Advances in Agricultural Systems Modeling

Transdisciplinary Research, Synthesis, and Applications

Volume 10

Lajpat R. Ahuja, Series Editor

Advances in Agricultural Systems Modeling

Transdisciplinary Research, Synthesis, and Applications

Volume 10

Lajpat R. Ahuja, Series Editor

Modeling Processes and Their Interactions in Cropping Systems

Challenges for the 21st Century

Lajpat R. Ahuja, Kurt C. Kersebaum, and Ole Wendroth, Editors

American Society of Agronomy

Crop Science SOCIETY OF AMERICA

Soil Science Society of America

WILEY

Copublication by American Society of Agronomy, Inc. / Crop Science Society of America, Inc. / Soil Science Society of America, Inc. and John Wiley & Sons, Inc.

Editorial Correspondence:
American Society of Agronomy, Inc.
Crop Science Society of America, Inc.
Soil Science Society of America, Inc.

5585 Guilford Road, Madison, WI 53711-58011, USA

agronomy.org • crops.org • soils.org

Registered Offices:
John Wiley & Sons, Inc., 111 River Street, Hoboken, NJ 07030, USA

For details of our global editorial offices, customer services, and more information about Wiley products, visit us at www.wiley.com.

Wiley also publishes its books in a variety of electronic formats and by print-on-demand. Some content that appears in standard print versions of this book may not be available in other formats.

Library of Congress Cataloging-in-Publication Data applied for

ISBN PRINT 0-89118-385-X 978-0-89118-385-3
ISBN Obook 0-89118-386-8 978-0-89118-386-0
ISBN epub 0-89118-387-6 978-0-89118-387-7
ISBN epdf 0-89118-388-4 978-0-89118-388-4
DOI DOI:10.1002/9780891183853

Cover design: Wiley
Cover image: © NRCS/SWCS photo

Set in 9/11pt Palatino by Straive, Pondicherry, India.

Contents

Preface   vii

1 — Water and Chemical Transport in the Soil Matrix
    and Macropores   1
    Lajpat R. Ahuja

2 — Heat Transport, Water Balance, Snowpack and
    Soil Freezing   33
    Gerald N. Flerchinger and Mark S. Seyfried

3 — Evapotranspiration, Transpiration, and Evaporation
    Processes and Modeling in the Soil–Residue–
    Canopy System   53
    Suat Irmak and Meetpal S. Kukal

4 — Advances and Improvements in Modeling Water
    Fluxes in Vegetation Canopy and their Relation
    to Photosynthesis   115
    Andree J. Tuzet, Dennis J. Timlin, and Alain Perrier

5 — Soil Organic Matter and Carbon–Nitrogen
    Transformations   155
    Marvin J. Shaffer and Lajpat R. Ahuja

6 — Equilibrium Soil Chemistry Submodels   179
    Donald L. Suarez and Todd H. Skaggs

7 — Pesticide Dissipation and Fate in Agricultural
    Settings: A Review of Research, 2000–2020   203
    R. Don Wauchope

8 — Coupling Machine Learning with APSIM Model Improves the Evaluation of Climate Extremes Impact on Wheat Yield in Australia   251

Puyu Feng, Bin Wang, De Li Liu, Qiang Yu, and Kelin Hu

9 — Simulating Maize Crop Processes in DSSAT: CSM-CERES-Maize vs. CSM-IXIM-Maize   277

Jon I. Lizaso, Vakhtang Shelia, and Gerrit Hoogenboom

10 — Simulation of Climate Change Effects on Evapotranspiration, Photosynthesis, Crop Growth, and Yield Processes in Crop Models   291

Kurt C. Kersebaum and Claudio O. Stöckle

11 — Quantifying and Modeling Management Effects on Soil Properties: Tillage, Reconsolidation, Crop Residues, and Crop Management   333

Lajpat R. Ahuja and Timothy R. Green

12 — Integrating Models with Field Experiments to Enhance Research: Cover Crop, Manure, Tillage, and Climate Change Impacts on Crops in a Humid Region   359

Gary Feng and Saseendran S. Anapalli

Index   393

Overview of Modeling processes and Their Interactions in Cropping Systems

Recently observed environmental trends regarding water and soil quality, climate change with extended drought and heat waves (Trnka et al., 2013, 2019), extreme weather events (Trnka et al., 2014), and irrigation water shortages due to urban use require continual adaptation and optimization of agricultural systems through changes in cropping and management. Moreover, the increasing world population and changing diets will raise global food demand by 100–110% from 2005 to 2050 (Tilman et al. 2011). To double food production by 2050, average annual growth rates in crop production of 2.4% (non-compounding rate) would be required (Ray et al. 2013). Considering the limited land resources and the increasing competition for productive land between food, fodder, and energy production, this increase in crop production must be achieved from the same or less land compared with today (Godfray et al., 2010; Zabel et al. 2014). The advancement of science and technology to achieve these changes requires cutting-edge field research, using a quantitative whole-systems approach.

Agricultural systems are composed of physical, biological, and chemical processes of soil, water, plant, climate, their complex interactions, and their management. The whole-system framework is designed for evaluating the impact of management practices and climate on soil water and nutrient dynamics, biomass production, and environmental quality. In this framework, process-level models integrate interdisciplinary knowledge with field research to evaluate management options and to create well-adapted agricultural management systems. The models help to: (a) integrate field research results in terms of the fundamental theory and concepts that are broadly applicable beyond the site-specific empirical relationships; (b) predict crop and soil process dynamics from knowledge of the fundamental drivers that determine the environment and plant growth under different climates, (c) transfer experimental findings to longer term weather conditions beyond the limited duration of field experiments and to soil types and climates diverging from the experimental plots, and (d) use extended results to develop broad-based precision management decision support tools or simple management guidelines for producers and other users, who may link them to economic and social considerations. The models also promote precision management of spatially variable soils (Kersebaum et al., 2005), help assimilate real-time information about soil and crop conditions obtained with remote or ground-level sensors (Wallor et al. 2019) and allow an in-depth analysis of complex policy or management decisions. As a result, the models are continually improved and serve as an evolving theoretical backbone of complex agricultural system research, decision support, and information transfer.

System modeling has been a vital step in many engineering disciplines. A lot more work is needed to bring agricultural system models to the level of physical and hydraulic system models and to enhance their structure and usability. However, agricultural system models have matured enough that, with some good data to serve as reference, they can be used for many practical applications in research and management (Ahuja et al., 2002; Ewert et al., 2015; Jones et al., 2003; Matthews & Stephens, 2002; Michalczyk et al., 2014; Stöckle et al., 2014). The effect of management practices regarding tillage or no-till management, different crops and cultivars, crop rotations, organic and mineral fertilizers, water supply, and natural or tile drainage on production and water quality can now be modeled (Anapalli et al., 2005, 2015; Andales et al., 2003; Ma et al., 2000; Malone et al., 2017; Wendroth et al., 2019; Yang et al., 2019). Their integration with field research will expose knowledge and conceptual gaps in the models that will then stimulate new research (Rötter et al., 2018). The International Service for National Agricultural Research (2004) made a case for applying modeling tools and processes in planning and implementation of natural resources research and making sense of the vast amount of data. The CGIAR Science Council (2005) noted in their research priorities:

> *Modeling and the ability to combine data from different sources, promises to revolutionize understanding of processes affecting management of natural resources. Thanks to strategic accumulation of data, tools, and modeling resources in the coming decade, one can expect the development of a more predictive approach to agriculture.*

Integration of field research with system models has the potential to significantly enhance the efficiency of agricultural research and raise agricultural science and technology to a higher level, leading to improved production and protection of the environment. Agricultural system models are also used in hydrological, climate, and ecological research and management.

The National Academies of Sciences, Engineering, and Medicine (2019) have released a new report that identifies the most promising scientific breakthroughs that are possible to be achieved in the next decade to increase the U.S. food and agriculture system's sustainability, competitiveness, and resilience in view of the following key challenges faced by agriculture:

- increasing nutrient use efficiency in crop production systems

- reducing soil loss and degradation

- mobilizing genetic diversity for crop improvement

- optimizing water use in agriculture

- improving food animal genetics

- developing precision livestock production systems

- early and rapid detection and prevention of plant and animal diseases

- early and rapid detection of foodborne pathogens

- reducing food loss and waste throughout the supply chain

The first four challenges involve cropping systems. The five breakthroughs for the next decade identified in the report are:

1. Transdisciplinary research and systems approach

2. Sensing technologies

3. Data science and informatics

4. Genomics and precision breeding

5. Animal, soil, and plant microbiomes

Process-based modeling of cropping systems, updated in this volume, will be an essential component of Breakthroughs 1, 3, and 5 and greatly facilitate Breakthroughs 2 and 4. The models are essential tools for evaluating and mitigating the effect of climate change and the management of increasingly limited water and other resources on the production of food, energy, fiber, and environmental quality (Anapalli et al., 2016; Gillette et al., 2018; Stöckle et al., 2010) and to identify yield gaps and analyze their causes to improve crop production for human nutrition (Schils et al., 2018; van Ittersum et al., 2013; van Wart et al., 2013).

Process-Level Components of Cropping Systems and Their Interactions

System models are composed of algorithms that describe several unique process components and their interactions. The primary components are the transport and dynamic status of soil water, mass, and energy (heat), including evapotranspiration; C/N ratio and nutrient dynamics; soil chemical equilibria; pesticide transport and metabolism dynamics; plant growth; the effects of management practices on the basic soil properties that control these processes; and the effect of weather conditions on all the above. How these processes and their interactions are simulated depends on the purpose of modeling and on the input data availability and uncertainty, which finally determines the performance of a system model at a specific scale. Evaluation and continual advances in knowledge of these processes and their interactions help improve the system models. In addition, certain special features of soil processes, such as the presence of active macropores, surface crusting, residue cover, transport of chemicals in runoff, and spatial variability of soil (see the Root Zone Water Quality Model [RZWQM]: Ahuja et al., 2000; Wallor et al., 2018; Wendroth & Robinson, 2008) need to be simulated well. Adequate estimates of the soil and plant parameters that control the above processes and special features are very important for simulations. However, their robustness must be tested under multiple site conditions to be suitable for extrapolation (Kersebaum et al., 2015). The evolution of system models coincides with basic science in agriculture. These unique process models also enhance applications in hydrology, climate, and ecology.

Modeling of soil–plant–climate–management processes is needed to enhance quantitative teaching of these processes at upper undergraduate and graduate levels and elevates education and the skill set of graduate students. The purpose of this volume in the Advances in Agricultural Systems Modeling series is to document the modeling of soil–plant–climate–management processes and their interactions in cropping systems and to serve as a textbook for graduate teaching and research. In 2000, we published a book on modeling processes and management effects in the RZWQM (Ahuja et al., 2000). Here, we update the modeling of processes beyond the RZWQM by introducing new concepts and knowledge of process modeling from other models and from new research in this area conducted during the past 20 years.

This publication will be the first comprehensive and updated state-of-the-art scientific textbook devoted to teaching modeling of the key soil–plant–climate–management processes at upper undergraduate and graduate levels; it also emphasizes new opportunities and paradigms for modern laboratory and field research to improve understanding and quantification of individual processes and their interactions.

Lajpat R. Ahuja
Kurt C. Kersebaum
Ole Wendroth

References

Ahuja, L. R., Ma, L., & Howell, T. A. (2002). Whole system integration and modeling—essential to agricultural science and technology in the 21st century. In L. R. Ahuja, L. Ma, & T. A. Howell (Eds.), *Agricultural system models in field research and technology transfer* (pp. 1–9Boca Raton, FL: CRC Press.

Ahuja, L. R., Rojas, K. W., Hanson, J. D., Shafer, M. J., & Ma, L. (2000). *Root Zone Water Quality Model: Modeling management effects on water quality and crop production*. Water Resources Publications.

Anapalli, S. S., Ahuja, L. R., Prasanna, P. H., Ma, L., Marek, G., Evett, S. R., & Howell, T. A. (2016). Simulation of crop evapotranspiration and crop coefficients with data in weighing lysimeters. *Agricultural Water Management*, 177, 274–283.

Anapalli, S. S., Nielsen, D. C., Ma, L., Ahuja, L. R., Vigil, M. F., Benjamin, J. G., & Halvorson, A. D. (2005). Effectiveness of RZWQM for simulating alternative Great Plains cropping systems. *Agronomy Journal*, 97, 1183–1193. https://doi.org/102134/agronj2005.0019

Anapalli, S. S., Trout, T. J., Ahuja, L. R., Ma, L., McMaster, G. S., Andales, A. A., Chaves, J., & Ham, J. (2015). Developing and normalizing average corn crop water production functions across years and locations using a system model. *Agricultural Water Management*, 157, 65–77. https://doi.org/101016/j.agwat.2014.09.002

Andales, A. A., Ahuja, L. R., & Peterson, G. A. (2003). Evaluation of GPFARM for dryland cropping systems in eastern Colorado. *Agronomy Journal*, 95, 1510–1524. https://doi.org/102134/agronj2003.1510

CGIAR Science Council. (2005). *System priorities for CGIAR research 2005–2015*. Science Council Secretariat.

Ewert, F., Rötter, R. P., Bindi, M., Webber, H., Trnka, M., Kersebaum, K. C., Olesen, J. E., Van Ittersum, M. K., Janssen, S., Rivington, M., Semenov, M. A., Wallach, D., Porter, J. R., Stewart, D., Verhagen, J., Gaiser, T., Palosuo, T., Tao, F., Nendel, C., … Asseng, S. (2015). Crop modelling for integrated assessment of risk to food production from climate change. *Environmental Modelling & Software*, 72, 287–303. https://doi.org/101016/j.envsoft.2014.12.003

Gillette, K., Malone, R. W., Kaspar, T. C., Ma, L., Parkin, T. B., Jaynes, D. B., Fang, Q. X., Hatfield, J. L., Feyereisen, G. W., & Kersebaum, K. C. (2018). N loss to drain flow and N$_2$O emissions from a corn–soybean rotation with winter rye. *Science of the Total Environment*, 618, 982–997. https://doi.org/101016/j.scitotenv.2017.09.054

Godfray, H. C. J., Beddington, J. R., Crute, I. R., Haddad, L., Lawrence, D., Muir, J. F., Pretty, J., Robinson, S., Thomas, S. M., & Toulmin, C. (2010). Food security: The challenge of feeding 9 billion people. *Science*, 327, 812–818.

International Service for National Agricultural Research. (2004). *Method in our madness: Making sense of ecoregional research with modelling tools and processes*. ISNAR.

Jones, J. W., Hoogenboom, G., Porter, C. H., Boote, K. J., Batchelor, W. D., Hunt, L. A., Huntd, P. W., Wilkense, U., Singhe, A. J., Gijsmana, J. T., & Ritchie, J. T. (2003). The DSSAT cropping system model. *European Journal of Agronomy*, 18, 235–265.

Kersebaum, K. C., Boote, K. J., Jorgenson, J. S., Nendel, C., Bindi, M., Frühauf, C., Gaiser, T., Hoogenboom, G., Kollas, C., Olesen, J. E., Rötter, R. P., Ruget, F., Thorburn, P. J., Trnka, M., & Wegehenkel, M. (2015). Analysis and classification of data sets for calibration and validation of agro-ecosystem models. *Environmental Modelling & Software*, 72, 402–417. https://doi.org/101016/j.envsoft.2015.05.009

Kersebaum, K. C., Lorenz, K., Reuter, H. I., Schwarz, J., Wegehenkel, M., & Wendroth, O. (2005). Operational use of agro-meteorological data and GIS to derive site specific nitrogen fertilizer recommendations based on the simulation of soil and crop growth processes. *Physics and Chemistry of the Earth*, 30, 59–67.

Ma, L., Ahuja, L. R., Ascough, J. C., II, Shaffer, M. J., Rojas, K. W., Malone, R. W., & Cameira, M. R. (2000). Integrating system modeling with field research in agriculture: Applications of the Root Zone Water Quality Model (RZWQM). *Advances in Agronomy*, 71, 233–292.

Malone, R. W., Kersebaum, K. C., Kaspar, T. C., Ma, L., Jaynes, D. B., & Gillette, K. (2017). Winter rye as a cover crop reduces nitrate loss to subsurface drainage in central Iowa as simulated by HERMES. *Agricultural Water Management*, 184, 156–169. https://doi.org/101016/j.agwat.2017.01.016

Matthews, R. B., & Stephens, W. (Eds.) (2002). *Crop–soil simulation models: Applications in developing countries*. CABI.

Michalczyk, A., Kersebaum, K. C., Roelcke, M., Hartmann, T., Yue, S. C., Chen, X. P., & Zhang, F. S. (2014). Model-based optimisation of nitrogen and water management for wheat–maize systems in the North China Plain. *Nutrient Cycling in Agroecosystems*, 98, 203–222.

National Academies of Sciences, Engineering, and Medicine. (2019). *Science breakthroughs to advance food and agricultural research by 2030.* National Academies Press. https://doi.org/1017226/25059

Ray, D. K., Mueller, N. D., West, P. C., & Foley, J. A. (2013). Yield trends are insufficient to double global crop production by 2050. *PLOS ONE, 8,* e66428. https://doi.org/1061371/journal.pone.0066428

Rötter, R. P., Appiah, M., Fichtler, E., Kersebaum, K. C., Trnka, M., & Hoffmann, M. P. (2018). Linking modelling and experimentation to better capture crop impacts of agroclimatic extremes: A review. *Field Crops Research, 221,* 152–156. https://doi.org/101016/j.fcr.2018.02.023

Schils, R., Olesen, J. E., Kersebaum, K. C., Rijk, B., Oberforster, M., Kalyada, V., Khitrykau, M., Gobinf, A., Kirchevg, H., Manolovag, V., Manolovg, I., Trnkahi, M., Hlavinkai, P., Palosuoj, T., Peltonen-Sainioj, P., Jauhiainenj, L., Lorgeouk, J., Marroul, H., Danalatosm, N., ... van Ittersum, M. (2018). Cereal yield gaps across Europe. *European Journal of Agronomy, 10,* 109–120. https://doi.org/101016/j.eja.2018.09.003

Stöckle, C. O., Kemanian, A. R., Nelson, R. L., Adam, J. C., Sommer, R., & Carlson, B. (2014). CropSyst model evolution: From field to regional to global scales and from research to decision support systems. *Environmental Modelling and Software, 62,* 361–369. https://doi.org/101016/j.envsoft.2014.09.006

Stöckle, C. O., Nelson, R. L., Higgins, S., Brunner, J., Grove, G., Boydston, R., Whiting, M., & Kruger, C. (2010). Assessment of climate change impact on eastern Washington agriculture. *Climate Change, 102,* 77–102.

Tilman, D., Balzer, C., Hill, J., & Befort, B. L. (2011). Global food demand and the sustainable intensification of agriculture. *Proceedings of the National Academy of Sciences, 108,* 20260–20264.

Trnka, M., Feng, S., Semenov, M. A., Olesen, J. E., Kersebaum, K. C., Rötter, R. P., Semerádová, D., Klem, K., Huang, W., Ruiz-Ramos, M., Hlavinka, P., Meitner, J., Balekpetr, J., Havlík, P., & Büntgen, U. (2019). Mitigation efforts will not fully alleviate the increase in the water scarcity occurrence probability in wheat-producing areas. *Science Advances, 5*(9), eaau2406. https://doi.org/101126/sciadv.aau2406

Trnka, M., Kersebaum, K. C., Eitzinger, J., Hayes, M., Hlavinka, P., Svoboda, M., Dubrovský, M., Semerádová, D., Wardlow, B., Pokorný, E., Možný, M., Wilhite, D., & Žalud, Z. (2013). Consequences of climate change for the soil climate in Central Europe and the Central Plains of the United States. *Climate Change, 120,* 405–418. https://doi.org/101007/s10584-013-0786-4

Trnka, M., Rötter, R. P., Ruiz-Ramos, M., Kersebaum, K. C., Olesen, J. E., Žalud, Z., & Semenov, M. A. (2014). Adverse weather conditions for European wheat production will become more frequent with climate change. *Nature Climate Change, 4,* 637–643. https://doi.org/101038/nclimate2242

van Ittersum, M. K., Cassman, K. G., Grassini, P., Wolf, J., Tittonell, P., & Hochman, Z. (2013). Yield gap analysis with local to global relevance: A review. *Field Crops Research, 143,* 4–17. https://doi.org/101016/j.fcr.2012.09.009

van Wart, J., Kersebaum, K. C., Peng, S., Milner, M., & Cassman, K. (2013). Estimating crop yield potential at regional to national scales. *Field Crops Research, 143,* 34–43. https://doi.org/101016/j.fcr.2012.11.018

Wallor, E., Kersebaum, K. C., Lorenz, K., & Gebbers, R. (2019). Soil state variables in space and time: The linkage between proximal soil sensing and process modelling. *Precision Agriculture, 20,* 313–334. https://doi.org/101007/s11119-018-9617-y

Wallor, E., Kersebaum, K. C., Ventrella, D., Bindi, M., Cammarano, D., Coucheney, E., Gaiser, T., Garofalo, P., Giglio, L., Giola, P., Hoffmann, M. P., Iocola, I., Lana, M., Lewan, E., Maharjan, G. R., Moriondo, M., Mula, L., Nendel, C., Pohankova, E., ... Trombi, G. (2018). The response of process-based agro-ecosystem models to within-field variability in site conditions. *Field Crops Research, 228,* 1–19. https://doi.org/101016/j.fcr.2018.08.021

Wendroth, O., Lascano, R. J., & Ma, L. (Eds.) (2019). *Bridging among disciplines by synthesizing soil and plant processes* (Advances in Agricultural Systems Modeling 8). ASA, SSSA, & CSSA.

Wendroth, O., & Robinson, D. A. (2008). Scaling processes in watersheds. In S. W. Trimble (Ed.), *Encyclopedia of water science* (2nd ed., Vol. 1, pp. 1024–1028). CRC Press.

Yang, W., Feng, G., Adeli, A., Kersebaum, K. C., Jenkins, J. N., & Li, P. F. (2019). Long-term effect of cover crop on rainwater balance components and use efficiency in the no-tilled and rainfed corn and soybean rotation system. *Agricultural Water Management, 219,* 27–39.

Zabel, F., Putzenlechner, B., & Mauser, W. (2014). Global agricultural land resources: A high resolution suitability evaluation and its perspectives until 2100 under climate change conditions. *PLOS ONE, 9,* e107522.

Lajpat R. Ahuja

Water and Chemical Transport in the Soil Matrix and Macropores

Abstract

In this chapter, we were especially interested in simulating water flow and chemical transport through macropores explicitly, as well as in the soil matrix. For this goal, we divided the water flow process into two phases: (a) infiltration into the soil matrix and macropores and macropore–matrix interaction during a rainfall or an irrigation, modeled by using the semi-analytical, vertical and radial, Green–Ampt approach, which has been theoretically related to Richards' equation; and (b) redistribution of water in the soil matrix following infiltration, computed by a mass-conservative numerical solution of the one-dimensional Richards equation. The chemical transport during both of these phases is based on the established concept of miscible displacement but is done in a simpler way using partial piston displacement followed by partial mixing in small depth increments. Again, the Green–Ampt approach during infiltration allowed explicit transport of chemicals through macropores and explicit interaction between chemical transport through macropores and the soil matrix.

Introduction

Water flow in soil is the driving force behind all the belowground processes of an agricultural system and significantly affects aboveground plant growth and surface hydrology. It determines how much water is stored in the root zone and how much is taken up by the roots at any given time and how surface-applied fertilizers and heat are transported through this zone. All the vital microbial and biochemical transformations of organic matter and nutrients are highly dependent on soil water content and temperature. Thus, the ability of a plant to grow is also highly dependent on water

Modeling Processes and Their Interactions in Cropping Systems: Challenges for the 21st Century, First Edition.
Edited by Lajpat R. Ahuja, Kurt C. Kersebaum, and Ole Wendroth.
© 2022 American Society of Agronomy, Inc. / Crop Science Society of America, Inc. / Soil Science Society of America, Inc.
Published 2022 by John Wiley & Sons, Inc.

flow. Furthermore, this flow largely determines how applied agricultural chemicals are transferred to surface runoff or leached below the root zone and thus how serious the impacts are on the quality of our surface and groundwater resources. This chapter is an update of Ahuja et al. (2000).

The theories of soil water movement and the related transport of chemicals and heat are now well developed for simple soil systems. One-dimensional vertical movement and retention of water in a partially or variably saturated soil profile is commonly described by a simpler form of the Richards equation that incorporates Darcy's Law and mass conservation and assumes that the role of the soil air or vapor phase and thermal gradients on the liquid flow process is negligible:

$$\frac{\partial \theta}{\partial t} = \frac{\partial}{\partial z}\left[K(h,z)\frac{\partial h}{\partial z} - K(h,z)\right] - S(z,t) \tag{1.1}$$

where θ is volumetric soil water content ($cm^3\,cm^{-3}$); t is time (h); z is soil depth (cm, assumed positive downward); h is the soil-water pressure head (cm), a function of θ; K is unsaturated hydraulic conductivity ($cm\,h^{-1}$), a function of h and z; and $S(z,t)$ is a sink term for root water uptake and tile drainage rates (h^{-1}).

Equation 1.1 can be used to described infiltration of water at the soil surface from rain or irrigation, with known rates or surface depths, and subsequent redistribution in the soil profile. For very dry portions of the soil profile, not commonly encountered in agriculture, vapor flow and the effect of soil temperatures on both liquid and vapor flow can be important. Equation 1.1 can then be modified to include these effects (Šimůnek et al., 2013). The modified equation has to be solved in conjunction with the soil heat flow equation.

The theories of chemical and heat transport in soil are based on the concept of miscible displacement of chemicals and heat by the flow of water. This concept is commonly approximated by a convection–dispersion equation. One-dimensional, miscible displacement of an absorptive and degradable chemical species through a soil column of constant water content and flux is simulated by

$$\frac{\partial}{\partial x}\left(D\frac{\partial C}{\partial x}\right) - v\frac{\partial C}{\partial x} - R\frac{\partial C}{\partial t} = \mu C - \gamma \tag{1.2}$$

where $C = C(x,t)$ is the chemical concentration at location x and time t; $D = \alpha_L\,|\,v\,|$ is the dispersion coefficient; α_L is the dispersivity; $v = q/\theta$ is the pore water velocity; θ is the volumetric water content; q is the flux density of the soil water (or the Darcy velocity); $R = 1 + \rho K_d/\theta$ is the retardation factor; ρ is the soil bulk density; K_d is the partition coefficient between solution and adsorbed phases at equilibrium; μ is the first-order decay constant; and γ is the zero-order production rate constant. For field conditions, this equation is extended to variable water contents and fluxes with space and time by solving in conjunction with the Richards equation. This equation can also be extended to nonlinear partition and nonlinear nonequilibrium or coupled sequential decay reactions (Šimůnek et al., 2013; see also Chapter 7). The heat transport equations are given in Chapter 2.

Generally for field conditions, Equations 1.1 and 1.2 have to be solved by the complex numerical methods. Simpler partially analytical approaches to these equations are available for certain conditions. However, real field soils have several complexities that play important roles which, at this stage, are not completely understood in mechanistic terms. Examples of these complexities are:

- the role of macropores—worm holes, decayed root channels, or structural cracks—in short-circuiting movement of water and chemical from soil surface to deeper layers. Šimůnek et al. (2013) approximately simulated macropore flow implicitly by an interacting dual-permeability (two-soil) system with the one-dimensional Richards equation applied to each soil.

- the role of surface aggregates in retarding downward leaching of surface-applied chemicals but increasing their transfer to runoff or to macropore flow.

- the role of interaggregate immobile pore space and chemical kinetics inside the soil matrix in retarding or enhancing transport of chemicals. This has been simulated by the Richards equation as a dual-porosity (mobile–immobile) system (Šimůnek et al., 2013).

- enormous changes in the basic soil hydraulic properties, macropores, and aggregation brought about by tillage and followed by temporal changes due to reconsolidation.

- the roles of a high water table and an installed drainage system, when incorporated in a one-dimensional model.

Most of the common cropping system models use simple sequential cascading schemes of water and chemical movement and retention in the root zone, following estimation of infiltration by a simple method such as the SCS Curve Number approach. In modeling water and chemical transport in the Root Zone Water Quality Model (RZWQM) (Ahuja et al., 2000), we utilized the existing theoretical understanding of simple soil systems but also incorporated the role of the above complexities in some simple ways, consistent with our limited knowledge of these complexities at this time. We were especially interested in simulating macropore water flow and chemical transport explicitly. For this goal, we divided the water flow process into two phases: (a) infiltration into the soil matrix and macropores and macropore–matrix interaction during a rainfall or an irrigation, modeled by using the semi-analytical, vertical and radial, Green–Ampt approach (Green & Ampt, 1911), which allows explicit flow through macropores and macropore–soil matrix interactions that the one-dimensional Richards equation does not, and has been theoretically related to the one-dimensional Richards equation (Ahuja, 1983; Morel-Seytoux & Khanji, 1974) and (b) redistribution of water in the soil matrix following infiltration, computed by a mass-conservative numerical solution of the one-dimensional Richards equation (Celia et al., 1990). Chemical transport in the soil matrix during both of these phases is based on the established concept of miscible displacement but is done in a simpler way using partial piston displacement followed by partial mixing in small depth increments. Again, the Green–Ampt approach during infiltration allowed explicit transport of chemicals through macropores and explicit interaction between chemical transport through macropores and the soil matrix.

This chapter describes the components of the model relevant to water flow and chemical transport through the soil matrix and macropores.

Description of the Soil Profile

In order to simulate the transport of water and chemicals, the soil profile must be well defined in its depth, horizon delineation, and physical and hydraulic properties. It can have either homogeneous soil properties with depth or up to a certain number of distinct soil horizons. The RZWQM uses soil horizons to define segments with like soil properties. It then creates two numerical grids, one a nonuniform layering system

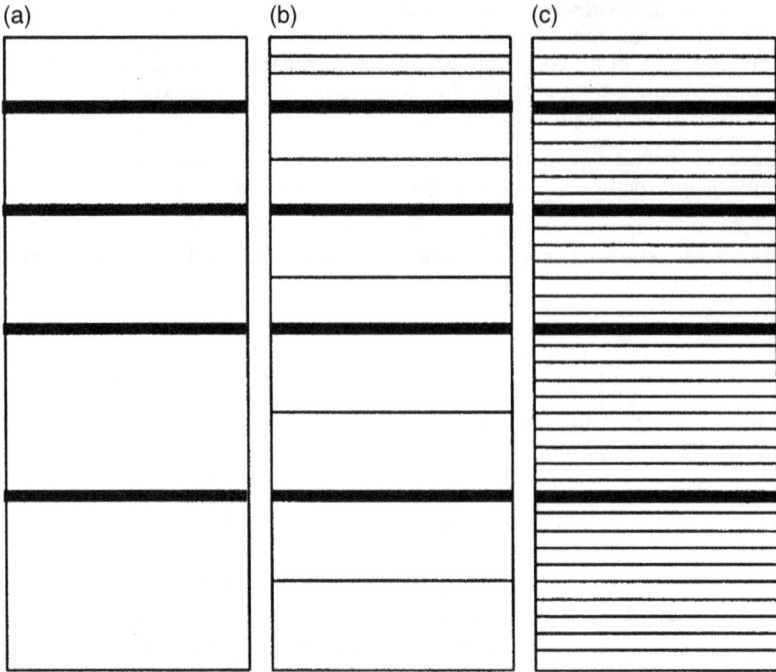

Figure 1.1 The three transport grids used by the Root Zone Water Quality Model: (a) soil profile delineated by horizons for defining hydraulic properties; (b) soil profile delineated by numerical layers for redistribution; and (c) soil profile delineated by 1-cm layers for infiltration.

for redistribution and the other a 1-cm grid used during infiltration (Figure 1.1). These grids are discussed further under Numerical Methods and Space Discretization.

Each soil horizon is characterized by its own soil physical and hydraulic properties. The physical properties are bulk density, particle density, porosity, and texture. Hydraulic properties are defined by the soil water content–matric suction relationship and the unsaturated hydraulic conductivity–matric suction relationship. These relationships are described by functional forms suggested by Brooks and Corey (1964) with slight modifications (Figure 1.2). The Brooks–Corey parameters have been compiled by Rawls et al. (1982) for 11 soil textural classes. These relationships are assumed non-hysteretic.

The soil water content versus the matric suction relationship is represented by

$$\theta(\tau) = \theta_s - A_1\tau; \quad \tau \leq \tau_b \qquad (1.3)$$

$$\theta(\tau) = \theta_r + B\tau^{-\lambda}; \quad \tau > \tau_b$$

where θ is volumetric soil water content ($cm^3\,cm^{-3}$), τ is matric suction head (cm, $\tau = |h|$, where h is the soil water pressure head), θ_s is the saturated soil water content ($cm^3\,cm^{-3}$), θ_r is the residual water content ($cm^3\,cm^{-3}$), τ_b is the air-entry or bubbling suction head (cm), λ is the pore size distribution index, and A_1, B, and λ are constants. The constant B is not an independent parameter; it is determined from other parameters by the condition of continuity at $\tau = \tau_b$. When A_1 is set equal to zero, Equation 1.3 reduces to the Brooks–Corey model.

$\theta(\tau)=0.473$: $\quad\quad\quad\quad\quad\quad \tau <= 12.0$ (cm)

$\theta(\tau)=0.473\ (\tau/12.0)^{-0.113}$: $\tau > 12.0$ (cm)

$K\,(\tau)=1.33$: $\quad\quad\quad\quad\quad\quad \tau <= 12.0$ (cm)

$K(\tau)=1.33\ (\tau/12.0)^{-2.39}$: $\tau > 12.0$ (cm)

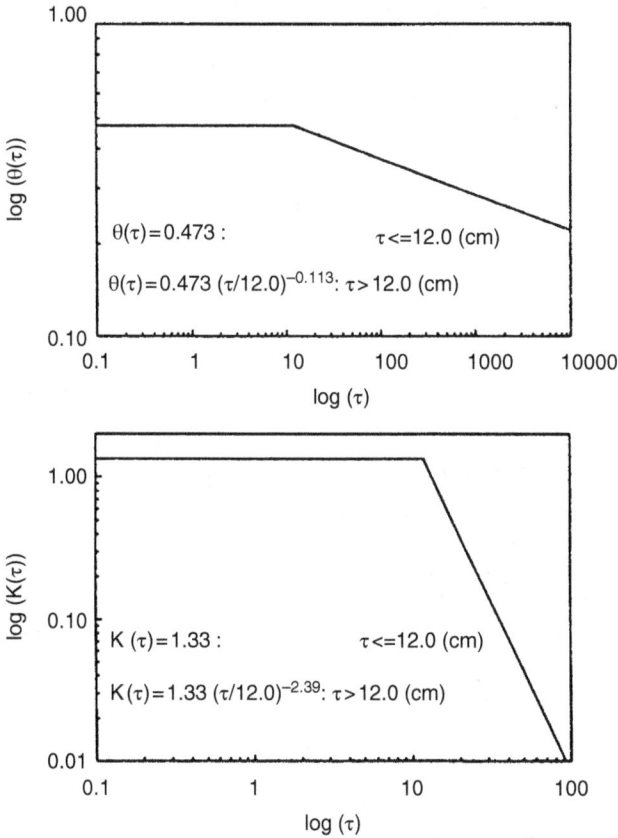

Figure 1.2 An example of the soil hydraulic property functions (soil water content–suction curve [θ–τ] and hydraulic conductivity–suction curve [K–τ]) used in the model. The values of the variables used in Equations 1.3–1.4 are θ_s = 0.473, A_1 = 0.0, θ_r = 0.0, B = 0.473/($12^{-0.113}$), λ = 0.113, τ_b = 12.0, K_s = 1.33, N_1 = 0.0, K_2 = 1.33/($12^{-2.39}$), N_2 = 2.39, and τ_{bK} = 12.0. Source: Based on Brooks and Corey 1964.

The hydraulic conductivity versus matric suction relation is expressed as

$$K(\tau) = K_s \tau^{-N_1}; \quad \tau \le \tau_{bK} \quad\quad\quad (1.4)$$
$$K(\tau) = K_2 \tau^{-N_2}; \quad \tau > \tau_{bK}$$

where K is hydraulic conductivity (cm h^{-1}), K_s is field-saturated hydraulic conductivity (cm h^{-1}), and τ_{bK} is the air-entry or bubbling suction head for this function (cm), which may be equal to τ_b introduced above, and N_1 and N_2 are constants. In the original Brooks–Corey equation, N_1 is zero, $K_2 = K_s$, and $N_2 = 2 + 3\lambda$.

Another commonly used model of soil hydraulic properties was proposed by van Genuchten (1980). This model smooths the water retention curve near saturation:

$$\theta(h) = \frac{\theta_r + (\theta_s - \theta_r)}{(1 + |h|^n)^m}; \quad h \le 0 \quad\quad\quad (1.3a)$$
$$\theta(h) = \theta_s; \quad\quad\quad\quad\quad h \ge 0$$

5

where h is the soil water pressure head ($-\tau$, cm), $\theta(h)$ (cm^3 cm^{-3}) is the volumetric water content (for $h < 0$), θ_r (cm^3 cm^{-3}) is the residual water content, θ_s (cm^3 cm^{-3}) is the saturated water content, m is $1 - (1/n)$ with $n > 1$, and α (cm^{-1}) and n are empirical parameters determining the shape of the curve. Equation 1.3a smooths the water retention curve around the air-entry value.

$$K(h) = K_s S_e^l \left[1 - \left(1 - S_e^{1/m}\right)^m\right]^2 \tag{1.4a}$$

where K_s (cm h^{-1}) is saturated hydraulic conductivity and S_e is the effective saturation, expressed as $S_e = (\theta - \theta_r)/(\theta_s - \theta_r)$. The value of the exponent l in the K function is about 0.5 as an average for many soils.

Vogel and Císlerová (1988) modified the van Genuchten equations to add flexibility in the properties near saturation (higher values). Kosugi (1996) proposed a lognormal distribution model for water retention.

Estimation of Soil Hydraulic Properties

When, as often happens, measured data for these hydraulic properties are not available, they are estimated from regression equations with simpler known properties of soil texture, bulk density, and soil-water content at a 333-cm (1/3-bar) suction head ($\theta_{1/3}$). The technique can also be applied if, instead of the 333-cm value, the soil-water content at 100-cm suction is known. If neither soil-water content value is known, the parameters for the hydraulic property functions are taken from Rawls et al. (1982) based on soil texture class and adjusted based on bulk density alone.

Similar-Media Scaling

If one value of the $\theta(\tau)$ relationship is known, the rest of the function values are estimated by the extended similar-media scaling technique (Ahuja et al., 1985). According to the scaling concept (Simmons et al., 1979; Warrick et al., 1977), the matric suction for a fixed degree of saturation, S [$S = (\theta - \theta_r)/(\theta_s - \theta_r)$], at the ith site, $\tau_i(S)$, is related to the mean τ value for a soil type, $\tau_m(S)$, by

$$\alpha_i \tau_i(S) = \tau_m(S) \tag{1.5}$$

where α_i is a scaling factor for the ith site that applies at all different values of S. Therefore, if $\tau_m(S)$ is known and one value of $\tau_i(S)$ is also known, we can obtain the rest of the unknown $\tau_i(S)$ function. In our model, we utilize the textural-class mean $\theta(\tau)$ or $\tau(S)$ functions of Rawls et al. (1982) as our reference $\tau_m(S)$ functions. We also assume that the soil texture, bulk density, and the 333- or 100-cm water content at the unknown ith site are known. The value of θ_s is assumed equal to porosity, which can be calculated from the soil bulk density. Thus we can estimate the entire $\tau_i(S)$ or $\theta_i(\tau)$ function. This scaling technique is illustrated in Figure 1.3.

A different technique is used in the model to estimate changes in soil hydraulic properties due to tillage and subsequent reconsolidation. For this purpose, the change in soil bulk density is assumed to be known. The effects of tillage on soil bulk density and hydraulic properties are discussed in Chapter 11.

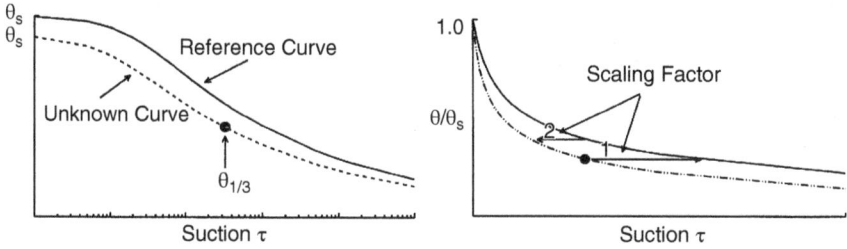

Figure 1.3 Scaling approach used to estimate the soil water content—suction relationship, $\theta(\tau)$, from two known points, θ at saturation (θ_s) and θ at 333-cm suction head ($\theta_{1/3}$), and a reference curve for the soil type or textural class.

One-Parameter Model

A one-parameter model is based on the strong linear correlation ($R^2 = 0.95$) observed between the slope b and intercept a of the log–log linear $h(\theta)$ form of the Brooks–Corey equation below the air-entry value (Ahuja & Williams, 1991; Gregson et al., 1987). Furthermore, the a versus b relationship of a large group of soils merged very nicely into one common relationship $a = p + qb$, with p and q nearly constant for all soils. This allows the determination of the $h(\theta)$ function from one known value, preferably the field capacity 333-cm soil water content. This model gave good results for several soils.

The saturated hydraulic conductivity (K_s) can be user defined or estimated from the effective porosity (ϕ_e), which is defined as

$$\phi_e = \theta_s - \theta_{1/3} \tag{1.6}$$

Recent work by Ahuja et al. (1989) with nine different soils indicated that the K_s of a soil layer is strongly related to ϕ_e as

$$K_s = 764.5\phi_e^{3.29} \tag{1.7a}$$

with $R^2 = 0.67$ and RMSE of log $K_s = 0.613$
Number of data points for Equation 1.7a = 473
Fitting to the texture mean values of Rawls, et al (1982) also gave good results:

$$K_s = 509.4\phi_e^{3.633} \tag{1.7b}$$

The measurement of saturated hydraulic conductivity is normally subject to a large error (an order of magnitude) because of the unknown effects of air entrapment and the presence of macropores. Considering this limitation, Equations 1.7a and 1.7b provide a good means to estimate K_s where measured data are not available. From knowledge of K_s and $\theta(\tau)$, the complete unsaturated hydraulic conductivity function, $K(\tau)$, is obtained by utilizing the approximate capillary-bundle approach described by Campbell (1974) and others. Once soil properties are defined, modeling of flow and transport processes begins.

Vertical Infiltration of Water into the Soil Matrix

The Green–Ampt equation is used to calculate infiltration rates into a homogeneous or layered profile divided into 1-cm increments (Childs & Bybordi, 1969; Green & Ampt, 1911; Hachum & Alfaro, 1980):

$$V = \overline{K}_s \frac{\tau_c + H_o + Z_{wf}}{Z_{wf}} \tag{1.8}$$

where V is the infiltration rate at any given time (cm h^{-1}), $\overline{K}_s$ is the effective average saturated hydraulic conductivity of the wetting zone (cm h^{-1}), τ_c is the capillary drive or suction head at the wetting front (cm), H_o is the depth of surface ponding (cm), if any, and Z_{wf} is the depth of the wetting front (cm).

If the infiltration rate calculated by Equation 1.8 is greater than the rainfall rate, it is set equal to the rainfall rate (Mein & Larson, 1973). If the converse is true, the excess rainfall is considered overland flow (runoff).

For a homogeneous soil or surface of a layered soil, the effective saturated hydraulic conductivity, $\overline{K}_s$, in Equation 1.8 is set equal to the field-saturated hydraulic conductivity of the soil or horizon. For a layered soil profile with hydraulic conductivity decreasing with depth, $\overline{K}_s$ is set equal to the harmonic mean effective saturated hydraulic conductivity of the wetted zone (Childs & Bybordi, 1969; Hachum & Alfaro, 1980). Thus, for such a layered soil, the effective saturated hydraulic conductivity changes as the wetted depth increases with time. If the saturated hydraulic conductivity of a subsoil layer in a layered soil is greater than that of the layers above, the conductivity of the previous layer continues to govern the flow in this subsoil layer (Hanks & Bowers, 1962).

Based on the work of Bouwer (1969), Morel-Seytoux and Khanji (1974), and Brakensiek and Onstad (1977), it has been found that $\overline{K}_s$ in Equation 1.8 may be reduced by about half due to entrapped air and the resulting viscous resistance. Thus, the actual infiltration rate may be less than that predicted by Equation 1.8. To account for the effects of entrapped air, $\overline{K}_s$ is divided by a viscous resistance correction factor of 2.0, changing Equation 1.8 to

$$V = \frac{\overline{K}_s}{2.0} \frac{\tau_c + H_o + Z_{wf}}{Z_{wf}} \tag{1.9}$$

To account for the effects of a surface crust, an adjustment is made to the saturated hydraulic conductivity, K_s, of the first horizon. Assuming the crust is 0.5 cm thick with a known conductivity, we approximate the value of the suction head at the crust–soil interface under steady-state conditions. Then, the saturated conductivity of the first horizon is assumed to be the conductivity associated with this head value. Because infiltration is governed by the smallest K_s of the wetted soil, this adjusted value can greatly decrease infiltration. Surface crusts are destroyed by tillage and are assumed to reform immediately after rainfall.

The capillary drive, τ_c, in Equation 1.9 varies from horizon to horizon, corresponding to the location of the wetting front. It is calculated from the unsaturated hydraulic conductivity–suction function, $K(\tau)$, of the wetting horizon as (Ahuja, 1983; Swartzendruber, 1987)

$$\tau_c = \frac{\int_0^{\tau_n} K(\tau)d\tau}{K_s} \tag{1.10}$$

where τ_n is the suction corresponding to the average initial soil water content of the soil horizon. The $K(\tau)$ function can be analytically integrated to solve for τ_c.

A simple and efficient scheme is then used to obtain cumulative infiltration with time. It involves the following steps:

1. An average infiltration rate, V, is calculated by Equation 1.9 using Z_{wf} corresponding to the middle of the 1.0-cm wetting increment.

2. The moisture deficit in this increment is calculated as the difference between the known field-saturated value for the layer (taken as 0.9 or a given fraction of porosity) and the initial moisture (the initial moisture of the wetting increment can change from the original value by lateral absorption below the wetting front from macropores, if present, as described below).

3. The time step in wetting this increment of soil is defined as the deficit divided by the average infiltration rate.

4. These incremental deficits and time steps are summed, thus calculating the cumulative infiltration and time.

The time step for infiltration is thus the time required to wet the last wetting 1-cm increment. The time step increases with time.

Flow into Tile Drains During Infiltration

Movement of water and chemicals below the wetting front during infiltration is assumed to occur according to unit gradient flow except in two cases. In the case of a constant pressure head bottom boundary condition, no transport occurs below the wetting front during infiltration. Unit gradient flow is also not assumed if a water table is present, but it is replaced by a different algorithm, discussed below. In all other cases, however, water contents and chemical concentrations below the current depth of wetting are updated according to the hydraulic conductivity of each 1-cm increment at the end of each time step. This allows for some redistribution to occur during rainfall events and results in a small amount of leakage below the lower boundary of the profile.

When a water table is present in the soil profile above the depth of tile drains, a flux out of the drain will occur. The drainage rate is calculated from soil conditions just prior to infiltration according to the Hooghoudt equation (as outlined below). In addition, if a positive leakage rate out the bottom of the profile exists, seepage from the water table will also occur. The water that flows to the drain and out the bottom of the profile at each infiltration time step is taken from the 1-cm layer at which the water table begins, with the requirement that this layer remain above the water content at 100-cm suction (Figure 1.4). If it should fall below this water content, the water table is moved down 1 cm and the water is taken from the next 1-cm layer. If the water table falls below the tile drain, no more drainage will occur during the infiltration event, although seepage will continue.

On the other hand, if the wetting front should catch up to the water table, completely saturating the profile, the rainfall event continues by taking time steps of five minutes.

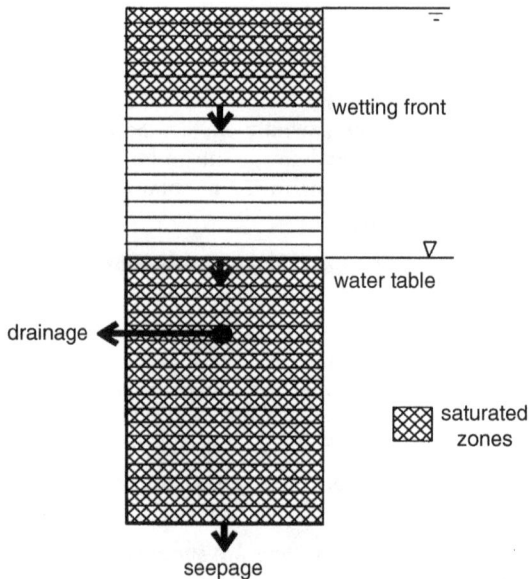

Figure 1.4 Movement of water during infiltration. The wetting front moves down in 1-cm increments as each 1 cm becomes saturated. The water table likewise moves down as water from the top of the water table is moved through the profile into drainage and/or seepage.

A new drainage rate is calculated based on the new depth of the water table, assumed to be at the surface. Then rainfall infiltrates at the drainage plus leakage rates, with any excess going into runoff. If the rainfall rate is such that it cannot satisfy drainage demands, the soil surface layers can become desaturated, and the water table will move back down.

Transport of a Chemical Within the Soil Matrix during Infiltration

We use a sequential partial-displacement and mixing approach in 1-cm increments to transport solutes in the spirit of the established phenomenon of the miscible displacement. However, the soil matrix is subdivided into micropore (immobile) and mesopore (mobile) zones, which introduces a form of preferential chemical transport in the soil matrix (the preferential flow in macropore channels outside the soil matrix is treated separately). The soil matrix porosity is divided into meso- and micropores based on either the user input values of these or the partitioning of the soil water retention curve at an arbitrary suction of 2,000 cm (Addiscott, 1977). Initially and during first wetting of a 1-cm depth increment, the water and chemicals in meso- and micropores are in equilibrium. During successive infiltration steps, the miscible displacement of solution in the saturated soil layers occurs only in the mesopores (mobile regions). This approach is similar to the layer model of Addiscott (1977), but the method of displacement involves only partial piston displacement as described below.

For each infiltration step, the soil solution is partially displaced sequentially across the 1-cm soil increments in the manner of piston displacement, followed by mixing in each increment. To simulate this partial displacement, we assume a two-stage process. In the first stage, half of the volumetric air space of the wetting increment is filled. Soil solution is displaced through the mesopores of the wetted increments by this amount and mixed before the second displacement step is taken. The second step then moves the remaining water and its solution through the mesopores of the wetted profile in the same manner. This two-stage process simulates miscible displacement and dispersion in the mesopores. In soils that are initially very dry and in which the dispersion is low, we have obtained better results with a one-stage displacement, but the two-stage process works well in most cases and no changes are needed in the program.

The model allows diffusion between meso- and micropore solutions for certain chemicals applied by the user, such as pesticides or tracers. The equation for this process is

$$\frac{\Delta C_{\text{sol}}}{\Delta t} = D_{\text{a}}(C_{\text{micr}} - C_{\text{sol}}) \tag{1.11}$$

where C_{sol} is the concentration of chemical in solution (mg L^{-1}) in the mesopores, C_{micr} is the concentration in solution of the micropores (mg L^{-1}), and D_{a} is the apparent diffusion coefficient or diffusion distance factor (user defined or calculated from a default database).

At the end of each time step, the exchange of chemical is calculated and the concentrations in each 1-cm increment appropriately adjusted.

For a soil-adsorbed chemical, such as some pesticides, a linear relationship and an instantaneous equilibrium are assumed to occur between the soil solution and adsorbed phases in both meso- and micropore regions. We also have an option in the model to allow a portion of a reactive chemical in the soil matrix to be subject to a first-order kinetic absorption–desorption between solid and solution phases. The equations for these processes are (Cameron & Klute, 1977)

$$C_{\text{ad}} = f K_{\text{d}} C_{\text{sol}} \tag{1.12}$$

$$\frac{\Delta C_{\text{ab}}}{\Delta t} = R_{\text{Kz}}[(1-f)K_{\text{d}}C_{\text{sol}} - C_{\text{ab}}] \tag{1.13}$$

where C_{ad} is the concentration of chemical on the soil surface (g g^{-1}) in equilibrium with the solution phase, C_{sol} is the concentration in the soil solution (mg L^{-1}), f is the fraction of sorption sites of soil in equilibrium, K_{d} is the partition coefficient between C_{sol} and C_{ad}, C_{ab} is the concentration in the solid phase (g g^{-1}) not in instantaneous equilibrium (kinetic pool), t is time, and R_{Kz} is a kinetic rate constant.

The parameters needed for the processes are either user supplied for each chemical or obtained from a default empirical database.

At the end of an infiltration event, when the water movement slows down, the meso- and micropore regions are allowed to equilibrate and the overall solute concentration is calculated. However, instantaneous equilibrium and kinetic pools are kept separate for redistribution.

If the water table is present, the flux within the water table will carry chemicals with it that may go out through the drain and into seepage. The transport of chemicals in this saturated zone is calculated using piston displacement in only the mesopores, similar to the method described below. After the initial infiltration and chemical transport into the soil matrix has occurred within a time step, the flux of water from the top of the water table to the bottom of the profile is displaced sequentially across the 1-cm increments, mixing with the chemical concentration in the mesopores of each increment. At the depth of the tile drain, the new mesopore chemical concentration is the concentration in the drainage water. Similarly, the concentration of chemicals in the seepage water is equal to the concentration in the mesopores in the bottom layer of the profile. To simplify the processes in the water table, two-stage displacement, diffusion between micro- and mesopores, and kinetic absorption–desorption processes are not simulated. The meso- and micropore regions below the water table are allowed to equilibrate at the end of the infiltration event.

Transfer of the Chemical to Overland Flow during Rainfall

Overland flow (also referred to as runoff) is calculated as the difference between the rainfall and infiltration in each time step. The model does not currently simulate the storage of runoff, but instead, in the absence of macropores, considers this water and its chemical solution as a loss to the system. The chemical in overland flow is extracted from the soil surface layers by raindrop impact. The extraction occurs from depths as great as 2 cm, but the contribution decreases exponentially with depth (Ahuja, 1986). This process can be modeled as an accelerated diffusion (Ahuja, 1990), but here we use a simpler approximation. We model this extraction process with a nonuniform mixing model (Ahuja, 1986; Heathman et al., 1986). The degree of mixing (M) between rainwater and soil solution in the micro- and mesopores is assumed to be complete (equal to unity) at the soil surface ($z = 0$ cm) and to decrease with depth as

$$M = \exp(-bz) \tag{1.14}$$

where b is a parameter that depends somewhat on the soil type, surface roughness, and cover conditions. Equation 1.14 can be integrated over a 1-cm increment and an average value of mixing, M_{ave}, calculated. This value may then be expressed in the same functional manner as Equation 1.14 as

$$M_{ave} = \exp(-Bz) \tag{1.15}$$

where B is a new exponent, different from b, and z is the depth of the midpoint of the increment. The value of B is kept constant for both of the top two soil compartments, which, with the experimentally determined value of B as 4.4 cm^{-1} (Ahuja, 1986), gives a degree-of-mixing value in the 1–2-cm layer of approximately $1/100$ of that in the top 1-cm increment. We compute chemical transfer to overland flow (then available to move into the macropores or run off the site) from micro- and mesopores by Equation 1.15 at the midpoint of the top two 1-cm intervals. During each infiltration step in which overland flow occurs, the chemical is first transferred to the rainfall solution, part of which then infiltrates into the soil. Thus, the water that infiltrates during the time interval has the same chemical concentration as the overland flow.

Transport of Overland Water and Chemical in Solution through Macropores

Two domains of flow, the soil matrix and macropore channels, interact through the walls of the macropores, which act as a common boundary condition or a source–sink term. Thus, in concept, our model is similar to the transient water flow models of Hoogmoed and Bouma (1980), Beven and Germann (1981) and Hetrick et al. (1982). However, the methods used here are different. The only source of water and chemicals transported through the macropores is the overland flow (rainfall excess) generated at the soil surface and the chemicals it picks up from the surface soil by mixing and raindrop impact. The solution flow in macropores is assumed very rapid and unaffected by pore tortuosity. The water solution is, however, subject to lateral absorption into the drier soil matrix. The reactive chemicals in solution are also subject to adsorption to or desorption from the macropore walls. The top soil horizons are assumed to have cylindrical macropore channels, although in the bottom horizons these may be replaced with planar cracks. Continuous macropores are idealized to be vertical and well dispersed within the soil matrix continuum. The continuity extends to the groundwater table or to a point above the depth of interest. However, a certain number of dead-end macropores are assumed to branch horizontally off the continuous pores in each soil horizon. The average volume fraction of continuous and dead-end macropores and the average size of voids (radius of cylindrical macropores and width and length of cracks) are assumed to be user supplied. From this information, the number of pores or total length of cracks per unit area of soil is calculated. The maximum flow-rate capacity (K_{mac}) of the macropores is then calculated using Poiseuille's law, assuming gravity flow (unit hydraulic head gradient).

For cylindrical holes:

$$K_{mac} = \frac{N_p \rho g \pi r_p^4}{8\eta} = \frac{P_{mac}\rho g r_p^2}{8\eta} \quad (cm\,h^{-1}) \tag{1.16}$$

For planar cracks:

$$K_{mac} = \frac{L_c \rho g w^3}{12\eta} = \frac{P_{mac}\rho g w^2}{12\eta} \quad (cm\,h^{-1}) \tag{1.17}$$

where ρ is the density of water (=1.0017 g cm^{-3}), g is the gravitational constant (=1.27 × 10^{10} cm h^{-2}), r_p is the radius of cylindrical holes (cm), w is the width of planar cracks (cm), η is the dynamic viscosity of water (=36.072 g h^{-1} cm^{-1}), N_p is the number of pores per unit area, L_c is the total length of cracks per unit area (cm), and P_{mac} is the continuous macroporosity as a fraction of the soil volume.

After ponding, the water and solutes available at the soil surface are allowed to flow into continuous macropores to the limit of macropore flow capacity. For each time step, the flow is sequentially routed downward through the continuous macropores in 1-cm depth increments. In each depth increment, the macropore flow is allowed to flow into the dead-end macropores and be absorbed by the soil matrix by radial or lateral infiltration from both continuous and dead-end parts of the macropores. The dead-end macropores can also store the solution. Time-dependent lateral absorption of the solution occurs only in the drier soil matrix below the transient wetting front generated by the continuing vertical infiltration into the matrix.

This absorption is computed by radial or lateral Green–Ampt-type equations. The transient radial infiltration rate from a cylindrical macropore, V_r, is calculated as

$$V_r = \frac{2\pi K_s \tau_c}{\ln(r_{wf}/r_p)} \tag{1.18}$$

where K_s is saturated hydraulic conductivity (cm h^{-1}), τ_c is a capillary drive term for the soil matrix in the depth increment in question (cm), r_{wf} is the wetted radius at any given time (cm), and r_p is the macropore radius (cm). The wetted radius, r_{wf}, is calculated from the quantity of water that has infiltrated.

Equation 1.18 is not applicable for the very first time step when $r_{wf} = r_p$. Then V_r is calculated as

$$V_r = 2\pi r_p \left[\frac{2K_s\tau_c(\theta_s - \theta_i)}{\Delta t_1/2} \right]^{1/2} \tag{1.19}$$

where $(\theta_s - \theta_i)$ is the initial volumetric soil water deficit in the depth increment, and Δt_1 is the first time step (h).

For planar cracks, the equation used for the lateral infiltration rate per unit length of crack is

$$V_r = \left[\frac{2\tau_c K_s(\theta_s - \theta_i)}{t} \right]^{1/2} \tag{1.20}$$

where all terms are as previously defined, and t is the cumulative time for lateral flow. Both of the above equations are constrained by an upper limit of lateral flow that depends on the initial deficit and average distance between holes or cracks.

Compaction along macropore walls may further influence the ability of the soil to absorb water and chemicals from macropores. To account for such effects, we multiplied the radial infiltration rate from the macropore by an appropriate reduction factor:

$$V_r^* = SFCT \times V_r \tag{1.21}$$

where V_r^* is the adjusted radial infiltration rate, V_r is the Green–Ampt radial infiltration rate, and SFCT is the absorption correction factor, a user-defined parameter.

Maximum absorption is achieved when the depth increment becomes saturated. The above routing continues until the available solution within a given time step is exhausted, a water table is encountered, or the lowest depth of interest is reached. Below the lowest depth, if the bottom boundary condition implies an outward flux, the remaining solution is allowed to drain away. If the bottom boundary is an impermeable layer, however, the remaining solution is added to overland flow. A record of the cumulative absorption and changing soil water content is maintained for each depth increment.

If a high water table is present, the macropores in the saturated region are assumed to be full. Thus, water moves down a macropore below the wetting front until it reaches a zone below which the soil is saturated. At that time, the remaining water may radially infiltrate into the 1-cm increments above the water table, thus raising the table and increasing the drainage flux to tiles according to Hooghoudt's equation (see below). Although this simplifying assumption ignores the radial infiltration rate by putting macropore flow directly into unsaturated increments, it allow us to simulate the macropore flow contribution to an increase in tile drainage.

Chemicals in the macropore flow are absorbed with water by the soil matrix in each depth increment. However, before water and solutes flow into the dead-end pores or are absorbed into the soil matrix, the reactive chemicals in the available macropore flow equilibrate with the wall of the pore, resulting in either a net adsorption to the wall or desorption from the wall. This mixing and equilibration of the macropore solution with the wall occurs in the wetted zone above the wetting front, even though there is no lateral water absorption into the soil matrix in this zone. The walls of pores are assumed to consist of 0.5 mm of soil. After absorption into the soil matrix, the chemicals are uniformly mixed and equilibrated with the soil and water in the mesopores within the depth increment.

Redistribution of Water and Chemicals after Infiltration

Between rainfall or irrigation events, the soil water is redistributed by using the Richards equation:

$$\frac{\partial \theta}{\partial z} = \frac{\partial}{\partial z}\left[K(h,z)\frac{\partial h}{\partial z} - K(h,z)\right] - S(z,t) \tag{1.22}$$

where θ is volumetric soil water content (cm^3 cm^{-3}), t is time (h), z is soil depth (cm, assumed positive downward), h is soil-water pressure head (cm), K is unsaturated hydraulic conductivity (cm h^{-1}), a function of h and z, and $S(z,t)$ is a sink term for root water uptake and tile drainage rates (h^{-1}).

The initial condition is given as

$$h = h(z); \quad t = 0, z \geq 0 \tag{1.23a}$$

The surface boundary condition is an evaporative flux (potential evaporation rate) until the surface pressure head falls below a minimum value (set at −20,000 cm), at which time a constant-head condition is used:

$$-K\frac{\partial h}{\partial z} + K = E; \quad z = 0, 0 < t < t_1, h(z = 0) > h_{min} \tag{1.23b}$$

$$h = h_{min}; \quad z = 0, t > t_1$$

where E is the potential evaporation rate on the soil surface after accounting for the effect of residue cover (cm h^{-1}) and the fraction of the soil surface not shaded by the canopy (see Chapter 3), h_{min} is the minimum value of soil-water pressure head (cm), set equal to −20,000 cm, and t_1 is the time up to which E can be sustained by the soil (h).

The bottom boundary condition can be specified as a unit gradient, constant flux, or constant pressure head:

$$\frac{\partial h}{\partial z} = 0; \quad z = z_w, t > 0$$

or

$$-K\frac{\partial h}{\partial z} + K = v_w; \quad z = z_w, t > 0$$

or

$$h = h(z_w); \quad z = z_w, t > 0 \tag{1.23c}$$

where z_w is the lower boundary of the soil profile (cm), and v_w is the leakage rate through the bottom of the profile (cm h^{-1}).

Before the process of redistribution begins, the chemical concentrations in solution and in the solid-phase equilibrium and kinetic pools are transformed from 1-cm depth increments to coarser and nonuniform depth increments used for computing redistribution. Thereafter, the chemicals in solution move with the Darcy flux of water from one depth increment to another, including upward movement due to evaporation. The chemical movement is done sequentially, starting from the bottom depth increment. At the end of each time step, chemical concentrations in the solution and solid phases are adjusted with respect to both the instantaneous equilibrium and kinetic pools.

Root Water Uptake

The sink term, $S(z,t)$, consists of both the distributed sink due to root uptake and a point sink arising from tile drainage. The root uptake part of the sink term, $S_r(z,t)$ (h^{-1}), is evaluated using the approach of Nimah and Hanks (1973):

$$-S_r(z,t) = \frac{[H_r + (RRES \times z) - h(z,t) - s(z,t)]R(z)K(h)}{\Delta x \Delta z} \tag{1.24}$$

where H_r is an effective root water pressure head (cm), RRES is a root resistance term and the product (RRES $\times$ z) accounts for gravity and friction loss in H_r (assumed = 1.05), $h(z,t)$ is the soil-water pressure head (cm), $s(z,t)$ is the osmotic pressure head (assumed = 0 cm), Δx is the distance from plant roots to where $h(z,t)$ is measured (assumed = 1 cm), Δz is the soil depth increment (cm), $R(z)$ is the proportion of the total root activity in the depth increment Δz as obtained from the plant growth model (could vary with t, and $K(h)$ is the hydraulic conductivity (cm h^{-1}).

The sum total of $S_r(z,t)$ throughout the root zone cannot exceed the potential transpiration demand. Equation 1.24 is solved iteratively by varying H_r until this demand is met, with the condition that H_r does not fall below h_{min}. After H_r reaches h_{min}, this value is assumed to hold steady, whereas the sum total of S_r for all depths (total root water uptake) falls below the potential demand.

Shani and Dudley (1996) used separated equations for root water uptake from the soil based on the soil water matric pressure head gradient, denoted as $S_h(z,t)$, and osmotic pressure head gradients, $S_o(z,t)$. The term for $S_h(z,t)$ still used the Nimah–Hanks equation (Equation 1.24) but without the osmotic term $s(z,t)$. The $S_o(z,t)$ term was based on the work of Maas and Hoffman (1977) as

$$-S_o(z,t) = \frac{S_{max}p(z,t)}{1 + [s(z,t)/s_o 50]^3}$$

where S_{max} is the potential transpiration at t, $s_o 50$ is the osmotic pressure head that causes a 50% yield loss, and $p(z,t)$ is a root distribution term similar to $R(z)$ in Equation 1.24.

The value of $S_r(z,t)$ was then obtained as a multiplicative function:

$$-S_r(z,t) = S_h(z,t)S_o(z,t) \tag{1.24a}$$

The integral of Equation 1.24a with respect to dz gives the total uptake from the profile at any time t. This equation gave a better simulation of the experimental data when osmotic pressure head was important (Shani & Dudley, 1996).

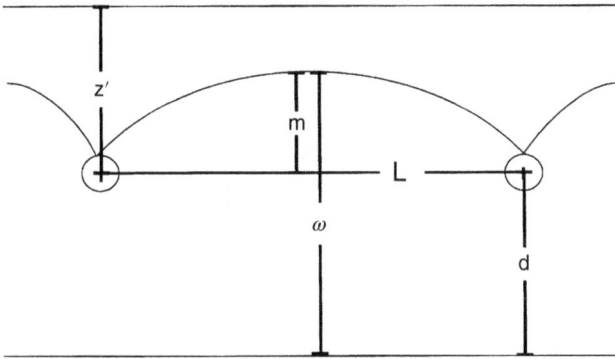

Figure 1.5 An illustration of a soil profile with a high water table and tile drains. The Root Zone Water Quality Model (RZWQM) simulates flux to the drains at the center point between the drains and reports the depth of the water table at this point. Variables represent distances used in calculating the flux to the drains in Equation 1.25.

Fluctuating Water Tables and Tile Drainage during Redistribution

If the soil profile contains a high, fluctuating water table, the depth of the water table is defined as the depth at which the pressure head first becomes non-negative and all heads below that depth are non-negative. It is allowed to fluctuate according to Equations 1.22, 1.23a, 1.23b, and 1.23c). Assuming that the pressure head value at the depth of each numerical node is known and varies linearly between nodes, the depth to the top of the water table ($h = 0$ cm) can be calculated.

Tile drainage is included in the term $S(z,t)$ as a point sink, $S_d(z',t)$ (h^{-1}). It is calculated from Hooghoudt's steady-state equation (Bouwer & van Schilfgaarde, 1963) as applied by Skaggs (1978). This equation is intended to correct for the two-dimensional effects of tile drainage by estimating this flux at the center point between two parallel drains, as shown in Figure 1.5. Thus, model estimates of the depth of the water table are given at the midpoint between drains. Depending on the depth of the water table, the equation for flux to the drain may be written as

$$S_d(z',t) = \frac{8K_e d_e m + 4K_e m^2}{CL^2 \Delta z} \qquad \omega > d$$
$$S_d(z',t) = 0, \qquad \omega \leq d \qquad (1.25)$$

where z' is the user-supplied depth of the drain (cm), ω is the distance from the water table to the bottom of the restricting layer (cm), d is the distance from the drain to the bottom of the restricting layer (cm), m is the water table height above the drain (cm), K_e is the user-supplied or model-calculated effective lateral hydraulic conductivity (cm h^{-1}), L is the user-supplied distance between drains (cm), C is the ratio of the average flux between drains to the flux midway between drains (set = 1.0), d_e is the equivalent depth from the drain to the bottom of the restricting layer (cm), and Δz is the soil depth increment at z' (cm).

The equivalent depth, d_e, used to correct drainage fluxes in Equation 1.25 for convergence near the drain is given by (Moody, 1967)

$$d_e = \frac{d}{1 + (d/L)[(8/\pi) \ln (d/r) - \alpha]}, \qquad 0 < \frac{d}{L} < 0.3$$

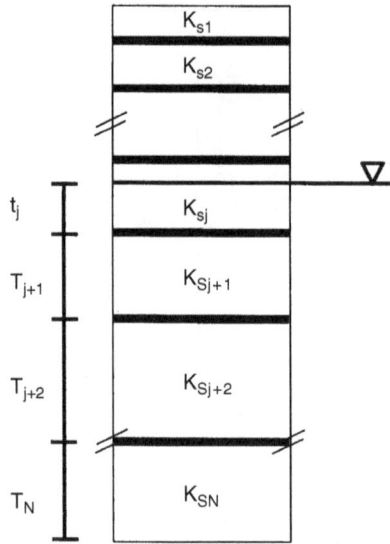

Figure 1.6 Variables used in calculating the effective saturated conductivity below the water table, where the conductivity, K_s, is defined for each horizon.

$$\alpha = 3.55 - \frac{1.6d}{L} + 2\left(\frac{d}{L}\right)^2 \tag{1.26}$$

$$d_e = \frac{L\pi}{8[\ln(L/r) - 1.15]}, \qquad \frac{d}{L} \geq 0.3$$

where r is the drain tube radius (cm) and all other terms are as previously defined. This equation assumes a completely open drain tube. To account for the additional loss of head as soil water approaches real tubes, which have only a finite number of openings, an effective radius, r_e, may be used in place of the real radius, r. This is defined as the radius such that a completely open drain tube with radius r_e will offer the same resistance to inflow as a real tube with radius r (Skaggs, 1978). In general, it is slightly less than the actual radius.

Figure 1.6 depicts the variables used to calculate the effective hydraulic conductivity in the lateral direction, K_e, which is given by

$$K_e = \frac{K_{sj}t_j + \sum_{i=j+1}^{N} K_{si}T_i}{t_j + \sum_{i=j+1}^{N} T_i} \tag{1.27}$$

where K_{si} is the saturated hydraulic conductivity in the ith horizon (cm h^{-1}), T_i is the thickness of the ith horizon (cm), t_j is the depth from the water table to the bottom of the jth horizon, where the water table is in the ith horizon (cm), and N is the number of horizons.

This flux is added to the root uptake at the depth of the drain to become the sink term at that depth

$$
\begin{aligned}
S(z,t) &= S_r(z',t) + S_d(z',t), & z = z' \\
S(z,t) &= S_r(z,t), & z \neq z'
\end{aligned}
\tag{1.28}
$$

Numerical Methods

Equations 1.22, 1.23a, 1.23b, and 1.23c are solved by a mass-conservative, mixed-form iterative finite-difference numerical solution (Celia et al., 1987; Celia et al., 1990). For this purpose, Equation 1.22 is written as

$$\left(\frac{1}{\Delta t}C^{n+1,m}\right)\delta^m - \frac{\partial}{\partial z}\left(K^{n+\alpha,m}\frac{\partial h^{n+1,m+1}}{\partial z}\right) = -S^n(z,t) - \frac{\partial k^{n+\alpha,m}}{\partial z} - \frac{\theta^{n+1,m}-\theta^n}{\Delta t} \qquad (1.29)$$

where the superscripts n, $n+1$, and $n+\alpha$ denote the time step level (with the value of α dependent on the scheme), Δt is the time step, and m is the iteration level. The variables $C^{n+1,m}$, δ^m, and $K^{n+\alpha,m}$ are defined as

$$C^{n+1,m} = \left.\frac{d\theta}{dh}\right|^{n+1,m}$$

$$\delta^m = h^{n+1,m+1} - h^{n+1,m} \qquad (1.30)$$

$$K^{n+\alpha,m} = K\left(\alpha h^{n+1,m} + |1-\alpha|h^n\right)$$

Equations 1.29 and 1.30 are then discretized by using either a fully implicit ($\alpha = 1$) or the Crank–Nicholson central ($\alpha = 2$) differencing scheme depending on the conditions in the soil. In soils with a high water table, we use $\alpha = 1$ and without we use $\alpha = 2$.

The initial guess of the solution of Equation 1.29 is made by taking a single Euler step (Burden & Faires, 1989, pp. 225–227):

$$\theta_i^{n+1,1} =$$

$$\left\{-K_{i-1/2}^n\left[\frac{h_i^n - h_{i-1}^n}{\frac{1}{2}\Delta z_i + \frac{1}{2}\Delta z_{i-1}}\right] + K_{i-1/2}^n + K_{i+1/2}^n\left[\frac{h_{i+1}^n - h_i^n}{\frac{1}{2}\Delta z_{i+1} + \frac{1}{2}\Delta z_i}\right] - K_{i+1/2}^n\right\}\frac{\Delta t}{\Delta z} + \theta_i^n \qquad (1.31)$$

$$K_{i\pm 1/2}^n = \sqrt{K(h_{i\pm 1}^n) \times K(h_i^n)}$$

and then determining $h_i^{n+1,1}$ from the soil water content–matric suction relationship.

The space mesh increments increase with depth from 1.0 to 10.0 cm, according to principles outlined below and shown in Figure 1.1b. The time increment increases with time from 10^{-5} to 1.0 hours and can be set back to aid convergence of Equation 1.29 when needed. If, during any time step, the scheme should fail to converge even after setting the time step back, and the cell by cell mass balance using the current solution would be off by more than 0.5cm, the solution is set to the initial guess from Equation 1.31. This prevents erroneous solutions that may occur due to divergence of the solution. The solution of the linear system of equations at each iteration step is obtained by the Thomas algorithm.

Space Discretization

The basic principles of the space discretization are the same as described by Bresler (1973) and Ahuja and Swartzendruber (1992). For a layered soil, the placement of the space mesh increments (the numerical layers) for the numerical solution of the Richards equation must follow predetermined criteria to achieve convergence. The rules for numerical layer formation are:

- All layer thicknesses must be an integer number of centimeters; no fractions are allowed. This requires that the horizon thicknesses must also be integer numbers.

- A boundary of a numerical layer must coincide with a horizon boundary. It follows that a small adjustment may have to be made to the horizon thickness to achieve a consistent set of numerical layer boundaries and horizon boundaries.

- To satisfy the solution technique used, numerical layer boundaries are centered between solution nodes. To adhere to this rule, the distance between nodes must be odd (i.e., 1, 3, 5, 7, ...) in value. This will allow layer thicknesses to remain integer values while increasing in thickness. To simulate flux boundary conditions, a fictitious node is created above the soil surface and another below the bottom of the profile.

- The first horizon must be greater than 2 cm thick.

When these rules are followed, the model will correctly move mass between the three grid systems inherent in the model. If mass imbalances appear during simulation, the grid system should be checked to ensure adherence to the numerical layer formation rules. The data-file generator program will generate correct numerical layers from horizon values.

Another factor affecting space discretization comes from the simulation of tillage. As a consequence of primary and secondary tillage, soil properties in some horizons will change. The model may create horizons in the soil profile to ensure that soil properties are changed only within the tillage zone. The depth of each type of tillage, however, varies with the instrument used. If the depth of the primary tillage is within ±1/3 of the thickness of the top soil horizon, we assume that the entire top horizon has been tilled. Otherwise, we create a new tilled horizon. The same procedure is applied for the secondary tillage. The secondary tillage generally results in creating a new horizon of its own, carved out of the original untilled or primary-tilled horizon. The depth of the secondary tilled horizon may be 2–3 cm or more depending on the implement used. When new tillage horizons are created, the numerical layers' thicknesses are adjusted to conform to the new horizon boundaries. Tillage operations are assumed to change only the soil bulk density and macroporosity, which in turn change the soil hydraulic and other properties. The decrease in soil bulk density of the tilled horizons is reversed by reconsolidation brought about by wetting and drying during subsequent rainfalls. The changes in soil physical and hydraulic properties due to tillage and reconsolidation are described in Chapter 11.

Exercises on Evaluation of the Water and Chemical Transport Model

Preferential Macropore Transport Studied with the RZWQM

The RZWQM components dealing with preferential water and chemical transport were used to study macropore flow and transport in a silty clay loam soil (Ahuja et al., 1993). Macroporosity of the soil was assumed to be 0.05% by volume, half of which was continuous and the rest discontinuous. Two rainfall sequences with two initial soil water contents, evaporation versus transpiration, macropore radius ranging from 1.0 to 0.125 mm, and three different chemicals were evaluated. During a five-weeks period, weekly rainfall of either 12.7 or 25.4 cm in one hour, with soil water redistribution and evaporation or transpiration occurring between storms, generated no macropore flow when the soil was initially dry (h = –15,000 cm). A slight amount of macropore flow was generated under the same rainfall when the soil was initially wet (h = –330 cm). Doubling the weekly rainfall amount and intensity generated macropore flow varying between 30 and 50% of rainfall depending on the initial and boundary conditions (Table 1.1). Chemicals transported with this flow were 0.05–8% of the

Table 1.1 Cumulative Amount of Water Entering the Macropores, Other Water Balance Components, and Amounts of Chemicals Transported in Macropores Under Different Conditions. Data for Only the Largest and Smallest Macropore Size are Presented.

Rainfall sequence—total amount for five rainfall events	Initial soil water pressure head	Upper boundary condition—evaporation (E) transpiration (T) or none (0.)[a]	Radius of macropores, if present	Flow entering the macropores (if applicable)	Runoff	Seepage below 150 cm	Actual evaporation or transpiration	Macropore flow as percent of rainfall	Amount of chemicals transported in macropores as a percent of the amount applied at the surface (40 µg/cm²)		
									Atrazine	Prometryn	Nitrate
(cm)	(cm)		(cm)	(cm)	(cm)	(cm)	(cm)	(%)	(%)	(%)	(%)
RAIN1 12.7	−15,000	0.	–	–	0.086	0.4×10^{-4}	0.	–	–	–	–
		E	–	–	0.0	0.4×10^{-4}	6.04	–	–	–	–
		E	0.10	0.	0.	0.4×10^{-4}	6.04	0.	0.	0.	0.
		E	0.0125	0.	0.	0.4×10^{-4}	6.04	0.	0.	0.	0.
		T	–	–	0.	0.4×10^{-4}	9.44	–	–	–	–
		T	0.10	0.	0.	0.4×10^{-4}	9.44	0.	0.	0.	0.
		T	0.0125	0.	0.	0.4×10^{-4}	9.44	0.	0.	0.	0.
	−330	0.	–	–	0.421	3.24	0.	–	–	–	–
		0.	0.10	0.426	0.	3.48	0.	3.35	0.13	0.125	0.57×10^{-3}
		0.	0.0125	0.426	0.	3.48	0.	3.35	0.13	0.125	0.57×10^{-3}
		E	–	–	0.006	0.50	8.87	–	–	–	–
		E	0.10	0.006	0.	0.50	8.87	0.05	0.01	0.35×10^{-2}	0.55×10^{-3}
		E	0.0125	0.006	0.	0.50	8.87	0.05	0.01	0.35×10^{-2}	0.55×10^{-3}
		T	–	–	0.006	0.22	13.82	–	–	–	–
		T	0.10	0.006	0.	0.22	13.82	0.05	0.01	0.35×10^{-2}	0.55×10^{-3}
		T	0.0125	0.006	0.	0.22	13.82	0.05	0.01	0.35×10^{-2}	0.57×10^{-3}
RAIN2 25.4	−15,000	0.	–	–	9.29	$.4 \times 10^{-4}$	0.	–	–	–	–
		0.	0.0125	10.67	0.	$.4 \times 10^{-4}$	0.	42.0	3.62	3.17	0.06
		E	–	–	7.38	$.4 \times 10^{-4}$	7.67	–	–	–	–
		E	0.10	8.43	0.	$.4 \times 10^{-4}$	9.05	33.2	4.62	2.67	0.13
		E	0.0125	8.48	0.	$.4 \times 10^{-4}$	9.05	33.4	4.65	2.70	0.13
		T	–	–	6.26	$.4 \times 10^{-4}$	13.14	–	–	–	–
		T	0.10	7.71	0.	$.4 \times 10^{-4}$	13.74	30.3	2.65	2.22	0.05
		T	0.0125	7.75	0.	$.4 \times 10^{-4}$	13.74	30.5	2.67	2.25	0.05

(continued)

Table 1.1 (*continued*)

Rainfall sequence-total amount for five rainfall events (cm)	Initial soil water pressure head (cm)	Upper boundary condition—evaporation (E) transpiration (T) or none (0.)[a]	Radius of macropores, if present (cm)	Flow entering the macropores (if applicable) (cm)	Runoff (cm)	Seepage below 150 cm (cm)	Actual evaporation or transpiration (cm)	Macropore flow as percent of rainfall (%)	Amount of chemicals transported in macropores as a percent of the amount applied at the surface (40 µg/cm²) Atrazine (%)	Prometryn (%)	Nitrate (%)
	−330	0.	–	–	11.03	4.81	0.	–	–	–	–
		0.	0.10	12.28	0.	13.61	0.	48.3	6.52	4.15	0.82
		0.	0.0125	12.33	0.	13.61	0.	48.5	6.57	4.17	0.82
		E	–	–	8.78	0.93	9.89	–	–	–	–
		E	0.10	9.97	0.	5.07	11.09	39.2	7.85	3.57	1.30
		E	0.0125	10.02	0.	5.07	11.09	39.4	7.90	3.60	1.32
		T	–	–	9.11	0.35	13.82	–	–	–	–
		T	0.10	10.88	0.	2.98	13.82	42.8	5.77	3.62	0.82

Note. Data for only the largest and smallest macropore size are presented.

[a] E, evaporation; T, transpiration; 0, none. Potential evaporation or transpiration imposed was 0.48 cm d^{-1}.

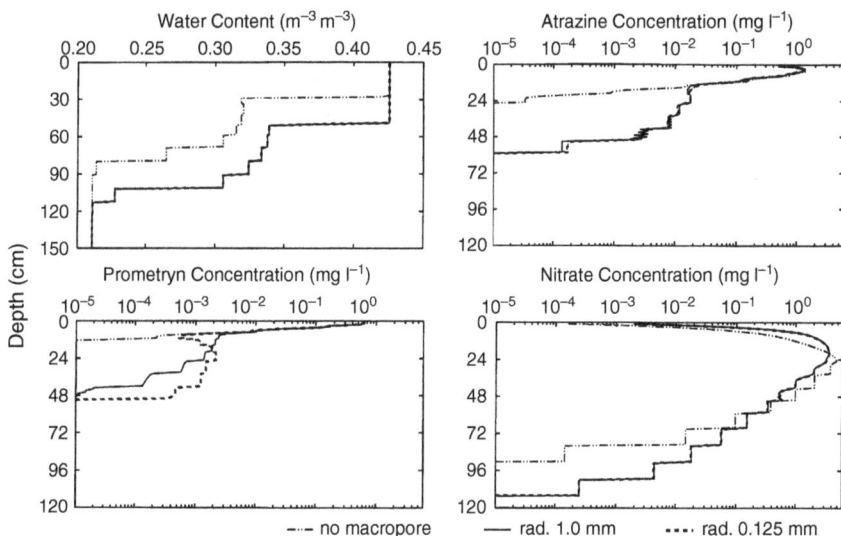

Figure 1.7 Water content and chemical concentration distributions in soil after application of 25.4 cm of rainfall, comparing results with no macropores versus macropores of two extreme sizes (1.0- and 0.125-mm radius). Initial soil water pressure head was −15,000 cm and the potential evaporation imposed at the soil surface was −4.8mm d^{-1}.

surface-applied amount, depending on conditions and the type of chemical. A moderately adsorbed chemical (atrazine) was the most susceptible to macropore transport, followed in order by a strongly adsorbed chemical (prometryn) and a mobile chemical (NO_3). The flow entering the macropores was partially absorbed by soil at progressively deeper depths; it increased the water content of the root zone and created a tail of low concentrations in the soil chemical content distributions (Figure 1.7). The macropore size had very little effect on macropore flow and transport, but the smallest size pores retarded the downward chemical movement by wall adsorption a little more than the largest size pores. Surface evaporation decreased macropore flow, soil water contents, and downward chemical movement but increased the chemical content of the macropore flow. Transpiration, on the other hand, decreased both macropore flow and its chemical content (Figure 1.8).

Macropore Transport of a Surface-Applied Bromide Tracer

This sub-model was further evaluated and refined by testing against experimental data from controlled studies in soil columns (Ahuja et al., 1995). The experiments consisted of eight treatment combinations, each in duplicate, of the following conditions: soil initially air dry versus soil initially wetted by rainfall; a 1.0-cm layer of dry aggregates on the surface versus no aggregates; and a 3-mm artificial macropore made along the column's vertical axis versus no macropore. A solution of SrBr^2 was atomized across the surface, followed by the application of simulated rainfall. The available data utilized were the amount of water and Br$^-$ in surface runoff or in macropore bottom outflow (seepage), soil water content distributions, and Br$^-$ content distributions with depth,

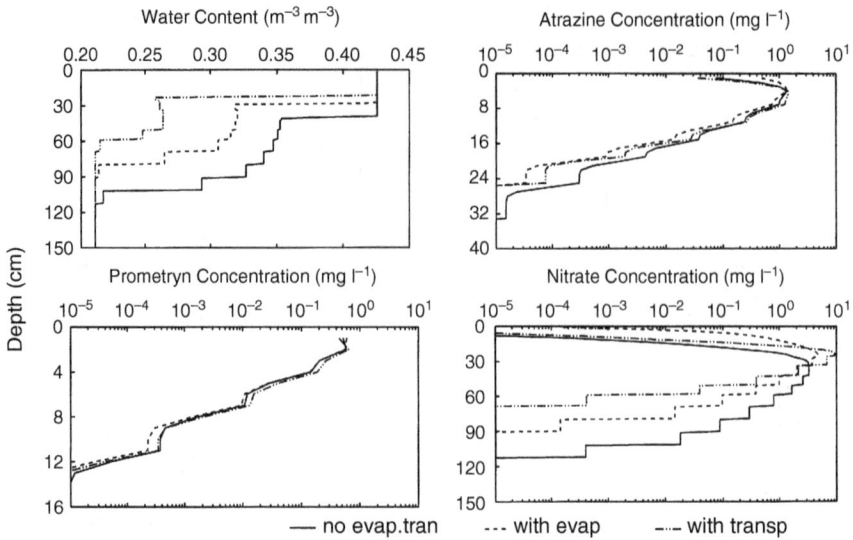

Figure 1.8 Comparison of soil moisture levels and chemical concentrations with no macropores and with and without evaporation and transpiration. Plots show conditions after application of 25.4 cm of rainfall. Initial soil water pressure head was −15,000 cm, evaporation and transpiration rates (when imposed) were −4.8 mm d^{-1}.

both above and below the main wetting front. Evaluation of the model simulations indicated that:

- The viscous resistance and entrapped air correction factor for reducing the Green–Ampt infiltration rates varied from about 2.0 for initially dry soil to about 3.0 for a prewetted soil.

- The effective lateral absorption of the macropore wall needed to be adjusted for compaction and variable water pressure around the pore circumference under small macropore flow rates.

- Better results were obtained if, in the wetted portion of the soil profile, the flow down the macropore mixed with about 0.5 mm of soil around the walls and its soil solution.

- Microporosity of surface aggregates was an important factor that held back chemical near the surface from downward leaching but kept this chemical available for mixing with rainfall, thus increasing the transfer to runoff or macropore flow.

With these refinements and calibration, the model simulations gave generally good descriptions of the data (Figures 1.9–1.14).

Other Evaluations

The sub-model has also been evaluated on field experimental data sets from the Netherlands and from Watkinsville, GA (Ahuja et al., 1996; Ma et al., 1995). For the Dutch data, the sub-model, with soil hydraulic properties estimated from bulk density and water content at $h = -333$ cm, gave good results for soil water content and Br$^-$ distributions. For two pesticides (atrazine and metribuzine), the distributions were reasonably reproduced when the kinetic option was used with the two-site model. For the

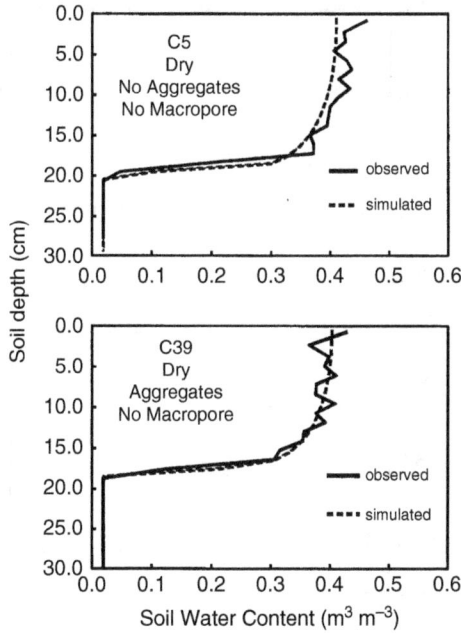

Figure 1.9 Comparison of observed and simulated soil water content distributions for initially dry columns without a macropore, with and without surface aggregates.

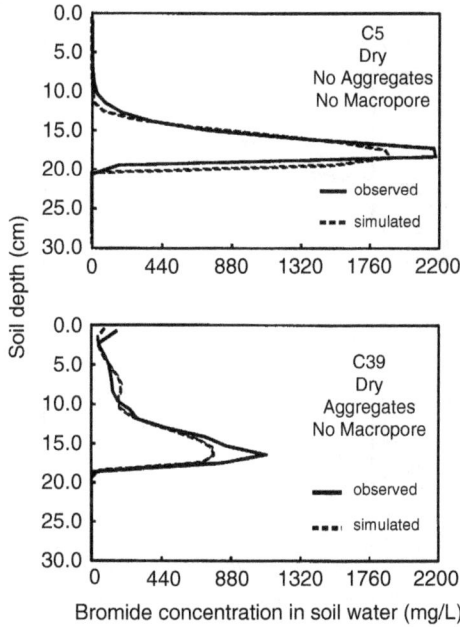

Figure 1.10 Comparison of observed and simulated soil Br⁻ concentration distributions for initially dry columns without a macropore, with and without surface aggregates.

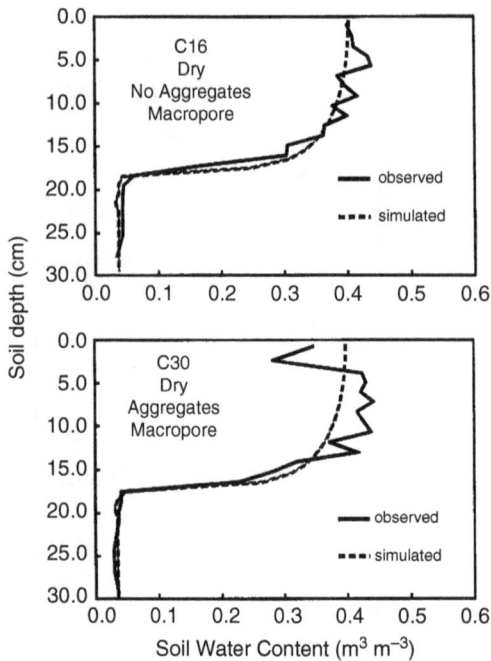

Figure 1.11 Comparison of observed and simulated soil water content distributions for initially dry columns with a macropore, with and without surface aggregates.

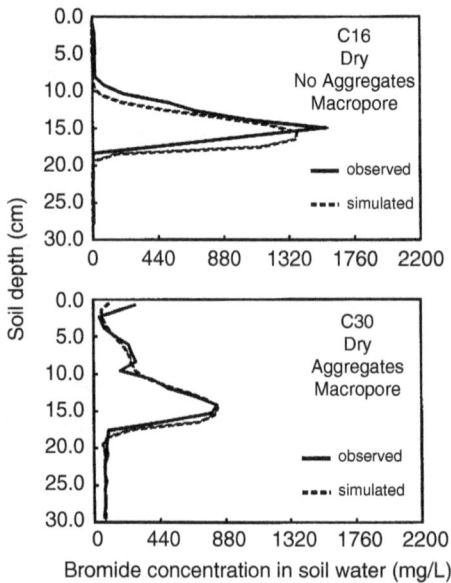

Figure 1.12 Comparison of observed and simulated soil Br⁻ concentration distributions for initially dry columns with a macropore, with and without surface aggregates.

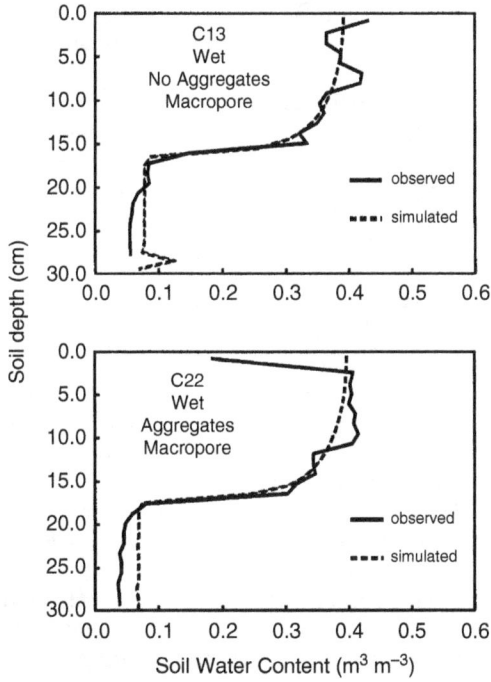

Figure 1.13 Comparison of observed and simulated soil water content distributions for initially wet columns with a macropore, with and without surface aggregates.

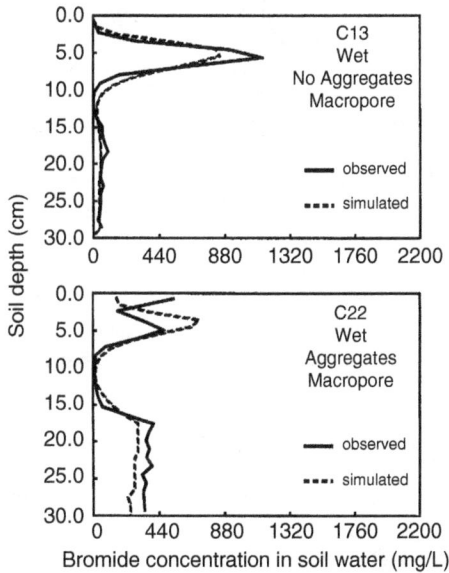

Figure 1.14 Comparison of observed and simulated soil Br⁻ concentration distributions for initially wet columns with a macropore, with and without surface aggregates.

27

Watkinsville site, the results for both soil water content and pesticides were good if a soil crust was included to match the peak runoff in one or two selected events. The pesticide concentrations in the runoff were also very well reproduced.

Using field data from the irrigation project of the Sorraia River in southern Portugal, RZWQM-simulated water content distributions during a growing season compared very favorably with observed distributions (Cameira et al., 1998). The model predicted the drying and wetting patterns seen in the top 30 cm before and after irrigations and came very close to the observed values of water content. The model also predicted very little flux below the compacted layer between 30 and 45 cm. On the same dates, predicted and observed NO_3–N concentrations were also compared. The NO_3–N concentrations were not predicted as accurately, particularly before and after the first irrigation event, but later in the season the simulated profile often came close to the range of observed values. It was demonstrated that the model did predict the downward movement of NO_3–N with water at rates comparable to those observed.

In another evaluation against soil-box data from Tifton, GA, the sub-model simulations compared very well with the experimental data for Br^- and pesticides in leaching and runoff when the kinetic option was used (Ma et al., 1996). This work demonstrated that the kinetics is extremely important in describing leaching and runoff losses from shallow soils and that the kinetics actually increased leaching of pesticides and amounts in runoff under these conditions.

Simulation of flow and transport in field soils under both dryland and irrigated corn production in eastern Colorado have further shown the ability of the RZWQM to predict these processes for a range of conditions (Buchleiter et al., 1995; Farahani et al., 1995a; 1995b, 1995c).

The tile drainage component of the RZWQM has been tested in a series of field studies in Iowa as summarized by Malone et al. (2013). The model simulated the total drainage, timing of peak flow, and seasonality fairly well. In addition, the tillage effects on tile flow were in agreement with experimental trends.

In the last two decades, extensive field testing and evaluation of the RZWQM have been done under different soil, climate, and management conditions for various applications, generally with good results. Water was the dominant factor in these studies, and the testing included parameterization of water and other important linked processes. Studies on the soil water management effects on crop water use and water quality under semiarid conditions are of special interest in this chapter (Anapalli et al., 2008, 2014, 2015a, 2015b, Bakhsh et al., 2004; Cameira et al., 2007; Hu et al., 2006; Ma et al., 2007, 2012). Recently, model use was extended to the humid conditions of the Mississippi Delta (Anapalli et al., 2016, 2018, 2021, 2019a, 2019b; Pinnamaneni et al., 2020a; 2020b; Yang et al., 2019). Guzman and Fox (2011) enhanced the RZWQM to simulate transport of soil fecal bacteria through soil macropores with good results. These studies indicate the robustness of the processes of water and chemical transport in the soil matrix and macropores described in this chapter.

Areas of Further Research

- Contribution of snow to soil water (infiltration) in conjunction with soil freezing, described in Chapter 2.

- Quantification of the effect of spatial variability of soil water properties on soil water distribution and water uptake from a field.

- Root growth distribution and its effect on water uptake distribution and on water and mass transport.

- Passive versus active plant uptake of N in conjunction with C and N and plant growth, described in the separate chapters on C and N (Chapter 5) and plant growth (Chapters 8, 9, and 10).

This chapter on soil water and chemical transport is a fundamental knowledge base that is essentially needed for advancing further research and best management in all other areas (chapters) in this book, especially for optimizing ET, transpiration and photosynthesis, surface soil and water management, climate change adaptations, heat waves, and pesticides (Chapters 3, 4, 7, 8, 10, 11, and 12).

References

Addiscott, T. M. (1977). A simple computer model for leaching in structured soils. *Journal of Soil Science*, 28, 554–563.

Ahuja, L. R. (1983). Modeling infiltration into crusted soils by the Green–Ampt approach. *Soil Science Society of America Journal*, 47, 412–418. https://doi.org/10.2136/sssaj1983.03615995004700030004x

Ahuja, L. R. (1986). Characterization and modeling of chemical transfer to runoff. *Advances in Soil Science*, 4, 149–188.

Ahuja, L. R. (1990). Modeling soluble chemical transfer to runoff with rainfall impact as a diffusion process. *Soil Science Society of America Journal*, 54, 312–321. https://doi.org/10.2136/sssaj1990.03615995005400020003x

Ahuja, L. R., Cassel, D. K., Bruce, R. R., & Barnes, B. B. (1989). Evaluation of spatial distribution of hydraulic conductivity using effective porosity data. *Soil Science*, 148, 404–411.

Ahuja, L. R., DeCoursey, D. G., Barnes, B. B., & Rojas, K. W. (1993). Characteristics of macropore transport studied with the ARS Root Zone Water Quality model. *Transactions of the ASAE*, 36, 369–380.

Ahuja, L. R., Johnsen, K. E., & Heathman, G. C. (1995). Macropore transport of a surface-applied bromide tracer: Model evaluation and refinement. *Soil Science Society of America Journal*, 59, 1234–1241. https://doi.org/10.2136/sssaj1995.03615995005900050004x

Ahuja, L. R., Johnsen, K. E., & Rojas, K. W. (2000). Water and chemical transport in soil matrix and macropores. In L. R. Ahuja, K. W. Rojas, J. D. Hanson, M. J. Shaffer, & L. Ma (Eds.), *Root Zone Water Quality Model: Modeling management effects on water quality and crop production* (pp. 13–50). Water Resources Publications.

Ahuja, L. R., Ma, Q. L., Rojas, K. W., Boesten, J. T. I., & Farahani, H. J. (1996). A field test of the Root Zone Water Quality Model: Pesticide and bromide behavior. *Pesticide Science*, 48, 101–108.

Ahuja, L. R., Naney, J. W., & Williams, R. D. (1985). Estimating soil water characteristics from simpler properties or limited data. *Soil Science Society of America Journal*, 49, 1100–1105. https://doi.org/10.2136/sssaj1985.03615995004900050005x

Ahuja, L. R., & Swartzendruber, D. (1992). Flow through crusted soils: Analytical and numerical approaches. In M. E. Sumner & B. A. Stewart (Eds.), *Soil crusting: Chemical and physical processes* (Advances in Soil Science (pp. 93–122). CRC Press.

Ahuja, L. R., & Williams, R. D. (1991). Scaling water characteristic and hydraulic conductivity based on Gregson–Hector–McGown approach. *Soil Science Society of America Journal*, 55, 308–319. https://doi.org/10.2136/sssaj1991.03615995005500020x

Anapalli, S. S., Ahuja, L. R., Nielsen, D. C., Trout, T. J., & Ma, L. (2008). Use of crop simulation models to evaluate limited irrigation management options for corn in a semiarid environment. *Water Resources Research*, 44, W00E02. https://doi.org.10.1029/2007WR006181

Anapalli, S. S., Ahuja, L. R., Ma, L., Nielsen, D. C., Trout, T. J., Andales, A. A., Chávez, J. L., & Ham, J. (2014). Enhancing the water stress factors for simulation of corn (*Zea mays* L.) in RZWQM2. *Agronomy Journal*, 106, 81–94. https://doi.org/10.2134/agronj2013.0300

Anapalli, S. S., Ahuja, L. R., Ma, L., Trout, T. J., McMaster, G. S., Nielsen, D. C., Hamd, J. M., Andales, A. A., Halvorsone, A. D., Chávezf, J. L., & Fang, Q. X. (2015a). Developing and generalizing average corn crop water production functions across years and locations using a system model. *Agricultural Water Management*, 157, 65–77. https://doi.org/10.1016/j.agwat.2014.09.002

Anapalli, S. S., Trout, T. J., Ahuja, L. R., Ma, L., McMaster, G. S., Nielsen, D. C., Andales, A. A., Chavez, J. L., & Ham, J. (2015b). Quantification of crop water stress factors from soil water measurements in limited irrigation experiments. *Agricultural Systems*, 137, 191–205. https://doi.org/10.1016/j.agsy.2014.11.005

Anapalli, S. S., Fisher, D. K., Reddy, K. N., Krutz, J. L., Pinnamaneni, S. R., & Sui, R. (2019a). Quantifying water and CO_2 fluxes and water use efficiencies across irrigated C_3 and C_4 crops in a humid climate. *Science of the Total Environment*, 663, 338–350. https://doi.org/10.1016/j.scitotenv.2018.12.471

Anapalli, S. S., Fisher, D. K., Reddy, K. N., Rajan, N., & Pinnamaneni, S. R. (2019b). Modeling evapotranspiration for irrigation water management in a humid climate. *Agricultural Water Management*, 225, 105731. https://doi.org/10.1016/j.agwat.2019.105731

Anapalli, S. S., Fisher, D. K., Reddy, K. N., Pettigrew, W. T., Sui, R., & Ahuja, L. R. (2016). Vulnerabilities and adapting irrigated and rainfed cotton to climate change in the Lower Mississippi Delta region. *Climate*, 4, 55. https://doi.org/10.3390/cli4040055

Anapalli, S. S., Pinnamaneni, S. R., Fisher, D. K., & Reddy, K. N. (2021). Vulnerabilities of irrigated and rainfed corn to climate change in a humid climate. *Climatic Change*, 164, 5. https://doi.org/10.1007/s10584-021-02999-0

Anapalli, S. S., Reddy, K. N., & Jagadamma, S. (2018). Conservation tillage impacts and adaptations in irrigated corn (*Zea mays* L.) production in a humid climate (includes modeling). *Agronomy Journal*, 110, 2673–2686. https://doi.org/10.2134/agronj2018.03.0195

Bakhsh, A., Hatfield, J. L., Kanwar, R. S., Ma, L., & Ahuja, L. R. (2004). Simulating nitrate drainage losses from a Walnut Creek Watershed field. *Journal of Environmental Quality*, 33, 114–123. https://doi.org/10.2134/jeq2004.1140

Beven, K. J., & Germann, P. (1981). Water flow in soil macropores: II. A combined flow model. *Journal of Soil Science*, 32, 15–29.

Bouwer, H. (1969). Infiltration of water into a nonuniform soil. *Journal of the Irrigation and Drainage Division, American Society of Civil Engineers*, 95(IR4), 451–462. https://doi.org/10.1061/JRCEA4.0000669

Bouwer, H., & van Schilfgaarde, J. (1963). Simplified method of predicting the fall of water table in drained land. *Transactions of the ASAE*, 6(288–291), 296. https://doi.org.10.13031/2013.40893

Brakensiek, D. L., & Onstad, C. A. (1977). Parameter estimation of the Green and Ampt infiltration equation. *Water Resources Research*, 13, 1009–1012.

Bresler, E. (1973). Simultaneous transport of solutes and water under transient unsaturated flow conditions. *Water Resources Research*, 9, 975–986.

Brooks, R. H., & Corey, A. T. (1964). *Hydraulic properties of porous media* (Hydrology paper 3). Colorado State University.

Buchleiter, G. W., Farahani, H. J., & Ahuja, L. R. (1995). Model evaluation of ground water contamination under center pivot irrigated corn in eastern Colorado. In C. Heatwole (Ed.), *Proceedings of the International Symposium on Water Quality Modeling, Orlando, FL* (pp. 41–50). American Society of Agricultural Engineers.

Burden, R. L., & Faires, J. D. (1989). *Numerical analysis* (4th ed.). PWS-Kent Publishing Company.

Cameira, M. R., Fernando, R. M., Ahuja, L. R., & Ma, L. (2007). Using RZWQM to simulate the fate of nitrogen in field soil–crop environment in the Mediterranean region. *Agricultural Water Management*, 90, 121–136. https://doi.org/10.1016/j.agwat.2007.03.002

Cameira, M. R., Sousa, P. L., Farahani, H. J., Ahuja, L. R., & Pereira, L. S. (1998). Simulation of water and nitrate movement under fertigation applied to level-basins. *European Journal of Agricultural Engineers*, 69, 331–341.

Cameron, D. R., & Klute, A. (1977). Convective–dispersive solute transport with a combined equilibrium and kinetic adsorption model. *Water Resources Research*, 13, 183–188.

Campbell, G. S. (1974). A simple method for determining unsaturated conductivity from moisture retention data. *Soil Science*, 117, 311–314.

Celia, M. A., Ahuja, L. R., & Pinder, G. F. (1987). Orthogonal collocation and alternating-direction procedures for unsaturated flow problems. *Advances in Water Resources*, 10, 178–187.

Celia, M. A., Bouloutaas, E. T., & Zarba, R. L. (1990). A general mass-conservative numerical solution for the unsaturated flow equation. *Water Resources Research*, 26, 1483–1496.

Childs, E. C., & Bybordi, M. (1969). The vertical movement of water in a stratified porous material. *Water Resources Research*, 5, 446–459.

Farahani, H. J., Ahuja, L. R., Buchleiter, G. W., & Peterson, G. A. (1995a). Mathematical modelling of irrigated and dryland corn production in eastern Colorado. In *Clean water–clean environment–21st century: Team Agriculture working to protect water resources: Conference proceedings, Kansas City, MO* (pp. 97–100). American Society of Agricultural Engineers.

Farahani, H. J., Ahuja, L. R., Peterson, G. A., Buchleiter, G. W., & Sherrod, L. A. (1995b). Modelling of dryland and irrigated corn production systems in eastern Colorado. In *Proceeding of the Workshop on Computer Applications in Water Management, Fort Collins, CO*. Fort Collins, CO: Great Plains Agricultural Council.

Farahani, H. J., Ahuja, L. R., Peterson, G. A., Sherrod, L. A., & Mrabet, R. (1995c). Root Zone Water Quality Model evaluation of dryland/no-till crop production in eastern Colorado. In C. Heatwole (Ed.), *Proceedings of the International Symposium on Water Quality Modeling, Orlando, FL* (pp. 11–20). American Society of Agricultural Engineers.

Green, W. H., & Ampt, G. A. (1911). Studies on soil physics: 1. Flow of air and water through soils. *The Journal of Agricultural Science*, 4, 1–24. https://doi.org/10.1017/S0021859600001441

Gregson, K., Hector, D. J., & McGowan, M. (1987). A one-parameter model for the soil water characteristic. *European Journal of Soil Science*, 38, 483–486. https://doi.org/10.1111/j.1365-2389.1987.tb02283.x

Guzman, J. A., & Fox, G. A. (2011). Implementation of biopore and soil fecal bacteria fate and transport routines in the Root Zone Water Quality Model (RZWQM). *Transactions of the ASABE*, 55, 73–84.

Hachum, A. Y., & Alfaro, J. F. (1980). Rain infiltration into layered soils: Prediction. *Journal of the Irrigation and Drainage Division, American Society of Civil Engineers*, 106, 311–321.

Hanks, R. J., & Bowers, S. A. (1962). Numerical solution of the moisture flow equation for infiltration into layered soils. *Soil Science Society of America Proceedings*, 26, 530–534. https://doi.org/10.2136/sssaj1962.03615995002600060007x

Heathman, G. C., Ahuja, L. R., & Baker, J. L. (1986). Test of a non-uniform mixing model for transfer of herbicides to surface runoff. *Transactions of the ASAE*, 39(450–455), 461.

Hetrick, D. M., Holdeman, J. T., & Luxmoore, R. J. (1982). *AGTEHM: Documentation of modifications to Terrestrial Ecosystem Hydrology Model (TEHM) for agricultural applications (ORNL/TM-7856)*. Oak Ridge National Laboratory.

Hoogmoed, W. B., & Bouma, J. (1980). A simulation model for predicting infiltration into cracked clay soil. *Soil Science Society of America Journal*, 44, 458–461. https://doi.org/10.2136/sssaj1980.03615995004400030003x

Hu, C., Saseendran, S. A., Green, T. R., Ma, L., Li, X., & Ahuja, L. R. (2006). Evaluating nitrogen and water management in a double-cropping system using RZWQM. *Vadose Zone Journal*, 5, 493–505. https://doi.org/10.2136/vzj2005.0004

Kosugi, K. (1996). Lognormal distribution model for unsaturated soil hydraulic properties. *Water Resources Research*, 32, 2697–2703.

Ma, L., Ahuja, L. R., Nolan, B. T., Malone, R. W., Trout, T. J., & Qi, Z. (2012). Root Zone Water Quality Model (RZWQM2): Model use, calibration and validation. *Transactions of the ASABE*, 55, 1425–1446.

Ma, L., Malone, R. W., Heilman, P., Karlen, D. L., Kanwar, R. S., Cambardella, C. A., Saseendran, S. A., & Ahuja, L. R. (2007). RZWQM simulation of long-term crop production, water and nitrogen balances in northeast Iowa. *Geoderma*, 140, 247–259.

Ma, Q., Ahuja, L. R., Rojas, K. W., Ferreira, V. F., & DeCoursey, D. G. (1995). Measured and RZWQM predicted atrazine movement in a field soil. *Transactions of the ASAE*, 38, 471–479.

Ma, Q., Ahuja, L. R., Wauchope, R. D., Johnsen, K. E., & Burgoa, B. (1996). Comparison of instantaneous equilibrium and equilibrium–kinetic sorption models for simultaneous leaching and runoff of pesticides. *Soil Science*, 161, 646–655.

Maas, E. V., & Hoffman, G. J. (1977). Crop salt tolerance: Current assessment. *Journal of the Irrigation and Drainage Division of the American Society of Civil Engineers*, 103, 115–134.

Malone, R. W., Logsdon, S., Shipitalo, M. J., Weatherington-Rice, J., Ahuja, L., & Ma, L. (2013). Tillage effect of macroporosity and herbicide transport in percolate. *Geoderma*, 116, 191–216. https://doi.org/10.1016/S0016-7061(03)00101-0

Mein, R. H., & Larson, C. L. (1973). Modeling infiltration during a steady rain. *Water Resources Research*, 9, 384–394.

Moody, W. T. (1967). Nonlinear differential equation of drain spacing. *Journal of the Irrigation and Drainage Division, American Society of Civil Engineers*, 92(IR2), 1–10. https://doi.org/10.1061/JRCEA4.0000420

Morel-Seytoux, H. J., & Khanji, J. (1974). Derivation of an equation of infiltration. *Water Resources Research*, 10, 795–800.

Nimah, M., & Hanks, R. J. (1973). Model for estimating soil–water–plant–atmospheric interrelation: I. Description and sensitivity. *Soil Science Society of America Proceedings*, 37, 522–527. https://doi.org/10.2136/sssaj1973.03615995003700040018x

Pinnamaneni, S. R., Anapalli, S. S., Reddy, K. N., & Fisher, D. K. (2020a). Effects of irrigation and planting geometry on cotton productivity and water use efficiency. *The Journal of Cotton Science*, 24, 2–10.

Pinnamaneni, S. R., Anapalli, S. S., Reddy, K. N., Fisher, D. K., & Quintana-Ashwell, N. E. (2020b). Assessing irrigation water use efficiency and economy of twin-row soybean in the Mississippi Delta. *Agronomy Journal*, 112, 4219–4231. https://doi.org/10.1002/agj2.20321

Rawls, W. J., Brakensiek, D. L., & Saxton, K. E. (1982). Estimation of soil water properties. *Transactions of the ASAE*, 25(1316–1320), 1328.

Shani, U., & Dudley, L. M. (1996). Modeling water uptake by roots under water and salt stress: Soil-based and crop response sink terms. In Y. Waisel, A. Eshel, & U. Kafkafi (Eds.), *Plant roots: The hidden half* (2nd ed., pp. 635–641). Marcel Dekker.

Simmons, C. S., Nielsen, D. R., & Biggar, J. W. (1979). Scaling of field-measured soil water properties. *Hilgardia*, 47, 77–154.

Šimůnek, J., Šejna, M., Saito, H., Sakai, M., & van Genuchten, M. T. (2013). *The HYDRUS-1D software package for simulating the one-dimensional movement of water, heat, and multiple solutes in variably-saturated media*. Department of Environmental Sciences, University of California.

Skaggs, R. W. (1978). *A water management model for shallow water table soils* (Report no. 134). Water Resources Research Institute of the University of North Carolina.

Swartzendruber, D. (1987). Rigorous derivation and interpretation of the Green and Ampt equation. In Y. S. Fok (Ed.), *Proceedings of the International Conference on Infiltration Development and Applications* (pp. 28–37). Water Resources Research Center, University of Hawaii.

van Genuchten, M. T. (1980). A closed-form equation for predicting the hydraulic conductivity of unsaturated soils. *Soil Science Society of America Journal*, 44, 892–898. https://doi.org/10.2136/sssaj1980.03615995004400050002x

Vogel, T., & Císlerová, M. (1988). On the reliability of unsaturated hydraulic conductivity calculated from the moisture retention curve. *Transport in Porous Media*, 3, 1–15.

Warrick, A. W., Mullen, G. J., & Nielsen, D. R. (1977). Scaling field-measured soil hydraulic properties using a similar media concept. *Water Resources Research*, 13, 355–362.

Yang, W., Feng, G., Adeli, A., Kersebaum, K. C., Jenkins, J. N., & Li, P. (2019). Long-term effect of cover crop on rainwater balance components and use efficiency in no-tilled and rainfed corn and soybean rotation system. *Agricultural Water Management*, 219, 27–39.

2

Gerald N. Flerchinger and
Mark S. Seyfried

Heat Transport, Water Balance, Snowpack and Soil Freezing

Abstract

Heat and water transfer processes that dictate the temperature and moisture regime of terrestrial ecosystems are the fundamental control for many biological and physical processes. This chapter focuses on mathematical algorithms for simulating surface fluxes, soil heat transfer, soil freezing, and snowmelt as implemented in the Simultaneous Heat and Water (SHAW) model. The SHAW model is a numerical representation of a vertical, one-dimensional system composed of a multispecies plant canopy, snow cover if present, plant residue, and the soil profile, which has been applied to a variety of important hydrological and ecosystems processes. Example model applications are presented to serve as a learning tool and to illustrate the ability of detailed physical representations for capturing varying site and climatic conditions on energy and water balance simulations.

Introduction

Heat and water transfer processes at the soil–atmosphere interface control biological, chemical, hydrological, and many physical processes in all terrestrial ecosystems. Temperature and moisture conditions in the near-surface soil environment resulting from these transfer processes drive many important plant and other biological processes, including plant germination and establishment, residue decomposition, insect population, and soil freezing. The strong coupling between the surface energy and water balance dictates many hydrological processes including evapotranspiration, snowmelt, and runoff from frozen soils. Understanding this coupling in the energy

Modeling Processes and Their Interactions in Cropping Systems: Challenges for the 21st Century, First Edition.
Edited by Lajpat R. Ahuja, Kurt C. Kersebaum, and Ole Wendroth.

and water balance is critical: snow and frozen soil are key factors influencing the winter hydrology in many areas, and evapotranspiration (ET) is often the largest outflow in the water balance. Land managers need to address the interactions between physical, chemical, and biological factors in the near surface but lack the information necessary. The ability to predict temperature and water movement within the soil–plant–atmosphere system enhances our ability to evaluate management options and enables better understanding of interactions between land surface processes and the atmosphere.

Permafrost and seasonally frozen soil strongly influence hydrology, runoff, erosion, and ecosystem processes on large areas of land around the world. Loss of permafrost with global warming is dramatically changing the hydrology and biogeochemical process in these regions (Walvoord & Kurylyk, 2016). In many areas, rain or snowmelt on seasonally frozen soil is the single leading cause of severe runoff and erosion events. As soils freeze, ice blocks the soil pores, greatly diminishing the permeability of the soil. This is aggravated by the tendency of water to migrate to the freezing front, causing elevated ice content and frost heave. A special section published in *Vadose Zone Journal* provided an overview of research related to these topics (Toride et al., 2013).

Several numerical models have been developed to simulate coupled heat and mass transfer in freezing and thawing soils dating back to Harlan (1974), who is typically credited for developing the first such model for freezing and thawing of porous media. Kurylyk and Watanabe (2013) gave a historic overview of modeling freezing–thawing, pointing out that development of these models has surged and waned over the years. Combining the dynamics of soil freezing–thawing into a soil–vegetation–atmosphere transfer (SVAT) model complete with one-dimensional heat and mass transfer through a plant canopy, snow, and surface residue did not occur until the late 1980s and early 1990s with development of the Simultaneous Heat and Water (SHAW) model (Flerchinger & Pierson, 1991; Flerchinger & Saxton, 1989) and the SOIL model (Johnsson & Lundin, 1991), which was subsequently developed into the COUP model (Jansson & Karlberg, 2004). Concern about the loss of permafrost to climate warming caused renewed interest in frozen soil modeling around the turn of the millennium (e.g., Zhang et al., 2003). Recognizing the need for representing the detailed dynamics of soil temperature, water, and freezing, model developers have incorporated these processes or the SVAT models themselves into complex ecosystem models, such as the Root Zone Water Quality Model (RZWQM; Flerchinger et al., 2000). Kelleners and Verma (2012) combined a frozen-soil, heat-mass transport model developed by Kelleners (2013) with a soil respiration model to simulate heat, water, and CO_2 exchange of vegetated areas influenced by snow and frozen soil. Increase in computational speed enabled development of two- and three-dimensional simulation of frozen soil processes (e.g., Painter, 2011) and spatially distributed hydrological simulation of permafrost and seasonal frozen soil processes (Zhang et al., 2013). These models that capture the detailed physics of heat and mass transfer at the soil–atmosphere interface can be used as powerful tools to assess the influence of management, climate, and vegetation shifts on terrestrial ecosystems.

This chapter describes routines used in the SHAW model to simulate energy and water fluxes within the plant–snow–residue–soil system. The SHAW model is a numerical representation of a vertical, one-dimensional system composed of a multi-species plant canopy, snow cover if present, plant residue, and the soil profile. The model simulates the surface energy balance, evaporation, transpiration, soil water, snow accumulation and melt, and heat and water fluxes using a detailed, physically based solution to the energy and mass balance equations. Model descriptions herein focus on surface fluxes, snow processes, soil freezing, and heat transfer within the soil profile. For more detail on energy and water transfer within the plant canopy and residue, see Flerchinger (2017). Example model applications are presented to illustrate the ability of the detailed physical representation to capture the influence of varying site and climatic conditions on energy and water balance simulations.

Surface Energy and Water Fluxes

The surface energy balance links heat and water vapor transfer at the land–atmosphere interface and fundamentally controls the thermal regime of terrestrial ecosystems. The surface energy balance may be written as

$$R_n + H + L_v E + G = 0 \tag{2.1}$$

where R_n is net all-wave radiation (W m^{-2}), H is sensible heat flux (W m^{-2}), $L_v E$ is latent heat flux (W m^{-2}), G is soil or ground heat flux (W m^{-2}), L_v is the latent heat of evaporation (J kg^{-1}), and E is total evapotranspiration from the soil surface and plant canopy (kg m^{-2} s^{-1}). These interrelated surface energy and water fluxes are computed for given soil and vegetation conditions from weather observations of air temperature, wind speed, relative humidity and solar radiation and are briefly addressed below.

Net Radiation

Net radiation of a surface consists of absorbed solar or shortwave radiation, absorbed long-wave radiation, and emitted long-wave radiation. A full radiation balance is computed for each layer within the plant canopy, residue, and surface (snow or soil) from the incoming atmospheric solar and long-wave radiation by computing the reflection, transmission, scattering, and absorption of each layer. Because snow is translucent, net solar radiation at the snow surface is distributed through the snow based on an extinction coefficient.

Solar radiation absorbed within the system is computed from the observed total incoming solar radiation. Total incoming solar radiation (S_t) consists of direct (or beam, S_b), and diffuse (S_d) components, which are absorbed and transmitted differently. Therefore, total solar radiation is separated into the two components based on the following empirical equation (Bristow et al., 1985):

$$\tau_d = \tau_t \left\{ 1 - \exp \left[\frac{0.6(1 - \tau_{t,\max}/\tau_t)}{\tau_{t,\max} - 0.4} \right] \right\} \tag{2.2}$$

where τ_d is the atmospheric diffuse transmission coefficient ($S_d/S_{b,o}$), τ_t is the atmospheric total transmission coefficient ($S_t/S_{b,o}$), $\tau_{t,\max}$ is the maximum clear-sky transmissivity of the atmosphere, and $S_{b,o}$ is total solar radiation incident on a horizontal surface at the outer edge of the atmosphere (W m^{-2}). At $\tau_{t,\max}$ (taken as 0.76), all solar radiation is assumed to reach the Earth's surface as beam radiation. At lower values of τ_t, more of the sun's radiation is blocked and scattered by clouds, dust, and aerosols, and a greater portion of the radiation that reaches the surface is diffuse. Hourly values for $S_{b,o}$ are calculated from the solar constant, S_o (~1,360 W m^{-2}), and the sun's altitude above the horizon, ϕ_s. Whereas the diffuse radiation incident on the surface is assumed independent of slope, direct solar radiation incident on a sloping surface is related to that on a horizontal surface by

$$S_s = S_b \frac{\sin \beta}{\sin \phi_s} \tag{2.3}$$

where β is the angle which the sun's rays make with the sloping surface and ϕ_s is computed based on the latitude of the site, the time of year, and the hour of the day.

Incoming long-wave radiation to the system is estimated from cloud cover, which in turn is estimated from S_t (see Flerchinger, 2017; Flerchinger et al., 2009). The net long-wave radiation balance is similar to net solar radiation except that emitted long-wave radiation must be considered using the Stefan–Boltzmann law and all long-wave radiation is transmitted diffusely.

The net radiation absorbed by each layer within the system depends not only on the incoming solar and long-wave radiation from above but also on the reflected, scattered, and emitted radiation from other layers within the plant canopy–snow–residue–soil system. Therefore, a complete radiation balance is performed by computing the direct and upward and downward diffuse radiation fluxes above and below each layer. These processes are described in Flerchinger (2017), Flerchinger et al. (2009), and Flerchinger and Yu (2007).

Sensible and Latent Heat Fluxes

Sensible and latent heat flux components of the surface energy balance are computed from temperature and vapor gradients between the canopy–residue–soil surface and the atmosphere using Monin–Obukuv stability theory. Sensible heat flux is calculated from

$$H = - \rho_a c_a \frac{T - T_a}{r_H} \tag{2.4}$$

where ρ_a, c_a, and T_a are the density (kg m^{-3}), specific heat (J kg^{-1} °C^{-1}), and temperature (°C) of the air at the measurement reference height z_{ref}, T is the temperature (°C) of the exchange surface, and r_H is the resistance to surface heat transfer (s m^{-1}) corrected for atmospheric stability. The exchange surface is either the top of the canopy, the residue layer, the snow surface, or the soil surface depending on the system profile. Latent heat flux is associated with the transfer of water vapor from the exchange surface to the atmosphere and is given by

$$L_v E = \frac{\rho_{vs} - \rho_{va}}{r_v} \tag{2.5}$$

where L_v is latent heat of vaporization (J kg^{-1}), E is vapor flux from the exchange surface (kg m^{-2}), ρ_{vs} and ρ_{va} are vapor density (kg m^{-3}) of the exchange surface and at the reference height z_{ref}, and the resistance value for vapor transfer, r_v, is taken to be equal to r_H. The resistance to convective heat transfer, r_H, is computed from

$$r_H = \frac{1}{u^* k} \left[\ln \left(\frac{z_{ref} - d + z_H}{z_H} \right) + \psi_H \right] \tag{2.6}$$

where u^* is the friction velocity (m s^{-1}) computed from

$$u^* = uk \left[\ln \left(\frac{z_{ref} - d + z_m}{z_m} \right) + \psi_m \right]^{-1} \tag{2.7}$$

k is von Karman's constant, d is the zero plane displacement, u is windspeed, z_H and z_m are the surface roughness parameters for the temperature and momentum profiles, and ψ_H and ψ_m are diabatic correction factors for heat and momentum, computed as a function of atmospheric stability. Corrections for atmospheric stability and a sparse vegetation canopy are described in Flerchinger (2017) and Flerchinger et al. (2015).

Ground Heat Flux

The net radiation and turbulent heat fluxes in Equation 2.1 are coupled to heat flux into the system through the surface temperature of the canopy, residue, snow, or soil. This surface temperature must satisfy the solution of the heat flux equations for the entire canopy–residue–soil profile, which is solved simultaneously and iteratively with the surface energy and water balance. Thus, G in Equation 2.1 is defined as the heat flux into the top of the canopy, snow, residue, or soil. In the examples presented here, we present soil heat flux, which is the heat flux at the soil surface. This is resolved by coupling the energy balance of the soil with that of the overlying material, which is controlled by the interface surface temperature.

Snow Processes

The energy balance for each layer within the snowpack is written as

$$\rho_{sp} c_i \frac{\partial T}{\partial t} + \rho_l L_f \frac{\partial w_{sp}}{\partial t} = \frac{\partial}{\partial z}\left(k_{sp}\frac{\partial T}{\partial z}\right) + \frac{\partial R_n}{\partial z} - L_s\left(\frac{\partial q_v}{\partial z} + \frac{\partial \rho_v}{\partial t}\right) \tag{2.8}$$

where ρ_{sp}, w_{sp}, and k_{sp} are the density (kg m^{-3}), volumetric liquid water content (m^3 m^{-3}), and thermal conductivity of the snow (W m^{-1} C^{-1}); c_i is the specific heat capacity of ice (J kg^{-1} C^{-1}); ρ_l is the density of water (kg m^{-3}); R_n is net downward radiation flux within the snow (W m^{-2}); L_f and L_s are latent heats of fusion and sublimation (J kg^{-1}); q_v is vapor flux (kg s^{-1} m^{-2}); and ρ_v is vapor density (kg m^{-3}) within the snow. The terms (W m^{-3}) in Equation 2.8 represent, from left to right, the change in specific heat stored due to a temperature increase, the latent heat required to melt snow, net thermal conduction into a layer, net radiation absorbed within the layer, and the net latent heat of sublimation associated with water vapor transfer. The complexities of heat transferred by liquid movement are not included in the snow energy balance equation. Instead, a mass balance of the snowpack is computed at the end of each time step to adjust the snowpack for melt, water percolation, and thermal advection.

Specific Heat

At temperatures below 0 °C, net energy absorbed by the snow results in a change in temperature. The volumetric specific heat of snow is computed from the density of the snow, ρ_{sp}, and the specific heat of ice, which is a function of temperature:

$$c_i = 92.96 + 7.37 T_K \tag{2.9}$$

where T_K is the temperature of the snow in Kelvin.

Latent Heat of Fusion

At 0 °C, the net energy absorbed by the snowpack results in the melting of ice and an increase in water content in the snowpack. The ice content of the snowpack is assumed constant during the hourly or daily time step and is adjusted for any melt at the end of the time step.

Thermal Conduction

Heat conduction within snow occurs primarily between and within ice crystals. Although the geometry of the snow crystals can influence conductivity, this is difficult to quantify. Nevertheless, the thermal conductivity of snow can be related to its density, and the following expression fits many empirically derived correlations:

$$k_{sp} = a_{sp} + b_{sp} \left(\frac{\rho_{sp}}{\rho_l} \right)^{c_{sp}}$$ (2.10)

where a_{sp}, b_{sp}, and c_{sp} are empirical coefficients with suggested values of 0.021 W $m^{-1}\,°C^{-1}$, 2.51 W $m^{-1}\,°C^{-1}$, and 2.0, respectively.

Radiation Absorption

Emission and absorption of long-wave radiation is considered only at the snow surface. However, snow is translucent, and solar radiation entering the surface of the snowpack is therefore attenuated and absorbed throughout the snowpack. The net solar radiation flux at a depth z (S_z) can be expressed as

$$S_z = (S_s + S_d)(1 - \alpha_{sp})e^{-vz}$$ (2.11)

where $(S_s + S_d)$ is the total solar radiation incident on the snow surface, α_{sp} is the albedo of the snow surface, and e is the base of the natural logarithm. The extinction coefficient, v, for radiation penetration through the snow is estimated from

$$v = 100\, C_v \left(\frac{\rho_{sp}}{\rho_l} \right) d_s^{-1/2}$$ (2.12)

where C_v is taken as 1.77 $mm^{1/2}\,cm^{-1}$ and d_s is the grain-size diameter of ice crystals (mm).

Latent Heat of Sublimation

Latent heat transfer within the snowpack arises from sublimation and vapor transfer through the snowpack in response to temperature gradients. Vapor density in snow is assumed equal to the saturated vapor density over ice; it therefore is independent of the snow water content and controlled solely by temperature. Warmer parts of the snowpack have a higher vapor density; vapor will therefore diffuse toward cooler parts, where, due to oversaturation, sublimation will occur and latent heat is released. Vapor flux through the snowpack is calculated as

$$q_v = D_e \frac{\partial \rho_v}{\partial z}$$ (2.13)

where D_e is the effective diffusion coefficient (m s^{-2}) for water vapor in snow, and ρ_v is the temperature-dependent vapor density within the snow. The net latent heat of sublimation for a layer in the snowpack is equal to the increase in vapor density minus the net transfer of vapor to that layer.

Density Changes and Outflow from Snow

Density and total water content (ice plus liquid) of the snow are assumed constant for each time step. After solving the snow energy balance, the thickness and density of each snow layer is adjusted for melt and metamorphosis of the snow at the end of the time step. Compaction and settling processes are included when computing changes in snow density. Liquid water is routed through the snowpack using attenuation and lag coefficients to estimate the outflow of water from the bottom of the snowpack. Detailed descriptions of these processes are given in Flerchinger (2017).

Heat and Water Relations During Soil Freezing and Thawing

The energy balance equation, considering convective heat transfer by liquid and latent heat transfer by vapor for a layer of freezing soil, is

$$C_s \frac{\partial T}{\partial t} - \rho_i L_f \frac{\partial \theta_i}{\partial t} = \frac{\partial}{\partial z}\left(k_s \frac{\partial T}{\partial z}\right) - \rho_l c_l \frac{\partial q_l T}{\partial z} - L_v \left(\frac{\partial q_v}{\partial z} + \frac{\partial \rho_v}{\partial t}\right) \tag{2.14}$$

where C_s and T are volumetric heat capacity (J m^{-3} °C^{-1}) and temperature (°C) of the soil, ρ_i is the density of ice (kg m^{-3}), θ_i is volumetric ice content (m^3 m^{-3}), k_s is soil thermal conductivity (W m^{-1} C^{-1}), ρ_l is the density of water, c_l is specific heat capacity of water (J kg^{-1} °C^{-1}), q_l is liquid water flux (m s^{-1}), q_v is water vapor flux (kg m^{-2} s^{-1}), and ρ_v is vapor density (kg m^{-3}) within the soil. The terms (W m^{-3}) in Equation 2.14 represent, from left to right, a specific heat term for the change in energy stored due to a temperature increase, the latent heat required to freeze water, net thermal conduction into a layer, net thermal advection into a layer due to water flux, and net latent heat evaporation within the soil layer.

Specific Heat

The volumetric heat capacity of soil is computed as the sum of the volumetric heat capacities of the soil constituents:

$$C_s = \sum \rho_j c_j \theta_j \tag{2.15}$$

where ρ_j, c_j, and θ_j are the density, specific heat capacity, and volumetric fraction of the jth soil constituent.

Latent Heat of Fusion

Due to matric and osmotic potentials, soil water exists in equilibrium with ice at temperatures below the normal freezing point of bulk water and across the entire range of soil freezing temperatures normally encountered. The total potential of the soil water with ice present is controlled by the vapor pressure over ice. Neglecting the influence of gas pressure and total water content (see Painter & Karra, 2014), the total water potential is given by the freezing point depression equation (Fuchs et al., 1978):

$$\phi = \pi + \psi = \frac{L_f}{g}\left(\frac{T}{T_K}\right) \tag{2.16}$$

where π is the soil water osmotic potential (m), ψ is the soil matric potential, g is the acceleration of gravity (m s^{-2}), T is the temperature difference from the freezing point of water (°C), and T_K is the absolute temperature (K). The osmotic potential in the soil is computed as

$$\pi = \frac{-cRT_K}{g} \tag{2.17}$$

where c is the solute concentration (mol kg^{-1}) in the soil solution. Given the osmotic potential, soil temperature defines the matric potential and, therefore, liquid water content. If the total water content is known, ice content and the latent heat term can be determined.

Thermal Conduction

Numerous algorithms exist for estimating soil thermal conductivity (Zhang & Wang, 2017). The SHAW model uses the theory presented by De Vries (1963). A fairly moist soil can be conceptualized as a continuous medium of liquid water, with granules of soil, crystals of ice, and pockets of air dispersed throughout. The same can be said for a dry soil, where air is the continuous medium. The thermal conductivity of such an idealized model is expressed as

$$k_s = \frac{\sum m_j k_j \theta_j}{\sum m_j \theta_j}$$

(2.18)

where m_j, k_j, and θ_j, are the weighting factor, thermal conductivity, and volumetric fraction of the jth soil constituent, that is, rock, sand, silt, clay, organic matter, water, ice, and air. The weighting factor of the continuous medium is 1.0. Weighting factors for the dispersed constituents (rock, sand, silt, clay, organic matter, and ice) are based on the theoretical temperature gradient in the dispersed constituent relative to that in the continuous medium. Values are computed from

$$m_j = \frac{2}{3}\left[1 + \left(\frac{k_j}{k_o} - 1\right)g_{a,j}\right]^{-1} + \frac{1}{3}\left[1 + \left(\frac{k_j}{k_o} - 1\right)\left(1 - 2g_{a,j}\right)\right]^{-1}$$

(2.19)

where k_o is the thermal conductivity of the continuous medium and $g_{a,j}$ is a particle shape factor relating to the axes of an ellipsoid. Values of $g_{a,j}$ range from 0.0 for disks to 1/3 for spheroids and 0.5 for needle-shaped particles. Particle shape factors for rock, sand, silt, clay, and organic matter are taken to be 0.333, 0.144, 0.144, 0.125, and 0.5, respectively. Kennedy and Sharratt (1998) identified problems with the shape factor set to 0.333 for ice, so the weighing factor for ice is taken as 1.0. The threshold liquid water content for water being the continuous medium is taken as 0.05 for sandy soils, 0.10 for loams, and 0.15 for clay soils; below this threshold, k_s is linearly interpolated to that computed for dry soil with zero liquid water.

Vapor Flux and Latent Heat of Vaporization

The net latent heat of vaporization occurring in a soil layer is computed from the rate of increase in vapor density minus the net vapor transfer into the layer. Vapor density in the soil is calculated, assuming equilibrium with the total water potential, as

$$\rho_v = h_r \rho_v' = \rho_v' \exp\left(\frac{M_w g}{RT_K}\phi\right)$$

(2.20)

where ρ_v is vapor density (kg m^{-3}), ρ_v' is saturated vapor density, h_r is relative humidity, M_w is the molecular weight of water (0.018 kg mol^{-1}), g is the acceleration of gravity (9.81 m s^{-2}), R is the universal gas constant (8.3143 J mol^{-1} K^{-1}), and ϕ is the total water potential (m).

Vapor transfer in the soil can be represented as the sum of the fluxes in response to a vapor density gradient related to a water potential gradient, q_{vp} (kg m^{-2}), and that due to a temperature gradient, q_{vT} (Campbell, 1985), where

heat transport, water balance, snowpack

$$q_v = q_{vp} + q_{vT} = -D_v \rho_v' \frac{dh_r}{\partial z} - \zeta D_v h_r s_v \frac{dT}{\partial z} \tag{2.21}$$

where D_v is vapor diffusivity ($m^2\ s^{-1}$) through the soil, h_r is relative humidity, s_v is the slope of the saturated vapor pressure curve ($d\rho_v'/dT$ in $kg\,m^{-3}\ ^\circ C^{-1}$), and ζ is an enhancement factor. Vapor diffusivity in the soil expressed as

$$D_v = D_v' b_v \theta_a^{c_v} \tag{2.22}$$

where D_v' is diffusivity of water vapor in air, θ_a is air porosity, and b_v and c_v are coefficients accounting for the tortuosity of the air voids.

Soil Water Flow

Darcy's equation can be used to describe steady-state one-dimensional water flux ($m\ s^{-1}$) through the soil:

$$q_l = -K \left[\frac{\partial \left(\psi + \psi_g \right)}{\partial z} \right] \tag{2.23}$$

where K is the unsaturated conductivity ($m\ s^{-1}$), ψ is the soil water matric potential (m), ψ_g is the gravitational potential, z is depth within the soil, and $\partial(\psi + \psi_g)/\partial z$ is the gradient in soil water potential.

The transient mass balance equation for water including the effects of freezing, thawing, and vapor flow within the soil can be written as

$$\frac{\partial \theta_l}{\partial t} + \frac{\rho_i}{\rho_l} \frac{\partial \theta_i}{\partial t} = \frac{\partial}{\partial z} \left[K \left(\frac{\partial \psi}{\partial z} + 1 \right) \right] + \frac{1}{\rho_l} \frac{\partial q_v}{\partial z} + U \tag{2.24}$$

The terms on the left-hand side of the equation represent the time rate of change of liquid water content and the time rate of change of ice content. The terms on the right-hand side are the gradient in water flux (i.e., the net flux of water into a layer of soil) and a source–sink term that can account for plant uptake or lateral flow into the soil layer. Equation 2.24 states that the net liquid water flux into a soil layer must equal the combined change in ice and water content of the soil. When the net flux into the layer is zero, any change in liquid water content must be offset by a change in ice content, adjusted for the difference in density. Although this change in density can result in expansion of the soil matrix, it is not the primary cause of frost heave. The relations among liquid water content, matric potential, and unsaturated hydraulic conductivity are determined from the soil moisture release curve. Users can choose between several formulations of the moisture release curve (Brooks & Cory, 1964; Campbell, 1974; Kosugi, 1996; van Genuchten, 1980).

As mentioned above, liquid water exists in the soil at temperatures below 0 °C due to negative water potentials. When ice is present in the soil, the soil matric potential is strongly influenced by temperature. As the temperature at the freezing front decreases, more and more water freezes, water potential becomes more negative, and liquid water content continues to drop, creating gradients in water potential and liquid water content. This drop in liquid water content at the freezing front has a similar effect as drying of the soil, and water will migrate from moist regions to the freezing front. This often results in elevated ice content, ice lenses, and frost heave; however, frost heave is not addressed by the SHAW model.

Example Applications

To illustrate surface energy, water, soil freezing and snow dynamics, examples from four sites on the USDA–ARS Reynolds Creek Experimental Watershed are presented: two sites with contrasting elevation and two with contrasting slope or aspect. The sites with contrasting elevation are described in Flerchinger et al. (2019) and those with contrasting slope or aspect are described in Seyfried et al. (2021). Site characteristics are given in Table 2.1.

Sites with Contrasting Elevation

The low-elevation Wyoming big sagebrush and the high-elevation mountain big sagebrush sites described by Flerchinger et al. (2019) are used for illustration. These sites are located at elevations of 1,425 and 2,111 m and are part of a network of eddy covariance sites along an elevation gradient within the watershed. A plot of measured and simulated daily soil temperature and soil heat flux for water year 2016 is presented in Figure 2.1. Not surprisingly, air and soil temperatures are generally warmer at the lower elevation site. However, wintertime soil temperatures are, in fact, colder at the lower elevation despite the warmer air temperatures. This phenomenon is examined more closely in Figures 2.2 and 2.3.

Colder winter temperatures and more precipitation at the higher elevation results in a more persistent snowpack than at the lower elevation (Figures 2.2 and 2.3). Snow depth at the higher elevation accumulated to more than 100 cm, but it scarcely reached 10 cm at the lower elevation. This has a tremendous impact on soil temperatures and the energy balance at the soil surface. Although the model did simulate some soil freezing at the high-elevation 5-cm soil depth, measured soil temperatures never dropped below 0 °C. Soil temperatures were very much insulated from the cold air temperature and remained very constant until the snow melted (Figure 2.2); indeed, soil heat flux remained nearly zero and constant for the entire winter. In contrast, the 5-cm soil temperature at the low elevation dropped to −5 °C and was much more responsive to variations in air temperature, as is also indicated by the diurnal variation in soil heat flux through much of the winter (Figure 2.3d). Final ablation of the snow can be readily identified by variations in soil temperature at both sites upon comparing Figure 2.2a with c and Figure 2.3a with c.

Table 2.1 Climate, Vegetation, and Soil Properties for the Four Study Sites (Soils Data Reflect Properties to 120 cm Depth)

Property	Low elevation	High elevation	North facing	South facing
Elevation, m	1,425	2,111	1,706	1,706
Mean annual temperature, °C	9.4	5.6	8.2	8.2
Slope, %	10	10	32	32
Aspect, ° N	0	0	22	160
Annual water balance				
Precipitation, mm	267	941	501	501
Measured evapotranspiration, mm	286	484	NA	NA
Evapotranspiration, mm[a]	261	471	490	467
Transpiration, mm[a]	97	324	302	298
Evaporation and sublimation[a]	164	147	188	169
Percolation, mm[a]	2	422	0	38
Overland runoff, mm[a]	0	0	0	0

Note. ET, evapotranspiration; NA, not available.
[a]Simulated values.

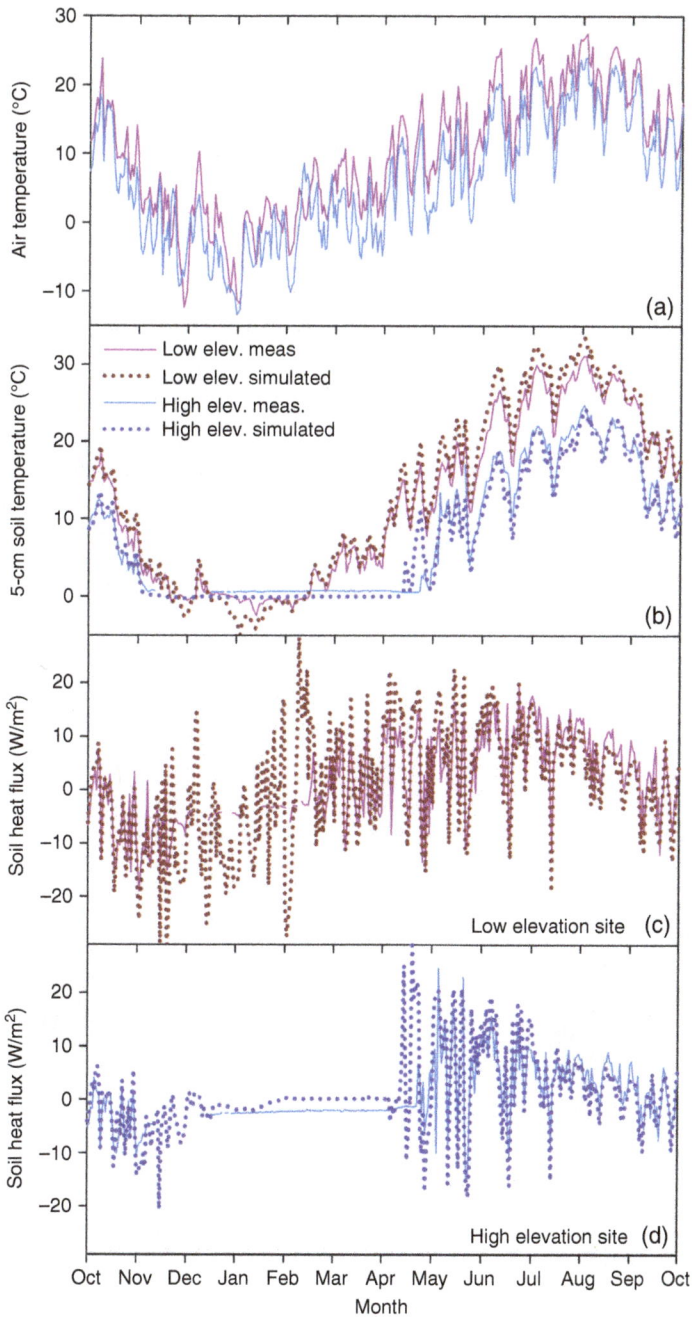

Figure 2.1 (a) Observed average daily air temperature at the high-elevation mountain big sagebrush site and the low-elevation Wyoming sagebrush site, (b) daily simulated and measured 5-cm soil temperature at the two sites, and soil heat flux at the soil surface for (c) the low-elevation and (d) the high-elevation site.

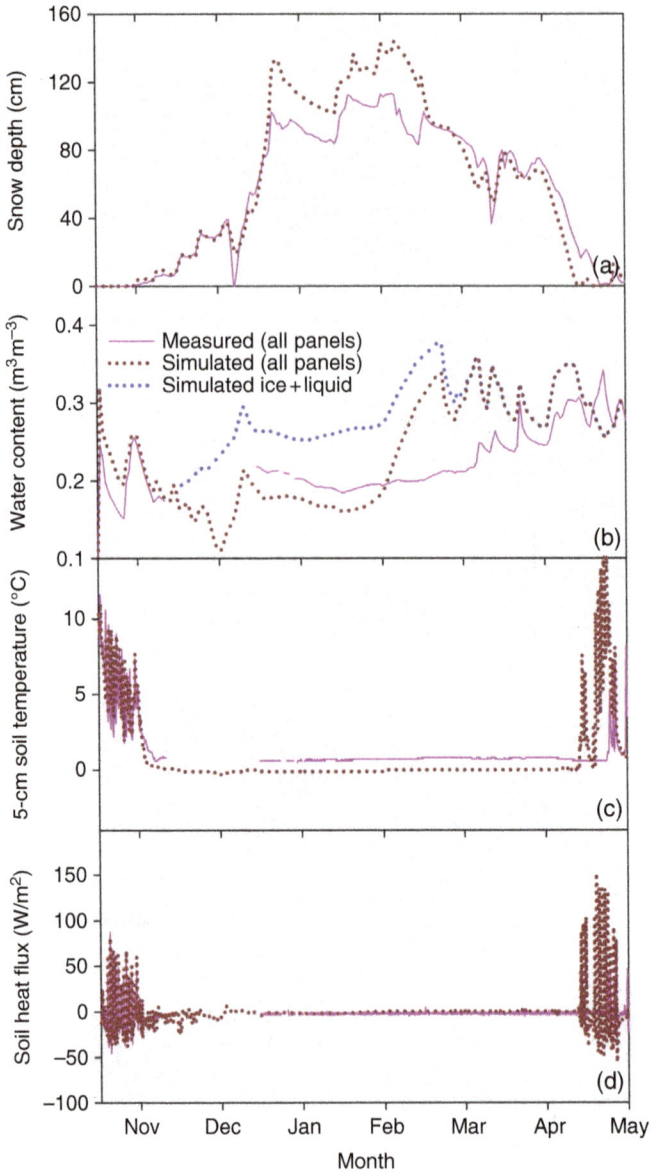

Figure 2.2 Measured and simulated hourly (a) snow depth, (b) 5-cm soil liquid water content and simulated total (ice + liquid) water content, (c) 5-cm soil temperature, and (d) surface soil heat flux for the high-elevation mountain big sagebrush site.

Liquid water content through the winter was simulated quite well at the lower elevation. The hard freeze resulted in moisture migration to the frozen soil from the unfrozen soil below, as indicated by the rise in simulated total water content upon freezing and the drop in the liquid water content of the 5-cm layer (Figure 2.3b).

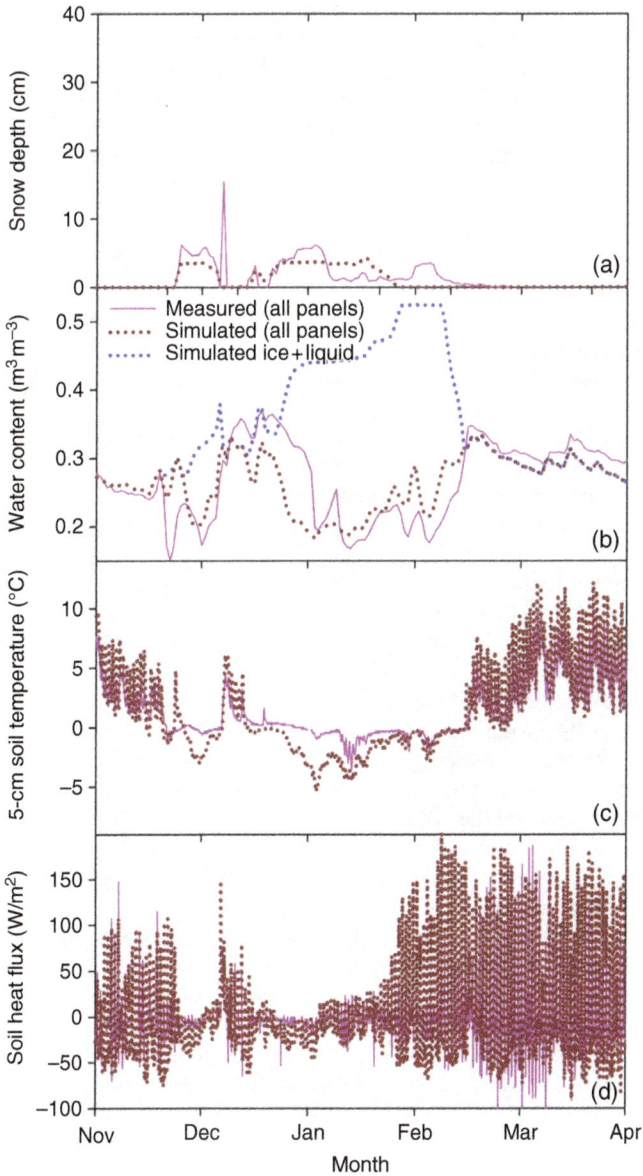

Figure 2.3 Measured and simulated hourly (a) snow depth, (b) 5-cm soil liquid water content and simulated total (ice + liquid) water content, (c) 5-cm soil temperature, and (d) surface soil heat flux for the low-elevation Wyoming big sagebrush site.

A summary of the annual water balance for the sites in Table 2.1 highlights the contrasts between the two sites. More than 98% of the precipitation at the lower elevation is consumed by ET, whereas it comprises only 50% at the high-elevation site. Simulated ET compares well with measured ET for both sites. The proportion of ET that supports transpiration and plant growth varies from around 40% at the lower elevation to nearly 70% at the higher elevation. This can be attributed to the fact that the lower elevation site has sparse vegetation cover and therefore more of the available water is lost to soil evaporation. The 422 mm of percolation beyond the root zone (>2 m) at the high elevation probably contributes to nearby perennial streamflow via subsurface flow. Negligible percolation was estimated at the lower elevation; the stream in the immediate vicinity of this site flows only in response to thunderstorms or rapid snowmelt events.

Sites with Contrasting Slope or Aspect

The sites described by Seyfried et al. (2021) are used to illustrate the effect of slope and aspect on the energy balance, snow accumulation, and soil freezing. The sites are situated at the same elevation on opposing slopes in a small watershed catchment. Both are on slopes of around 32%, one being nearly south-facing and the other facing north-northwest.

The annual soil temperature response of the north- and south-facing slopes for water year 2014, presented in Figure 2.4, illustrates the consistently warmer regime of the south-facing slope. This is a direct result of the increased solar radiation incident on the south-facing slope compared with the north-facing slope (Figure 2.4d). The variability of incident radiation on the steep north-facing slope is small because it receives little to no direct radiation, is dominated by diffuse radiation, and therefore is not influenced significantly by cloud cover.

The difference in temperature regime between the two slopes has a large influence on winter conditions. Snowpack on the south-facing slope is much more intermittent through the winter (Figures 2.5a and 2.6a). Even though the soil is bare for much of the winter and lacks the insulating snowpack of the north-facing slope, it experiences less soil freezing than the north-facing slope. Indeed, during this year, the 20-cm depth of the south-facing soil did not experience freezing temperatures, while the north-facing slope was frozen much of the winter, as evidenced by the separation of the simulated liquid water content line from the total water (ice + liquid) content line beginning in early December. Exposure of the soil to solar radiation has a large influence on soil heat flux into the soil profile (Figures 2.5e and 2.6e).

The model captured differences in the winter dynamics between the two slopes quite well, although it did underestimate snow depth on the north-facing slope. It appears from Figure 2.5a that snow accumulation during many of the early season (December and January) snow events was underestimated on the north-facing slope, perhaps due to not accounting for wind redistribution of the snow.

The warmer temperature regime of the south-facing slope persists through the summer months, as illustrated in Figure 2.7, where the south-facing slope is approximately 5 °C warmer. Although the diurnal temperature amplitude on the north-facing slope is underestimated at the 5-cm depth by the model (Figure 2.7a), the model performed well at the 5-cm south-facing soil depth and captured the contrast in temperature between the two sites at the 20-cm depth.

Despite the large difference in temperature between the two sites, the primary difference in the simulated water balance is the 38 mm of percolation from the south-facing site (Table 2.1). The simulated percolation occurred primarily by February prior to complete snow ablation on the north-facing slope. Because both sites use nearly all post-melt available water for ET, this translated to nearly 38 mm more simulated ET

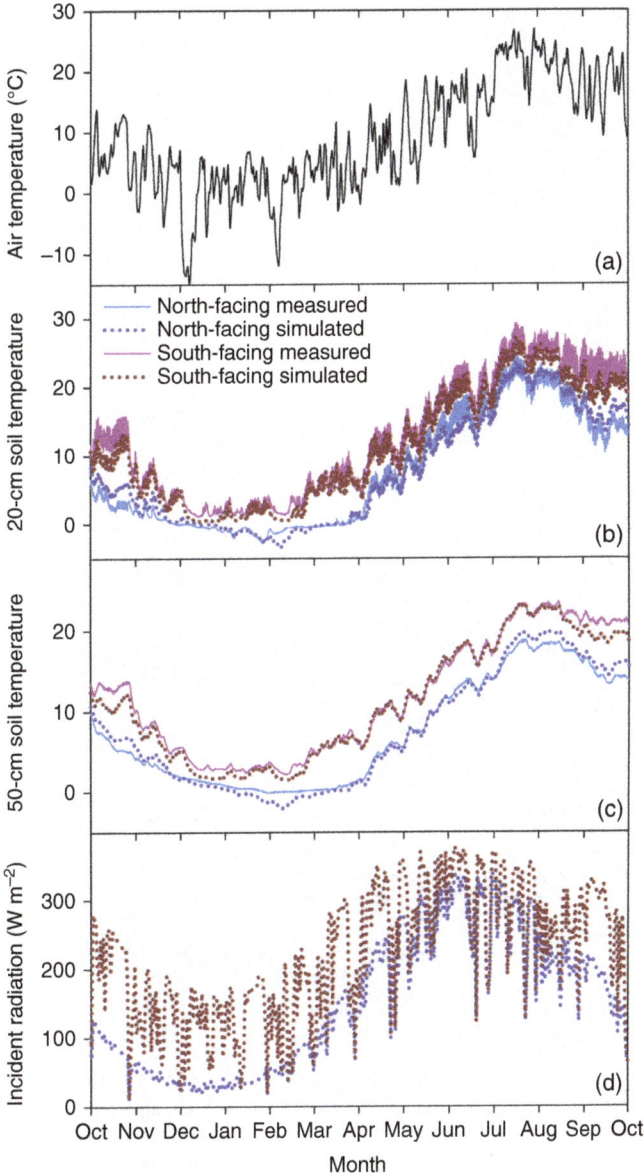

Figure 2.4 (a) Observed average daily air temperature and hourly simulated and measured (b) 20-cm soil temperature, (c) 50-cm soil temperature, and (d) incident solar radiation on the local slope for opposing north-facing and south-facing slopes.

from the north-facing site than the south-facing site. The additional ET on the north-facing slope occurred in late May–early June after ET on the south-facing slope started to subside; simulated ET during this two-weeks period was 83 mm on the north-facing slope and 46 mm on the south-facing slope.

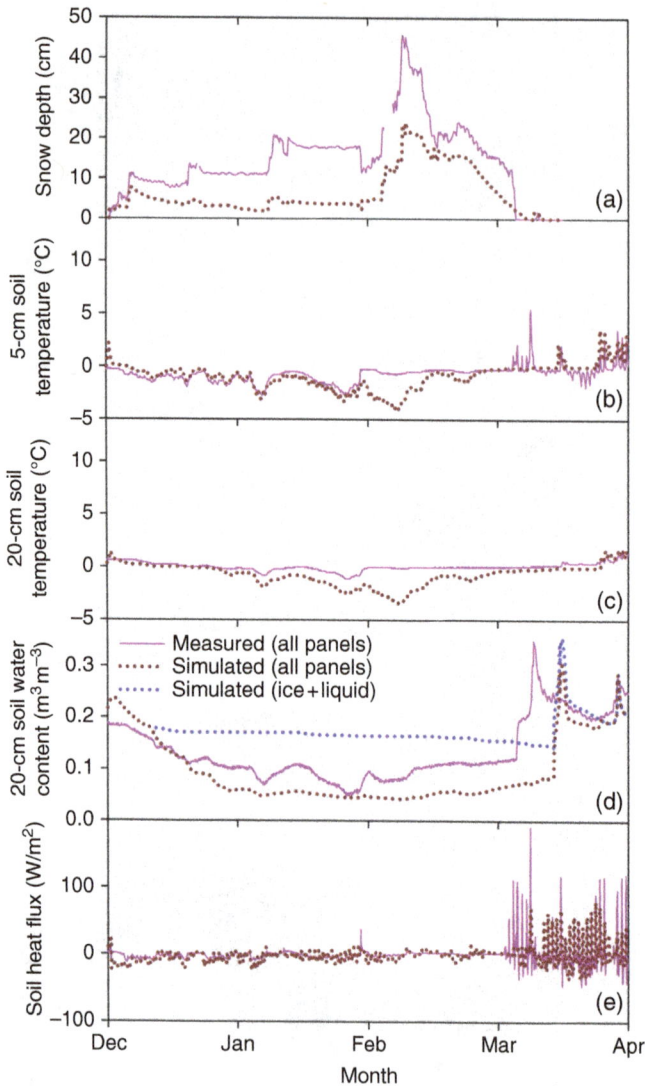

Figure 2.5 Measured and simulated hourly (a) snow depth, (b) 5-cm soil temperature water content, (c) 20-cm soil temperature, (d) 20-cm soil liquid water content and simulated total (ice + liquid) water content, and (e) surface soil heat flux on a steep north-facing slope.

Future Research and Modeling Needs

Opportunities abound in modeling heat and water fluxes in the near-surface soil environment, particularly in the areas of coupling these processes with ecosystem and ecohydrologic phenomena controlled by the microclimatic temperature and moisture regime and application to and understanding of cold region hydrologic and ecosystem

heat transport, water balance, snowpack

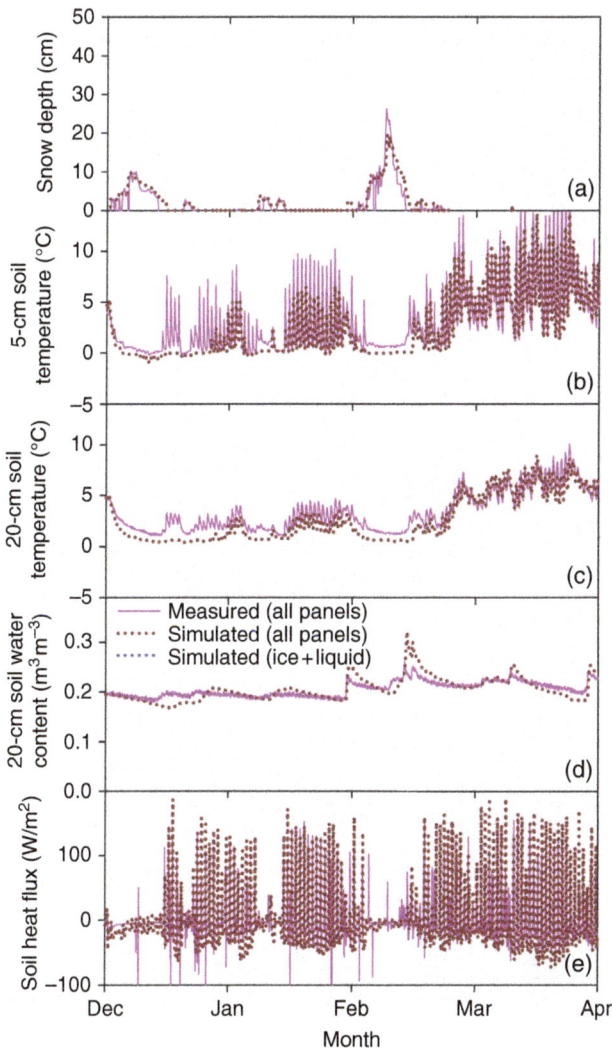

Figure 2.6 Measured and simulated hourly (a) snow depth, (b) 5-cm soil temperature water content, (c) 20-cm soil temperature, (d) 20-cm soil liquid water content and simulated total (ice + liquid) water content (ice content was zero so the line is obscured by the liquid water content line), and (e) surface soil heat flux on a steep south-facing slope.

processes. Advances will most efficiently be made when these efforts are combined with field or laboratory studies.

The utility of modeling heat and water fluxes at the soil–atmosphere interface arises when these processes are used to assist management decisions or to further our understanding of ecosystem and environmental processes. Countless such applications exist and include water use and planning, ecosystem productivity, carbon and nutrient cycling, population dynamics, weed control, and impacts of climate shifts, to name

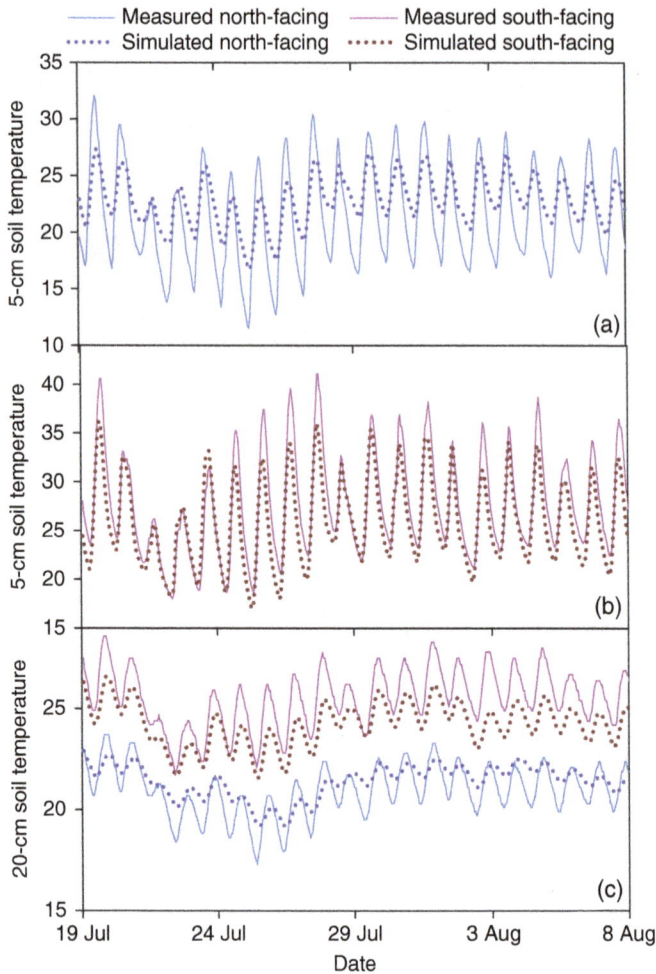

Figure 2.7 Hourly simulated and measured (a) 5-cm soil temperature on a steep north-facing slope, (b) 5-cm soil temperature on a steep south-facing slope, and (c) 20-cm soil temperature on both slopes during the summer period 19 July–8 August.

a few; Vereecken et al. (2016) gave a broad review of applications for modeling soil processes. Modeling the impacts of permafrost degradation on hydrology and carbon release is a particularly challenging area of research that has attracted recent attention (Painter et al., 2013; Walvoord & Kurylyk, 2016). Other applications include determination of the effects of heat, water, and frost on geomorphology (Anderson et al., 2013) and soil development (e.g., Finke & Hutson, 2008).

Despite decades of research on quantifying the hydraulic and thermal properties of frozen soils, there still remain gaps and uncertainties. Uncertainties linger in accurately computing liquid water content and freezing point depression of frozen soils (Kurylyk & Watanabe, 2013; Painter & Karra, 2014). This may be partly attributed to the difficulties and expense in accurately measuring liquid water content, in the case of nuclear magnetic resonance, or uncertainties in less expensive measurement techniques, such as time domain reflectometry (Kurylyk & Watanabe, 2013; Seyfried & Murdock, 1996). Further work is also needed to quantify infiltration and the hydraulic conductivity of frozen soils (Watanabe et al., 2013; Watanabe & Osada, 2017), frozen soil thermal conductivity (He et al., 2021), and frost heave (Groenevelt & Grant, 2013; Rempel, 2012).

References

Anderson, R. S., Anderson, S. P., & Tucker, G. E. (2013). Rock damage and regolith transport by frost: An example of climate modulation of the geomorphology of the critical zone. *Earth Surface Processes and Landforms*, 38, 299–316.

Bristow, K. L., Campbell, G. S., & Saxton, K. E. (1985). An equation for separating daily solar radiation irradiation into direct and diffuse components. *Agricultural and Forest Meteorology*, 35, 123–131.

Brooks, R. H., & Cory, A. T. (1964). *Hydraulic properties of porous media* (Hydrology Paper 3). Civil Engineering Department, Colorado State University.

Campbell, G. S. (1974). A simple method for determining unsaturated conductivity from moisture retention data. *Soil Science*, 117, 311–314.

Campbell, G. S. (1985). *Soil physics with BASIC: Transport models for soil–plant systems*. Elsevier.

De Vries, D. A. (1963). Thermal properties of soils. In W. R. van Wijk (Ed.), *Physics of plant environment* (pp. 210–235). North-Holland.

Finke, P. A., & Hutson, J. L. (2008). Modeling soil genesis in calcareous loess. *Geoderma*, 145, 462–479.

Flerchinger, G. N. (2017). *The Simultaneous Heat and Water (SHAW) model: Technical documentation* (Tech. Rep. NWRC 2017–02). Boise, ID: USDA–ARS, Northwest Watershed Research Center. https://www.ars. usda.gov/ARSUserFiles/20520500/SHAW/ShawDocumentation.30.pdf

Flerchinger, G. N., Aiken, R. M., Rojas, K. W., & Ahuja, L. R. (2000). Development of the Root Zone Water Quality Model (RZWQM) for over-winter conditions. *Transactions of the American Society of Agricultural Engineering*, 43, 59–68.

Flerchinger, G. N., Fellows, A. W., Seyfried, M. S., Clark, P. E., & Lohse, K. A. (2019). Water and carbon fluxes along a climate gradient in a sagebrush ecosystem. *Ecosystems*, 23, 246–263. http://dx.doi.org/10.1007/ s10021-019-00400-x

Flerchinger, G. N., & Pierson, F. B. (1991). Modeling plant canopy effects on variability of soil temperature and water. *Agricultural and Forest Meteorology*, 56, 227–246.

Flerchinger, G. N., Reba, M. L., Link, T. E., & Marks, D. (2015). Modeling temperature and humidity profiles within forest canopies. *Agricultural and Forest Meteorology*, 213, 251–262.

Flerchinger, G. N., & Saxton, K. E. (1989). Simultaneous heat and water model of a freezing snow–residue–soil system: I. Theory and development. *Transactions of American Society of Agricultural Engineering*, 32, 565–571.

Flerchinger, G. N., Xaio, W., Marks, D., Sauer, T. J., & Yu, Q. (2009). Comparison of algorithms for incoming atmospheric long-wave radiation. *Water Resources Research*, 45, W03423. http://dx.doi.org/10.1029/ 2008WR007394

Flerchinger, G. N., Xaio, W., Sauer, T. J., & Yu, Q. (2009). Simulation of within-canopy radiation exchange. *NJAS–Wageningen Journal of Life Sciences*, 57, 5–15.

Flerchinger, G. N., & Yu, Q. (2007). Simplified expressions for radiation scattering in canopies with ellipsoidal leaf angle distributions. *Agricultural and Forest Meteorology*, 144, 230–235.

Fuchs, M., Campbell, G. S., & Papendick, R. I. (1978). An analysis of sensible and latent heat flow in a partially frozen unsaturated soil. *Soil Science Society of America Journal*, 42, 379–385.

Groenevelt, P. H., & Grant, C. D. (2013). Heave and heaving pressure in freezing soils: A unifying theory. *Vadose Zone Journal*, 12(1). http://dx.doi.org/10.2136/vzj2012.0051

Harlan, R. L. (1974). Analysis of coupled heat–fluid transport in partially frozen soil. *Water Resources Research*, 9, 1314–1323. http://dx.doi.org/10.1029/WR009i005p01314

He, H., Flerchinger, G. N., Kojima, Y., Dyck, M., & Lv, J. (2021). A review and evaluation of 39 thermal conductivity models for frozen soils. *Geoderma*, 382, 114694.

Jansson, P. E., & Karlberg, L. (2004). *Coupled heat and mass transfer model for soil–plant–atmosphere systems*. Department of Civil and Environmental Engineering, Royal Institute of Technology. https://www.coupmodel. com/documentation

Johnsson, H., & Lundin, L. C. (1991). Surface runoff and soil–water percolation as affected by snow and soil frost. *Journal of Hydrology*, 122, 141–159. https://doi.org/10.1016/0022-1694(91)90177-J

Kelleners, T. J. (2013). Coupled water flow and heat transport in seasonally frozen soils with snow accumulation. *Vadose Zone Journal*, 12(4). http://dx.doi.org/10.2136/vzj2012.0162

Kelleners, T. J., & Verma, A. K. (2012). Coupled water flow and heat transport in seasonally frozen soils with snow accumulation. *Vadose Zone Journal*, 11(3). http://dx.doi.org/10.2136/vzj2011.0205

Kennedy, I., & Sharratt, B. (1998). Model comparisons to simulate soil frost depth. *Soil Science*, 163, 636–645.

Kosugi, K. (1996). Lognormal distribution model for unsaturated soil hydraulic properties. *Water Resources Research*, 32, 2697–2703. http://dx.doi.org/10.1029/96WR01776

Kurylyk, B. L., & Watanabe, K. (2013). The mathematical representation of freezing and thawing processes in variably-saturated, non-deformable soils. *Advances in Water Resources*, 60, 160–177.

Painter, S. L. (2011). Three-phase numerical model of water migration in partially frozen geological media: Model formulation, validation, and applications. *Computational Geosciences*, 15, 69–85. http://dx.doi.org/10.1007/s10596-010-9197-z

Painter, S. L., & Karra, S. (2014). Constitutive model for unfrozen water content in subfreezing unsaturated soils. *Vadose Zone Journal*, 13(4). http://dx.doi.org/10.2136/vzj2013.04.0071

Painter, S. L., Moulton, J. D., & Wilson, C. J. (2013). Modeling challenges for predicting hydrologic response to degrading permafrost. *Hydrogeology Journal*, 21, 221–224.

Rempel, A. W. (2012). Hydromechanical processes in freezing soils. *Vadose Zone Journal*, 11(4). http://dx.doi.org/10.2136/vzj2012.0045

Seyfried, M. S., Flerchinger, G. N., Bryden, S., Link, T. E., Marks, D., & McNamara, J. (2021). Slope and aspect controls on soil climate: Field documentation and implications for large-scale simulation of critical zone processes. *Vadose Zone Journal*, 2021;e20158. http://dx.doi.org/10.1002/vzj2.20158.

Seyfried, M. S., & Murdock, M. D. (1996). Calibration of time domain reflectometry for measurement of liquid water in frozen soils. *Soil Science*, 161, 87–89. http://dx.doi.org/10.1097/00010694-199602000-00002

Toride, N., Watanabe, K., & Hayashi, M. (2013). Special section: Progress in modeling and characterization of frozen soil processes. *Vadose Zone Journal*, 12(1). http://dx.doi.org/10.2136/vzj2013.01.0001

van Genuchten, M. T. (1980). A closed-form equation for predicting the hydraulic conductivity of unsaturated soils. *Soil Science Society of America Journal*, 44, 892–898. http://dx.doi.org/10.2136/sssaj1980.03615995004400050002x

Vereecken, H., Schnepf, A., Hopmans, J. W., Javaux, M., Or, D., Roose, T., Vanderborght, J., Young, M. H., Amelung, W., Aitkenhead, M., Allison, S. D., Assouline, S., Baveye, P., Berli, M., Brüggemann, N., Finke, P., Flury, M., Gaiser, T., Govers, G., … Young, I. M. (2016). Modeling soil processes: Review, key challenges, and new perspectives. *Vadose Zone Journal*, 15(5). http://dx.doi.org/10.2136/vzj2015.09.0131

Walvoord, M. A., & Kurylyk, B. L. (2016). Hydrologic impacts of thawing permafrost: A review. *Vadose Zone Journal*, 15(6). http://dx.doi.org/10.2136/vzj2016.01.0010

Watanabe, K., Kito, T., Dun, S., Wu, J. Q., Greer, R. C., & Flury, M. (2013). Water infiltration into a frozen soil with simultaneous melting of the frozen layer. *Vadose Zone Journal*, 12(1). http://dx.doi.org/10.2136/vzj2011.0188

Watanabe, K., & Osada, Y. (2017). Simultaneous measurement of unfrozen water content and hydraulic conductivity of partially frozen soil near 0 °C. *Cold Regions Science and Technology*, 142, 79–84.

Zhang, Y., Chen, W., & Cihlar, J. (2003). A process-based model for quantifying the impact of climate change on permafrost thermal regimes. *Journal of Geophysical Research*, 108(D22), 4695. http://dx.doi.org/10.1029/2002JD003354

Zhang, Y., Cheng, G., Li, X., Han, X., Wang, L., Li, H., Chang, X., & Flerchinger, G. N. (2013). Coupling of a simultaneous heat and water model with a distributed hydrological model and evaluation of the combined model in a cold region watershed. *Hydrological Processes*, 27, 3762–3776. http://dx.doi.org/10.1002/hyp.9514

Zhang, N., & Wang, Z. (2017). Review of soil thermal conductivity and predictive models. *International Journal of Thermal Sciences*, 117, 172–183.

3

Suat Irmak and
Meetpal S. Kukal

Evapotranspiration, Transpiration, and Evaporation Processes and Modeling in the Soil–Residue–Canopy System

Abstract

Accurate quantification of evapotranspiration (ET) in irrigated and rainfed agriculture under different soil, water, and crop management practices and climatic conditions has been, and will always be, paramount for effective management of water resources. This is especially important for enhancing crop and water productivity for the world's rapidly increasing population. While changing global climatic conditions further exacerbate this challenge, scientific investigations and discoveries will be necessary to address these large-scale challenges and develop practical solutions at the scale of management. ET, which is a combination of evaporative losses from the plant canopy through transpiration and evaporation from the plant–soil interface and other surfaces in agroecosystems, is one of the most complex processes in terrestrial system water vapor and energy exchange and hydrologic processes. Its accurate quantification is important for various applications, including irrigation and water management, crop growth modeling, climate change studies, assessing the impact of climate change on agricultural productivity and water resources, assessing the impact of land use change on agroecosystem hydrology and water balance, plant physiology and agronomy studies, groundwater resources replenishment, agricultural water rights transfers, and others. This chapter focuses on recent developments and improvements in modeling of ET, transpiration (T), and soil evaporation (E) processes occurring from the soil–residue–canopy system in the light of changing climate conditions. While

Modeling Processes and Their Interactions in Cropping Systems: Challenges for the 21st Century, First Edition.
Edited by Lajpat R. Ahuja, Kurt C. Kersebaum, and Ole Wendroth.
© 2022 American Society of Agronomy, Inc. / Crop Science Society of America, Inc. / Soil Science Society of America, Inc.
Published 2022 by John Wiley & Sons, Inc.

various models have been introduced over the decades, ET models must be capable of responding to the changes to which the agricultural ecosystems have been subjected to in the past and will be in the future. We describe the formulations and applications (across agricultural, horticultural, and natural landscapes) of the most-used single-source and multisource combination ET models, namely the Shuttleworth–Wallace (SW) and Penman–Monteith models and their recent refinements and applications. Major complements and critiques of these models are also discussed, especially when compared with our current theoretical level of understanding. In general, ET models are limited in their application to comprehensively reflect the physiological and bio-physical changes in plants in response to elevated CO_2 concentrations, the implications of which are discussed. Alternative approaches that compensate for or bypass these limitations are discussed as well, with recommendations for future lines of research and applications.

Terrestrial ET is the transfer of water from the land surface to the atmosphere, entailing a phase change from liquid to gas. The energy required to evaporate a given water mass is equivalent to 600 times that required to raise its temperature by 1 K, and further, 2,400 times that required to achieve the same for air (Seneviratne et al., 2010). On average, terrestrial latent heat accompanying ET (λE) uses ~60% (range of 48–88%) of net radiation (Trenberth et al., 2009), making it a dominant component of the surface energy balance. From a water cycle standpoint, it accounts for 67% of the mean 700 mm year^{-1} of precipitation received by the land surface (Chahine, 1992; Oki & Kanae, 2006). In semi-arid and arid regions, the percentage of evaporative losses can reach ~80–90%, depending on numerous environmental and terrestrial landscape characteristics, including water availability as well as management (Irmak, 2010; Irmak & Mutiibwa, 2010). In addition to being a major determinant of the terrestrial energy and water budget, ET also influences atmospheric processes to a large extent (Kanemasu et al., 1992; Pielke et al., 1998; Rind et al., 1992). The extent of this influence is clear from the fact that the northern hemisphere would be 15–25 °C warmer if the cooling effect of ET was assumed as absent ($\lambda E = 0$) (Shukla & Mintz, 1982).

Evapotranspirational processes have also been recognized to be the most challenging to quantify among the various water and energy balance components, especially when achieving the accuracy required for various applications. The progress of understanding ET, T, and soil E processes emerged from improved measurement techniques (soil-water budget, surface energy budget, lysimetry, eddy covariance [EC] [correlation], Bowen ratio, scintillometry, sap flow gauges, stable isotopes, etc.), allowing us to observe and question weaknesses in existing ideas and hypotheses and to further innovate (Shuttleworth, 2007). The progress made in the last 30–35 year has led to a recognition that ET is a coupled process that involves feedback between the surface and the overlying atmosphere. In practical situations where ET information is desired, ET measurement instrumentation and resources are usually insufficient, thus creating the potential and need for the utilization of ET models. This task becomes extremely difficult when spatiotemporal complexity arising from heterogeneity of ground cover is accounted for.

The progress in evaporation modeling has relied on hypotheses and concepts to aid in the operational determination of ET, such as the concepts of potential ET (Thornthwaite, 1948), crop potential ET (Anonymous., 1956; Dingman, 1992; Jensen,

1968; Penman, 1956, 1963 World Meteorological Organization, 1963), and reference ET (Allen et al., 1998; ASCE Environmental and Water Resources Institute, 2005). Although these are distinct concepts and formulations, they are usually confused among themselves or applied as identical concepts (Dinpashoh et al., 2011; Mardikis et al., 2005; Tanner & Pelton, 1960; Vicente-Serrano et al., 2010; Xu & Singh, 2002). It is critical that the subtle differences among the terminologies and formulations are understood prior to application. The distinctions among these concepts should be realized so caution can be practiced in clearly discerning various concepts and definitions of terrestrial evaporation, which are concisely described here:

- *Potential evapotranspiration* was first defined by Thornthwaite (1948) and Penman (1948) as the atmospheric evaporation demand, namely the amount of water that can be transferred to the air from land or water, representing the transport of water from the Earth back to the atmosphere, the reverse of precipitation (Thornthwaite, 1948). It is the maximum value of E that would occur under ideal conditions.

- Penman (1956, 1963) defined *potential transpiration* as a measure of the transpiration rate from an extensive short green vegetation cover completely shading the ground and adequately supplied with water. This definition was referenced and redefined by Anonymous. (1956) as "the rate of water vapor loss from a short grass canopy under the following conditions: grown in a large surface, during an active growth stage, completely covering the soil, of homogeneous height, in optimal water and nutritional status" and by the World Meteorological Organization (1963) as "the quantity of water vapor which could be emitted by a surface of pure water, per unit surface area and unit time, under the existing atmospheric conditions." In all these definitions of potential transpiration, all stresses, including water and nutrient stresses, advection, disease and weed pressure, insects, and other environmental limitations are assumed to be under ideal conditions. In other words, they do not impede plant growth and development for a plant transpiring at a potential rate in response to atmospheric evaporative demand. Transpiration is regulated, in part, by soil moisture availability and plant stomatal behavior with respect to its response to environmental variables.

- Jensen (1968) defined *crop potential evapotranspiration* as "the rate of evapotranspiration for a given crop at a given stage of growth when water is not limiting and other factors such as insects, diseases, and nutrients have not materially restricted plant development." A major difference here was that Jensen (1968) considered the limitations such as the evaporating surface and crop growth conditions that could affect ET. He also defined potential evapotranspiration (ET_p) in his paper as "the upper limit of evapotranspiration that occurs with a well-watered agricultural crop that has an aerodynamically rough surface such as alfalfa with 30–50 cm of top growth," which is the first description of ET_p in agronomy and is far more specific than that of Thornthwaite (1948).

- A summarized description of the ideal conditions for maximum transfer of water transpiring from a crop or the "potential" level was presented by Dingman (1992): "the rate at which evapotranspiration would occur from a large area completely and uniformly covered with growing vegetation which has access to an unlimited supply of soil water, and without advection or heating effects."

- Considering numerous meanings associated with the potential ET term, it was imperative to define a standardized term for the sake of consistency and a common understanding of an evaporating surface in different applications. A standardization effort was imperative to report evaporative demand of the atmosphere independent of crop type, development, and management as well as different locations and seasons. Allen et al. (1998), based on definitions by Doorenbos and Pruitt (1977), Jensen (1974), Jensen et al. (1971), and Wright and Jensen (1972), defined (grass) reference evapotranspiration (ET_o) as "the rate of evapotranspiration from a hypothetical crop with an assumed crop height (12 cm) and a fixed surface resistance ($70 \, s \, m^{-1}$) and surface albedo (0.23) which would closely resemble evapotranspiration from an extensive surface of green grass cover of uniform height, actively growing, completely shading the ground and not short of water." The American Society of Civil Engineers (ASCE) Environmental and Water Resources Institute (EWRI) in 2005 extended the reference surface concept to a 0.50-m-tall alfalfa crop by defining an alfalfa-reference ET term (ET_r). The two evaporating surfaces vary in their canopy resistance, aerodynamic resistance, and timescale for the tall and short reference crops (ASCE Environmental and Water Resources Institute, 2005). Jensen and Allen (2016) provided a detailed description of this standardization effort to consistently define reference surface ET, including updates since Allen et al. (1998).

Depending on what definitions or concepts of evaporation are suitable to particular applications, numerous models are currently available to estimate E (or ET). Significant progress has been made in developing ET models, and our modeling capabilities have generally improved with time as measurement capabilities have advanced. Combination-based ET models are superior and more physically sound than models that are temperature or radiation based (Arellano & Irmak, 2016). McMahon et al. (2016) compiled a list of 38 single-source and multisource combination-based evaporation models (Table 3.1), developed since as early as 1948 (Penman, 1948) until as recent as 2015 (Tegos et al., 2015; Valiantzas, 2015), that allow estimation of potential, reference crop, open water, deep lake, and pan evaporation from near-surface standard meteorological data (Table 3.1). Among these models, 20 single-source and eight multisource models estimate either ET_p, ET_o, or actual evapotranspiration (ET_c). Amid these alternatives, this chapter focuses on the most widely applied single-source and multisource combination models (highlighted in bold in Table 3.1) that account for both energy balance and aerodynamic aspects. The chapter also focuses on recent developments, their limitations, and recommendations that have been proposed by the scientific community during the last two decades to allow model refinement, improved application, and potential for future use.

Various ET models (Table 3.1) rely on different definitions, have contrasting data requirements, and account for different environmental and surface variables, resulting in vast differences in their performance. In general, not all ET models perform with sufficient accuracy for estimating evaporative losses in all climatic regions for all soils, crops, and management practices. Our agroecosystems, their environmental conditions, and their management are changing in ways that necessitate accounting for them in our ET estimation processes. These requirements mean that certain ET models should be preferable than others to be able to account for the respective changes. Two primary elements of these changes are presented below:

1. **Increased no-till (NT) or reduced-till adoption:** Owing to evidence of positive impacts of NT on reducing water and wind erosion, improving water retention, drainage, organic matter content, and C sequestration, in addition to profitability from labor and fuel savings, adoption of NT and reduced-till management is increasing in the United States and globally (Derpsch et al., 2010 Kassam et al., 2015 Pittelkow et al., 2015). Azzari et al. (2019) used satellite imagery and georeferenced surveys to develop tillage intensity maps from 2005 to 2016 across the

Table 3.1 Combination-Based Evaporation (*E*) (or Evapotranspiration [ET]) Models Categorized by Date and Application

Model year	Reference	Potential E (ET$_p$)	Reference ET(ET$_o$)	Actual ET (ET$_c$)	Shallow water E	Lake E	Pan E
		Single-source combination models					
1948	Penman (1948)			✓			
1955	Kohler et al. (1955)					✓	
1956	**Penman (1956)**			✓			
	Lascano and Van Bavel (2007)		✓	✓			
	Budyko (1956/1958)	✓					
1960	Slatyer and McIlroy (1961)	✓		✓			
1964	Sellers (1964)	✓					
1965	Monteith et al. (1965)	✓		✓			
	Keijman (1981)	✓					
1966	van Bavel (1966)	✓					
1967	Kohler and Parmele (1967)					✓	
1974	Keijman (1974)					✓	
1982	Wright (1982)		✓				
1983	Shi et al. (2008)			✓			
1988	Paw and Gao (1988)			✓			
1989	Granger and Gray (1989)			✓			
1991	Milly (1991)			✓			
1994	Linacre (1994)						✓
1996	Vardavas and Fountoulakis (1996)					✓	
1998	**Allen et al. (1998)**		✓				
1999	Todorovic (1999)		✓				
2000	**ASCE Standardization of Reference Evapotranspiration Task Committee (2000); Allen et al. (2005)**		✓				
2001	Finch (2001)					✓	
2006	Valiantzas (2006)		✓		✓		
	Rotstayn et al. (2006)						✓
2008	McJannet et al. (2008)					✓	
2009	**Shuttleworth and Wallace (2009)**		✓			✓	
2013	Valiantzas (2013)		✓				
2015	Tegos et al. (2015)	✓					
	Valiantzas (2015)		✓				
		Multisource combination models					
1985	**Shuttleworth and Wallace (1985)**			✓			
1988	Choudhury and Monteith (1988)			✓			
1994	Wessel and Rouse (1994)			✓			
1996	**Farahani and Ahuja (1996)**			✓			
1997	Brenner and Incoll (1997)			✓			
	Hough and Jones (1997)			✓			
2012	Lhomme et al. (2012)			✓			
2013	Lhomme et al. (2013)			✓			

Note. This list is a part of a larger compendium of evaporation models compiled by McMahon et al. (2016).
Note. The models discussed here are highlighted in bold.

U.S. Corn Belt (Figure 3.1a). They found that the low-intensity tillage area significantly increased for most counties at rates ranging from 1 to 6% per year (Figure 3.1a), with the highest increases in Nebraska (Figure 3.1c), Kansas (Figure 3.1b), and southern Illinois.

Careful measurements of micrometeorological dynamics across conventional and conservation tillage systems have shown that ET$_c$ is a strong function of residue management. Irmak et al. (2019) measured ET$_c$ across disk-till (DT) and NT

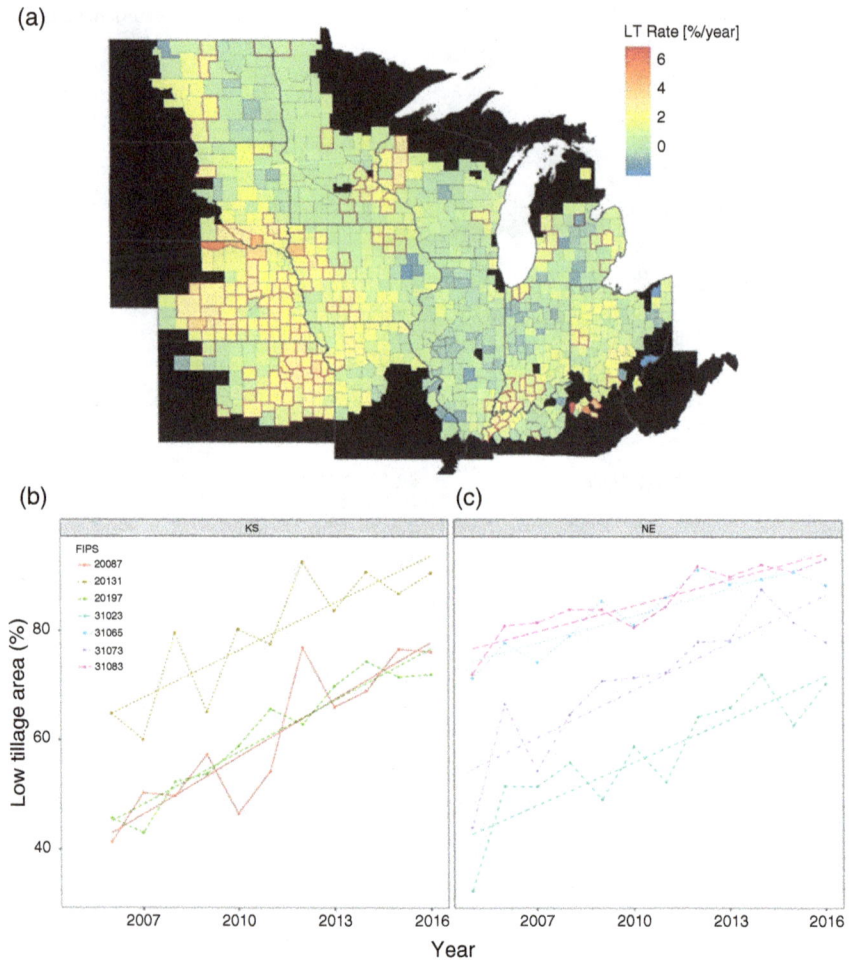

Figure 3.1 (a) Rate of change of low-intensity tillage fractional area at the county level, **(b)** linear regressions of low-intensity tillage fractional area in selected counties in Kansas, and **(c)** same as (b) but for Nebraska. The rates were estimated using linear regressions of fractional area estimates against time for only counties with at least six years of data. Counties with a red border had a statistically significant regression (p < 0.01). Reproduced from Azzari et al. (2019) / with permission of Elsevier.

tillage systems (Figure 3.2) in two carefully managed and monitored producers' fields (Figure 3.3a), which have been under these tillage management practices for more than 17 year. They found that, on a three-year total (2011–2013) basis, the DT maize ET_c (2091 mm) was 92 mm higher than the NT maize ET_c (1999 mm) owing to reduced E from residue coverage under NT maize. Also, NT maize had less pre-anthesis water use than DT maize and greater post-anthesis water use than DT maize in all three growing seasons (Figure 3.3b). The post-anthesis ET_c being greater under the NT system was primarily due to a higher leaf area index (LAI) in the NT field (Irmak et al., 2019). These ET_c differences translated into differences in crop coefficients [both single (K_c) and basal (K_{cb})]: NT maize showed

Figure 3.2 Side-by-side no-till (NT) (upper picture) and disk-till (DT) (bottom picture) research fields that contrasts the surface residue cover. The residue cover in the NT field is close to 95–100%, whereas the DT field had about 15–20% residue cover after disking, which was done once in spring before planting and once in fall after harvest. These DT and NT fields are a part of the Nebraska Water and Energy Flux Measurement, Modeling and Research Network (Irmak, 2010), which has been continuously operating more than 10 flux towers since 2004 to measure surface energy balance variables, including evapotranspiration; soil heat flux; sensible heat flux; all radiation envelopes; surface albedo; soil water content (every 0.30 m down to 1.8 m); soil temperature; irrigation timing and amount; fertilizer type, amount, and application date and method; grain yield; stomatal resistance, photosynthetic photon flux density, and canopy temperature (for some vegetation surfaces); and other variables on an hourly basis year-round for irrigated and rainfed grasslands; DT, NT, and strip-till maize–soybean rotation; vineyard; popcorn; cover crops; alfalfa; riparian vegetation; black turtle bean; irrigated and rainfed winter wheat; grain sorghum; and other surfaces. Photos by S. Irmak.

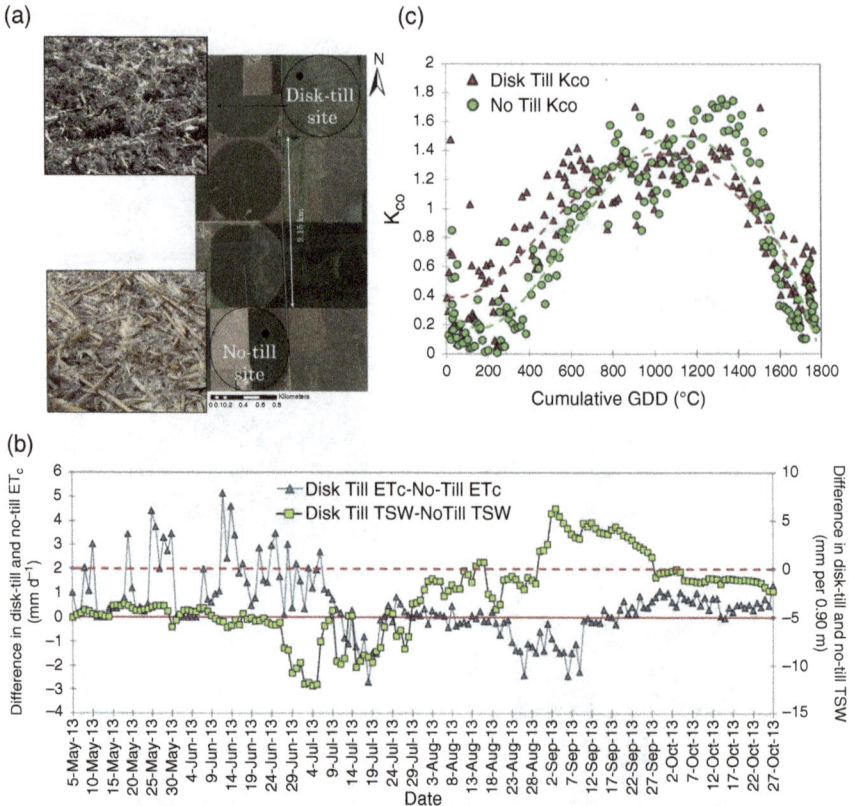

Figure 3.3 (a) Location and surface cover conditions (bare soil and residue cover) of the two experimental sites—a disk-till maize field and a no-till maize field that are 2.15 km from each other, both equipped with a Bowen Ratio Energy Balance System—for which all the critical site characteristics and management practices (except tillage) were identical across the two fields, (b) differences in daily disk-till maize evapotranspiration (ET_c) and no-till ET_c and daily disk-till total soil water (TSW) and no-till TSW during the 2013 growing season (the solid red horizontal line represents no ET_c difference, and the broken red horizontal line represents no TSW difference), and (c) seasonal trends in daily grass-reference single crop coefficients (K_{co}) for disk-till and no-till maize as a function of cumulative growing degree days (GDD) during the 2013 growing season. (a) Reproduced from Irmak et al. (2019)/with permission of Elsevier. (b) Reproduced from Irmak et al. (2019)/with permission of Elsevier. (c) Reproduced from Irmak, S., & Kukal, M. S, (2019)/with permission of Elsevier.

lower K_c and K_{cb} during pre-anthesis and higher K_c and K_{cb} during post-anthesis, evident from Figure 3.3c (Irmak & Kukal, 2019). Thus, ET models have to be capable of replicating these revealed water use differences across tillage systems via measurements. Irmak and Kukal (2019) reported that differences in ET_c between the two tillage systems were also responsible for modification of the field-scale microclimate. The difference in DT and NT maize ET_c was negatively related to differences in DT measured and NT measured air temperature, vapor pressure deficit, wind speed, and sensible heat flux. The difference in DT and NT maize ET_c was positively related to DT measured and NT measured net radiation and total soil water (TSW) in the root zone.

2. **Elevated ambient CO_2:** The CO_2 concentration in the atmosphere has increased from 354 mg kg^{-1} in 1990 to 411 mg kg^{-1} in 2019 (Earth System Research Laboratory, 2020). These levels are further expected to increase by mid-century, with 500 mg kg^{-1} for a reduced-emission scenario of representative concentration pathway (RCP) 4.5 (Thomson et al., 2011) and 540 mg kg^{-1} for a high-emission scenario of RCP 8.5 (Riahi et al., 2011). The CO_2 fertilization and enrichment effect leads to increased photosynthesis and a reduction in stomatal conductance (g_s) and leaf-level T, primarily in C_3 crop species (Drake et al., 1997 Kimball et al., 2002 Leakey et al., 2009). Kimball (2016) reviewed 27 year of Free-Air CO_2 Enrichment (FACE) experiments and found that elevated CO_2 decreased ET_c of both C_3 and C_4 plants by about 10% (Figure 3.4).

In the light of these changes in surface and atmospheric characteristics, ET models need to stand up to the challenge of accurately representing these dynamics in their respective frameworks to produce reliable estimates of the evolution of water use characteristics of terrestrial surfaces. For example, when the majority of crop fields are under NT or reduced-till management, E has to be considered from two distinct sources (bare and residue-covered soil fractions), making well-characterized and developed multisource models necessary. Similarly, models that allow for a variable stomatal resistance (r_s) in their formulations are advantageous for application in a changing climate.

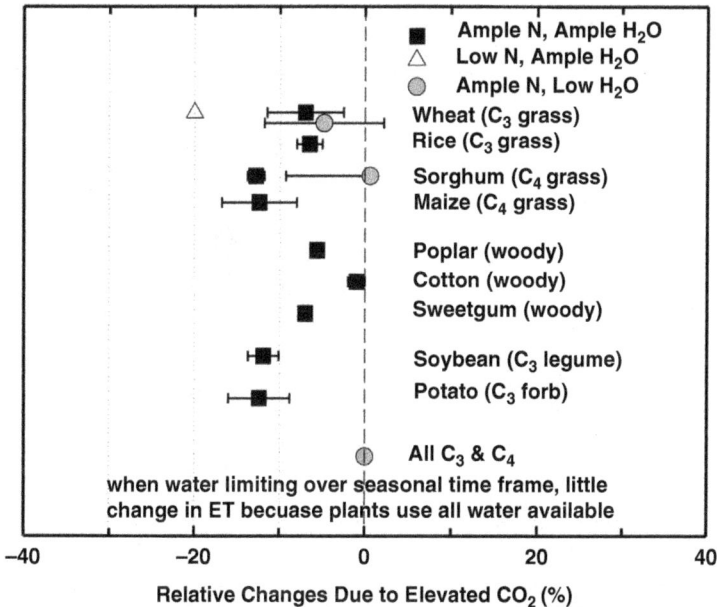

Current Opinion in Plant Biology

Figure 3.4 Evapotranspiration (ET) responses to elevated CO_2 (+200 mg kg^{-1} from free-air CO_2 enrichment studies) at ample and limited levels of soil water and N as compiled from multiple research studies. Reproduced from Kimball (2016) / with permission of Elsevier.

Descriptions of ET, *T*, and *E* Modeling Frameworks

Penman–Monteith Approach

The Penman–Monteith (PM) model (presented as a schematic in Figure 3.5) builds on the original Penman combination equation (Penman, 1948), which was the first attempt to integrate an aerodynamic approach and a surface energy model driven by net radiation (R_n). The integrated approach was developed to use surface radiation, temperature, and humidity to estimate E from a saturated surface. The PM model (Monteith, 1965) extends the application of the Penman approach to vegetated surfaces by incorporating resistance factors (Figure 3.5).

The PM model (Allen et al., 1998) is

$$\text{ET}_{PM} = \frac{1}{\lambda} \frac{\Delta(R_n - G) + \rho_a c_a(\text{VPD}/r_a)}{\Delta + \gamma(1 + r_s/r_a)} \tag{3.1}$$

where ET_{PM} is Penman–Monteith evapotranspiration (mm d^{-1}), R_n is net daily radiation at the vegetated surface (MJ m^{-2} d^{-1}), G is soil heat flux (MJ m^{-2} d^{-1}), ρ_a is the mean air density at constant pressure (kg m^{-3}), c_a is the specific heat of the air (MJ kg^{-1} °C^{-1}), r_a is aerodynamic resistance to water vapor transport (s m^{-1}) for neutral conditions of stability, r_s is a surface resistance term (s m^{-1}), VPD is vapor pressure deficit (kPa), λ is the latent heat of vaporization (MJ kg^{-1}), Δ is the slope of the saturation vapor pressure curve (kPa °C^{-1}) at air temperature, and γ is the psychrometric constant (kPa °C^{-1}).

The r_a is computed as (Allen et al., 1998):

$$r_a = \frac{\ln\left[(z_m - d)/z_{0m}\right] \ln\left[(z_h - d)/z_{0h}\right]}{k^2 u_z} \tag{3.2}$$

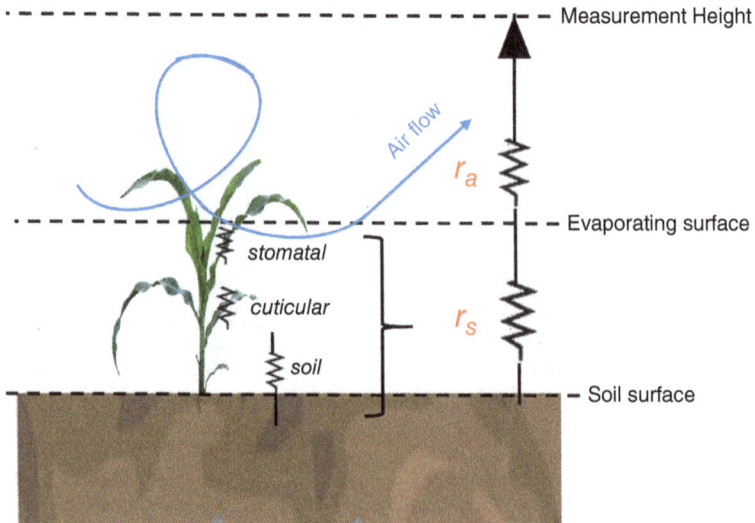

Figure 3.5 A simplified representation of the (bulk) surface (r_s) and aerodynamic (r_a) resistance for water vapor flow in the Penman–Monteith approach. Adapted from Allen et al. (1998).

where $d = 0.66h_c$, z_m and z_h are the heights at which wind speed and relative humidity (RH) are measured; z_{0m} and z_{0h} are roughness lengths for momentum and heat and vapor transfer, respectively, and are defined as $z_{0m} = 0.123h_c$ and $z_{0h} = 0.1z_{0m}$; u_z is the wind speed measured at height z (ms^{-1}); k is the von Karman constant (0.41); and h_c is the reference plant height. All heights are in meters.

When this r_a formulation is applied to a reference surface (h_c of 0.12 m), it is simplified as

$$r_a = \frac{208}{u_z} \tag{3.3}$$

A further application-oriented development in this direction was the adoption of the characteristics of a hypothetical reference crop by the ASCE Standardization of Reference Evapotranspiration Task Committee to formulize the ASCE-EWRI standardized reference ET equation (ASCE Environmental and Water Resources Institute, 2005):

$$\text{ET}_{\text{ASCE}} = \frac{0.408\Delta(R_n - G) + \gamma[C_n/(T_{\text{mean}} + 273)]u_2\text{VPD}}{\Delta + \gamma(1 + C_d u_2)} \tag{3.4}$$

where ET_{ASCE} is either short reference (clipped, cool-season grass) ET (ET_o) or tall-reference (full-cover alfalfa) ET (ET_r) (mm d^{-1}), R_n is net radiation (MJ m^{-2} d^{-1}), G is soil heat flux (MJ m^{-2} d^{-1}) and is assumed to be zero for a daily time step (at hourly timestep: 10 and 50% of R_n for short reference during day and night, respectively, and 4 and 20% of R_n for tall reference during day and night, respectively), γ is the psychometric constant (kPa °C^{-1}), T_{mean} is daily average air temperature (°C), u_2 is wind speed at 2-m height (m s^{-1}), Δ is the slope of the saturation vapor pressure curve at air temperature (kPa °C^{-1}), and C_n and C_d are constants that change with the reference surface and time step. Coefficients C_n and C_d are 900 °C mm s^{3}Mg^{-1} d^{-1} and 0.34 s m^{-1} for a grass-reference surface and 1600 °C mm s^3 Mg^{-1} d^{-1} and 0.38 s m^{-1} for an alfalfa-reference surface, respectively, on a daily timescale (ASCE Environmental and Water Resources Institute, 2005). At hourly time steps, ASCE Environmental and Water Resources Institute (2005) specifies values of hourly C_n as 37 and 66 °C mm s^3 Mg^{-1} d^{-1} for short-reference and tall-reference, respectively. Similarly, hourly C_d was specified as 0.24 and 0.96 s m^{-1} for day and night, respectively, for short reference and 0.25 and 1.7 s m^{-1} for day and night, respectively, for tall reference.

Estimating actual crop evapotranspiration (ET_c) requires the most common application, that is, a two-step approach. While the first step is determining ET_{ASCE} (Equation 3.4), the second step is multiplying ET_{ASCE} by a crop coefficient (K_c), which must be the K_c associated with the specific reference surface used in determining ET_{ASCE} (grass or alfalfa). The K_c values represent the plant cover or LAI as the crop develops, resulting in a decrease in bulk surface resistance.

The crop coefficient approach can also be used in a more refined manner than using a mean K_c to represent crop growth. Under this framework, referred to as the dual-K_c approach, the K_c is divided into two components. One component is called a basal crop coefficient (K_{cb}) and represents primarily the T component of ET and a small E component from soil that is visually dry at the surface. The other component is an evaporation coefficient (K_e), which adjusts for increased evaporation from wet soil immediately following a wetting event, that is, rain or irrigation (Allen et al., 1998; Jensen & Allen, 2016). The effects of water stress on crop water use can be incorporated into the ET_c determination via the use of a dimensionless coefficient K_s, which is a function of the available soil water:

$$ET_c = (K_s K_{cb} + K_e)ET_{ASCE} \tag{3.5}$$

The most widely used and easy to formulate K_c is the K_c curve provided by FAO-56 (Allen et al., 1998). The FAO-56 K_c curve divides the growing season into four parts corresponding to crop phenology, namely: initial, crop development, midseason, and late season periods. Three tabulated values are selected to represent average K_c (or K_{cb}) during the initial, midseason, and end of late seasons, and linear progressions can be estimated between them. The tabulated values in FAO-56 [$K_{cb(table)}$] are experimentally measured (typically via lysimetry or energy balance). The values of K_{cb} and K_e are calculated as

$$K_{cb} = K_{cb(table)} + [0.04(u_2 - 2) - 0.004(RH_{min} - 45)]\left(\frac{h_p}{3}\right)^{0.3} \tag{3.6}$$

$$K_e = K_r(K_{c\_max} - K_{cb}) \leq f_{ew} K_{c\_max} \tag{3.7}$$

where u_2 is mean daily wind speed at 2-m height over a grass surface (m s^{-1}), RH_{min} is mean daily minimum relative humidity (%), h_p is mean plant height (m), f_{ew} is the fraction of soil contributing to the majority of evaporation, K_r is a reduction factor based on soil water content (θ), and K_{c_max} is the maximum evaporation factor, which is a function of the energy available for ET at the soil surface.

Evapotranspiration from a Sparse Canopy

The Shuttleworth–Wallace (SW) model (Shuttleworth & Wallace, 1985) was proposed in 1985 as a compartment model to allow partitioning of ET by treating the bare substrate (soil) and closed canopy as distinct sources of evaporation in a one-dimensional model. Canopy interactions have been accepted to be considered as a single source (Monteith, 1965), and thus the model is a pragmatic solution to physical modeling rigor and convenience for field applications. A one-dimensional model necessitates that all model elements are averaged across a sufficiently persistent horizontal area. Another assumption for the SW model development is the hypothetical "mean canopy flow" (Thom, 1972), which can be represented using temperature, humidity, and wind speed. The fulfillment of this hypothesis requires sufficient aerodynamic mixing in the canopy and has been tested, especially in row crops (Black et al., 1970; Szeicz et al., 1973).

The SW model is based on a two-component (crop and soil) structure, as depicted from the schematic in Figure 3.6, which was reconstructed from the original work of Shuttleworth and Wallace (1985). The model framework argues that the air adjacent to the soil surface is not saturated, except under wet soil surface conditions, and thus introduces the concept of surface resistance at the substrate surface (soil), that is, r_s^s, in addition to the bulk stomatal resistance (r_s^c). Thus, the two surface resistances in the model are r_s^c and r_s^s. Similarly, the aerodynamic resistance is represented as $r_a^a + r_a^c$ in a closed canopy with no soil evaporation and $r_a^a + r_a^s$ to describe soil evaporation.

The total evaporation from the crop–soil system (λE) is depicted in the following form (Equation 3.8), where PM_c and PM_s are evaporation terms analogous to the Penman–Monteith combination equation when applied for computing evaporation

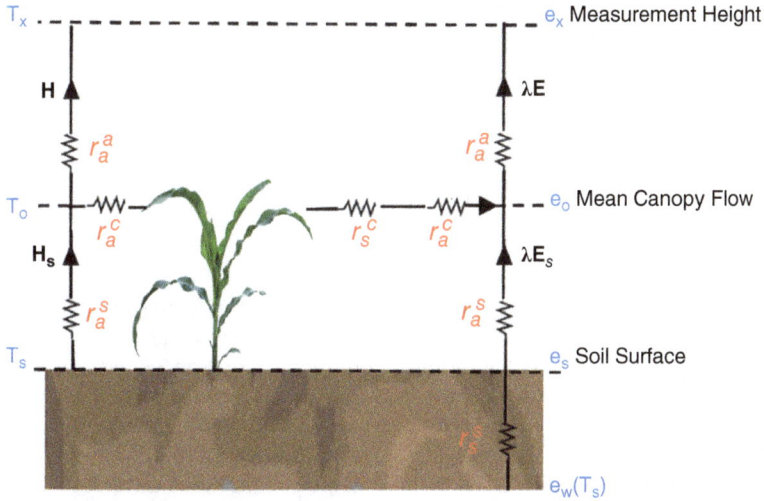

Figure 3.6 Schematic representation of a one-dimensional description of energy partition for sparse crops (partial canopy) in the Shuttleworth and Wallace (SW) model. Adapted from Shuttleworth and Wallace (1985).

from a closed (full) canopy (Equation 3.9) and from bare soil (Equation 3.10), respectively. The two terms represent defined asymptotic limits (closed canopy vs. bare soil):

$$\lambda E = C_c PM_c + C_s PM_s \tag{3.8}$$

$$PM_c = \frac{\Delta A + \left[(\rho c_p D - \Delta r_a^c A_s)/(r_a^a + r_a^c) \right]}{\Delta + \gamma \left[1 + r_s^c/(r_a^a + r_a^c) \right]} \tag{3.9}$$

$$PM_s = \frac{\Delta A + \left[\rho c_p D - \Delta r_a^s (A - A_s) \right]/(r_a^a + r_a^s)}{\Delta + \gamma \left[1 + r_s^s/(r_a^a + r_a^s) \right]} \tag{3.10}$$

where ρ is the density of air (kg m^{-3}), c_p is specific heat at constant pressure (J kg^{-1} K^{-1}), D is vapor pressure at reference height (mbar [1 mbar = 0.1 kPa]) defined as $e_w(T_x) - e_x$, Δ is the mean rate of change of saturation vapor pressure with air temperature (mbar K^{-1}), γ is the psychometric constant (mbar K^{-1}), and A and A_s are total available energy and available energy at the soil surface, respectively, and are defined as

$$A = \lambda E + H = R_n - S - P - G \tag{3.11}$$

$$A_s = \lambda E_s + H_s = R_n^s - G \tag{3.12}$$

where R_n and R_n^s are the incoming net radiation and incoming net radiation at the soil surface, respectively, λE_s is the latent heat flux from the substrate (soil), H is the sensible heat flux, S and P are the physical and biochemical energy storage terms, and G is the soil heat flux.

The coefficients C_c and C_s are defined as

$$C_c = \left[1 + \frac{R_c R_a}{R_s(R_c + R_a)} \right]^{-1} \tag{3.13}$$

$$C_s = \left[1 + \frac{R_s R_a}{R_c(R_s + R_a)}\right]^{-1} \tag{3.14}$$

Parameters R_a, R_s, and R_c are defined as

$$R_a = (\Delta + \gamma)r_a^a \tag{3.15}$$

$$R_s = (\Delta + \gamma)r_a^s + \gamma r_s^s \tag{3.16}$$

$$R_c = (\Delta + \gamma)r_a^c + \gamma r_s^c \tag{3.17}$$

The asymptotic nature of Equation 3.8 implies that if there is no soil evaporation, r_s^s and R_s are infinite, PM_s is 0, and thus, C_c is unity. Moreover, if there is no sensible heat flux from the soil (H_s), A_s is 0. Under these conditions, Equation 3.8 reduces to the PM evaporation under closed (full) canopy conditions and no soil interaction. In contrast, if bare soil conditions exist (no canopy presence), r_s^c and R_c are infinite and thus A is equal to A_s. In this case, Equation 3.8 reduces to PM evaporation from the soil only. Within these two asymptotic limits, a wide range of conditions can exist where both canopy and soil surface coexist that can represent any given instance within the crop growing season. In such cases, total evaporation (λE) computed from Equation 3.8 can be substituted in Equation 3.18 to deduce D_o, which can further be used in Equations 3.19 and 3.20 to quantify the component fluxes λE_s (latent heat flux from the substrate) and λE_c (latent heat flux from the plant canopy), respectively:

$$D_o = D + [\Delta A - (\Delta + \gamma)\lambda E]\frac{r_a^a}{\rho c_p} \tag{3.18}$$

$$\lambda E_s = \frac{\Delta A_s + \rho c_p D_o/r_a^s}{\Delta + \gamma\left(1 + r_s^s/r_a^s\right)} \tag{3.19}$$

$$\lambda E_c = \frac{\Delta(A - A_s) + \rho c_p D_o/r_a^c}{\Delta + \gamma\left(1 + r_s^c/r_a^c\right)} \tag{3.20}$$

Evapotranspiration from a Residue-Covered System

The extended SW model (Farahani & Ahuja, 1996), as represented by a schematic in Figure 3.7, further enables the SW model framework with the capability of defining a partially covered soil and partitioning evaporation into bare soil and residue-covered fractions.

The total flux of latent heat (λE) above the canopy (at the measurement height) is the sum of latent heat from the canopy (λT), bare soil evaporation flux (λE_s), and residue-covered soil evaporation flux (λE_r):

$$\lambda E = T + C_s E_s + C_r E_r \tag{3.21}$$

Following Shuttleworth and Wallace (1985), when the VPD at the height of the canopy air stream (D_o) is known, λT and λE_s can be expressed as

$$\lambda T = \frac{\Delta(R_n - G) - R_{nsub} + \rho c_p D_o/r_a^c}{\Delta + \gamma\left(1 + r_s^c/r_a^c\right)} \tag{3.22}$$

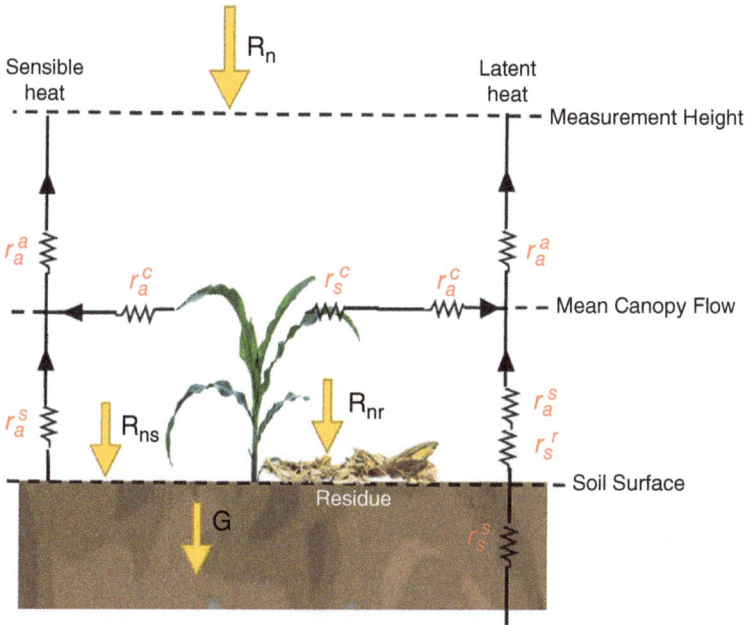

Figure 3.7 A schematic representation of the resistances in the extended Shuttleworth–Wallace) model applicable to partial canopy- or residue-covered fields. Adapted from Farahani and Ahuja (1996).

$$\lambda E_s = \frac{\Delta(R_{ns} - G_s) + \rho c_p D_o / r_a^s}{\Delta + \gamma\left(1 + r_s^s / r_a^s\right)} \tag{3.23}$$

where R_{nsub} is the net radiation below the canopy and is determined as

$$R_{nsub} = C_s R_{ns} + C_r R_{nr} \tag{3.24}$$

By a similar logic as followed in Equations 3.22 and 3.23, λE_r can be formulated as

$$\lambda E_r = \frac{\Delta(R_{nr} - G) + \rho c_p D_o / r_a^s}{\Delta + \gamma\left[1 + \left(r_s^s + r_s^r\right) / r_a^s\right]} \tag{3.25}$$

and D_o can be formulated by introducing flux equations for λE and H into the energy balance equation above the canopy:

$$D_o = D + [\Delta(R_n - G) - (\Delta + \gamma)\lambda E] \frac{r_a^a}{\rho c_p} \tag{3.26}$$

Inserting the formula for D_o into Equations 3.22, 3.23, and 3.25, the expression for λE changes to

$$\lambda E = CC\, PM_c + CS\, PM_s + CR\, PM_r \tag{3.27}$$

where

$$PM_c = \frac{\Delta(R_n - G) + \left(\rho c_p D - \Delta r_a^c R_{nsub}\right) / \left(r_a^a + r_a^c\right)}{\Delta + \gamma\left[1 + r_s^c / \left(r_a^a + r_a^c\right)\right]} \tag{3.28}$$

$$PM_s = \frac{\Delta(R_n - G) + \{\rho c_p D - \Delta r_a^s[(R_n - G) - (R_{ns} - G_s)]\}/(r_a^a + r_a^s)}{\Delta + \gamma[1 + r_s^s/(r_a^a + r_a^s)]} \tag{3.29}$$

$$PM_r = \frac{\Delta(R_n - G) + \{\rho c_p D - \Delta r_a^s[(R_n - G) - (R_{nr} - G_r)]/(r_a^a + r_a^s)\}}{\Delta + \gamma[1 + (r_s^s + r_s^r)/(r_a^a + r_a^s)]} \tag{3.30}$$

The coefficients CC, CS, and CR are given as

$$CC = \left[1 + \frac{R_a R_c (C_s R_r + C_r R_s)}{R_s R_r (R_c + R_a)}\right]^{-1} \tag{3.31}$$

$$CS = \left[1 + \frac{R_a (C_s R_r + C_r R_s)}{R_r C_s (R_a + R_s)}\right]^{-1} \tag{3.32}$$

$$CR = \left[1 + \frac{R_a (C_s R_r + C_r R_s)}{R_s C_r (R_a + R_r)}\right]^{-1} \tag{3.33}$$

where R_a, R_s, R_c, and R_r are defined as

$$R_a = (\Delta + \gamma)r_a^a \tag{3.34}$$

$$R_s = (\Delta + \gamma)r_a^s + \gamma r_s^s \tag{3.35}$$

$$R_c = (\Delta + \gamma)r_a^c + \gamma r_s^c \tag{3.36}$$

$$R_r = (\Delta + \gamma)r_a^s + \gamma(r_s^s + r_s^r) \tag{3.37}$$

One-Step Estimation of Crop Water Requirements: Matt–Shuttleworth Approach

Shuttleworth (2006) initiated the debate among relative merits of estimating crop water requirements by (a) a two-step estimation method involving the ET_{ASCE} and accounting for intercrop differences using K_c or (b) a one-step estimation based on the PM approach, accounting for crop-to-crop differences by using crop-specific values of the surface and aerodynamic resistances. He argued that due to the location of aerodynamic and surface resistances relative to meteorological controls with the PM model, K_c values are a complex mixture of both the physiology of the crop that they represent and the climate within which the K_c values are derived and/or used. Generally, it is claimed that K_c is conservative to climate conditions and only sensitive to crop characteristics (Wright, 1981), which was contradicted by Wallace (1995). Wallace (1995) proposed that using a crop-height-dependent aerodynamic resistance and carrying out field studies to define effective surface resistances should be the preferable approach over complexities presented by the two-step approach. He also identified and proposed solutions of two outstanding issues that limited the application of the one-step approach, that is, the issues of a reference height and the lack of a surface resistance tabulation.

In the Matt–Shuttleworth (M–S) approach (Shuttleworth, 2006; Shuttleworth & Wallace, 2009), the first step is to select the surface resistance $(r_s)_c$ of a well-watered crop equivalent to the FAO crop coefficient for the target crop from Shuttleworth and Wallace (2009). These values were presented by Shuttleworth (2006) as interim sources of crop-specific surface resistance estimates, pending field investigations.

Next, climatological resistance (r_{clim}) (s m^{-1}) needs to be computed using

$$r_{clim} = 86,400 \frac{\rho_a c_a \text{VPD}}{\Delta R_n} \qquad (3.38)$$

where ρ_a is the mean air density (kg m^{-3}) at constant pressure, c_a is the specific heat of the air (MJ kg^{-1} °C^{-1}), VPD is the vapor pressure deficit (kPa), Δ is the slope of the vapor pressure curve (kPa °C^{-1}) at air temperature, and R_n is the net radiation (MJ m^{-2} d^{-1}) at the vegetated surface.

Next, the ratio of VPD at 50- and 2-m heights is computed:

$$\frac{\text{VPD}_{50}}{\text{VPD}_2} = \frac{302(\Delta + \gamma) + 70\gamma u_2}{208(\Delta + \gamma) + 70\gamma u_2} + \frac{1}{r_{clim}} \left\{ \left[\frac{302(\Delta + \gamma) + 70\gamma u_2}{208(\Delta + \gamma) + 70\gamma u_2} \left(\frac{208}{u_2} \right) - \frac{302}{u_2} \right] \right\} \qquad (3.39)$$

where u_2 is the mean daily wind speed (m s^{-1}) at 2-m height.

The next step is to calculate the aerodynamic coefficient (s m^{-1}) for crop height (h), that is, r_c^{50}:

$$r_c^{50} = \frac{1}{(0.41)^2} \ln \left(\frac{50 - 0.67h}{0.123h} \right) \ln \left(\frac{50 - 0.67h}{0.0123h} \right) \frac{\ln \left[(2 - 0.08)/0.0148 \right]}{\ln \left[(50 - 0.08)/0.0148 \right]} \qquad (3.40)$$

The final step is to calculate ET$_c$:

$$\text{ET}_c = \frac{1}{\lambda} \frac{\Delta R_n + \left[\rho_a c_a u_2 (\text{VPD}_2)/r_c^{50} \right] (\text{VPD}_{50}/\text{VPD}_2)}{\Delta + \gamma \left[1 + (r_s)_c u_2/r_c^{50} \right]} \qquad (3.41)$$

Extension of the Matt–Shuttleworth Approach to Dual Crop Coefficients

The translation of the FAO-56 methodology into a one-step approach by Shuttleworth (2006) motivated Lhomme et al. (2015) to transform the two-source model of ET into a PM-type equation by including foliage transpiration resistance and soil evaporation resistance combined in a bulk surface resistance term.

Canopy ET$_c$ is the sum of foliage evaporation (ET$_f$) and soil surface evaporation (ET$_s$):

$$\text{ET}_c = (K_{cb} + K_e)\text{ET}_0 = \text{ET}_f + \text{ET}_s \qquad (3.42)$$

These component evaporations are expressed as a function of their respective surface resistances ($r_{s,f}$ and $r_{s,s}$):

$$\text{ET}_f = \frac{1}{\lambda} \frac{\Delta R_{n,f} + \rho c_p D_m/r_{a,f}}{\Delta + \gamma(1 + r_{s,f}/r_{a,f})} \qquad (3.43)$$

$$\text{ET}_s = \frac{1}{\lambda} \frac{\Delta (R_{n,s} - G) + \rho c_p D_m/r_{a,s}}{\Delta + \gamma(1 + r_{s,s}/r_{a,s})} \qquad (3.44)$$

where D_m is the saturation deficit at canopy source height ($z_m = d + z_{0,m}$), $R_{n,f}$ and $R_{n,s}$ are radiation loads for foliage and soil surface, and $r_{a,f}$ and $r_{a,s}$ are air resistances within the canopy (usually considered negligible to produce the consequent equations).

The two surface resistances are retrieved by equating Equation 3.43 with $K_{cb}ET_o$ and Equation 3.44 with K_eET_o:

$$r_{s,f} = \frac{\rho c_p}{\gamma} \frac{D_m}{K_{cb}\lambda ET_0} \tag{3.45}$$

$$r_{s,s} = \frac{\rho c_p}{\gamma} \frac{D_m}{K_e\lambda ET_0} \tag{3.46}$$

The value of D_m can be inferred from the saturation deficit at reference height (D_a) as

$$D_m = D_a + \frac{[\Delta(R_n - G) - \lambda ET_c(\Delta + \gamma)]r_a}{\rho c_p} \tag{3.47}$$

In this equation, D_a and r_a should be replaced by those calculated at the blending height (Shuttleworth, 2006). The bulk surface resistances can be expressed as the parallel sum of two component resistances:

$$\frac{1}{r_{s,v}} = \frac{1}{r_{s,f}} + \frac{1}{r_{s,s}} \tag{3.48}$$

Finally, the bulk surface resistance should be used in the following equation to infer evaporative flux:

$$\lambda E = \frac{\Delta(R_n - G) + \rho c_p D_a/r_a}{\Delta + \gamma(1 + r_{s,v}/r_a)} \tag{3.49}$$

Lhomme et al. (2015) recommended that $r_{s,f}$ should be calculated with the standard climatic conditions of K_{cb}, whereas $r_{s,s}$ should instead be calculated with actual conditions of the crop. If there is no soil evaporation, $K_e = 0$ and $r_{s,s}$ trends to infinity.

Notations used in the descriptions of the various ET models are provided in the Appendix.

Applications of ET Models

Agricultural and Agroecosystem Applications

The literature is rife with applications and evaluations of the two-step approach and SW model for their performance to simulate ET, E, and T in a wide variety of agricultural and horticultural ecosystems. Li et al. (2013) compared the performance of the PM model, the original SW model, and an adjusted version of the SW model to account for film mulch (MSW model) to estimate maize ET_c and soil E against EC and microlysimetric measurements in a temperate region in China. They found that MSW performed the best, underestimating 30-minutes maize ET_c and E by 2 and 7%, respectively, during the experimental period (two growing seasons), followed by PM (underestimating ET_c by 6%), and SW (overestimating ET_c by 17% and E by 241%). The SW model was shown to be effective to simulate ET_c accurately in a sparse sorghum canopy in an arid region in Japan (Kato et al., 2004). Gardiol et al. (2003) demonstrated the better performance of the SW model over the PM model when estimating ET_c under an entire range of maize covers (bare soil to full canopy cover) in two contrastingly dense canopies (22,000 and 91,000 plants ha^{-1}) when compared with the soil water balance ET_c in Argentina. Wei et al. (2019) evaluated four ET models (SW, PM, Priestly–Taylor, and Flint–Childs models) to estimate spring maize ET_c at an arid site against 30-minutes EC data. They found that the SW model had the best performance among all four models. They also evaluated the physical mechanisms underlying the performance behaviors. The main reasons for the outperforming of the SW model were its physically rigorous structure and the extinction coefficient parameter (defines light transmission in the canopy), which is sensitive and has a significant impact on the performance of the model, being well constrained. They also concluded that good fitting of SW ET_c values to observations can counterbalance its greater complexity relative to other models.

A recent development (Kimball et al., 2019) by the Agricultural Model Inter-Comparison and Improvement Project (AgMIP) serves as the most comprehensive assessment of crop models (29 in total) to simulate daily maize ET_c during 2006–2013 at an EC site in Ames, IA. Figure 3.8 presents these ET comparisons across 29 models during 2011 when a "blind" test was performed, whereby the modelers were given only weather, phenology, management, and soils information but no crop response data. Within these 29 models, the PM (FAO-56) and SW models form the basis for 13 (CropSyst, DSSAT-CERES family; Expert-N family, IXIM, JULES, MCWLA, MONICA, SARRAH, and SIMPLACE LINTUL5) and two (RZWQM2 and STICS_ETP) crop models, respectively. Four out of six models that best predicted mid-season T during a typical year (2011) were based on the FAO-56 PM model, including the best predicting model. Only one out of seven best early-season E predicting models during a typical year were FAO-56 PM based. During a dry year (2012), however, no PM-based model was among the top seven best performing models in estimating early-season E, while only one PM-based model predicted mid-season T well. A SW-based model (RZWQM2) was among the best for the dry year in predicting mid-season T. When total seasonal ET_c was estimated, four out of seven models that performed well across all eight year of investigation were PM based, while no SW-based model performed well.

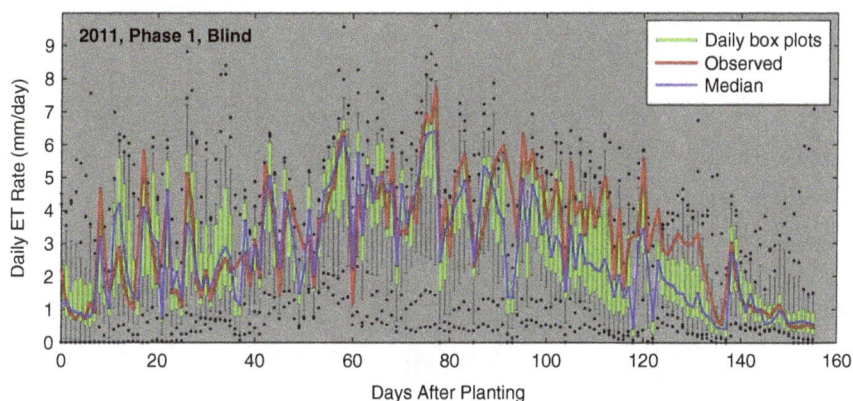

Figure 3.8 Boxplots of daily maize evapotranspiration (ET_c), where the lower and upper limits of the box indicate the 25th and 75th percentile of ET values simulated by 29 maize growth models included in Kimball et al. (2019), the lower and upper whiskers indicate the 10th and 90th percentiles, and the points are outliers. Observed values and the median values from the 29 models are also shown. The simulated values in this plot came from Phase 1 (a "blind" test). Reproduced from Kimball et al. (2019) / with permission of Elsevier.

The extended SW model has been applied to estimate E and T in partial-canopy and residue-covered systems across diverse cropping systems. Odhiambo and Irmak (2011) evaluated the performance of the extended SW model to estimate soybean ET_c from a partially residue-covered, subsurface drip-irrigated field in south-central Nebraska during the 2007 and 2008 growing seasons (Figures 3.9a and 3.9b, respectively)

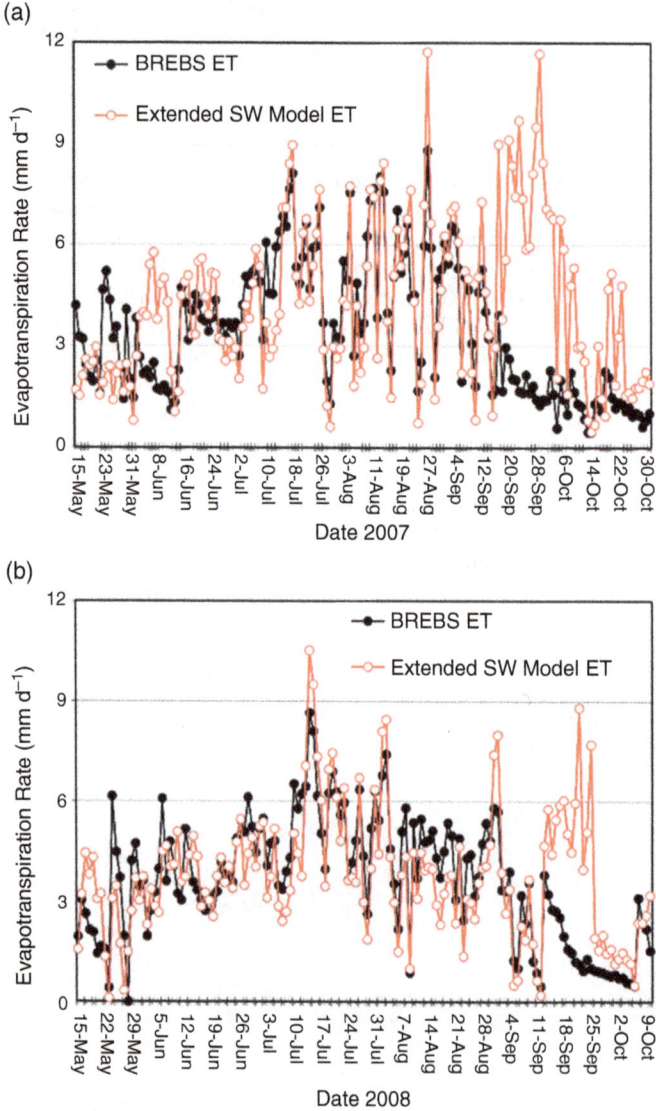

Figure 3.9 Daily soybean evapotranspiration rate or ET_c (mm d^{-1}) estimated by the extended Shuttleworth–Wallace (SW) model compared with Bowen Ratio Energy Balance (BREBS) measurements (a) during the 2007 growing season and (b) during the 2008 growing season; (c) regression plots of extended SW model ET_c estimates against BREBS ET_c measurements for various growth stages in the 2007 and 2008 growing seasons. Adopted from Odhiambo and Irmak (2011).

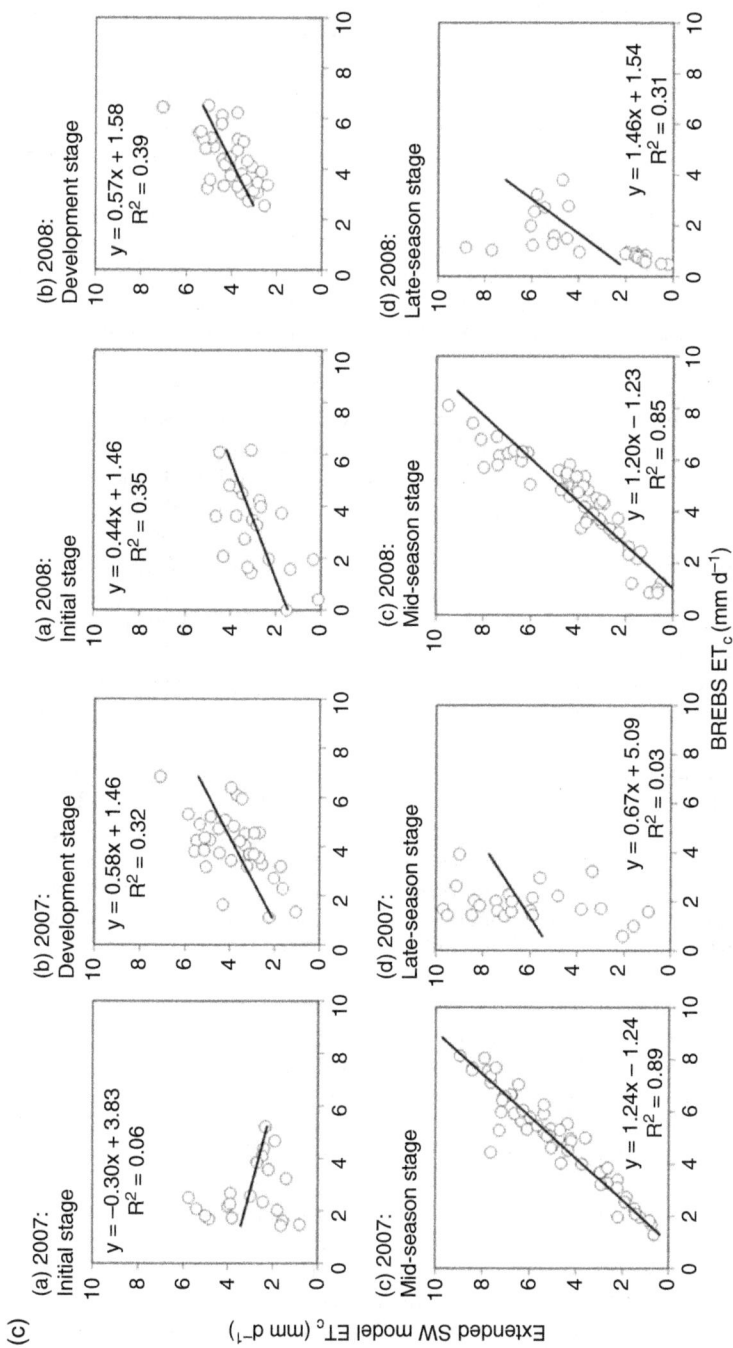

(c)

(a) 2007:
Initial stage

$y = -0.30x + 3.83$
$R^2 = 0.06$

(b) 2007:
Development stage

$y = 0.58x + 1.46$
$R^2 = 0.32$

(a) 2008:
Initial stage

$y = 0.44x + 1.46$
$R^2 = 0.35$

(b) 2008:
Development stage

$y = 0.57x + 1.58$
$R^2 = 0.39$

(c) 2007:
Mid-season stage

$y = 1.24x - 1.24$
$R^2 = 0.89$

(d) 2007:
Late-season stage

$y = 0.67x + 5.09$
$R^2 = 0.03$

(c) 2008:
Mid-season stage

$y = 1.20x - 1.23$
$R^2 = 0.85$

(d) 2008:
Late-season stage

$y = 1.46x + 1.54$
$R^2 = 0.31$

Extended SW model ET_c (mm d^{-1})

BREBS ET_c (mm d^{-1})

Figure 3.9 (Continued)

compared with Bowen ratio energy balance system (BREBS)-measured ET_c. They found that daily ET_c during mid-season (T dominant) was in a better agreement with the BREBS ET_c measurements than during both initial and late-season ET_c (E dominant), as evident from the regression plots shown in Figure 3.9c. Estimating initial and late-season ET_c is a challenging and ongoing issue for most models, including the PM model and all its derivatives and including two-step approaches. In investigating two-step PM methods' performance in estimating ET_c for a subsurface drip-irrigated maize canopy, Irmak et al. (2008b) found that the two-step ETc calculation procedures work well during the mid-season when the crop canopy is fully developed (Stage II, mid-season). In initial (Stage I) and late-season (Stage III) stages, all two-step methods revealed negative coefficients of determination (R^2) and poor Nash–Sutcliffe model efficiency (NSE) values between the daily predicted and measured ET_c. They found significant sensible heat losses, indicating low rates of transpiration in both Stages I and III. The time-dependent crop coefficients were found to be incapable of scaling the ET_o to represent the daily ET_c fluctuations in Stages I and III. Irmak and Mutiibwa (2009a) observed that the two-step PM (standardized ASCE-EWRI-PM) approach overestimated the measured ET_c by approximately 10% in the $1.2 < LAI < 2.7$ range and then consistently underestimated by 10–20% until LAI reached ~5.0. They observed that some of the largest underestimation by the ASCE-EWRI-PM model occurred in the early season during partial canopy and late in the growing season during leaf maturity and senescence due to the inability of the K_c values to capture the reduced ET_c (and transpiration) due to leaf senescence and maturity. These commonly observed early- and late-season discrepancies between measured and two-step approach estimated ET_c indicates that additional research is necessary to improve the ability of K_c values to account for phenological and physiological changes and associated plant behavior to better estimate early- and late-season evaporative losses.

In addition to being evaluated as a stand-alone application, the extended SW model has also been applied extensively via the Root Zone Water Quality Model (RZWQM) (Farahani & DeCoursey, 2000). The Root Zone Water Quality Model 2 (RZWQM2) is a widely used process-based agricultural system model to simulate agricultural management practice impacts on soil water, crop production, and water quality (Ahuja et al., 2000). The crop transpiration in this model is limited by potential ET, which is represented by the extended SW model (Farahani & Ahuja, 1996; Farahani & DeCoursey, 2000). The specifics of this model have already been discussed above. The model integrates simulated processes of soil surface residue dynamics, tillage, soil water, and soil C and N as derived from RZWQM with crop-specific growth modules of the DSSAT 4.0 crop model (Jones et al., 2003). The RZWQM2 and its predecessors have been used extensively for maize growth and response to water, N, and other environmental variables simulated under various climatic and soil and crop management conditions in the U.S. Great Plains (Anapalli et al., 2005; Ma et al., 2003; Saseendran et al., 2004; 2008, 2009, 2010).

Anapalli et al. (2016) used RZWQM2 to simulate maize ET during 1990, 2006, and 2007 and compared the resulting simulated K_c with lysimeter-measured K_c, observing a fair amount of year-to-year variation (Figure 3.10). They suggested that the high degree of variation necessitates that longer term averages of K_c values need to be considered if the use of the K_c approach is to continue. Models such as RZWQM2 have the capability of simulating these long-term K_c values, which can be cultivar- and site-specific.

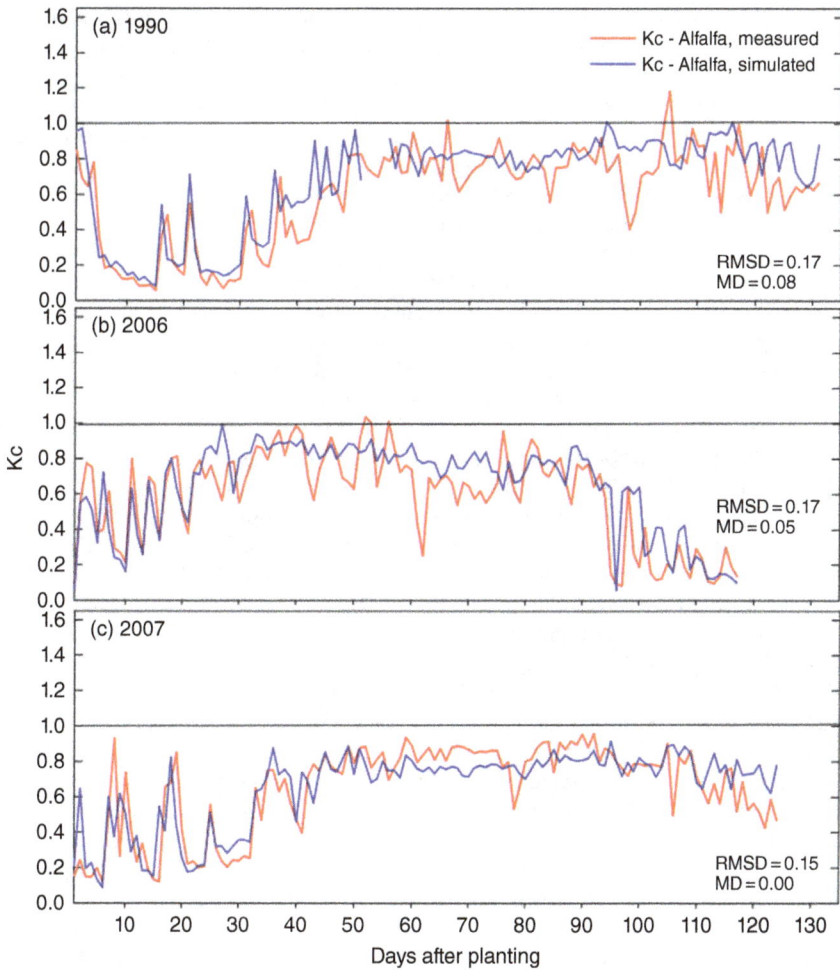

Figure 3.10 Comparison between measured and simulated alfalfa-reference maize crop coefficients (K_c) during 1990, 2006, and 2007. Reproduced from Anapalli et al. (2016) / with permission of Elsevier.

Estimating ET_c for water-stressed plants is a challenging task essentially for all models due to the complex and unpredictability of occurrence (timing) and magnitude of water stress (Irmak, 2015; Irmak et al., 2000. This challenge is further exacerbated when plants are subject to both water and nutrient (primarily N) stress simultaneously; it is an extremely difficult task to accurately distinguish and quantify the magnitude of each stress and its quantitative impact on plant physiological functions. Plant physiological functions, growth and development, grain production, ET_c, and water productivity can exhibit substantial variability or be greatly variable as a function of management practices, climatic and environmental factors, water availability, N availability, heat and wind stress, and numerous other factors. Plant response to abiotic stresses is a strong function of climatic conditions, primarily precipitation, but additionally maximum air temperature, shortwave radiation, VPD, and wind speed, all of which are drivers of ET and thus influence the water stress magnitude and timing. Because all these variables exhibit extreme variation throughout the growing season, differences in the occurrences of the amount and timing of water stress result in complex interactions between environmental factors and biotic stress (Irmak et al., 2000). Thus, developing a model to account for all these dynamic variables is an extremely difficult task. This challenge can potentially be addressed through long-term and carefully established field research under different climatic and soil-water and N management conditions. Simultaneous measurement and analyses of all these dynamic variables would enable the development of a dynamic model to better simulate plants' response to stressed conditions. Because different hybrids, cultivars, and varieties respond to varying levels of stress in dissimilar manners and degrees, quantification of a plant's response to water, N, heat, and other stresses should be carried out for different hybrids and lines, which further exacerbates the difficulty in developing a complex and comprehensive dynamic modeling effort to accurately simulate plants' response to stress causing limited growth and productivity. Saseendran et al. (2014, 2015) improved the simulation of limited water scenarios in maize production in RZWQM2 (via the DSSAT-CSM-CERES MAIZE module [Jones et al., 2003; Lizaso et al., 2011]) by modifying the water stress factor. The improved water stress factor accounted for limited water impacts on photosynthesis and potential root water uptake and heating of the canopy from the unused energy of potential soil evaporation. The improvement was shown to be superior to the previous water stress factor across multiple water-level experiments at three maize sites. Water stress parameterization in RZWQM2 using potential ET (ET_p or PET) estimated from the SW versus ASCE-EWRI-PM approaches was evaluated against lysimeter-measured maize ET at Bushland, TX, by Zhang et al. (2018), who found that fully irrigated maize ET_c was simulated within 3% of the lysimeter ET_c. Under unirrigated conditions, ET_c was underestimated by ~11%, with the SW-based water stress factor performing slightly better. They concluded that ET_c simulations need improvement under irrigated and unirrigated conditions.

The extended SW model, via the use of RZWQM2, has been used to evaluate irrigation strategies and their effects on ET and soil water besides growth and yield. The applications include evaluating irrigation amounts and irrigation capacities (Ma et al., 2012; Saseendran et al., 2009), evaluating irrigation water management decision support (Anapalli et al., 2005; Chen et al., 2020a; Saseendran et al., 2008, 2009), evaluating water dynamics as causes for lower NT yields than conventional tillage yields (Anapalli et al., 2018; Ding et al., 2020), identifying cultivars for climate change mitigation (Ma et al., 2017), identifying soil texture impacts on ET (Kozak et al., 2005), developing crop coefficients (Anapalli et al., 2016), estimating drainage and leaching (Craft et al., 2018; Gillette et al., 2018; Jeong et al., 2019; Jiang et al., 2018; Ma et al., 2007; Smith et al., 2019, and evaluating cover crop impacts on the soil water balance (Yang et al., 2019a).

More recently, Elfarkh et al. (2020) incorporated a thermal stress index into the SW model for large-scale ET_c mapping for complex surfaces. They introduced a new approach to characterize spatial soil resistance and vegetation resistance to water vapor transfer based on two thermal-based proxy variables. They derived the land surface temperature and normalized difference vegetation index (NDVI) from Landsat data, which were combined with the minimum and maximum temperatures for soil and vegetation, simulated by a surface energy balance model, to compute the soil and the vegetation temperatures. Based on these temperatures, they determined two thermal proxies (for soil and vegetation) and related them to soil and vegetation resistances. The proposed approach was initially evaluated at a local scale by comparing the results to measurements by an EC system installed over an area planted with olive trees intercropped with wheat. In a second step, they applied the new approach across a large area that contained mixed vegetation (tall and short) to derive soil and vegetation resistances to estimate ET_c using the SW model. They used a large-aperture scintillometer installed along a line transect of 1.4 km and spanning the total area to validate the SW model ET_c. The comparison confirmed the ability of the proposed approach to provide satisfactory SW ET_c maps with a root mean squared error (RMSE), mean bias error, and R^2 equal to 0.08 mm h^{-1}, 0.06 mm h^{-1}, and .80, respectively. The new approach proposed and evaluated by Elfarkh et al. (2020) can serve as a dynamic link between both vegetation and soil resistances with plant water stress indices, which can improve the estimation of spatial ET_c using the SW model under water-stressed environments.

Horticultural Applications

In addition to agronomic row crop applications, the SW model has also been applied to controlled greenhouse conditions for horticultural crops. The ET_c, T, and E were estimated for tomato (Gong et al., 2019) and cucumber (Huang et al., 2020) using the SW model, and both studies showed satisfactory accuracy against microlysimeter and sap flow measurements. Viticulture research has also relied on the application of PM and SW models. Zhang et al. (2008) compared PM and SW models to estimate vineyard ET_c against the BREBS method in an arid and water-limited region of China. Similar to the results of a bigger body of literature, the SW model performed better than the PM model, including under conditions immediately following wetting events.

Zhao et al. (2015) studied the use of the dual crop coefficient method in FAO-56 (FAO dual-K_c) and the SW model in estimating ET_c, T, and E of a vineyard in an arid region of Northwest China. They conducted continuous measurements of ET_c with EC, T with sap flow, and soil evaporation E with micro-lysimeters to validate the performance of the two approaches. The FAO dual-K_c method in partitioning ET_c was acceptable when using the site-specific basal crop coefficient obtained from sap flow, with the slope and intercept of linear regression of 0.96 (dimensionless) and −0.13 mm d^{-1} (R^2 = .81) for ET_c, 0.92 (dimensionless) and −0.07 mm d^{-1} (R^2 = .76) for E, and 0.93 (dimensionless) and 0.16 mm d^{-1} (R^2 = .80) for T, respectively. Results indicated that the SW model can better estimate ET_c but overestimated T and underestimated E when using site-specific soil surface resistance, with the slope and intercept of linear regression of 0.98 and 0.28 mm d^{-1} (R^2 = .79) for ET_c, 0.49 and 0.42 mm d^{-1} (R^2 = .46) for E, and 1.10 and 0.38 mm d^{-1} (R^2 = .81) for T, respectively. They concluded that both approaches can estimate ET_c with good accuracy, but the FAO dual-K_c method had higher accuracy in estimating E and T separately.

Other horticultural crops have been subjected to the application of SW models and its variants in the literature such as a cherry orchard (Li et al., 2010), apple orchard (Liu et al., 2015), and an olive orchard (Riveros-Burgos et al., 2019). Chen et al. (2020b) used random forest (RF) and support vector regression (SVR) to predict cabbage T based on meteorological, crop, and soil factors. Under scenarios of sufficient meteorological data and lacking soil and vegetation data, RF was recommended, whereas SVR was recommended when meteorological data are absent. Tang et al. (2018) found that SVR and a genetic algorithm accurately estimated maize ET_c under no mulch and partial plastic film mulch conditions when both crop and meteorological inputs were provided.

Applications in Natural Landscapes

In addition to agricultural applications, modeling of water use dynamics has immense applications across a variety of natural landscapes. Li et al. (2011) used the SW model to parameterize a transpiration model coupled with models of stomatal conductance (Ball–Berry model) and Farquhar and Collatz photosynthesis models (Collatz et al., 1991) to understand the ecophysiological dynamics of C_3 and C_4 species in a Central Asian desert ecosystem. They found that the SW model explained 61% (C_3) and 52% (C_4) of measured T (EC) variance, suggesting promising performance in arid biomes such as northern Australia, the western United States, and northwestern Mexico. Hu et al. (2009) used the SW model to estimate ET partitioning (E/ET) at four grassland sites in China using Monte Carlo simulations for model parameter estimations. Model estimates were in good agreement with EC measurements (Figure 3.11), except underestimations by 3–11% on rainy days due to the model's inability to represent rainfall interception. The inability of an EC system to accurately estimate evaporative losses on rainy days could be the cause of the discrepancy. Advection due to surface discontinuities and high wind speeds were probably responsible for the SW model's unsatisfactory performance in estimating EC-measured ET of an extremely arid forest in China (Gao et al., 2016). Fisher et al. (2005) carried out a comparison of five PET models, including the SW and PM models, to estimate 30-minutes PET in an Ameriflux tower-equipped Sierra Nevada forest ecosystem. Their findings suggested that both PM and SW models are equally effective, with slightly better performance of the SW model. The better performance of the SW model was more significant in a wet year due to

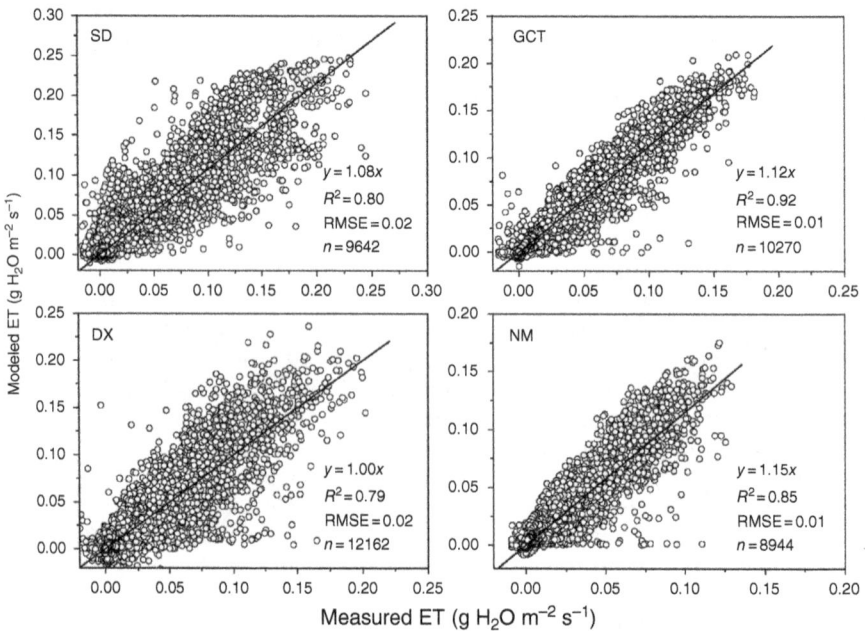

Figure 3.11 Comparisons between the Shuttleworth–Wallace-modeled and measured evapotranspiration (ET) at half-hourly time scale for four grassland ecosystems. Reproduced from Hu et al. (2009) / with permission of Elsevier.

increased soil evaporation, which is not accounted for in the PM model as a source. They also investigated the sensitivity of the three assumed variables in the SW model—surface resistance of the substrate, roughness length of a bare substrate, and extinction coefficient of the crop for net radiation—by multiplying each parameter by 0.5, 2, and 10. They concluded that the SW model was not highly sensitive to these parameters, with the simulated PET differing only by less than 5%.

The applications of the SW model are not limited to the field or point scale but have also been extended to large spatial scales such as the Yellow River Basin (Zhou et al., 2007) and Mekong River Basin (Zhou et al., 2006). Adopting spatiotemporal variation in vegetation from Advanced Very High Resolution Radiometer (AVHRR) satellite platform-derived NDVI, Climatic Research Unit-derived (New et al., 1999) meteorological data, and vegetation parameters from a land cover dataset, Zhou et al. (2007) calculated SW PET for the Yellow River Basin during the 20th century. They suggested that the SW model, being physically based and accounting for changing surface conditions, produces sound PET estimates.

The SW model has been integrally included and modified for modeling ecosystem processes and their responses to changing climate conditions. An improved version of the SW model, the Shuttleworth–Wallace–Hu (SWH) model (Hu et al., 2013, 2017) has been proposed that estimates and partitions ET using meteorological and remote sensing data by improving canopy stomatal and soil surface conductances. The improved model estimates r_s by coupling the Ball–Berry stomatal conductance model (Ball, 1988) and a light use efficiency-based gross primary productivity model into the SW model. The ET estimates from SWH model performed well when validated against global flux tower sites (Hu et al., 2017). Recently, Wu et al. (2020) used Monte Carlo techniques for optimization of SWH model parameters and incorporation of their spatial variation using 163 flux tower sites. The LAI and clay fraction were strongly correlated with model parameters, and their inclusion resulted in greatly improved model performance. The SWH model was applied at a spatial scale across the Yellow River Basin (Jiang et al., 2020) to identify spatiotemporal variations of E, T, and ET. The SWH model was also concluded to be the most superior among a suite of satellite-based ET models investigated in Northwest China (Sun et al., 2020).

The PM and SW models have also served as inspirations and fundamental concept engines for multiple layer models. The structure of the PM-type ET models varies from single-source (PM) to two-layer (Shuttleworth & Wallace, 1985), three-source (Mu et al., 2011), and four-layer (Choudhury & Monteith, 1988) structures. Ershadi et al. (2015) examined how these model structures compare when used for ET estimation at 20 globally distributed FLUXNET tower sites across all major biomes. They found that no single model structure produced top-ranked results across all the biomes. In croplands, a three-source model structure was optimal, while single-source and two-layer model structures were better in short canopies (grassland and shrubland) and forest canopies, respectively. They also investigated the relative success of various r_a and r_s parameterization schemes and concluded that the models were sensitive to them. The best performing parameterizations were biome-specific lookup table r_s values proposed by Mu et al. (2011) and r_a proposed by Thom (1975). They suggested, considering these findings, that a multi-model ensemble tailored for each biome might be the most appropriate approach for a global ET product. Zhao et al. (2020) suggested, with evidence from EC observations, that more complex resistance parameterizations result in greater uncertainty and bias in ET estimates. While some studies have assessed the performance of these single- and multilayer modeling approaches, their performance has been compared for only very limited sites and surfaces. Additional detailed studies can aid in improving these models' physically based parametrization, perhaps toward more dynamic parameterization and modeling approaches, which could enhance their accuracy for estimating evaporative losses for different surfaces rather than performing well only for certain vegetation surfaces and poorly for others.

Comparative Suitability of ET Models

The FAO-56 approach (Allen et al., 1998), further updated by ASCE Environmental and Water Resources Institute (2005) and the ASCE-EWRI Task Committee on Revision of Manual 70 (Jensen & Allen, 2016) has emerged as the most widely used crop ET modeling framework. The strengths of the FAO-56 approach stem from its operational nature and clear definitions as well as the simplistic nature of its applicability in practice. First, the approach provided a clear (although hypothetical) definition of a reference surface, which was meant to reduce confusion among the ET_{ref} and PET concepts (Xiang et al., 2020), both of which can be easily confused. Second, the basis of ET_{ref} in the PM equation makes it superior for its analytical description of energy exchange between a crop surface and the atmosphere compared with statistical formulations such as FAO-24 (Katerji & Rana, 2014). Last, the approach is easily operational and practical because it is based on input variables that are measurable via standard meteorological stations.

Despite the strengths of the FAO-56 approach, its focus on simplicity and operational nature has resulted in several possible sources of error, as compiled by Katerji and Rana (2014):

1. The PM equation is largely adopted at a daily time step due to the ease associated with using the 24-hours average of weather variables. Under the daily time step, the weather variables do not represent the conditions when ET actually occurs. The accuracy trade-offs associated with the use of a daily-time-step PM equation have been quantified (Allen et al., 2006; Althoff et al., 2019; Berengena & Gavilán, 2005; Doorenbos & Pruitt, 1977; Gavilán et al., 2008; Irmak, 2011; Irmak et al., 2005; Itenfisu et al., 2003; Jensen et al., 1990; Ji et al., 2017; Katerji & Rana, 2014), but deeming the performance satisfactory or unsatisfactory is dependent on the allowable buffer for error in the intended applications, which can vary widely. Nevertheless, as mentioned by Allen et al. (2006), formulation of the combination equation (combined energy balance and aerodynamic components) theoretically requires weather inputs on a nearly instantaneous basis. Thus, an ideal alternative could be to use high-temporal-resolution weather variables that refer to only the part of the day when the largest amount of water is lost. Thus, conceptually, the use of an hourly time step PM equation would be preferred over a daily time step calculation as it keeps radiation and aerodynamic parameters synchronized in time. Sinclair and Ghanem (2020) and Ghanem et al. (2020) also pointed out that VPD in the PM equation is erroneous due to the use of a daily average value, which instead should be weighted for the daily cycle of transpiration. The approach outlined by Allen et al. (2006) allows hourly parameterization of surface resistance (r_s) to estimate hourly ET_{ref} and eventually the sum-of-hourly, daily ET_{ref}. Furthermore, Irmak et al. (2005) investigated the differences between sum-of-hourly and daily estimates of ET_{ref} (ET_o) using the ASCE-EWRI-PM method in a humid coastal region (Fort Pierce and Bradenton on the east and west coasts of Florida, respectively), semiarid temperate regions (Bushland, TX, and North Platte, NE), two Mediterranean-type regions along the west coast of California (Santa Rosa and Santa Barbara), and Twitchell Island, CA, and found that the daily approach

estimated higher than the hourly approach at all locations except one, and agreement between the computational time steps was best in humid regions. The greatest differences between daily and hourly approaches were in locations where strong, dry, hot winds cause advective increases in ET_{ref}. Three locations showed considerable signs of advection. Some of the differences between the time steps were attributed to uncertainties in predicting soil heat flux and to the difficulty of the daily approach to effectively account for abrupt diurnal changes in wind speed, air temperature, and VPD. They suggested a potential improvement in accuracy when using the standardized ASCE-EWRI-PM procedure applied hourly rather than daily. The hourly application helps to account for abrupt changes in atmospheric conditions on ET_{ref} estimation in advective and other environments when hourly climate data are available. Irmak (2011) further emphasized the importance of sum-of-hourly rather than daily ET modeling applications to better account for nocturnal evaporative losses, which can be a sizeable portion of the daily ET in many agroecosystems.

2. The two-step approach assumes a fixed canopy resistance for the reference surface, which is a simplification of a much more complex process of vegetation water vapor exchange with the surrounding environment. The experimental observations of stomatal resistance varying with climatic variables (meteorological and CO_2), for instance, as shown by Irmak and Mutiibwa (2010) (Figure 3.12), are inconsistent with the hypothesis that all weather effects are included in ET_{ref} (Damour et al., 2010; Katerji & Rana, 2006; Lecina et al., 2003; Perez et al., 2006; Rana et al., 1994 Steduto et al., 2003). Moreover, the constant hourly canopy

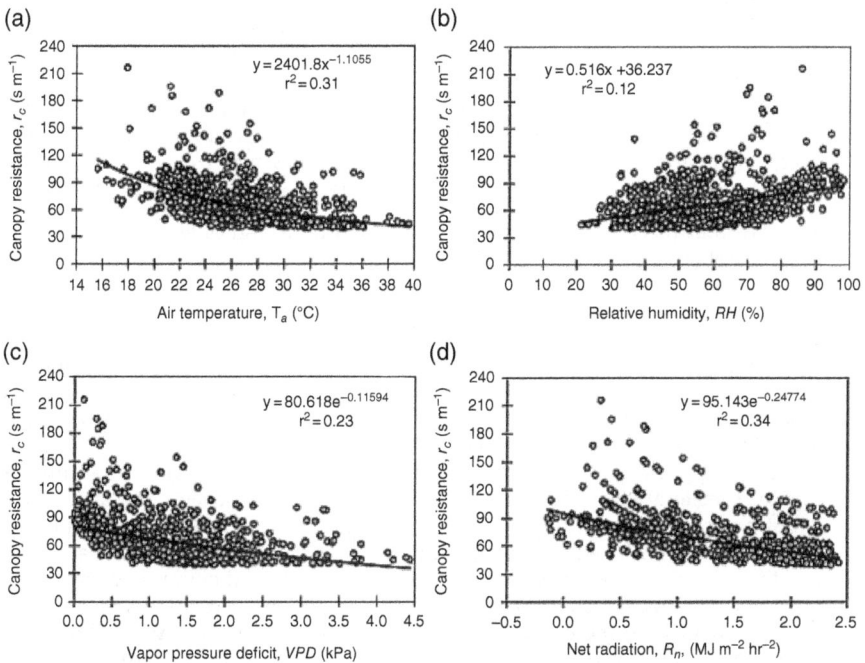

Figure 3.12 Relationship between canopy resistance (r_c) and primary microclimatic variables: (a) air temperature (T_a), (b) relative humidity (RH), (c) vapor pressure deficit (VPD), and (d) net radiation (R_n) ($n = 755$ for each case). Adapted from Irmak and Mutiibwa (2010).

(surface) resistance (r_c) (Allen et al., 1998, 2006) proposition is not deterministic and is derived based on literature-based average values. Katerji et al. (1983) demonstrated that the minimum alfalfa r_s was ~100 s m^{-1} only on the upper leaf surface, even when R_s was greater than 400 W m^{-2}. These values can certainly be higher on the lower leaf surface as well as under low R_s.

3. The consideration of only 50% of the leaf surface contributing actively to T (equivalent to LAI = 1.42 for a 0.12-m reference crop height) is inconsistent with the observations of Monteith et al. (1965), where the complete leaf surface in barley contributed to transpiration until LAI was half its maximum. Additionally, Farahani and Bausch (1995) assumed that complete maize LAI was actively transpiring until 50% of maximum LAI (4.0), and only 50% of LAI was active for LAI > 2.0. This scheme was also followed by Gardiol et al. (2003). The total LAI was considered when calculating ET in snap beans (Black et al., 1970), soybean, and grain sorghum (Brun et al., 1972), maize (Perrier, 1976), and alfalfa crops (Katerji & Perrier, 1985). Although a functional relationship between LAI and LAI contributing to T has not yet been found, there is ample evidence to revisit the assumption of only 50% of the leaf surface contributing to T.

Katerji and Rana (2014) also highlighted two shortcomings of the K_c values proposed by Allen et al. (1998) in the FAO-56:

1. The first weakness is the ambiguous origin of the K_c values listed within FAO-56, a case also brought forward by Shuttleworth and Wallace (2009). The research at the source of these listed values (Doorenbos & Kassam, 1979 Doorenbos & Pruitt, 1977 Pruitt, 1986 Snyder et al., 1989; Wright, 1981, 1982 occurred at times that preceded the introduction of our current ET$_o$ formulations. In fact, Allen et al. (1998) criticized the ET$_o$ determinations that were actually used in these research studies, which were also conducted in the same period. Moreover, averaging K_c values across heterogeneous data that originate from extremely different climatic, soil, crop management, water management and other management regimes is a substantial compromise (Doorenbos & Kassam, 1979). Thus, caution should be practiced when approaching these K_c data, especially when using them in local estimates of crop water use under different water management practices.

2. The idea that K_c is conservative to climate conditions and only sensitive to crop characteristics has not been confirmed observationally. In fact, the opposite has been shown by Shuttleworth (2006), who stated that the K_c values listed within FAO-56 implicitly depend on the climatic conditions that prevailed when these particular K_c were measured via field studies. The transferability challenges of K_c values from developmental climates to diverse target climates and agricultural practices has been recognized (Farahani et al., 2007 Lazzara & Rana, 2010). This inherent link among K_c and local climate prevents K_c being confidently. Irmak et al. (2008a) stated that the technical information on the transferability of the K_c values for applications under different climatic conditions is lacking. They argued that the transferability of K_c values for the same vegetation surface between regions is not a straightforward process, and substantial errors can be introduced when using K_c values developed for a specific climate, soil, water, and crop management condition in another location that has different climate and environmental conditions and management practices. Questions that should receive greater focus in this regard include the evaluation of the temporal and spatial variability of K_c values, quantification of differences among K_c values obtained with different ET$_{ref}$ methods, and transferability of K_c values among climatic regions. Djaman and Irmak (2013) found that ET$_c$ was overestimated by the two-step approach

compared with the soil-water balance approach, supporting similar observations by Irmak et al. (2008a) and Howell et al. (2000). They attributed these overestimation patterns to K_c uncertainty in plant conditions during the initial and ending periods of the growing season, especially for rainfed and stressed treatments. Upon investigating sensible and latent heat losses relative to the available energy in a maize field, Irmak et al. (2008a) observed significant sensible heat losses during both initial (prior to canopy closure) and late (during crop maturity and senescence) stages, for which the two-step approach may not be appropriate.

Significant variability has been observed in daily measured K_c values (Anderson et al., 2017; Facchi et al., 2013; Irmak & Kukal, 2019; Irmak et al., 2013a, 2013b, 2015; Ji et al., 2017; Payero & Irmak, 2011, 2013), making it challenging to fit a growing-stage-average K_c curve such as proposed in FAO-56 (Figure 3.13). Both wetting events and crop stress can lead to significant deviation from proposed average K_c curves, traditionally used to estimate ET_c. This is especially problematic for humid and subhumid regions and frequently (sprinkler) irrigated fields. These issues can be somewhat relieved by using the dual crop coefficient approach to represent these processes using K_{cb}, K_s, and K_e. Nevertheless, the initial, midseason, and late-season K_{cb} values are subject to change as a function of soil, crop, and irrigation management, hybrid characteristics, tillage, plant density, etc. Moreover, the fixed duration of each of these windows, especially if represented using calendar days, makes it challenging to account for phenological impacts resulting from water-limited conditions (Djaman & Irmak, 2013).

Derivation of crop coefficients under non-humid conditions is problematic and hinders accurate application of the FAO-56 two-step approach. Shuttleworth and Wallace (2009) demonstrated this by using both the one-step M–S approach and FAO-56 two-step approach to estimate the ET_c of irrigated sugarcane, cotton, and pasture across five important irrigation districts under arid and windy conditions in Australia. The growth-stage-specific estimates under the two approaches differed by −27 to 19% for well-watered sugarcane and by −36 to 14% for well-watered cotton across all five sites. When these estimates were aggregated across the entire growing seasons, M–S estimates were 3–15% higher, 4–6% higher, and 0.5–2.5% higher for sugarcane, cotton, and pasture crops, respectively, than the FAO-56 two-step approach. The M–S approach, due to its recognition of the crop height dependency of aerodynamic resistance, produces better estimates of day-to-day variations in ET_c in response to day-to-day weather variability. To demonstrate this, Shuttleworth and Wallace (2009) presented K_c calculated using ET_c determined by the M–S approach during Stage 3 sugarcane growth. Figure 3.14 shows the K_c variation that would need to be reproduced by the FAO-56 approach, contrary to the assumed constant K_c suggested for Stage 3 in the FAO-56 approach.

Shuttleworth and Wallace (2009) argued that the practical application of the rigorous M–S approach is preferable to the two-step approach, which is relatively unattractive for the following reasons:

1. The M–S approach is consistent with our current understanding of the physical and physiological controls on the evaporation processes, unlike the FAO-56 two-step approach.

2. Due to the realistic representation of plant control via surface resistance, it is possible to capture future crop responses to weather, crop development, leaf area, and soil conditions. The use of a surface resistance in the M–S approach will also allow

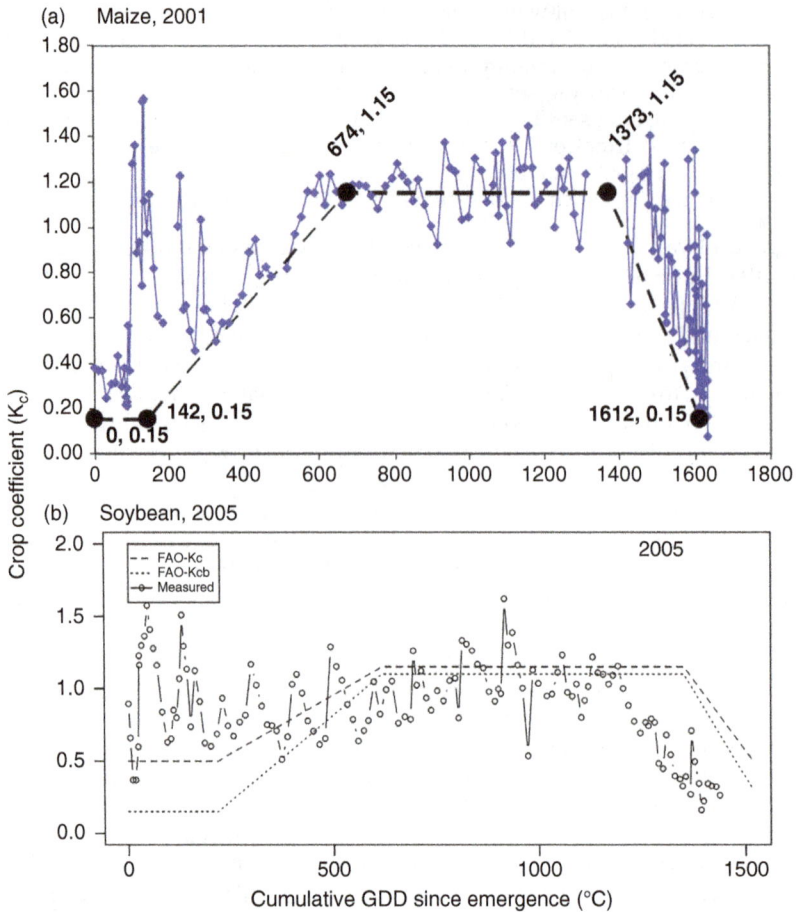

Figure 3.13 Crop coefficient (K_c) curves for (a) maize during 2001 and (b) soybean during 2005 measured at North Platte, NE, as a function of cumulative growing degree days (GDD) from crop emergence. The K_c values were calculated based on a grass-reference evapotranspiration (ET_o), which was calculated with the FAO-56 method. The broken line in (a) represents the K_c function proposed in FAO-56 with growth stage lengths for Kimberly, ID. The FAO K_c and FAO K_{cb} curves in (b) were constructed based on the values of the average K_c and basal crop coefficients (K_{cb}) given in FAO-56 for soybean for the central United States. Reproduced from (a) Payero and Irmak, 2011 / IntechOpen / Licensed under CC BY 3.0 and (b) Reproduced from Payero and Irmak, 2013 / with permission of Elsevier.

future improvements in ET_c, T, and E estimations via two- or three-source resistance-based models (e.g., Farahani & Ahuja, 1996; Shuttleworth & Wallace, 1985).

3. The initial hindrances for the use of the M–S approach have been eliminated. First, methods exist to use standard 2-m-height climate station data for applying the M–S approach. Second, pending measured surface resistance from field investigations, an interim surface resistance determination methodology has been proposed using the FAO-56 K_c, based on crop height and a preferred temperature corresponding to K_c development (20 °C), for example, presented in Table 3.2 for selected crops. The use of these determinations will yield ET_c estimates at least as accurate as those

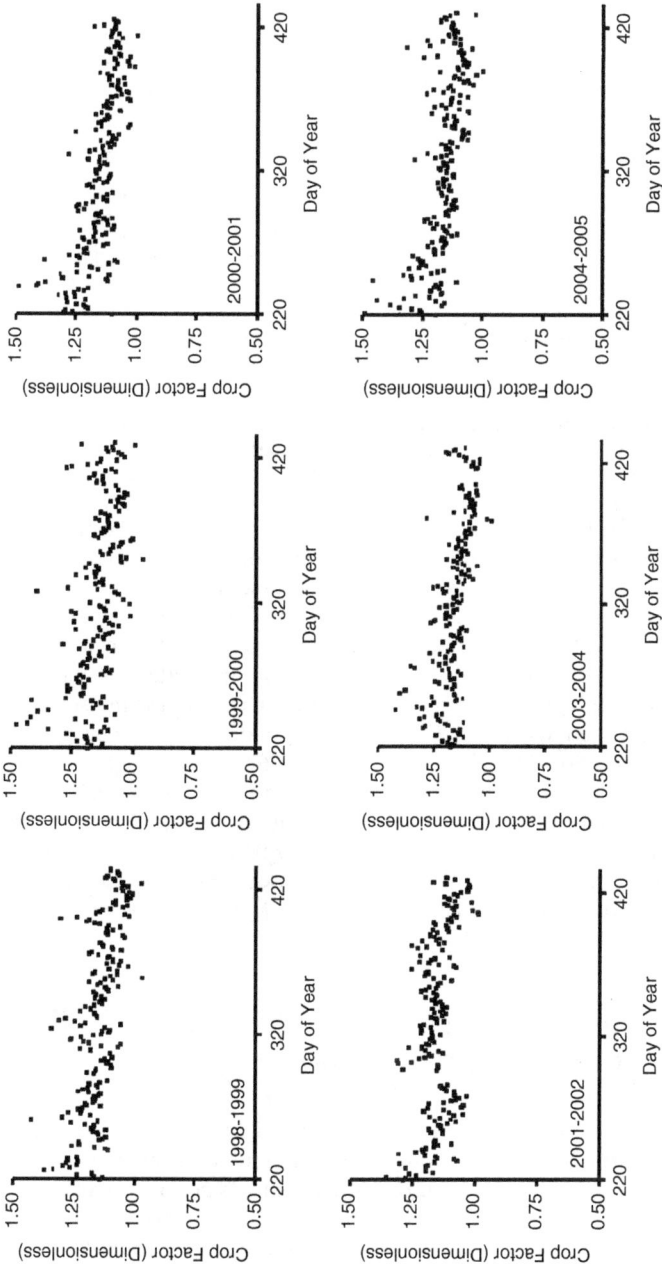

Figure 3.14 Values of the crop coefficient (crop factor) for sugarcane during Stage 3 growth calculated as the ratio between the daily evapotranspiration rate calculated using the Matt-Shuttleworth approach divided by the daily value of reference crop evaporation. Reprinted from Shuttleworth and Wallace (2009) with permission.

Table 3.2 Values of Equivalent Surface Resistance for Selected Crops Calculated with FAO-56 Crop Coefficient (K_c) Values and Maximum Crop Height Derived from Allen et al. (1998, Table 12)

| Irrigated crop | Assumed K_c | Assumed crop height m | Surface resistance for selected values of T^{prefa} | | | Suggested surface resistance and range |
			10° C	20° C	30° C	
Reference crop (grass)	1.00	0.12	70	70	70	70
Maize (grain)	1.20	2.00	62	64	66	64 ± 2
Sorghum (grain)	1.05	1.50	90	100	115	100 ± 2
Rice	1.20	1.00	49	46	40	46 ± 5
Sugarcane	1.25	3.00	61	63	66	63 ± 3
Cotton	1.18	1.35	60	60	61	60 ± 1
Legumes	1.15	0.55	49	44	36	44 ± 6
Alfalfa	0.95	0.70	111	127	151	127 ± 20
Roots and tubers	1.10	0.68	65	66	67	66 ± 1

Note. This table is derived from a larger database presented by Shuttleworth and Wallace (2009).
[a] T^{pref} is the air temperature during the period when the value of FAO-56 K_c was derived.

estimated from the FAO-56 approach but will have improved representation of atmospheric aridity and sub-seasonal variability in weather.

Using these arguments, Shuttleworth (2014) proposed that direct one-step application of the PM model should be preferred by FAO, and thus the FAO-56 document can be updated with the M-S approach by deriving tabulated values of surface resistance from Allen et al. (1998, Table 12). Efforts have also been made for operational application of a one-step PM model to be used with standard weather station datasets to estimate ET in irrigated soybean, sweet sorghum, and tomato (Katerji & Rana, 2006; Rana et al., 2012; Rana & Katerji, 2009). A major impediment in the effective application of the M-S approach stems from the ambiguous (essentially unknown) nature of the meteorological conditions that prevailed when the K_c values provided in Allen et al. (1998) were calibrated. This situation forced Shuttleworth (2006) to knowingly assume that $ET_o = E_{PT}$ (Priestley–Taylor estimate of the ET rate) as a default value of r_{clim} (55 s m^{-1}), representing subhumid conditions. This assumption does not hold under other climate conditions, although the resulting error may be minor (Lhomme et al., 2014). However, there is a provision in the M-S framework to fine-tune estimates of r_s^c if additional information on weather conditions during K_c calibrations are known (Shuttleworth, 2014).

Agreement Among the Well-Established ET, E, and T Measurements

Evaluation of ET, T, and E model performance will always be a function of the gold standard measurements used in the process to measure and partition ET. Kool et al. (2014) reviewed 52 ET partitioning studies, out of which 20 studies measured all components—E, T, and ET—to evaluate how various measurements performed and compared with each other. The most commonly used approaches were micro-lysimeters for E and stem heat balance for T determination. From these studies, they compiled eight different combinations for E + T measurements and how they agreed with different ET measurements (Figure 3.15). They observed that: (a) combinations that use EC for E result in smaller E + T compared with ET, and (b) combinations that use EC for ET

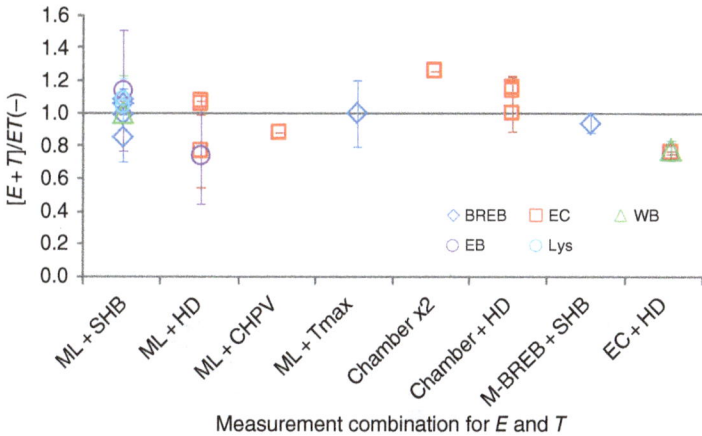

Figure 3.15 An evaluation of eight independent measurement combinations of evaporation from the soil surface (E) and transpiration (T). The combinations of E and T were compared with evapotranspiration (ET) measurements using the ratio $[E + T]/ET$ with the objective of deriving trends that could indicate the accuracy of the methods. Each point represents a single study. The different colors represent different methods for determining ET. Abbreviations: ML, microlysimeter; SHB, stem heat balance; HD, heater dissipation; CHPV, compensated heat-pulse velocity; (M)-BREB, (micro)-Bowen ratio energy balance; EC, eddy covariance; EB, energy balance; Lys., weighing lysimeter; WB, water balance. Reproduced from Kool et al. (2014) / with permission of Elsevier.

result in larger $E + T$ values compared with ET. This led to the conclusion that EC systems underestimate fluxes, as was shown by Twine et al. (2000). On the other hand, several combinations use BREB for ET, resulting in good agreement with $E + T$. Other combinations had limited data to make conclusive observations, although most of the studies achieved 90% agreement for ET = $E + T$, implying that effective and accurate partitioning of fluxes is possible with current technology (Kool et al., 2014). The uncertainties in the various approaches used need to be quantified in order to take account of the uncertainties of the modeling performance evaluation process (Flo et al., 2019; Irmak et al., 2014; Mauder et al., 2013 Sharma et al., 2021).

Implications of Failure to Account for Elevated CO_2 in PET Formulations

The PM-based PET models and their derivatives fail to capture the biological processes that are otherwise accounted for in climate models (Milly & Dunne, 2016). Due to zero or fixed ($70\,s\,m^{-1}$) stomatal conductance (g_s) in PM models, the reductions in g_s from increasing CO_2 concentrations (Roderick et al., 2015) are ignored. The daytime g_s values have been shown to be 40% lower under non-water-stressed conditions in the future than the historical period (Figure 3.16), as obtained from the GFDL-ESM2M model (Milly & Dunne, 2016). Direct and indirect leaf-level responses to elevated CO_2 are responsible for the positive r_s vs. CO_2 relationship. The direct mechanism is an increase in r_s from elevated CO_2 due to partial stomatal closure (Field et al., 1995), while the indirect mechanism is an increase in r_s from higher VPD due to elevated CO_2 (Novick et al., 2016). Ficklin and Novick (2017) showed that increased future VPD can reduce g_s by ~9–51%. Both r_s and VPD have opposing controls on

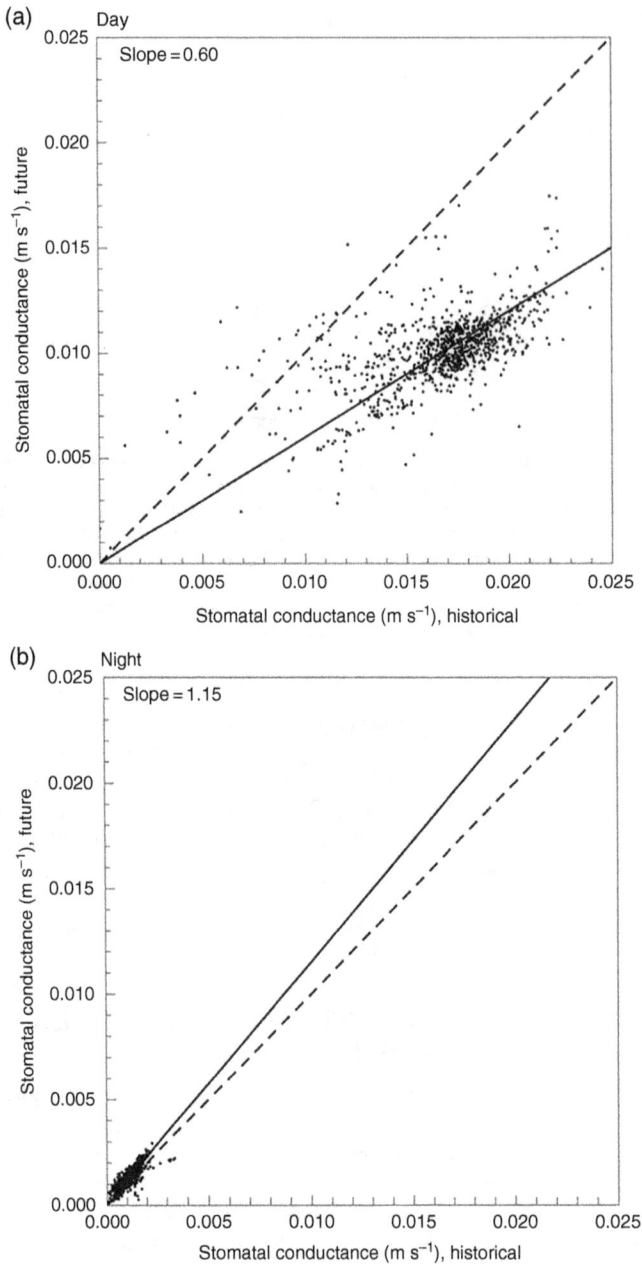

Figure 3.16 Future vs. historical stomatal conductance for the GFDL-ESM2M climate model for (a) daytime and (b) nighttime. Each point represents a time average for one non-water-stressed grid cell-month. Dashed lines are 1:1 lines and solid lines are least-squares fits through the origin. Reproduced from Milly and Dunne (2016) / with permission of Springer Nature.

non-water-stressed ET, with an increase in r_s causing suppressed ET and an increase in VPD causing higher ET. However, Yang et al. (2019b) showed that these opposing effects cancel out among all the models, refuting the common perception of "warming leads to drying" (Cook et al., 2014; Dai, 2013; Feng & Fu, 2013 Fu & Feng, 2014; Lin et al., 2015; Naumann et al., 2018 Park et al., 2018; Scheff & Frierson, 2015; Sherwood & Fu, 2014; Trenberth et al., 2014.

Due to the concerns mentioned above, future drought predictions are subject to inconsistency when obtained using climate models vs. traditional offline impact models and drought indices (Milly & Dunne, 2016; 2017; Yang et al., 2020). These PM-based PET estimates overestimate the trends obtained from anthropogenic climate change experiments, and thus the observed continental drying is actually weaker than previously thought (Swann et al., 2016). Milly and Dunne (2016) showed that commonly used PET estimators such as the PM model for an open water surface (Shuttleworth, 1993) or a reference crop (Allen et al., 1998) cannot replicate the non-water-stressed ET obtained from anthropogenic climate change experiments. The simplistic resistance network of the PM model relative to the climate models, ignorance of vegetation water interception, and time steps have also been cited as secondary factors for failure of the PM-based models (Milly & Dunne, 2016).

According to Milly and Dunne (2016), the bias in predicting the change in non-water-stressed ET from climate models observed when using the PM-based PET is reduced to some extent with the use of the SW model (Figure 3.17), primarily due to accounting for both vegetation and ground as sources. The RZWQM2 model, with its plant growth simulated by the DSSAT4.0-CERES module, accounts for the increased r_s in the SW model as a consequence of elevated CO_2, resulting in decreased root water uptake and alleviation of crop water stress (Leakey et al., 2009). Ko et al. (2010a) calibrated and validated RZWQM2 for spring wheat growth and soil-water simulations in the FACE experiments at Maricopa, AZ. As a result, soil moisture was simulated with reasonable accuracy, with root mean squared difference ranging from 0.016 to

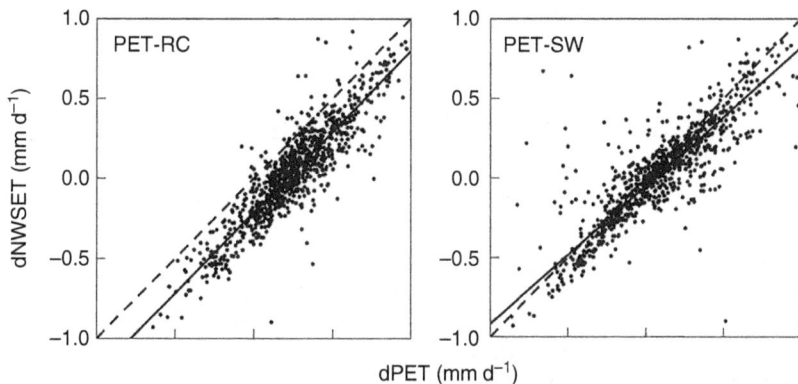

Figure 3.17 Scatterplot of change in non-water-stressed evapotranspiration (ET) from the GFDL-ESM2M climate model (dNWSET) against change in potential ET (dPET). The value of dPET is computed by the reference-crop Penman–Monteith method (PET-RC) and the Shuttleworth–Wallace method (PET-SW). Each point represents a difference between historical and future 20-year averages for one non-water-stressed grid cell-month. Dashed lines are 1:1 lines through the origin, and solid lines are Theil–Sen slope estimators. Reproduced from Milly and Dunne (2016) / with permission of Springer Nature.

$0.072 \text{ cm}^3 \text{ cm}^{-3}$. The ability of RZWQM2 (or the SW model) to account for elevated CO_2 impact on r_s makes it suitable to investigate the climate change impacts on the water requirement of crops. Perhaps more important, the SW provides more a realistic estimate of evaporative losses under changing climatic conditions by accounting for plants' response (resistance) to elevated CO_2 and adjusting ET accordingly. Thus, the model has been used for this purpose for cotton (Chen et al., 2019), maize (Durand et al., 2018; Islam et al., 2012; Ko et al., 2012; Ma et al., 2017, wheat (Ko et al., 2010b, 2012 Ko & Ahuja, 2013), and millet (Ko et al., 2012).

It is necessary to incorporate the dynamism of r_s resulting from elevated CO_2 in our ET calculations to obtain realistic future assessments that account for radiative, aerodynamic, and biological controls on the hydrologic cycle. Yang et al. (2019b) inverted the PM model with climate model outputs of actual ET and meteorological variables under non-stressed conditions to examine long-term changes in r_s. They used 16 Couple Model Intercomparison Project Phase 5 (CMIP5) models under historical (1861–2005) and RCP 8.5 (2006–2100) experiments. They found that mean r_s increased from ~55 s m^{-1} in the historical period (CO_2 concentration of ~300 mg kg^{-1}) to 80 s m^{-1} in the future (2071–2100) period (CO_2 concentration of ~807 mg kg^{-1}). The mean sensitivity of r_s to a change in the CO_2 concentration was ~0.09% (mg kg^{-1})$^{-1}$ at 300 mg kg^{-1}, which was similar across space and models. The relative sensitivity [~0.09% (mg kg^{-1})$^{-1}$] documented by Yang et al. (2019b) is identical to that synthesized from actual global observations [~0.08% (mg kg^{-1})$^{-1}$] by Ainsworth and Rogers (2007). Thus, the assumption of zero or a constant r_s will no longer hold.

Yang et al. (2019b) also proposed a simple approach to account for CO_2 impacts on r_s in the PM model under non-water-stressed conditions:

$$r_{s\_elevated} = r_{s\_baseline}[1 + S_{r_s}(CO_{2\_elevated} - CO_{2\_baseline})] \tag{3.50}$$

where $r_{s\_elevated}$ and $r_{s\_baseline}$ are stomatal resistances at elevated and baseline CO_2 concentrations, respectively, $CO_{2\_elevated}$ and $CO_{2\_baseline}$ are elevated and baseline CO_2 concentrations; and S_{r_s} is the relative sensitivity of r_s to a change in CO_2 concentration. In the CMIP5 model ensemble, $r_{s\_baseline}$ at the $CO_{2\_baseline}$ of 300 mg kg^{-1} is ~55 s m^{-1} and S_{r_s} is ~0.09% (mg kg^{-1})$^{-1}$. Using these parameters, $r_{s\_elevated}$ can be quantified at any given future $CO_{2\_elevated}$.

Modeling of Biological Control on ET: Surface Resistance

Multiple sensitivity analyses of ET models have shown the critical importance of the resistance terms used to represent the biological control on evaporation. Lagos et al. (2013) performed a sensitivity analysis of the model parameters in a four-layer ET model (Lagos et al., 2009) for maize in eastern Nebraska. The model parameters evaluated were extinction coefficient, VPD, soil temperature, crop height, and the aerodynamic, boundary layer, soil surface, residue, and surface canopy resistances. They found that the simulated hourly ET was most sensitive to changes in surface canopy resistance, soil surface resistance, and residue surface resistance. Zhao et al. (2020) suggested that calibration of physical and empirical parameters used in resistance estimation can result in improved performance. Rana and Katerji (1998) found that crop ET was most sensitive to r_c, among available energy, VPD, and r_a, especially for tall crops under water stress, which represents the majority of dryland maize grown in the

United States. Thus, to accurately model ET or T in these critical ecosystems, it is necessary that r_c is modeled using climate, canopy, and soil water content and to avoid simplistic models of r_c, an incomplete representation that does not account for critical variables that influence r_c and in turn ET, which can result in unrealistic quantification of evaporative losses. Irmak et al. (2006) conducted extensive sensitivity analyses for the ASCE-EWRI-PM ET_o model, with a fixed resistance term, in semiarid, Mediterranean-type, coastal humid, inland humid, and semi-humid climates and an island region. In general, ET_o was most sensitive to VPD at all locations, while the sensitivity of ET_o to the same variable showed significant variation from one location to another and also exhibited interannual variation in sensitivity to the same variables at the same location. After VPD, ET_o was most sensitive to u_2 in semiarid and windy regions. The R_s was the dominant driving force of ET_o at humid locations during the summer months. In a Mediterranean climate, the sensitivity of ET_o to u_2 was minimal during the summer months. At the semi-arid and island locations, ET_o was more sensitive to maximum temperature than R_s in summer months, whereas it was equally sensitive to maximum temperature and R_s at the semi-humid location. The ET_o was not sensitive to T_{min} at any of the locations. The change in ET_o was linearly related to change in climate variables with $R^2 \geq .96$ in most cases, except T_{min}, at all sites. Increase in ET_o with respect to increases in climate variables changed considerably by month. The sensitivity coefficient for R_s was higher during the summer months and lower during the winter months, and the opposite was observed for VPD, except for the island site. The decrease in the sensitivity coefficients for R_s corresponding to an increase in the sensitivity coefficient for VPD is due to a decrease in the energy term in favor of the increase in significance of the aerodynamic term of the standardized ASCE-EWRI-PM equation in summer versus winter months. While sensitivity analyses of ASCE-EWRI-PM can provide valuable information about the behavior of PM-type equations' response to climate variables, using a fixed resistance term would not provide a complete description of these combination-based energy balance equations' behavior with one resistance term because the equations' response to a unit increase or decrease in climate variables would differ, even substantially in certain cases. Thus, sensitivity analyses need to be conducted for varying resistances under varying climate variables simultaneously for more realistic quantification of the sensitivity of evaporative loss.

Due to increased uncertainty owing to the presence of numerous resistance terms in ET models, models that do not explicitly require any resistance parameterizations are attractive (Zhao et al., 2020). These models include the 3T model (Paw & Daughtry, 1984 Qiu, 1996), the Priestley-Taylor method (Priestley & Taylor, 1972), the triangle or trapezoidal method (Long & Singh, 2012; Price, 1990, and the surface renewal method (Paw et al., 1995). Although these methods require relatively less input information, they have shown comparable accuracy (Ershadi et al., 2014; Long & Singh, 2012; Xiong et al., 2015; Zhou et al., 2014. However, in order to appropriately represent the physiological controls on ET, attempts to parameterize bulk surface resistance (r_s) under actual crop and environmental conditions have justly received attention. Zhu et al. (2014) used a Bayesian method to simultaneously parameterize the soil surface and spring maize canopy resistances in the SW model in arid China and found good agreement with EC and microlysimeter measurements. They attributed the underestimations by the SW model to micro-scale advection over heterogeneous land surfaces and interception-driven evaporation following wetting events. Mutiibwa and Irmak (2013) conducted sensitivity and elasticity analyses for r_s and found that r_s estimates were most sensitive to uncertainties introduced in parameter a_1 of the photosynthetic photon flux density (PPFD) sub-function (i.e., $r_s = a_0(\text{PPFD})^{a_1}(1 - a_2\text{VPD})^{a_3}\left(1 - a_4(298 - T_a)^{a_5} + r_{s\_min}^{1/\exp(\text{LAI})}\right)$ (Irmak & Mutiibwa, 2009b; Mutiibwa & Irmak, 2013) due to its exponential impact on r_s in the Jarvis-type parameterizations. They found that for accurate estimates of r_s, uncertainties in parameter a_1 should not exceed the range of ±2% so that the error in estimated r_s is kept between −3.5 and 3.6%. They observed that the relative change in r_s due to uncertainties in parameters a_2 and a_3 of the VPD subfunction was a linear function and less sensitive than the PPFD subfunction. The sensitivity of r_s to uncertainties in air temperature subfunction parameters (a_4 and a_5) was higher than that of VPD subfunction parameters but less than that of PPFD subfunction parameters. The uncertainty in parameters a_4 and a_5 should range within ±10%, and the calibration of these parameters should be determined with greater precision than the VPD subfunction parameters.

Some major empirical models have been developed to capture the response of plant stomata to environmental conditions, and are briefly presented here:

The Jarvis–Stewart equation:

$$g_s = \frac{1}{r_s} = g_0 \times f_D(\text{VPD}) \times f_T(T_a) \times f_s(R_s) \times f_w(\theta) \tag{3.51}$$

where g_s is stomatal conductance (inverse of r_s); g_0 is the maximum g_s, which is plant function type-specific; $f_D, f_T, f_s,$ and f_w range from 0 to 1 and are stress factors associated with VPD, T_a, R_s, and θ, respectively. This model may be subject to underestimation under conditions of high RH due to linear correlation among g_s to RH (Wang et al., 2009). Rather, a non-linear function of RH or VPD may alleviate this bias (Leuning, 1995; Wang et al., 2009).

Katerji and Perrier (KP) model:
The KP model (Katerji & Perrier, 1983) is based on the prospect that the ratio of r_c to r_a is linearly related to the ratio of r^* (climatic resistance term) to r_a:

$$\frac{r_c}{r_a} = a\frac{r*}{r_a} + b \tag{3.52}$$

where a and b are empirical coefficients that have high transferability (Rana et al., 1997) and have been experimentally determined for alfalfa, rice, grass, lettuce, sweet and grain sorghum, sunflower, soybean, clementine orchard, and sloping grassland (Katerji & Rana, 2006), including under water-stress conditions (Rana et al., 1997, 2001). Having no physical meaning, the coefficients and the model have been limited to applicability only for periods when the Bowen ratio varies between −0.3 and 0.3 (Pereira et al., 1999).

The parameter r^* is defined as

$$r* = \frac{\Delta + \gamma}{\Delta\gamma}\frac{\rho c_p\text{VPD}}{R_n - G} \tag{3.53}$$

Todorovic model:
The Todorovic model (Todorovic, 1999) is applied by calculating the difference between the actual canopy temperature and the canopy temperature under wet conditions (t):

$$t = \frac{\gamma}{\Delta}\frac{D}{(\Delta + \gamma)} \tag{3.54}$$

Next, the ratio (X) of r_c to climatological resistance (r^*) is calculated as

$$X = \frac{r_c}{r*} = \frac{a + \sqrt{b^2 - 4ac}}{2a} \tag{3.55}$$

where a, b, and c are defined as

$$a = \frac{\Delta + \gamma \Delta}{\Delta + \gamma} \gamma VPD \tag{3.56}$$

$$b = -\gamma \frac{r*}{r_a} t \tag{3.57}$$

$$c = -(\Delta + \gamma)t \tag{3.58}$$

Todorovic (1999) defined r^* slightly differently than in the KP model:

$$r^* = \frac{\rho c_p VPD}{\gamma (R_n - G)} \tag{3.59}$$

Finally, r_c is calculated as

$$r_c = Xr^* \tag{3.60}$$

Li et al. model:
The model proposed by Li et al. (2009) is a biophysically improved version of the Todorovic model. They incorporated a term C in the definition of t, for which the Todorovic model assigned a value of $C = 1$:

$$t = \frac{\gamma}{\Delta} \frac{DC}{(\Delta + \gamma)} \tag{3.61}$$

$$C = \frac{(\Delta/\lambda)(1/r^*) + 1/r_a}{(1 + \gamma/\Delta)(1/r_c) + (\gamma/\Delta)(1/r_a)} \tag{3.62}$$

Following Leuning (1995) to parameterize the response of g_s to VPD, Li et al. (2009) proposed C as $(1 + VPD/VPD_0)$, where VPD_0 accounts for the response of t to VPD and was considered 1.5 kPa for winter wheat in the North China Plain. However, VPD_0 is subject to variation with crops and climate conditions and has no physical meaning.

Shuttleworth one-step model:
Shuttleworth (2006) proposed calculation of r_c as a function of weather variables and K_c values from FAO-56 as a preferred alternative to the use of fixed K_c, weaknesses of which are discussed above (under Comparative Suitability of ET Models). The one-step approach was presented in detail above, it being a focus of this chapter.

Upscaling leaf r_s using PPFD:
Stating inaccurate performance of the two-step approach resulting from fixed r_c as their justification and motivation, Irmak et al. (2008c) proposed a methodology for using measured, leaf-level, porometer-measured stomatal resistance (r_s) to upscaled canopy resistance (r_c) using PPFD, LAI for sunlit and shaded leaves, and direct and diffuse solar radiation. The process requires several steps building up to r_c, and the approach is presented in detail in Irmak et al. (2008c). The approach was developed for maize and was consequently validated for soybean after reparameterization (Irmak et al., 2013a;

Mutiibwa & Irmak, 2011 and for riparian vegetation species, including common reed, peachleaf willow, and cottonwood (Kabenge & Irmak, 2012). The resulting ET estimates from the r_c scaled up from r_s approach were superior to the fixed resistance model, especially under high VPD conditions (Irmak & Mutiibwa, 2009a). Irmak and Mutiibwa (2010) and Irmak et al. (2013a) also developed simple empirical models for non-stressed maize and soybean to calculate r_c using weather variables.

Ball–Berry equation:

$$g_s = \frac{1}{r_s} = m \frac{A}{C_s} \frac{e}{e_s} P_{atm} + b \tag{3.63}$$

where parameter m depends on plant functional type and θ; A is net C assimilation and is estimated as the minimum of W_c, W_j, and W_e, which are assimilation rates limited by Rubisco enzyme, light, and export capacity, respectively (Ball et al., 1987; Collatz et al., 1991; Sellers et al., 1996a, 1996b); C_s is the CO_2 partial pressure at the leaf surface; e is the water vapor pressure at the leaf surface; e_s is the saturation vapor pressure inside the leaf; P_{atm} is the atmospheric pressure; and b is the minimum g_s when A is 0.

Medlyn equation:
Although empirical models of g_s (Jarvis–Stewart and Ball–Berry equations) are attractive due to confident parameter estimation and their already widespread use, they can only summarize our current understanding of stomata. Instead, in order to gain insights into controls on gas exchange, mechanistic and optimality approaches are promising (Buckley, 2017). The Medlyn equation (Medlyn et al., 2011) presents a combination of optimal and empirical approaches to model g_s:

$$g_s = \frac{1}{r_s} = g_0 + \left(1 + \frac{g_1}{\sqrt{D}}\right) \frac{A}{C_a} \tag{3.64}$$

where g_0 and g_1 are fitted model parameters, A is the net C assimilation rate, C_a is the atmospheric CO_2 concentration at the leaf surface, and D is leaf-to-air VPD.

Canopy resistance r_c can be modeled by scaling up r_s (Ding et al., 2014; Irmak et al., 2008c; Kim & Verma, 1991; Mutiibwa & Irmak, 2011 Rochette et al., 1991; Zhang et al., 2011).

Transitioning from Meteorologically Based to Plant-Based Estimation of Crop Transpiration

Sinclair and Ghanem (2020) and Ghanem et al. (2020) argued that the use of "assumed" constant values of r_a and r_c in the reference ET approach ignores any unique sensitivity of ET to r_a and r_c. Moreover, they argued that K_c has no mechanistic meaning and exhibits great variation across crop species, the point in the growing season, cultural practices (sowing dates, irrigation practices, etc.), weather conditions, the percentage of ground covered by crops, plant health, drought conditions, etc. (Hanson & May, 2006; Marin et al., 2016; Rosa et al., 2012; Spinelli et al., 2017; Wherley et al., 2015; Yang et al., 2014. They also argued that variables used in the FAO-56 two-step approach such as R_n and r_a are never or rarely measured and are only empirical constructs based on R_s and u_2, respectively.

As an alternative to meteorological approximations, Sinclair and Ghanem (2020) and Ghanem et al. (2020) presented a plant-based predictor of T ($g\,m^{-2}\,d^{-1}$), which avoids the assumptions, empiricisms, and unknowns of the FAO-56 (and ASCE-EWRI-PM) two step approach and is instead based on explicit, plant-specific inputs. The approach is based on fundamental physiological and physical characteristics of the canopy water use efficiency expression given by de Wit (1958) and Tanner and Sinclair (1983):

$$T = I_o \left[1 - \exp\left(\frac{-g\mathrm{LAI}}{\sin(\beta)} \right) \right] \frac{(\mathrm{RUE})(\mathrm{VPD})}{k_d} \qquad (3.65)$$

where I_o is incident photosynthetically active radiation ($MJ\,m^{-2}\,d^{-1}$), g is leaf shadow projection defined by leaf and incident light angles, LAI is leaf area index (unitless), β is weighted daily sun angle, RUE is radiation use efficiency ($g\,MJ^{-1}$), VPD is vapor pressure deficit (Pa), and k_d is a species-specific constant (Pa) that is defined mostly using constant (and conservative) parameters, except two variables, the first denoting fractional conversion of photosynthate to plant mass depending on the synthesized composition of carbohydrate, protein, and oil in the plant products, and the second being a dimensionless ratio based on the leaf internal CO_2 concentration and atmospheric CO_2 concentration. The current value of k_d for major crop species is approximately 9 Pa for C_4 species (maize, sorghum, and sugarcane), 6 Pa for C_3 cereals (wheat and rice), and 5 Pa for C_3 legumes (soybean and peanut).

Sinclair and Ghanem (2020) and Ghanem et al. (2020) suggested values of RUE under unstressed conditions as $3.6\,g\,MJ^{-1}$ for maize and sugarcane, $3.0\,g\,MJ^{-1}$ for wheat and rice, and $2.4\,g\,MJ^{-1}$ for soybean. Extensive research has documented field-determined RUE for various crop species (Campbell et al., 2001; Edwards et al., 2005; Hatfield, 2014; Kemanian et al., 2004 Kiniry et al., 2005; Kukal & Irmak, 2020a, 2020b Lindquist et al., 2005; Narayanan et al., 2013; Purcell et al., 2002; Singer et al., 2011 Van Roekel & Purcell, 2014). Unlike the daily mean VPD used in the FAO-56 reference ET equation, VPD in Equation 3.65 is weighted on a daily basis to reflect the distribution of transpiration through the daily cycle. Howell (1990) claimed that daytime VPD should be preferred to predict transpiration over mean daily VPD. The weighted VPD is calculated as the difference among weighted saturated vapor pressure (e_s) and actual vapor pressure (e_a), where e_s is calculated as

$$e_s = e_{s(T_{max})}\theta_w + e_{s(T_{min})}(1 - \theta_w) \qquad (3.66)$$

where θ_w is a weighting parameter ranging between 0 and 1. A θ_w value >0.50 gives more weight to a daytime VPD. Sinclair and Ghanem (2020) and Ghanem et al. (2020) suggested a θ_w of 0.75 based on the literature (Abbate et al., 2004; Ghanem et al., 2020; Tanner, 1981 Tanner & Sinclair, 1983). The daily value of $g/\sin(\beta)$ is 0.7 for latitudes ~30° and a leaf angle of ~50° (Sinclair, 2006), although recalculation is suggested for other latitudes. The plant-based approach for T determination has been demonstrated in a case study for turfgrasses, as presented in Figure 3.18 (Wherley et al., 2015). Being physically based, and operating on explicitly defined, readily observable, mostly conservative (varying by photosynthetic pathway) inputs, this approach has merits over meteorological approximations such as the FAO-56 approach. However, direct comparisons of the two approaches should inform these assessments in the future.

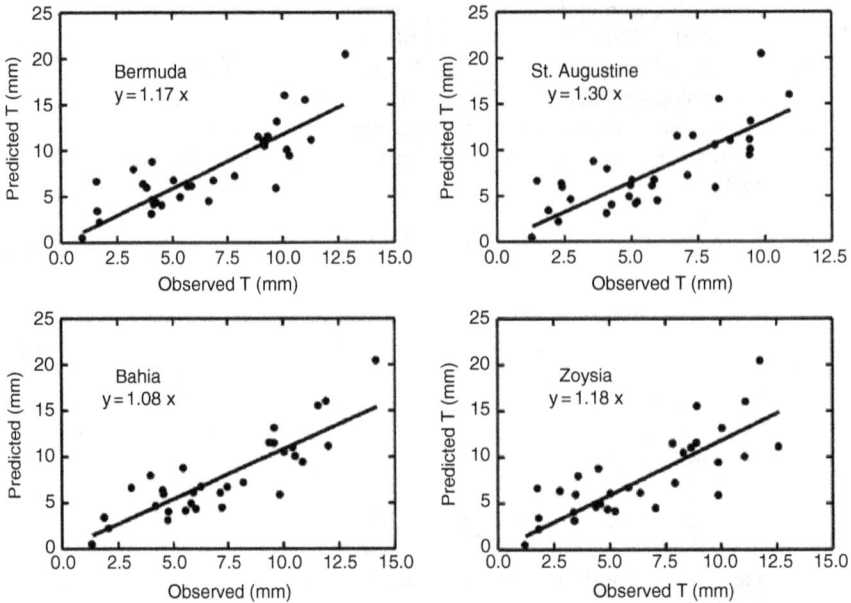

Figure 3.18 Physiological prediction of transpiration amounts versus observed transpiration for all one-to three-days periods studied during three year of measurements for four turfgrasses. Linear regressions through the origin are shown for each species ($p < .0001$). Reproduced from Sinclair et al. (2014) / with permission of John Wiley & Sons.

Concluding Remarks and Future Recommendations

1. The metadata for ET, T, and E reported in various platforms, including journal articles, are very important and immensely beneficial for discerning the accuracy and representativeness of the ET data and their derivatives such as K_c in different applications (Allen et al., 2011). Such data and information are critical to better understand the behavior of different ET, T, and E models under different environmental, climatic, and management conditions. These metadata at least should include the description of the vegetation, its aerodynamic fetch, water management and background soil moisture, types of equipment and calibration checks, photographs of the measured vegetation–equipment combinations, and independent assessments of measured ET using models or other means. However, in many of the ET-related studies and subsequent published papers, especially in non-agricultural communities, adequate or sufficient information about the detailed conditions of the evaporating surface is not provided. The scientific community needs to be encouraged to provide such important information for enabling and enhancing the reproducibility of the study and to better understand the surface characteristics from which the ET data are being reported.

2. Recommendations of research on ET models' suitability in varied (aridity) ecosystems should be heeded for application. For example, windy and semi-arid regions should preferably use the M–S approach over the FAO-56 PM approach, which is

more suitable for humid conditions. There is ample opportunity and need for evaluating the M–S approach for a variety of crops and environmental conditions and to expand empirical evidence on the greater suitability of the M-S approach over the FAO-56 PM approach under certain conditions, especially water-stressed and arid to semi-arid conditions as well as partial canopy conditions, in which the performance of the PM models seems to decline. Practical applications where adequate room for uncertainty exists can prefer convenient but robust models like FAO-56, but research investigations should transition toward more physically sound models.

3. The PM equation, being only applicable to full crop cover conditions and mostly for well-watered conditions, is not theoretically appropriate to improve ET estimates under partial cover conditions (Farahani et al., 2007). Instead, two-source models (e.g., SW model) should be preferred and further refined to estimate ET from partial cover canopies. Additionally, NT or reduced-till systems should be subject to the use of approaches (e.g., the Farahani & Ahuja, 1996 approach in the SW model) that explicitly account for residue cover to accurately represent the modified surface–atmosphere interactions from a residue-covered soil surface.

4. It is critical to conduct field studies that address the knowledge gap of effective hourly and all-day average values for r_c and r_a for different crops equivalent to that for crop coefficients (Shuttleworth, 2007).

5. The use of the PM equation under deficit irrigation management requires better quantification of the effect of stress on stomatal control besides the use of the K_s parameter. Farahani et al. (2007) restated the suggestions of Monteith (1985) to investigate the one-step PM approach for estimating ET in light of impacts of water stress on surface resistance.

6. Future research should focus on the intersection of direct and indirect human intervention and evaporation. With improved plant productivity via selective breeding, change in nutrient availability, and elevated CO_2 concentrations, modified plant physiological controls and plant phenology will affect transpiration patterns, investigation of which requires coupling between biogeochemical and evaporation models (Shuttleworth, 2007).

7. Practical use of the FAO-56 two-step approach (or ET_{ref}) to estimate ET_c is widespread, accepted, and an ideal solution to the weaknesses that existed in ET models until its proposal. This approach has been serving the water management community and in other applications well, mostly due to its nature in that it does not require crop-plant-specific resistance values. Thus, the use of ET_{ref} is expected to continue, mostly due to its operational nature and simplicity. The use of ET_{ref} is especially promising when one realizes the inaccuracies, limited availability and uncertainties in historical and current weather measurements, and crop and irrigation information (Pereira et al., 2015). Although, as a community, we are at a more advanced level of understanding ET processes and advanced models exist, these shortcomings will continue to hinder applications of more complex models (such as one-step approach and multilayer approaches). To motivate users to adopt complex approaches that are closer to our understanding of ET processes, effort has to be placed on the development of computer models or case studies to exemplify their operational use. For instance, it can be argued that the presence of these support resources (e.g., examples in FAO-56 and the ASCE-EWRI Manual of Practice 70) have had a reasonable role to play in its understanding and

rapid application. Similar or even more effective materials coupled with well-coordinated educational programs can be developed to educate professionals involved in ET, water management, agronomy, climate science, plant physiology, hydrology, and related disciplines to highlight the importance of one-step approach estimates of ET for better accuracy and the realistic nature of this approach compared with the two-step PM approach.

8. The ET community has been moving away from the original comprehensive, physically based and realistic approach of one-step estimation of evaporative losses for different surfaces, especially since the development of the FAO-56 and ASCE-EWRI PM methods. This is primarily due to the lack of availability of resistance terms for various evaporating surfaces. Thus, well-coordinated and comprehensive field research should be carried out in the future to measure such resistances for as many different vegetation and other surfaces as possible. Additionally, good quality and robust modeling efforts can be expanded to model the resistance terms for different surfaces as a function of environmental, climatic, crop, soil, water, and other variables that impact the magnitude of resistances. Surfaces for which resistance terms for application of one-step ET methods are needed should include agricultural cropping systems, horticultural plant species, vegetables, fruits, forestry, landscape, forestry, water bodies, and other surfaces to once again enable and enhance the use of the one-step approach for more accurate evaporative loss quantification.

9. The database of K_c values needs to be continuously expanded, especially for new varieties of crops or natural vegetation, management practices, and environments or untested conditions. To do so, qualified and well-operated measuring systems such as lysimetry, Bowen ratio, EC, and soil-water balance measurements should be continued (Pereira et al., 2015).

10. The use of the FAO-56 and ASCE-EWRI-PM models and their derivatives is widespread but has several drawbacks. Both successful and unsuccessful studies of ET using these methods should be documented.

11. Accurate and robust ET, T, and E estimates are becoming more and more important as we work toward the goal of higher crop water productivity. Assessments of ET, T, and E form the very basis of evaluating water management practices, of which crop water production functions are highly instrumental. Development of T-based (rather than ET) crop water production functions is highly desired due to T being the determinant of crop biomass and/or grain yield production. Furthermore, to evaluate management strategies that potentially minimize non-beneficial E losses, robust E quantification will be imperative. Thus, modeling frameworks that allow robust partitioning of T and E will be especially useful in this context.

12. The FAO-56 and ASCE-EWRI-PM models and the procedures outlined in these publications to estimate different variables needed to estimate water use have been suggested as "standardized" methods. However, in many ET and related publications, these methods have been misrepresented as "standard" methods, which is significantly different and false. The scientific communities are free to use any ET model for their studies that they deem appropriate for specific applications. Thus, when the FAO-56 or ASCE-EWRI PM models are not utilized in any given scientific publication, study, report, etc., this cannot be (and should not be) evaluated as a weakness or wrong practice. In fact, standardized methods are consistent, well-documented, and well-known and thus allow fair comparisons.

Appendix
Notations and Symbols

Notations for certain parameters have been defined more than once, using either different units or slight variations, to communicate their use exactly as defined in their original sources.

A	total energy flux leaving the complete crop ($W\,m^{-2}$)
A_s	total energy flux leaving the substrate ($W\,m^{-2}$)
b	value of minimum g_s when A is 0
c_a	specific heat of the air ($MJ\,kg^{-1}\,°C^{-1}$, $J\,kg^{-1}\,K^{-1}$)
C_a	atmospheric CO_2 concentration at the leaf surface
C_d	denominator coefficient for the ASCE–EWRI standardized PM equation
C_n	numerator coefficient for the ASCE–EWRI standardized PM equation
$CO_{2\_baseline}$	baseline CO_2 concentration
$CO_{2\_elevated}$	elevated CO_2 concentration
c_p	specific heat of the air ($MJ\,kg^{-1}\,°C^{-1}$, $J\,kg^{-1}\,K^{-1}$)
C_s	fraction of unit ground area covered by soil (decimal)
C_s	CO_2 partial pressure at the leaf surface
C_r	fraction of unit ground area covered by residue (decimal)
d	zero plane displacement height (m)
D	vapor pressure deficit (kPa, mbar)
D_a	vapor pressure deficit at reference height (Pa)
D_m	vapor pressure deficit at canopy source height (Pa)
D_o	vapor pressure deficit at canopy source height
DT	disk-till
e	water vapor pressure at the leaf surface
E	evaporation (mm)
e_a	actual vapor pressure
E_{PT}	Priestley–Taylor estimate of evapotranspiration
e_s	saturation vapor pressure inside the leaf
ET	evapotranspiration (mm)
ET_{ASCE}	reference evapotranspiration rate defined by the American Society of Civil Engineers ($mm\,d^{-1}$)
ET_c	actual crop evapotranspiration
ET_f	foliage evaporation ($mm\,d^{-1}$)
ET_o	grass reference evapotranspiration
ET_p	potential evapotranspiration
ET_{PM}	Penman-Monteith evapotranspiration ($mm\,d^{-1}$)
ET_r	alfalfa reference evapotranspiration
ET_{ref}	reference evapotranspiration
ET_s	soil surface evaporation ($mm\,d^{-1}$)
FACE	free-air CO_2 enrichment
f_{ew}	fraction of soil contributing to the majority of evaporation
G	soil heat flux ($MJ\,m^{-2}\,d^{-1}$)
g	leaf shadow projection defined by leaf and incident light angles
GDD	growing degree days
G_r	heat flux into residue-covered soil ($W\,m^{-2}$)
g_s	stomatal conductance
G_s	heat flux into bare soil ($W\,m^{-2}$)
H	sensible heat flux from the complete crop ($W\,m^{-2}$)
$h_{p,}$	h mean plant or crop height (m)

H_s	sensible heat flux from the substrate (W m^{-2})
h_c	reference plant height (m)
I_o	incident photosynthetically active radiation (MJ m^{-2} d^{-1})
k	von Karman constant (0.41)
K_c	single crop coefficient
K_{cb}	basal crop coefficient
K_{c_max}	maximum evaporation factor
k_d	species-specific constant (Pa)
K_e	coefficient to adjust for increased evaporation from the soil
K_r	reduction factor based on soil water content
K_s	adjustment coefficient for water stress
LAI	leaf area index (m^2 m^{-2})
M–S	Matt–Shuttleworth
NT	no-till
P	biochemical storage of energy in the crop below reference height (W m^{-2})
P_{atm}	atmospheric pressure
PET	potential evapotranspiration
PM	Penman–Monteith
PM$_c$	Penman–Monteith combination equation for evaporation from a closed canopy
PM$_r$	Penman–Monteith combination equation for evaporation from residue-covered soil
PM$_s$	Penman–Monteith combination equation for evaporation from bare substrate
PPFD	photosynthetic photon flux density
r^*	climatic resistance but defined differently in Katerji & Perrier and Todorovic models
r_a	aerodynamic resistance (s m^{-1})
$r_a{}^a$, $r_{a,s}$	aerodynamic resistance between the substrate and canopy source height (s m^{-1})
$r_a{}^c$	bulk boundary layer resistance of the vegetative elements in the canopy (s m^{-1})
$r_{a,f}$	bulk boundary-layer resistance of the foliage for water vapor (s m^{-1}) $r_a{}^s$ aerodynamic resistance between the substrate and canopy source height (s m^{-1})
$r_c{}^{50}$	aerodynamic coefficient at blending height of 50 m (s m^{-1})
r_{clim}	climatic resistance (s m^{-1})
RCP	representative concentration pathway
RH$_{min}$	daily minimum relative humidity
R_n	net radiation into the complete crop (MJ m^{-2} d^{-1}, W m^{-2})
$R_{n,f}$	net radiation of the foliage (W m^{-2})
R_{nr}	net radiation at the residue-covered ground (W m^{-2})
$R_n{}^s$, R_{ns}, $R_{n,s}$	net radiation flux into the (bare) substrate (W m^{-2})
R_{nsub}	net radiation below the canopy (W m^{-2})
r_s	stomatal resistance
$r_{s\_baseline}$	stomatal resistance at baseline CO$_2$ concentration
$r_{s\_elevated}$	stomatal resistance at elevated CO$_2$ concentration
R_s	incoming shortwave radiation (W m^{-2})
r_s, $r_s{}^c$, $r_{s,f}$	bulk stomatal resistance
$(r_s)_c$, r_c, $r_{s,v}$	crop surface resistance (s m^{-1})
$r_s{}^r$	resistance to soil evaporative flux through the surface residue (s m^{-1})
$r_s{}^s$, $r_{s,s}$	surface resistance at the substrate surface (soil)

RUE	radiation use efficiency $(\mathrm{g\,MJ^{-1}})$
S	physical storage of energy in the atmosphere and crop below reference height $(\mathrm{W\,m^{-2}})$
S_{rs}	relative sensitivity of r_s to change in CO_2 concentration
SW	Shuttleworth–Wallace
T	transpiration $(\mathrm{mm,\,g\,m^{-2}\,d^{-1}})$
T_a	daily average air temperature $(^\circ \mathrm{C})$
T_{mean}	daily average air temperature $(^\circ \mathrm{C})$
T^{pref}	air temperature during the period when the value of FAO-56 K_c is derived
TSW	total soil water
u_2	wind speed at 2-m height $(\mathrm{m\,s^{-1}})$
u_z	wind speed measured at height z $(\mathrm{m\,s^{-1}})$
VPD	vapor pressure deficit (kPa, mbar)
VPD_2	vapor pressure deficit at 2-m height (kPa)
VPD_{50}	vapor pressure deficit at 50-m height (kPa)
W_c	assimilation rate limited by Rubisco enzyme
W_e	assimilation rate limited by export capacity
W_j	assimilation rate limited by light
X	ratio of canopy resistance to climatological resistance
z_{0h}	roughness length for heat and vapor transfer (m)
z_{0m}	roughness length for momentum transfer (m)
z_h	height at which relative humidity is measured (m)
z_m	height at which wind speed is measured (m)
β	weighted daily sun angle
γ	psychrometric constant $(\mathrm{kPa\,^\circ C^{-1},\,mbar\,K^{-1}})$
Δ	slope of saturation vapor pressure curve at the corresponding air temperature $(\mathrm{kPa\,^\circ C^{-1},\,mbar\,K^{-1}})$
θ	soil water content
θ_w	weighting parameter in the plant-based estimation of transpiration in Sinclair and Ghanem (2020) and Ghanem et al. (2020)
λ	latent heat of vaporization $(\mathrm{MJ\,kg^{-1}})$
λE	latent heat flux from the complete crop $(\mathrm{W\,m^{-2}})$
λE_c	latent heat flux from the plant canopy $(\mathrm{W\,m^{-2}})$
λE_s	latent heat flux from the (bare) substrate $(\mathrm{W\,m^{-2}})$
λE_r	evaporation rate from the residue-covered soil $(\mathrm{W\,m^{-2}})$
λT	latent heat flux from the plant canopy $(\mathrm{W\,m^{-2}})$
ρ_a	ρ mean air density at constant pressure $(\mathrm{kg\,m^{-3}})$

References

Abbate, P. E., Dardanelli, J. L., Cantarero, M. G., Maturano, M., Melchiori, R. J. M., & Suero, E. E. (2004). Climatic and water availability effects on water-use efficiency in wheat. *Crop Science*, 44, 474–483. https://doi.org/10.2135/cropsci2004.4740

Ahuja, L., Rojas, K. W., Hanson, J. D., Shaffer, M. J., & Ma, L. (Eds.) (2000). *Root Zone Water Quality Model: Modelling management effects on water quality and crop production.* Water Resources Publications.

Ainsworth, E. A., & Rogers, A. (2007). The response of photosynthesis and stomatal conductance to rising [CO_2]: Mechanisms and environmental interactions. *Plant, Cell & Environment*, 30, 258–270.

Allen, R. G., Pereira, L. S., Howell, T. A., & Jensen, M. E. (2011). Evapotranspiration information reporting: II. Recommended documentation. *Agricultural Water Management*, 98, 921–929.

Allen, R. G., Pereira, L. S., Raes, D., & Smith, M. (1998). *Crop evapotranspiration: Guidelines for computing crop water requirements* (FAO Irrigation and Drainage Paper 56). FAO.

Allen, R. G., Pruitt, W. O., Wright, J. L., Howell, T. A., Ventura, F., Snyder, R., Itenfisu, D., Steduto, P., Berengena, J., & Yrisarry, J. B. (2006). A recommendation on standardized surface resistance for hourly calculation of reference ET$_o$ by the FAO56 Penman-Monteith method. *Agricultural Water Management*, 81, 1–22. https://doi.org/10.1016/j.agwat.2005.03.007

Allen, R. G., Walter, I. A., Elliott, R. L., Howell, T. A., Iitenfisu, D., Jensen, M. E., & Snyder, R. L. (2005). *The ASCE standardized reference evapotranspiration equation*. Environmental and Water Resources Institute, American Society of Civil Engineers.

Althoff, D., Filgueiras, R., Dias, S. H. B., & Rodrigues, L. N. (2019). Impact of sum-of-hourly and daily timesteps in the computations of reference evapotranspiration across the Brazilian territory. *Agricultural Water Management*, 226, 105785.

Anapalli, S. S., Ahuja, L. R., Gowda, P. H., Ma, L., Marek, G., Evett, S. R., & Howell, T. A. (2016). Simulation of crop evapotranspiration and crop coefficients with data in weighing lysimeters. *Agricultural Water Management*, 177, 274–283.

Anapalli, S. S., Ma, L., Nielsen, D. C., Vigil, M. F., & Ahuja, L. R. (2005). Simulating planting date effects on corn production using RZWQM and CERES-Maize models. *Agronomy Journal*, 97, 58–71. https://doi.org/10.2134/agronj2005.0058

Anapalli, S. S., Reddy, K. N., & Jagadamma, S. (2018). Conservation tillage impacts and adaptations in irrigated corn production in a humid climate. *Agronomy Journal*, 110, 2673–2686. https://doi.org/10.2134/agronj2018.03.0195

Anderson, R. G., Alfieri, J. G., Tirado-Corbalá, R., Gartung, J., McKee, L. G., Prueger, J. H., Wang, D., Ayars, J. E., & Kustas, W. P. (2017). Assessing FAO-56 dual crop coefficients using eddy covariance flux partitioning. *Agricultural Water Management*, 179, 92–102.

Anonymous. (1956). Proceedings of the informal meeting on physics in agriculture [Special issue]. *NJAS–Wageningen Journal of Life Sciences*, 4. https://doi.org/10.18174/njas.v4i1.17766

Arellano, M. G., & Irmak, S. (2016). Reference (potential) evapotranspiration. I: Comparison of temperature, radiation, and combination-based energy balance equations in humid, subhumid, arid, semiarid, and Mediterranean-type climates. *Journal of Irrigation and Drainage Engineering*, 142(4), 4015065. https://doi.org/10.1061/(ASCE)IR.1943-4774.0000978

ASCE Environmental and Water Resources Institute. (2005). *The ASCE standardized reference evapotranspiration equation*. American Society of Civil Engineers.

ASCE Standardization of Reference Evapotranspiration Task Committee. (2000). *ASCE's standardized reference evapotranspiration equation*. In National Irrigation Symposium, Phoenix, AZ. American Society of Civil Engineers.

Azzari, G., Grassini, P., Edreira, J. I. R., Conley, S., Mourtzinis, S., & Lobell, D. B. (2019). Satellite mapping of tillage practices in the North Central US region from 2005 to 2016. *Remote Sensing of Environment*, 221, 417–429. https://doi.org/10.1016/j.rse.2018.11.010

Ball, J. T. (1988). *An analysis of stomatal conductance* (Doctoral dissertation). Stanford University.

Ball, J. T., Woodrow, I. E., & Berry, J. A. (1987). A model predicting stomatal conductance and its contribution to the control of photosynthesis under different environmental conditions. In J. Biggins (Ed.), *Progress in photosynthesis research* (Vol. 4, pp. 221–224). Springer. https://doi.org/10.1007/978-94-017-0519-6_48

Berengena, J., & Gavilán, P. (2005). Reference evapotranspiration estimation in a highly advective semiarid environment. *Journal of Irrigation and Drainage Engineering*, 131(2), 147–163.

Black, T. A., Tanner, C. B., & Gardner, W. R. (1970). Evapotranspiration from a snap bean crop. *Agronomy Journal*, 62, 66–69. https://doi.org/10.2134/agronj1970.00021962006200010022x

Brenner, A. J., & Incoll, L. D. (1997). The effect of clumping and stomatal response on evaporation from sparsely vegetated shrublands. *Agricultural and Forest Meteorology*, 84, 187–205.

Brun, L. J., Kanemasu, E. T., & Powers, W. L. (1972). Evapotranspiration from soybean and sorghum fields. *Agronomy Journal*, 64, 145–148. https://doi.org/10.2134/agronj1972.00021962006400020005x

Buckley, T. N. (2017). Modeling stomatal conductance. *Plant Physiology*, 174, 572–582.

Budyko, M. I. (1956/1958). *The heat balance of the Earth's surface* (N. A. Stepanova, Trans.). U.S. Department of Commerce.

Campbell, C. S., Heilman, J. L., Mcinnes, K. J., Wilson, L. T., Medley, J. C., Wu, G., & Cobos, D. R. (2001). Seasonal variation in radiation use efficiency of irrigated rice. *Agricultural and Forest Meteorology*, 110, 45–54. https://doi.org/10.1016/S0168-1923(01)00277-5

Chahine, M. T. (1992). The hydrological cycle and its influence on climate. *Nature*, 359, 373–380.

Chen, H., Huang, J. J., & McBean, E. (2020a). Partitioning of daily evapotranspiration using a modified Shuttleworth–Wallace model, random forest and support vector regression, for a cabbage farmland. *Agricultural Water Management*, 228, 105923. https://doi.org/10.1016/j.agwat.2019.105923

Chen, X., Qi, Z., Gui, D., Sima, M. W., Zeng, F., Li, L., Li, X., Gu, Z., & Gu, Z. (2020b). Evaluation of a new irrigation decision support system in improving cotton yield and water productivity in an arid climate. *Agricultural Water Management*, 234, 106139.

Chen, X., Qi, Z., Gui, D., Gu, Z., Ma, L., Zeng, F., & Li, L. (2019). Simulating impacts of climate change on cotton yield and water requirement using RZWQM2. *Agricultural Water Management*, 222, 231–241.

Choudhury, B. J., & Monteith, J. L. (1988). A four-layer model for the heat budget of homogeneous land surfaces. *Quarterly Journal of the Royal Meteorological Society*, 114, 373–398.

Collatz, G. J., Ball, J. T., Grivet, C., & Berry, J. A. (1991). Physiological and environmental regulation of stomatal conductance, photosynthesis and transpiration: A model that includes a laminar boundary layer. *Agricultural and Forest Meteorology*, 54, 107–136.

Cook, B. I., Smerdon, J. E., Seager, R., & Coats, S. (2014). Global warming and 21st century drying. *Climate Dynamics*, 43, 2607–2627.

Craft, K. J., Helmers, M. J., Malone, R. W., Pederson, C. H., & Schott, L. R. (2018). Effects of subsurface drainage systems on water and nitrogen footprints simulated with RZWQM2. *Transactions of the ASABE*, 61, 245–261.

Dai, A. (2013). Increasing drought under global warming in observations and models. *Nature Climate Change*, 3, 52–58.

Damour, G., Simonneau, T., Cochard, H., & Urban, L. (2010). An overview of models of stomatal conductance at the leaf level. *Plant, Cell & Environment*, 33, 1419–1438.

de Wit, C. T. (1958). *Transpiration and crop yield* (Rep. 64.6). Institute of Biological and Chemical Research on Field Crops and Herbage.

Derpsch, R., Friedrich, T., Kassam, A., & Li, H. (2010). Current status of adoption of no-till farming in the world and some of its main benefits. *International Journal of Agricultural and Biological Engineering*, 3(1). https://doi.org/10.3965/j.issn.1934-6344.2010.01.0-0

Ding, J., Hu, W., Wu, J., Yang, Y., & Feng, H. (2020). Simulating the effects of conventional versus conservation tillage on soil water, nitrogen dynamics, and yield of winter wheat with RZWQM2. *Agricultural Water Management*, 230, 105956.

Ding, R., Kang, S., Du, T., Hao, X., & Zhang, Y. (2014). Scaling up stomatal conductance from leaf to canopy using a dual-leaf model for estimating crop evapotranspiration. *PLoS One*, 9(4), e95584.

Dingman, S. L. (1992). *Physical hydrology*. Prentice Hall.

Dinpashoh, Y., Jhajharia, D., Fakheri-Fard, A., Singh, V. P., & Kahya, E. (2011). Trends in reference crop evapotranspiration over Iran. *Journal of Hydrology*, 399, 422–433.

Djaman, K., & Irmak, S. (2013). Actual crop evapotranspiration and alfalfa- and grass-reference crop coefficients of maize under full and limited irrigation and rainfed conditions. *Journal of Irrigation and Drainage Engineering*, 139, 433–446.

Doorenbos, J., & Kassam, A. H. (1979). *Yield response to water* (FAO Irrigation and Drainage Paper 33). FAO.

Doorenbos, J., & Pruitt, W. O. (1977). *Guidelines for predicting crop water requirements* (FAO Irrigation and Drainage Paper 24). FAO.

Drake, B. G., Gonzàlez-Meler, M. A., & Long, S. P. (1997). More efficient plants: A consequence of rising atmospheric CO_2? *Annual Review of Plant Physiology and Plant Molecular Biology*, 48, 609–639. https://doi.org/10.1146/annurev.arplant.48.1.609

Durand, J.-L., Delusca, K., Boote, K., Lizaso, J., Manderscheid, R., Weigel, H. J., Ruane, A. C., Anapalli, S. S., Ahuja, L., Basso, B., Baron, C., Bertuzzi, P., Ripoche, D., Biernath, C., Priesak, E., Deryng, D., Ewert, F., Gaiser, T., Gayler, S., … Ahuja, L. (2018). How accurately do maize crop models simulate the interactions of atmospheric CO_2 concentration levels with limited water supply on water use and yield? *European Journal of Agronomy*, 100, 67–75.

Earth System Research Laboratory. (2020). *Trends in atmospheric carbon dioxide*. Global Monitoring Laboratory, ESRL–NOAA. Retrieved November 17, 2020 https://www.esrl.noaa.gov/gmd/ccgg/trends/weekly.html

Edwards, J. T., Purcell, L. C., & Karcher, D. E. (2005). Soybean yield and biomass responses to increasing plant population among diverse maturity groups: II. Light interception and utilization. *Crop Science*, 45, 1778–1785. https://doi.org/10.2135/cropsci2004.0570

Elfarkh, J., Er-Raki, S., Ezzahar, J., Chehbouni, A., Aithssaine, B., Amazirh, A., Khabba, S., & Jarlan, L. (2020). Integrating thermal stress indexes within Shuttleworth–Wallace model for evapotranspiration mapping over a complex surface. *Irrigation Science*, 39, 45–61. https://doi.org/10.1007/s00271-020-00701-3

Ershadi, A., McCabe, M. F., Evans, J. P., Chaney, N. W., & Wood, E. F. (2014). Multi-site evaluation of terrestrial evaporation models using FLUXNET data. *Agricultural and Forest Meteorology*, 187, 46–61.

Ershadi, A., McCabe, M. F., Evans, J. P., & Wood, E. F. (2015). Impact of model structure and parameterization on Penman–Monteith type evaporation models. *Journal of Hydrology*, 525, 521–535.

Facchi, A., Gharsallah, O., Corbari, C., Masseroni, D., Mancini, M., & Gandolfi, C. (2013). Determination of maize crop coefficients in humid climate regime using the eddy covariance technique. *Agricultural Water Management*, 130, 131–141.

Farahani, H. J., & Ahuja, L. R. (1996). Evapotranspiration modeling of partial canopy/residue-covered fields. *Transactions of the ASAE*, 39, 2051–2064. https://doi.org/10.13031/2013.27708

Farahani, H. J., & Bausch, W. C. (1995). Performance of evapotranspiration models for maize: Bare soil to closed canopy. *Transactions of the ASAE*, 38, 1049–1059. https://doi.org/10.13031/2013.27922

Farahani, H. J., & DeCoursey, D. G. (2000). Potential evaporation and transpiration processes in the soil-residue-canopy system. In J. D. Hanson, K. W. Rojas, & L. Ahuja (Eds.), *Root Zone Water Quality Model: Modelling management effects on water quality and crop production* (pp. 51–80). Water Resources Publications.

Farahani, H. J., Howell, T. A., Shuttleworth, W. J., & Bausch, W. C. (2007). Evapotranspiration: Progress in measurement and modeling in agriculture. *Transactions of the ASABE*, 50, 1627–1638.

Feng, S., & Fu, Q. (2013). Expansion of global drylands under a warming climate. *Atmospheric Chemistry and Physics*, 13, 10081–10094. https://doi.org/10.5194/acp-13-10081-2013

Ficklin, D. L., & Novick, K. A. (2017). Historic and projected changes in vapor pressure deficit suggest a continental-scale drying of the United States atmosphere. *Journal of Geophysical Research: Atmospheres*, 122, 2061–2079. https://doi.org/10.1002/2016JD025855

Field, C. B., Jackson, R. B., & Mooney, H. A. (1995). Stomatal responses to increased CO_2: Implications from the plant to the global scale. *Plant, Cell & Environment*, 18, 1214–1225.

Finch, J. W. (2001). A comparison between measured and modelled open water evaporation from a reservoir in south-east England. *Hydrological Processes*, 15, 2771–2778.

Fisher, J. B., DeBiase, T. A., Qi, Y., Xu, M., & Goldstein, A. H. (2005). Evapotranspiration models compared on a Sierra Nevada forest ecosystem. *Environmental Modelling & Software*, 20, 783–796.

Flo, V., Martinez-Vilalta, J., Steppe, K., Schuldt, B., & Poyatos, R. (2019). A synthesis of bias and uncertainty in sap flow methods. *Agricultural and Forest Meteorology*, 271, 362–374.

Fu, Q., & Feng, S. (2014). Responses of terrestrial aridity to global warming. *Journal of Geophysical Research: Atmospheres*, 119, 7863–7875.

Gao, G., Zhang, X., Yu, T., & Liu, B. (2016). Comparison of three evapotranspiration models with eddy covariance measurements for a *Populus euphratica* Oliv. forest in an arid region of northwestern China. *Journal of Arid Land*, 8, 146–156.

Gardiol, J. M., Serio, L. A., & Della Maggiora, A. I. (2003). Modelling evapotranspiration of corn (*Zea mays*) under different plant densities. *Journal of Hydrology*, 271, 188–196.

Gavilán, P., Estévez, J., & Berengena, J. (2008). Comparison of standardized reference evapotranspiration equations in southern Spain. *Journal of Irrigation and Drainage Engineering*, 134(1), 1–12.

Ghanem, M. E., Kehel, Z., Marrou, H., & Sinclair, T. R. (2020). Seasonal and climatic variation of weighted VPD for transpiration estimation. *European Journal of Agronomy*, 113, 125966.

Gillette, K., Malone, R. W., Kaspar, T. C., Ma, L., Parkin, T. B., Jaynes, D. B., Fang, Q. X., Hatfield, J. L., Feyereisen, G. W., & Kersebaum, K. C. (2018). N loss to drain flow and N_2O emissions from a corn–soybean rotation with winter rye. *Science of the Total Environment*, 618, 982–997. https://doi.org/10.1016/j.scitotenv.2017.09.054

Gong, X., Liu, H., Sun, J., Gao, Y., & Zhang, H. (2019). Comparison of Shuttleworth–Wallace model and dual crop coefficient method for estimating evapotranspiration of tomato cultivated in a solar greenhouse. *Agricultural Water Management*, 217, 141–153.

Granger, R. J., & Gray, D. M. (1989). Evaporation from natural nonsaturated surfaces. *Journal of Hydrology*, 111, 21–29.

Hanson, B. R., & May, D. M. (2006). Crop coefficients for drip-irrigated processing tomato. *Agricultural Water Management*, 81, 381–399.

Hatfield, J. L. (2014). Radiation use efficiency: Evaluation of cropping and management systems. *Agronomy Journal*, 106, 1820–1827. https://doi.org/10.2134/agronj2013.0310

Hough, M. N., & Jones, R. J. A. (1997). The United Kingdom Meteorological Office rainfall and evaporation calculation system: MORECS version 2.0—An overview. *Hydrology and Earth System Sciences*, 1, 227–239. https://doi.org/10.5194/hess-1-227-1997

Howell, T. A. (1990). Relationships between crop production and transpiration, evapotranspiration, and irrigation. In R. J. Lascano & R. E. Sojka (Eds.), *Irrigation of agricultural crops* (2nd ed., pp. 391–434). Agronomy Monograph 30). ASA, CSSA, and SSSA.

Howell, T. A., Evett, S. R., Schneider, A. D., Dusek, D. A., & Copeland, K. S. (2000, 14–16 Nov.). Irrigated fescue grass ET compared with calculated reference grass ET. In National Irrigation Symposium: *Proceedings of the 4th Decennial Symposium, Phoenix, Arizona* (pp. 228–242). American Society of Agricultural Engineers.

Hu, Z., Li, S., Yu, G., Sun, X., Zhang, L., Han, S., & Li, Y. (2013). Modeling evapotranspiration by combing a two-source model, a leaf stomatal model, and a light-use efficiency model. *Journal of Hydrology*, 501, 186–192. http://dx.doi.org/10.1016/j.jhydrol.2013.08.006

Hu, Z., Wu, G., Zhang, L., Li, S., Zhu, X., Zheng, H., Zhang, L., Sun, X., & Yu, G. (2017). Modeling and partitioning of regional evapotranspiration using a satellite-driven water–carbon coupling model. *Remote Sensing*, 9(1), 54. https://doi.org/10.3390/rs9010054

Hu, Z., Yu, G., Zhou, Y., Sun, X., Li, Y., Shi, P., Yanfen, W., Xia, S., Zemei, Z., Lia, Z., & Zhang, L. (2009). Partitioning of evapotranspiration and its controls in four grassland ecosystems: Application of a two-source model. *Agricultural and Forest Meteorology*, 149, 1410–1420.

Huang, S., Yan, H., Zhang, C., Wang, G., Acquah, S. J., Yu, J., Li, L., Ma, J., & Darko, R. O. (2020). Modeling evapotranspiration for cucumber plants based on the Shuttleworth–Wallace model in a Venlo-type greenhouse. *Agricultural Water Management*, 228, 105861.

Irmak, S. (2010). Nebraska water and energy flux measurement, modeling, and research network (NEBFLUX). *Transactions of the ASABE*, 53, 1097–1115.

Irmak, S. (2011). On the dynamics of nocturnal, daytime and sum-of-hourly evapotranspiration and other surface energy fluxes over a non-stressed maize canopy. *Journal of Irrigation and Drainage Engineering*, 137, 475–490.

Irmak, S. (2015). Inter-annual variation in long-term center pivot-irrigated maize evapotranspiration (ET) and various water productivity response indices: I. Grain yield, actual and basal ET, irrigation–yield production functions, ET–yield production functions, and yield response factors. *Journal of Irrigation and Drainage Engineering*, 141, 04014068. https://doi.org/10.1061/(ASCE)IR.1943-4774.0000825

Irmak, S., Djaman, K., & Sharma, V. (2015). Winter wheat (*Triticum aestivum* L.) evapotranspiration and single (normal) and basal crop coefficients. *Transactions of the ASABE*, 58, 1047–1066.

Irmak, S., Haman, D. Z., & Bastug, R. (2000). Determination of crop water stress index for irrigation timing and yield estimation of corn. *Agronomy Journal*, 92, 1221–1227. https://doi.org/10.2134/agronj2000.9261221x

Irmak, S., Howell, T. A., Allen, R. G., Payero, J. O., & Martin, D. L. (2005). Standardized ASCE–Penman–Monteith: Impact of sum-of-hourly vs. 24-hr-time step computations at reference weather station sites. *Transactions of the ASAE*, 48, 1063–1077.

Irmak, S., Irmak, A., Howell, T. A., Martin, D. L., Payero, J. O., & Copeland, K. S. (2008a). Variability analyses of alfalfa-reference to grass-reference evapotranspiration ratios in growing and dormant seasons. *Journal of Irrigation and Drainage Engineering*, 134, 147–159.

Irmak, S., Istanbulluoglu, E., & Irmak, A. (2008b). An evaluation of evapotranspiration model complexity against performance in comparison with Bowen ratio energy balance measurements. *Transactions of the ASABE*, 51, 1295–1310.

Irmak, S., Mutiibwa, D., Irmak, A., Arkebauer, T. J., Weiss, A., Martin, D. L., & Eisenhauer, D. E. (2008c). On the scaling up leaf stomatal resistance to canopy resistance using photosynthetic photon flux density. *Agricultural and Forest Meteorology*, 148, 1034–1044.

Irmak, S., & Kukal, M. S. (2019). Disk-till vs. no-till maize grass- and alfalfa-reference single (average) and basal (dual) crop coefficients. *Agricultural Water Management*, 226, 105815. https://doi.org/10.1016/j.agwat.2019.105815

Irmak, S., Kukal, M. S., Mohammed, A. T., & Djaman, K. (2019). Disk-till vs. no-till maize evapotranspiration, microclimate, grain yield, production functions and water productivity. *Agricultural Water Management*, 216, 177–195. https://doi.org/10.1016/j.agwat.2019.02.006

Irmak, S., & Mutiibwa, D. (2009a). On the dynamics of evaporative losses from Penman–Monteith with fixed and variable canopy resistance during partial and complete maize canopy. *Transactions of the ASABE*, 52, 1139–1153.

Irmak, S., & Mutiibwa, D. (2009b). On the dynamics of stomatal resistance: Relationships between stomatal behavior and micrometeorological variables and performance of Jarvis-type parameterization. *Transactions of the ASABE*, 52, 1923–1939.

Irmak, S., & Mutiibwa, D. (2010). On the dynamics of canopy resistance: Generalized linear estimation and relationships with primary micrometeorological variables. *Water Resources Research*, 46, W08526. https://doi.org/10.1029/2009WR008484

Irmak, S., Mutiibwa, D., Payero, J., Marek, T., & Porter, D. (2013a). Modeling soybean canopy resistance from micrometeorological and plant variables for estimating evapotranspiration using one-step Penman–Monteith approach. *Journal of Hydrology*, 507, 1–18. https://doi.org/10.1016/j.jhydrol.2013.10.008

Irmak, S., Odhiambo, L. O., Specht, J. E., & Djaman, K. (2013b). Hourly and daily single and basal evapotranspiration crop coefficients as a function of growing degree days, days after emergence, leaf area index, fractional green canopy cover, and plant phenology for soybean. *Transactions of the ASABE*, 56, 1785–1803.

Irmak, S., Payero, J. O., Martin, D. L., Irmak, A., & Howell, T. A. (2006). Sensitivity analyses and sensitivity coefficients of standardized daily ASCE–Penman–Monteith equation. *Journal of Irrigation and Drainage Engineering*, 132, 564–578.

Irmak, S., Skaggs, K. E., & Chatterjee, S. (2014). A review of the Bowen ratio surface energy balance method for quantifying evapotranspiration and other energy fluxes. *Transactions of the ASABE*, 57, 1657–1674.

Islam, A., Ahuja, L. R., Garcia, L. A., Ma, L., Saseendran, A. S., & Trout, T. J. (2012). Modeling the impacts of climate change on irrigated corn production in the central Great Plains. *Agricultural Water Management*, 110, 94–108.

Itenfisu, D., Elliott, R. L., Allen, R. G., & Walter, I. A. (2003). Comparison of reference evapotranspiration calculations as part of the ASCE standardization effort. *Journal of Irrigation and Drainage Engineering*, 129, 440–448.

Jensen, M. E. (1968). Water consumption by agricultural plants. In T. T. Kozlowski (Ed.), *Plant water consumption and response: Water deficits and plant growth* (Vol. II, pp. 1–22). Academic Press.

Jensen, M. E. (1974). *Consumptive use of water and irrigation water requirements*. American Society of Civil Engineers.

Jensen, M. E., & Allen, R. G. (Eds.) (2016). *Evaporation, evapotranspiration, and irrigation water requirements* (2nd ed.). American Society of Civil Engineers. https://doi.org/10.1061/9780784414057

Jensen, M. E., Burman, R. D., & Allen, R. G. (1990). *Evapotranspiration and irrigation water requirements* (Manual of Practice no. 70). American Society of Civil Engineers.

Jensen, M. E., Wright, J. L., & Pratt, B. J. (1971). Estimating soil moisture depletion from climate, crop and soil data. *Transactions of the ASAE*, 14, 954–959.

Jeong, H., Pittelkow, C. M., & Bhattarai, R. (2019). Simulated responses of tile-drained agricultural systems to recent changes in ambient atmospheric gradients. *Agricultural Systems*, 168, 48–55.

Ji, X. B., Chen, J. M., Zhao, W. Z., Kang, E. S., Jin, B. W., & Xu, S. Q. (2017). Comparison of hourly and daily Penman–Monteith grass- and alfalfa-reference evapotranspiration equations and crop coefficients for maize under arid climatic conditions. *Agricultural Water Management*, 192, 1–11. https://doi.org/10.1016/j.agwat.2017.06.019

Jiang, Q., Qi, Z., Madramootoo, C. A., & Singh, A. K. (2018). Simulating hydrologic cycle and crop production in a subsurface drained and sub-irrigated field in southern Quebec using RZWQM2. *Computers and Electronics in Agriculture*, 146, 31–42.

Jiang, Z.-Y., Yang, Z.-G., Zhang, S.-Y., Liao, C.-M., Hu, Z.-M., Cao, R.-C., & Wu, H.-W. (2020). Revealing the spatio-temporal variability of evapotranspiration and its components based on an improved Shuttleworth-Wallace model in the Yellow River Basin. *Journal of Environmental Management*, 262, 110310.

Jones, J. W., Hoogenboom, G., Porter, C. H., Boote, K. J., Batchelor, W. D., Hunt, L. A., Wilkens, P. W., Singh, U., Gijsman, A. J., & Ritchie, J. T. (2003). The DSSAT cropping system model. *European Journal of Agronomy*, 18, 235–265.

Kabenge, I., & Irmak, S. (2012). Evaporative losses from a common reed-dominated peachleaf willow and cottonwood riparian plant community. *Water Resources Research*, 48, W09513. https://doi.org/10.1029/2012WR011902

Kanemasu, E. T., Verma, S. B., Smith, E. A., Fritschen, L. J., Wesely, M., Field, R. T., Kustas, W. P., Weaver, H., Stewart, J. B., Gurney, R., Panin, G., & Gurney, R. (1992). Surface flux measurements in FIFE: An overview. *Journal of Geophysical Research: Atmospheres*, 97(D17), 18547–18555.

Kassam, A., Friedrich, T., Derpsch, R., & Kienzle, J. (2015). Overview of the worldwide spread of conservation agriculture. *Field Actions. Science Reports*, 8. https://journals.openedition.org/factsreports/3966

Katerji, N., & Perrier, A. (1983). Modelisation de l'evapotranspiration reelle ETR d'une parcelle de luzerne: Role d'un coefficient cultural. *Agronomie*, 3, 513–521.

Katerji, N., & Perrier, A. (1985). Détermination de la résistance globale d'un couvert végétal à la diffusion de vapeur d'eau et de ses différentes composantes: Approche théorique et vérification expérimentale sur une culture de luzerne. *Agricultural and Forest Meteorology*, 34, 105–120.

Katerji, N., Perrier, A., & Oulid-Aïssa, A. K. (1983). Exploration au champ et interprétation de la variation horizontale et verticale de la résistance stomatique: Cas d'une culture de luzerne (*Médicago sativa* L.). *Agronomie*, 3, 847–856.

Katerji, N., & Rana, G. (2006). Modelling evapotranspiration of six irrigated crops under Mediterranean climate conditions. *Agricultural and Forest Meteorology*, 138, 142–155.

Katerji, N., & Rana, G. (2014). FAO-56 methodology for determining water requirement of irrigated crops: Critical examination of the concepts, alternative proposals and validation in Mediterranean region. *Theoretical and Applied Climatology*, 116, 515–536.

Kato, T., Kimura, R., & Kamichika, M. (2004). Estimation of evapotranspiration, transpiration ratio and water-use efficiency from a sparse canopy using a compartment model. *Agricultural Water Management*, 65, 173–191.

Keijman, J. Q. (1974). The estimation of the energy balance of a lake from simple weather data. *Boundary-Layer Meteorology*, 7, 399–407.

Keijman, J. Q. (1981). Theoretical background of some methods for the determination of evaporation. In *Evaporation in relation to hydrology* (Proceedings and Information 28, pp. 12–24). Committee for Hydrological Research TNO.

Kemanian, A. R., Stöckle, C. O., & Huggins, D. R. (2004). Variability of barley radiation-use efficiency. *Crop Science*, 44, 1662–1672. https://doi.org/10.2135/cropsci2004.1662

Kim, J., & Verma, S. B. (1991). Modeling canopy photosynthesis: Scaling up from a leaf to canopy in a temperate grassland ecosystem. *Agricultural and Forest Meteorology*, 57, 187–208.

Kimball, B. A. (2016). Crop responses to elevated CO_2 and interactions with H_2O, N, and temperature. *Current Opinion in Plant Biology*, 31, 36–43.

Kimball, B. A., Boote, K. J., Hatfield, J. L., Ahuja, L. R., Stockle, C., Archontoulis, S., Barongh, C., Bassoi, B., Bertuzzij, P., Constantink, J., Derynglm, D., Dumontn, B., Durando, J.-L., Ewertpq, F., Gaiserq, T., Gaylerr, S., Hoffmannos, M. P., Jiang, Q., … Constantin, J. (2019). Simulation of maize evapotranspiration: An intercomparison among 29 maize models. *Agricultural and Forest Meteorology*, 271, 264–284.

Kimball, B. A., Kobayashi, K., & Bindi, M. (2002). Responses of agricultural crops to free-air CO_2 enrichment. *Advances in Agronomy*, 77, 293–368.

Kiniry, J. R., Simpson, C. E., Schubert, A. M., & Reed, J. D. (2005). Peanut leaf area index, light interception, radiation use efficiency, and harvest index at three sites in Texas. *Field Crops Research*, 91, 297–306. https://doi.org/10.1016/j.fcr.2004.07.021

Ko, J., Ahuja, L., Kimball, B., Anapalli, S., Ma, L., Green, T. R., Ruane, A. C., Wall, G. W., Pinter, P., & Bader, D. A. (2010a). Simulation of free air CO_2 enriched wheat growth and interactions with water, nitrogen, and temperature. *Agricultural and Forest Meteorology*, 150, 1331–1346.

Ko, J., Ahuja, L., Kimball, B., Anapalli, S., Ma, L., Green, T. R., Ruane, A. C., Wall, G. W., Pinter, P., & Bader, D. A. (2010b). Simulation of free air CO_2 enriched wheat growth and interactions with water, nitrogen, and temperature. *Agricultural and Forest Meteorology*, 150, 1331–1346.

Ko, J., & Ahuja, L. R. (2013). Global warming likely reduces crop yield and water availability of the dryland cropping systems in the US central Great Plains. *Journal of Crop Science and Biotechnology*, 16, 233–242.

Ko, J., Ahuja, L. R., Saseendran, S. A., Green, T. R., Ma, L., Nielsen, D. C., & Walthall, C. L. (2012). Climate change impacts on dryland cropping systems in the central Great Plains, USA. *Climatic Change*, 111, 445–472.

Kohler, M. A., Nordenson, T. J., & Fox, W. E. (1955). *Evaporation from pans and lakes* (Weather Bureau Research Paper 38). U.S. Government Printing Office.

Kohler, M. A., & Parmele, L. H. (1967). Generalized estimates of free-water evaporation. *Water Resources Research*, 3, 997–1005.

Kool, D., Agam, N., Lazarovitch, N., Heitman, J. L., Sauer, T. J., & Ben-Gal, A. (2014). A review of approaches for evapotranspiration partitioning. *Agricultural and Forest Meteorology*, 184, 56–70.

Kozak, J. A., Ahuja, L. R., Ma, L., & Green, T. R. (2005). Scaling and estimation of evaporation and transpiration of water across soil textures. *Vadose Zone Journal*, 4, 418–427. https://doi.org/10.2136/vzj2004.0119

Kukal, M. S., & Irmak, S. (2020a). Interrelationships among water use efficiency and light use efficiency in four row crop canopies. *Agrosystems, Geosciences, and Environment*, 3, e20110. https://doi.org/10.1002/agg2.20110

Kukal, M. S., & Irmak, S. (2020b). Canopy light interactions, use and efficiency in four row crops under optimal growth conditions. *Agricultural and Forest Meteorology*, 284, 107887. https://doi.org/10.1016/j.agrformet.2019.107887

Lagos, L. O., Martin, D. L., Verma, S. B., Irmak, S., Irmak, A., Eisenhauer, D., & Suyker, A. (2013). Surface energy balance model of transpiration from variable canopy cover and evaporation from residue-covered or bare soil systems: Model evaluation. *Irrigation Science*, 31, 135–150.

Lagos, L. O., Martin, D. L., Verma, S. B., Suyker, A., & Irmak, S. (2009). Surface energy balance model of transpiration from variable canopy cover and evaporation from residue-covered or bare-soil systems. *Irrigation Science*, 28, 51–64.

Lascano, R. J., & Van Bavel, C. H. M. (2007). Explicit and recursive calculation of potential and actual evapotranspiration. *Agronomy Journal*, 99, 585–590.

Lazzara, P., & Rana, G. (2010). The use of crop coefficient approach to estimate actual evapotranspiration: A critical review for major crops under Mediterranean climate. *Italian Journal of Agrometeorology*, 2, 25–39.

Leakey, A. D. B., Ainsworth, E. A., Bernacchi, C. J., Rogers, A., Long, S. P., & Ort, D. R. (2009). Elevated CO_2 effects on plant carbon, nitrogen, and water relations: Six important lessons from FACE. *Journal of Experimental Botany*, 60, 2859–2876.

Lecina, S., Martínez-Cob, A., Pérez, P. J., Villalobos, F. J., & Baselga, J. J. (2003). Fixed versus variable bulk canopy resistance for reference evapotranspiration estimation using the Penman–Monteith equation under semiarid conditions. *Agricultural Water Management*, 60, 181–198.

Leuning, R. (1995). A critical appraisal of a combined stomatal-photosynthesis model for C_3 plants. *Plant, Cell & Environment*, 18, 339–355. https://doi.org/10.1111/j.1365-3040.1995.tb00370.x

Lhomme, J.-P., Boudhina, N., & Masmoudi, M. M. (2014). On the Matt–Shuttleworth approach to estimate crop water requirements. *Hydrology and Earth System Sciences*, 18, 4341–4348.

Lhomme, J. P., Boudhina, N., Masmoudi, M. M., & Chehbouni, A. (2015). Estimation of crop water requirements: Extending the one-step approach to dual crop coefficients. *Hydrology and Earth System Sciences*, 19, 3287–3299.

Lhomme, J.-P., Montes, C., Jacob, F., & Prevot, L. (2012). Evaporation from heterogeneous and sparse canopies: On the formulations related to multi-source representations. *Boundary-Layer Meteorology*, 144, 243–262.

Lhomme, J. P., Montes, C., Jacob, F., & Prévot, L. (2013). Evaporation from multi-component canopies: Generalized formulations. *Journal of Hydrology*, 486, 315–320.

Li, L., Luo, G., Chen, X., Li, Y., Xu, G., Xu, H., & Bai, J. (2011). Modelling evapotranspiration in a central Asian desert ecosystem. *Ecological Modelling*, 222, 3680–3691.

Li, L., Yu, Q., Su, Z., & van der Tol, C. (2009). A simple method using climatic variables to estimate canopy temperature, sensible and latent heat fluxes in a winter wheat field on the North China Plain. *Hydrological Processes*, 23, 665–674. https://doi.org/10.1002/hyp.7166

Li, S., Kang, S., Zhang, L., Ortega-Farias, S., Li, F., Du, T., Tong, L., Wang, S., Ingman, M., & Guo, W. (2013). Measuring and modeling maize evapotranspiration under plastic film-mulching condition. *Journal of Hydrology*, 503, 153–168.

Li, X., Yang, P., Ren, S., Li, Y., Liu, H., Du, J., Li, P., Wang, C., & Ren, L. (2010). Modeling cherry orchard evapotranspiration based on an improved dual-source model. *Agricultural Water Management*, 98, 12–18.

Lin, L., Gettelman, A., Feng, S., & Fu, Q. (2015). Simulated climatology and evolution of aridity in the 21st century. *Journal of Geophysical Research: Atmospheres*, 120, 5795–5815.

Linacre, E. T. (1994). Estimating U.S. Class A pan evaporation from few climate data. *Water International*, 19, 5–14. https://doi.org/10.1080/02508069408686189

Lindquist, J. L., Arkebauer, T. J., Walters, D. T., Cassman, K. G., & Dobermann, A. (2005). Maize radiation use efficiency under optimal growth conditions. *Agronomy Journal*, 97, 72–78. https://doi.org/10.2134/agronj2005.0072

Liu, C., Sun, G., McNulty, S. G., & Kang, S. (2015). An improved evapotranspiration model for an apple orchard in northwestern China. *Transactions of the ASABE*, 58, 1253–1264.

Lizaso, J. I., Boote, K. J., Jones, J. W., Porter, C. H., Echarte, L., Westgate, M. E., & Sonohat, G. (2011). CSM-IXIM: A new maize simulation model for DSSAT version 4.5. *Agronomy Journal*, 103, 766–779. https://doi.org/10.2134/agronj2010.0423

Long, D., & Singh, V. P. (2012). A two-source trapezoid model for evapotranspiration (TTME) from satellite imagery. *Remote Sensing of Environment*, 121, 370–388.

Ma, L., Ahuja, L. R., Islam, A., Trout, T. J., Saseendran, S. A., & Malone, R. W. (2017). Modeling yield and biomass responses of maize cultivars to climate change under full and deficit irrigation. *Agricultural Water Management*, 180, 88–98.

Ma, L., Malone, R. W., Heilman, P., Jaynes, D. B., Ahuja, L. R., Saseendran, S. A., Kanwar, R. S., & Ascough, J. C., II (2007). RZWQM simulated effects of crop rotation, tillage, and controlled drainage on crop yield and nitrate-N loss in drain flow. *Geoderma*, 140, 260–271.

Ma, L., Nielsen, D. C., Ahuja, L. R., Malone, R. W., Saseendran, S. A., Rojas, K. W., Hanson, J. D., & Benjamin, J. G. (2003). Evaluation of RZWQM under varying irrigation levels in eastern Colorado. *Transactions of the ASAE*, 46, 39–49. https://doi.org/10.13031/2013.1254

Ma, L., Trout, T. J., Ahuja, L. R., Bausch, W. C., Saseendran, S. A., Malone, R. W., & Nielsen, D. C. (2012). Calibrating RZWQM2 model for maize responses to deficit irrigation. *Agricultural Water Management*, 103, 140–149.

Mardikis, M. G., Kalivas, D. P., & Kollias, V. J. (2005). Comparison of interpolation methods for the prediction of reference evapotranspiration: An application in Greece. *Water Resources Management*, 19, 251–278.

Marin, F. R., Angelocci, L. R., Nassif, D. S. P., Costa, L. G., Vianna, M. S., & Carvalho, K. S. (2016). Crop coefficient changes with reference evapotranspiration for highly canopy–atmosphere coupled crops. *Agricultural Water Management*, 163, 139–145.

Mauder, M., Cuntz, M., Drüe, C., Graf, A., Rebmann, C., Schmid, H. P., Schmidt, M., & Steinbrecher, R. (2013). A strategy for quality and uncertainty assessment of long-term eddy-covariance measurements. *Agricultural and Forest Meteorology*, 169, 122–135.

McJannet, D. L., Webster, I. T., Stenson, M. P., & Sherman, B. S. (2008). *Estimating open water evaporation for the Murray–Darling basin*. CSIRO Land and Water.

McMahon, T. A., Finlayson, B. L., & Peel, M. C. (2016). Historical developments of models for estimating evaporation using standard meteorological data. *Wiley Interdisciplinary Reviews: Water*, 3, 788–818.

Medlyn, B. E., Duursma, R. A., Eamus, D., Ellsworth, D. S., Prentice, I. C., Barton, C. V. M., Crous, K. Y., de Angelis, P., Freeman, M., & Wingate, L. (2011). Reconciling the optimal and empirical approaches to modelling stomatal conductance. *Global Change Biology*, 17, 2134–2144.

Milly, P. C. D. (1991). A refinement of the combination equations for evaporation. *Surveys in Geophysics*, 12, 145–154.

Milly, P. C. D., & Dunne, K. A. (2016). Potential evapotranspiration and continental drying. *Nature Climate Change*, 6, 946–949. https://doi.org/10.1038/nclimate3046

Milly, P. C. D., & Dunne, K. A. (2017). A hydrologic drying bias in water-resource impact analyses of anthropogenic climate change. *Journal of the American Water Resources Association*, 53, 822–838.

Monteith, J. L. (1965). Evaporation and environment. *Symposia of the Society for Experimental Biology*, 19, 205–234.

Monteith, J. L. (1985, 16–17 Dec.). Evaporation from land surfaces: Progress in analysis and prediction since 1948. In *Proceedings of the National Conference on Advances in Evapotranspiration*, Chicago, IL (ASAE Publ. 14-85, pp. 4–12). American Society of Agricultural Engineers.

Monteith, J. L., Szeicz, G., & Waggoner, P. E. (1965). The measurement and control of stomatal resistance in the field. *Journal of Applied Ecology*, 2, 345–355. https://doi.org/10.2307/2401484

Mu, Q., Zhao, M., & Running, S. W. (2011). Improvements to a MODIS global terrestrial evapotranspiration algorithm. *Remote Sensing of Environment*, 115, 1781–1800.

Mutiibwa, D., & Irmak, S. (2011). On the scaling up soybean leaf level stomatal resistance to canopy resistance for one-step estimation of actual evapotranspiration. *Transactions of the ASABE*, 54, 141–154.

Mutiibwa, D., & Irmak, S. (2013). Transferability of Jarvis-type models developed and re-parameterized for maize to estimate stomatal resistance of soybean: Analyses on model calibration, validation, performance, sensitivity, and elasticity. *Transactions of the ASABE*, 56, 409–422.

Narayanan, S., Aiken, R. M., Vara Prasad, P. V., Xin, Z., & Yu, J. (2013). Water and radiation use efficiencies in sorghum. *Agronomy Journal*, 105, 649–656. https://doi.org/10.2134/agronj2012.0377

Naumann, G., Alfieri, L., Wyser, K., Mentaschi, L., Betts, R. A., Carrao, H., Spinoni, J., Vogt, J., & Feyen, L. (2018). Global changes in drought conditions under different levels of warming. *Geophysical Research Letters*, 45, 3285–3296.

New, M., Hulme, M., & Jones, P. (1999). Representing twentieth-century space–time climate variability: I. Development of a 1961–90 mean monthly terrestrial climatology. *Journal of Climate*, 12, 829–856. https://doi.org/10.1175/1520-0442(1999)012<0829:RTCSTC>2.0.CO;2

Novick, K. A., Ficklin, D. L., Stoy, P. C., Williams, C. A., Bohrer, G., Oishi, A. C., Papuga, S. A., Blanken, P. D., Noormets, A., Sulman, B. N., Scott, R. L., Wang, L., & Sulman, B. N. (2016). The increasing importance of atmospheric demand for ecosystem water and carbon fluxes. *Nature Climate Change*, 6, 1023–1027.

Odhiambo, L., & Irmak, S. (2011). Performance of extended Shuttleworth–Wallace model for estimating and partitioning of evapotranspiration in a partial residue-covered subsurface drip-irrigated soybean field. *Transactions of the ASABE*, 54, 915–930. https://doi.org/10.13031/2013.37117

Oki, T., & Kanae, S. (2006). Global hydrological cycles and world water resources. *Science*, 313, 1068–1072.

Park, C.-E., Jeong, S.-J., Joshi, M., Osborn, T. J., Ho, C.-H., Piao, S., Chen, D., Liu, J., Yang, H., Park, H., Kim, B.-M., & Park, H. (2018). Keeping global warming within 1.5 °C constrains emergence of aridification. *Nature Climate Change*, 8, 70–74.

Paw, K. T., & Gao, W. (1988). Applications of solutions to non-linear energy budget equations. *Agricultural and Forest Meteorology*, 43(2), 121–145.

Paw, K. T., & Daughtry, C. S. T. (1984). A new method for the estimation of diffusive resistance of leaves. *Agricultural and Forest Meteorology*, 33, 141–155. https://doi.org/10.1016/0168-1923(84)90066-2

Paw, K. T., Qiu, J., Su, H.-B., Watanabe, T., & Brunet, Y. (1995). Surface renewal analysis: A new method to obtain scalar fluxes. *Agricultural and Forest Meteorology, 74*, 119–137. https://doi.org/10.1016/0168-1923(94)02182-J

Payero, J. O., & Irmak, S. (2011). Daily crop evapotranspiration, crop coefficient and energy balance components of a surfaceirrigated maize field. In G. Gerosa (Ed.), *Evapotranspiration: From measurements to agricultural and environmental applications* (pp. 59–78). InTech.

Payero, J. O., & Irmak, S. (2013). Daily energy fluxes, evapotranspiration and crop coefficient of soybean. *Agricultural Water Management, 129*, 31–43. https://doi.org/10.1016/j.agwat.2013.06.018

Penman, H. L. (1948). Natural evaporation from open water, bare soil and grass. *Proceedings of the Royal Society of London, Series A, 193*, 120–145.

Penman, H. L. (1956). Estimating evaporation. *Eos, Transactions of the American Geophysical Union, 37*(1), 43–50.

Penman, H. L. (1963). *Vegetation and hydrology (Technical Communication 53).* Commonwealth Agricultural Bureaux.

Pereira, L. S., Allen, R. G., Smith, M., & Raes, D. (2015). Crop evapotranspiration estimation with FAO56: Past and future. *Agricultural Water Management, 147*, 4–20.

Pereira, L. S., Perrier, A., Allen, R. G., & Alves, I. (1999). Evapotranspiration: Concepts and future trends. *Journal of Irrigation and Drainage Engineering, 125*, 45–51. https://doi.org/10.1061/(ASCE)0733-9437(1999)125:2(45)

Perez, P. J., Lecina, S., Castellvi, F., Martínez-Cob, A., & Villalobos, F. J. (2006). A simple parameterization of bulk canopy resistance from climatic variables for estimating hourly evapotranspiration. *Hydrological Processes, 20*, 515–532.

Perrier, A. (1976). *Etude et essai de modélisation des échanges de masse et d'énergie au niveau des couverts végétaux: Profils microclimatiques, évapotranspiration et photosynthèse nette.* Université Pierre et Marie Curie.

Pielke, R. A., Avissar, R., Raupach, M., Dolman, A. J., Zeng, X., & Denning, A. S. (1998). Interactions between the atmosphere and terrestrial ecosystems: Influence on weather and climate. *Global Change Biology, 4*, 461–475.

Pittelkow, C. M., Liang, X., Linquist, B. A., van Groenigen, K. J., Lee, J., Lundy, M. E., van Gestel, N., Six, J., Venterea, R. T., & van Kessel, C. (2015). Productivity limits and potentials of the principles of conservation agriculture. *Nature, 517*, 365–368.

Price, J. C. (1990). Using spatial context in satellite data to infer regional scale evapotranspiration. *IEEE Transactions on Geoscience and Remote Sensing, 28*, 940–948.

Priestley, C. H. B., & Taylor, R. J. (1972). On the assessment of surface heat flux and evaporation using large-scale parameters. *Monthly Weather Review, 100*, 81–92.

Pruitt, W. O. (1986). *Traditional methods: Evapotranspiration research priorities for the next decade* (ASAE Paper 86-2629). American Society of Agricultural Engineers.

Purcell, L. C., Ball, R. A., Reaper, J. D., & Vories, E. D. (2002). Radiation use efficiency and biomass production in soybean at different plant population densities. *Crop Science, 42*, 172–177. https://doi.org/10.2135/cropsci2002.1720

Qiu, G. Y. (1996). *A new method for estimation of evapotranspiration* (Doctoral dissertation). United Graduate School of Agricultural Sciences, Tottori University.

Rana, G., & Katerji, N. (1998). A measurement-based sensitivity analysis of the Penman–Monteith actual evapotranspiration model for crops of different height and in contrasting water status. *Theoretical and Applied Climatology, 60*, 141–149.

Rana, G., & Katerji, N. (2009). Operational model for direct determination of evapotranspiration for well-watered crops in Mediterranean region. *Theoretical and Applied Climatology, 97*, 243–253.

Rana, G., Katerji, N., Lazzara, P., & Ferrara, R. M. (2012). Operational determination of daily actual evapotranspiration of irrigated tomato crops under Mediterranean conditions by one-step and two-step models: Multiannual and local evaluations. *Agricultural Water Management, 115*, 285–296.

Rana, G., Katerji, N., Mastrorilli, M., & El Moujabber, M. (1994). Evapotranspiration and canopy resistance of grass in a Mediterranean region. *Theoretical and Applied Climatology, 50*, 61–71.

Rana, G., Katerji, N., Mastrorilli, M., El Moujabber, M., & Brisson, N. (1997). Validation of a model of actual evapotranspiration for water stressed soybeans. *Agricultural and Forest Meteorology, 86*, 215–224.

Rana, G., Katerji, N., & Perniola, M. (2001). Evapotranspiration of sweet sorghum: A general model and multi-local validity in semiarid environmental conditions. *Water Resources Research, 37*, 3237–3246.

Riahi, K., Rao, S., Krey, V., Cho, C., Chirkov, V., Fischer, G., Kindermann, G., Nakicenovic, N., & Rafaj, P. (2011). RCP 8.5: A scenario of comparatively high greenhouse gas emissions. *Climatic Change, 109*, 33. https://doi.org/10.1007/s10584-011-0149-y

Rind, D., Rosenzweig, C., & Goldberg, R. (1992). Modelling the hydrological cycle in assessments of climate change. *Nature*, 358, 119–122.

Riveros-Burgos, C., Ortega-Farias, S., López-Olivari, R., & Chávez, J. L. (2019). Parameterization of a clumped model to directly simulate actual evapotranspiration over a superintensive drip-irrigated olive orchard. *Journal of Hydrometeorology*, 20, 935–946.

Rochette, P., Pattey, E., Desjardins, R. L., Dwyer, L. M., Stewart, D. W., & Dube, P. A. (1991). Estimation of maize (*Zea mays* L.) canopy conductance by scaling up leaf stomatal conductance. *Agricultural and Forest Meteorology*, 54, 241–261.

Roderick, M. L., Greve, P., & Farquhar, G. D. (2015). On the assessment of aridity with changes in atmospheric CO_2. *Water Resources Research*, 51, 5450–5463.

Rosa, R. D., Paredes, P., Rodrigues, G. C., Alves, I., Fernando, R. M., Pereira, L. S., & Allen, R. G. (2012). Implementing the dual crop coefficient approach in interactive software: 1. Background and computational strategy. *Agricultural Water Management*, 103, 8–24.

Rotstayn, L. D., Roderick, M. L., & Farquhar, G. D. (2006). A simple pan-evaporation model for analysis of climate simulations: Evaluation over Australia. *Geophysical Research Letters*, 33, L17715. https://doi.org/10.1029/2006GL027114

Saseendran, S. A., Ahuja, L. R., Ma, L., Nielsen, D. C., Trout, T. J., Andales, A. A., Chavez, J. L., & Ham, J. (2014). Enhancing the water stress factors for simulation of corn in RZWQM2. *Agronomy Journal*, 106, 81–94. https://doi.org/10.2134/agronj2013.0300

Saseendran, S. A., Ahuja, L. R., Ma, L., Timlin, D., Stöckle, C. O., Boote, K. J., & Hoogenboom, G. (2008). Current water deficit stress simulations in selected agricultural system models. In L. R. Ahuja, V. R. Reddy, S. A. Saseendran, & Q. Yu (Eds.), *Response of crops to limited water: Understanding and modeling water stress effects on plant growth processes*, Advances in Agricultural Systems Modeling 1 (pp. 1–38). ASA, CSSA, & SSSA. https://doi.org/10.2134/advagricsystmodel1.c1

Saseendran, S. A., Nielsen, D. C., Lyon, D. J., Ma, L., Felter, D. G., Baltensperger, D. D., Hoogenboom, G., & Ahuja, L. R. (2009). Modeling responses of dryland spring triticale, proso millet and foxtail millet to initial soil water in the High Plains. *Field Crops Research*, 113, 48–63.

Saseendran, S. A., Nielsen, D. C., Ma, L., Ahuja, L. R., & Halvorson, A. D. (2004). Modeling nitrogen management effects on winter wheat production using RZWQM and CERES-Wheat. *Agronomy Journal*, 96, 615–630. https://doi.org/10.2134/agronj2004.0615

Saseendran, S. A., Nielsen, D. C., Ma, L., Ahuja, L. R., & Vigil, M. F. (2010). Simulating alternative dryland rotational cropping systems in the central Great Plains with RZWQM2. *Agronomy Journal*, 102, 1521–1534. https://doi.org/10.2134/agronj2010.0141

Saseendran, S. A., Trout, T. J., Ahuja, L. R., Ma, L., McMaster, G. S., Nielsen, D. C., Andales, A. A., Chávez, J. L., & Ham, J. (2015). Quantifying crop water stress factors from soil water measurements in a limited irrigation experiment. *Agricultural Systems*, 137, 191–205.

Scheff, J., & Frierson, D. M. W. (2015). Terrestrial aridity and its response to greenhouse warming across CMIP5 climate models. *Journal of Climate*, 28, 5583–5600.

Sellers, P. J., Randall, D. A., Collatz, G. J., Berry, J. A., Field, C. B., Dazlich, D. A., Zhang, C., Collelo, G. D., & Bounoua, L. (1996a). A revised land surface parameterization (SiB2) for atmospheric GCMs: I. Model formulation. *Journal of Climate*, 9, 676–705.

Sellers, P. J., Tucker, C. J., Collatz, G. J., Los, S. O., Justice, C. O., Dazlich, D. A., & Randall, D. A. (1996b). A revised land surface parameterization (SiB2) for atmospheric GCMs: II. The generation of global fields of terrestrial biophysical parameters from satellite data. *Journal of Climate*, 9, 706–737.

Sellers, W. D. (1964). Potential evapotranspiration in arid regions. *Journal of Applied Meteorology*, 3, 98–104. https://doi.org/10.1175/1520-0450(1964)003<0098:PEIAR>2.0.CO;2

Seneviratne, S. I., Corti, T., Davin, E. L., Hirschi, M., Jaeger, E. B., Lehner, I., Orlowsky, B., & Teuling, A. J. (2010). Investigating soil moisture–climate interactions in a changing climate: A review. *Earth-Science Reviews*, 99, 125–161.

Sharma, K., Irmak, S., & Kukal, M. S. (2021). Propagation of soil moisture sensing uncertainty into estimation of total soil water, evapotranspiration and irrigation decision-making. *Agricultural Water Management*, 243, 106454. https://doi.org/10.1016/j.agwat.2020.106454

Sherwood, S., & Fu, Q. (2014). A drier future? *Science*, 343, 737–739.

Shi, T., Guan, D., Wang, A., Wu, J., Jin, C., & Han, S. (2008). Comparison of three models to estimate evapotranspiration for a temperate mixed forest. *Hydrological Processes*, 22, 3431–3443.

Shukla, J., & Mintz, Y. (1982). Influence of land-surface evapotranspiration on the Earth's climate. *Science*, 215, 1498–1501.

Shuttleworth, W. J. (1993). Evaporation. In D. R. Maidment (Ed.), *Handbook of hydrology (Chapter 4)*. McGraw-Hill.

Shuttleworth, W. J. (2006). Towards one-step estimation of crop water requirements. *Transactions of the ASABE, 49*, 925–935.

Shuttleworth, W. J. (2007). Putting the "vap" into evaporation. *Hydrology and Earth System Sciences, 11*, 210–244. https://doi.org/10.5194/hess-11-210-2007

Shuttleworth, W. J. (2014). Comment on "Technical note: On the Matt–Shuttleworth approach to estimate crop water requirements" by Lhomme et al. (2014). *Hydrology and Earth System Sciences, 18*, 4403–4406.

Shuttleworth, W. J., & Wallace, J. S. (1985). Evaporation from sparse crops: An energy combination theory. *Quarterly Journal of the Royal Meteorological Society, 111*, 839–855.

Shuttleworth, W. J., & Wallace, J. S. (2009). Calculating the water requirements of irrigated crops in Australia using the Matt–Shuttleworth approach. *Transactions of the ASABE, 52*, 1895–1906.

Sinclair, T. R. (2006). A reminder of the limitations in using Beer's law to estimate daily radiation interception by vegetation. *Crop Science, 46*, 2343–2347. https://doi.org/10.2135/cropsci2006.01.0044

Sinclair, T. R., & Ghanem, M. E. (2020). Plant-based predictions of canopy transpiration instead of meteorological approximations. *Crop Science, 60*, 1133–1141. https://doi.org/10.1002/csc2.20067

Singer, J. W., Meek, D. W., Sauer, T. J., Prueger, J. H., & Hatfield, J. L. (2011). Variability of light interception and radiation use efficiency in maize and soybean. *Field Crops Research, 121*, 147–152. https://doi.org/10.1016/j.fcr.2010.12.007

Slatyer, R. O., & McIlroy, I. C. (1961). Evaporation and the principle of its measurement. In *Practical meteorology*. CSIRO.

Smith, W., Qi, Z., Grant, B., VanderZaag, A., & Desjardins, R. (2019). Comparing hydrological frameworks for simulating crop biomass, water and nitrogen dynamics in a tile drained soybean–corn system: Cascade vs computational approach. *Journal of Hydrology X, 2*, 100015.

Snyder, R. L., Lanini, B. J., Shaw, D. A., & Pruitt, W. O. (1989). *Using reference evapotranspiration (ET₀) and crop coefficients to estimate crop evapotranspiration (ETc) for trees and vines (Leaflet 21428)*. University of California Cooperative Extension.

Spinelli, G. M., Shackel, K. A., & Gilbert, M. E. (2017). A model exploring whether the coupled effects of plant water supply and demand affect the interpretation of water potentials and irrigation management. *Agricultural Water Management, 192*, 271–280.

Steduto, P., Todorovic, M., Caliandro, A., & Rubino, P. (2003). Daily reference evapotranspiration estimates by the Penman–Monteith equation in southern Italy: Constant vs. variable canopy resistance. *Theoretical and Applied Climatology, 74*, 217–225.

Sun, S. K., Li, C., Wang, Y. B., Zhao, X. N., & Wu, P. T. (2020). Evaluation of the mechanisms and performances of major satellite-based evapotranspiration models in Northwest China. *Agricultural and Forest Meteorology, 291*, 108056.

Swann, A. L. S., Hoffman, F. M., Koven, C. D., & Randerson, J. T. (2016). Plant responses to increasing CO_2 reduce estimates of climate impacts on drought severity. *Proceedings of the National Academy of Sciences, 113*, 10019–10024.

Szeicz, G., Van Bavel, C. H. M., & Takami, S. (1973). Stomatal factor in the water use and dry matter production by sorghum. *Agricultural Meteorology, 12*, 361–389.

Tang, D., Feng, Y., Gong, D., Hao, W., & Cui, N. (2018). Evaluation of artificial intelligence models for actual crop evapotranspiration modeling in mulched and non-mulched maize croplands. *Computers and Electronics in Agriculture, 152*, 375–384.

Tanner, C. B. (1981). Transpiration efficiency of potato. *Agronomy Journal, 73*, 59–64. https://doi.org/10.2134/agronj1981.00021962007300010014x

Tanner, C. B., & Pelton, W. L. (1960). Potential evapotranspiration estimates by the approximate energy balance method of Penman. *Journal of Geophysical Research, 65*, 3391–3413.

Tanner, C. B., & Sinclair, T. R. (1983). Efficient water use in crop production: Research or re-search? In H. M. Taylor, W. R. Jordan, & T. R. Sinclair (Eds.), *Limitations to efficient water use in crop production* (pp. 1–27). ASA, CSSA, and SSSA. https://doi.org/10.2134/1983.limitationstoefficientwateruse.c1

Tegos, A., Malamos, N., & Koutsoyiannis, D. (2015). A parsimonious regional parametric evapotranspiration model based on a simplification of the Penman–Monteith formula. *Journal of Hydrology, 524*, 708–717.

Thom, A. S. (1972). Momentum, mass and heat exchange of vegetation. *Quarterly Journal of the Royal Meteorological Society, 98*, 124–134.

Thom, A. S. (1975). Momentum, mass, and heat exchange of plant communities. *Vegetation and the Atmosphere, 1*, 57–109.

Thomson, A. M., Calvin, K. V., Smith, S. J., Kyle, G. P., Volke, A., Patel, P., Delgado-Arias, S., Bond-Lamberty, B., Wise, M. A., Clarke, L. E., & Clarke, L. E. (2011). RCP4.5: A pathway for stabilization of radiative forcing by 2100. *Climatic Change*, 109, 77.

Thornthwaite, C. W. (1948). An approach toward a rational classification of climate. *Geographical Review*, 38, 55–94.

Todorovic, M. (1999). Single-layer evapotranspiration model with variable canopy resistance. *Journal of Irrigation and Drainage Engineering*, 125, 235–245.

Trenberth, K. E., Dai, A., van der Schrier, G., Jones, P. D., Barichivich, J., Briffa, K. R., & Sheffield, J. (2014). Global warming and changes in drought. *Nature Climate Change*, 4, 17–22.

Trenberth, K. E., Fasullo, J. T., & Kiehl, J. (2009). Earth's global energy budget. *Bulletin of the American Meteorological Society*, 90, 311–324.

Twine, T. E., Kustas, W. P., Norman, J. M., Cook, D. R., Houser, P., Meyers, T. P., Pruegerg, J. H., Starks, P. J., & Wesely, M. L. (2000). Correcting eddy-covariance flux underestimates over a grassland. *Agricultural and Forest Meteorology*, 103, 279–300.

Valiantzas, J. D. (2006). Simplified versions for the Penman evaporation equation using routine weather data. *Journal of Hydrology*, 331, 690–702.

Valiantzas, J. D. (2013). Simplified forms for the standardized FAO-56 Penman–Monteith reference evapotranspiration using limited weather data. *Journal of Hydrology*, 505, 13–23.

Valiantzas, J. D. (2015). Simplified limited data Penman's ET_0 formulas adapted for humid locations. *Journal of Hydrology*, 524, 701–707.

Van Bavel, C. H. M. (1966). Potential evaporation: The combination concept and its experimental verification. *Water Resources Research*, 2, 455–467.

Van Roekel, R. J., & Purcell, L. C. (2014). Soybean biomass and nitrogen accumulation rates and radiation use efficiency in a maximum yield environment. *Crop Science*, 54, 1189–1196. https://doi.org/10.2135/cropsci2013.08.0546

Vardavas, I. M., & Fountoulakis, A. (1996). Estimation of lake evaporation from standard meteorological measurements: Application to four Australian lakes in different climatic regions. *Ecological Modelling*, 84, 139–150.

Vicente-Serrano, S. M., Beguería, S., & López-Moreno, J. I. (2010). A multiscalar drought index sensitive to global warming: The standardized precipitation evapotranspiration index. *Journal of Climate*, 23, 1696–1718.

Wallace, J. S. (1995). Calculating evaporation: Resistance to factors. *Agricultural and Forest Meteorology*, 73, 353–366.

Wang, S., Yang, Y., Trishchenko, A. P., Barr, A. G., Black, T. A., & McCaughey, H. (2009). Modeling the response of canopy stomatal conductance to humidity. *Journal of Hydrometeorology*, 10, 521–532.

Wei, G., Zhang, X., Ye, M., Yue, N., & Kan, F. (2019). Bayesian performance evaluation of evapotranspiration models based on eddy covariance systems in an arid region. *Hydrology and Earth System Sciences*, 23, 2877–2895. https://doi.org/10.5194/hess-23-2877-2019

Wessel, D. A., & Rouse, W. R. (1994). Modelling evaporation from wetland tundra. *Boundary-Layer Meteorology*, 68, 109–130.

Wherley, B., Dukes, M. D., Cathey, S., Miller, G., & Sinclair, T. (2015). Consumptive water use and crop coefficients for warm-season turfgrass species in the southeastern United States. *Agricultural Water Management*, 156, 10–18.

World Meteorological Organization. (1963). *Sites for wind-power installations* (WMO 156, Technical Note 63). WMO.

Wright, J. L. (1981). Crop coefficients for estimates of daily crop evapotranspiration [southern Idaho] (ASAE Publ. 23-81). American Society of Agricultural Engineers.

Wright, J. L. (1982). New evapotranspiration crop coefficients. *Proceedings of the American Society of Civil Engineers, Journal of the Irrigation and Drainage Division*, 108(IR2), 57–74.

Wright, J. L., & Jensen, M. E. (1972). Peak water requirements of crops in southern Idaho. *Proceedings of the American Society of Civil Engineers, Journal of the Irrigation and Drainage Division*, 98(IR2), 193–201.

Wu, G., Cai, X., Keenan, T. F., Li, S., Luo, X., Fisher, J. B., Cao, R., Lic, F., Purdy, A. J., Zhao, W., Sun, X., & Zhao, W. (2020). Evaluating three evapotranspiration estimates from model of different complexity over China using the ILAMB benchmarking system. *Journal of Hydrology*, 590, 125553.

Xiang, K., Li, Y., Horton, R., & Feng, H. (2020). Similarity and difference of potential evapotranspiration and reference crop evapotranspiration: A review. *Agricultural Water Management*, 232, 106043.

Xiong, Y. J., Zhao, S. H., Tian, F., & Qiu, G. Y. (2015). An evapotranspiration product for arid regions based on the three-temperature model and thermal remote sensing. *Journal of Hydrology*, 530, 392–404.

Xu, C.-Y., & Singh, V. P. (2002). Cross comparison of empirical equations for calculating potential evapotranspiration with data from Switzerland. *Water Resources Management*, 16, 197–219.

Yang, F., Zhang, Q., Wang, R., & Zhou, J. (2014). Evapotranspiration measurement and crop coefficient estimation over a spring wheat farmland ecosystem in the Loess Plateau. *PLOS ONE*, 9(6), e100031.

Yang, W., Feng, G., Adeli, A., Kersebaum, K. C., Jenkins, J. N., & Li, P. (2019a). Long-term effect of cover crop on rainwater balance components and use efficiency in the no-tilled and rainfed corn and soybean rotation system. *Agricultural Water Management*, 219, 27–39.

Yang, Y., Roderick, M. L., Zhang, S., McVicar, T. R., & Donohue, R. J. (2019b). Hydrologic implications of vegetation response to elevated CO_2 in climate projections. *Nature Climate Change*, 9, 44–48.

Yang, Y., Zhang, S., Roderick, M. L., McVicar, T. R., Yang, D., Liu, W., & Li, X. (2020). Comparing Palmer Drought Severity Index drought assessments using the traditional offline approach with direct climate model outputs. *Hydrology and Earth System Sciences*, 24, 2921–2930. https://doi.org/10.5194/hess-24-2921-2020

Zhang, B., Kang, S., Li, F., & Zhang, L. (2008). Comparison of three evapotranspiration models to Bowen ratio-energy balance method for a vineyard in an arid desert region of Northwest China. *Agricultural and Forest Meteorology*, 148, 1629–1640.

Zhang, B., Liu, Y., Xu, D., Cai, J., & Li, F. (2011). Evapotranspiration estimation based on scaling up from leaf stomatal conductance to canopy conductance. *Agricultural and Forest Meteorology*, 151, 1086–1095.

Zhang, H., Malone, R. W., Ma, L., Ahuja, L. R., Anapalli, S. S., Marek, G. W., Gowda, P. H., Evett, S. R., & Howell, T. A. (2018). Modeling evapotranspiration and crop growth of irrigated and non-irrigated corn in the Texas High Plains using RZWQM. *Transactions of the ASABE*, 61, 1653–1666.

Zhao, P., Li, S., Li, F., Du, T., Tong, L., & Kang, S. (2015). Comparison of dual crop coefficient method and Shuttleworth–Wallace model in evapotranspiration partitioning in a vineyard of Northwest China. *Agricultural Water Management*, 160, 41–56.

Zhao, W. L., Qiu, G. Y., Xiong, Y. J., Paw, K. T., Gentine, P., & Chen, B. Y. (2020). Uncertainties caused by resistances in evapotranspiration estimation using high-density eddy covariance measurements. *Journal of Hydrometeorology*, 21, 1349–1365.

Zhou, M. C., Ishidaira, H., Hapuarachchi, H. P., Magome, J., Kiem, A. S., & Takeuchi, K. (2006). Estimating potential evapotranspiration using Shuttleworth–Wallace model and NOAA-AVHRR NDVI data to feed a distributed hydrological model over the Mekong River basin. *Journal of Hydrology*, 327, 151–173.

Zhou, M. C., Ishidaira, H., & Takeuchi, K. (2007). Estimation of potential evapotranspiration over the Yellow River basin: Reference crop evaporation or Shuttleworth–Wallace? *Hydrological Processes*, 21, 1860–1874.

Zhou, S., Medlyn, B., Sabaté, S., Sperlich, D., & Prentice, I. C. (2014). Short-term water stress impacts on stomatal, mesophyll and biochemical limitations to photosynthesis differ consistently among tree species from contrasting climates. *Tree Physiology*, 34, 1035–1046.

Zhu, G. F., Li, X., Su, Y. H., Zhang, K., Bai, Y., Ma, J. Z., Li, C. B., Hu, X. L., & He, J. H. (2014). Simultaneously assimilating multivariate data sets into the two-source evapotranspiration model by Bayesian approach: Application to spring maize in an arid region of northwestern China. *Geoscientific Model Development*, 7, 1467–1482.

4

Andree J. Tuzet,
Dennis J. Timlin,
and Alain Perrier

Advances and Improvements in Modeling Water Fluxes in Vegetation Canopy and their Relation to Photosynthesis

Abstract

Water fluxes from vegetation are a key component of the global water cycle and a link among the terrestrial cycles of water, carbon, and energy. Accurate estimates of these fluxes are important for hydrological, meteorological, and agricultural research and applications, such as quantifying surface energy and water balances, forecasting weather, scheduling irrigation, and regulating agricultural production. In addition, quantification of these fluxes is fundamental to elucidating how hydrological cycles respond to climate change. However, the study of water fluxes from vegetation is challenging because of its complex interactions across the soil–vegetation–atmosphere interface. Modeling, as well as scientific observation and experimentation, is a useful approach to fully integrate these interactions and increase our understanding of the relationships. It requires prioritization of the processes involved for creation of a simplified reality.

This chapter gives a general outline of the problem and a historical overview of water fluxes in vegetation from ancient times through the end of the 20th century. This historical account is far from exhaustive; rather, it attempts to show, as succinctly as possible, the currents of thought and the essential references and observations that have made it possible to specify the representation of the water cycle that we have adopted today. A description of the main components of

Modeling Processes and Their Interactions in Cropping Systems: Challenges for the 21st Century, First Edition.
Edited by Lajpat R. Ahuja, Kurt C. Kersebaum, and Ole Wendroth.

soil–plant–atmosphere continuum models is then given. This chapter outlines the conceptualization and the mathematical formulation of energy balance. The energy balance, which is the basis of all surface exchanges, only defines these exchanges in terms of surface water availability. Knowledge of this surface water availability necessarily requires good management of the soil water balance and therefore an analysis of the coupling between the energy balance and the water balance. To refine this approach, an overview of the models currently available to describe the dynamics of water flow in plants, water uptake by roots, stomatal conductance, CO_2 assimilation, and transpiration is presented. Finally, the authors reflect on the response of plants to water stress and on the modification of agricultural practices that should allow better adaptation to climate change.

Introduction

Water is the source of all life. It is a key compound for sustaining life as it is a foundational building block that supports biochemistry on the planet. It is therefore not surprising that water remains the primary constituent (60–95%) of all living things. Furthermore, during the Devonian period, the origin of plants on Earth was almost certainly triggered by the drop in atmospheric CO_2 concentration and the presence of water, increasing the efficiency with which CO_2 could be captured for photosynthesis. The spread of herbaceous plants permitted a rich development of terrestrial and complex vegetative cover with a progressive adaptation to their environmental conditions. All mineral or vegetative surfaces were subjected to rainfall that induced strong chemical and physical weathering of soils, even in deeper sediments. This resulted in changes to soil depth as weathered or deposited sediment that influenced hydrological water fluxes. Since then, vegetation has played an important and primary role in soil development and protection. Vegetation can protect soils against erosive factors. It can act as a hydrological catchment regulator by intercepting raindrops and increasing water infiltration in soils. It can provide mechanical protection by reducing the energy of raindrops and "splash" effects. It can also play a role in thermal regulation by reducing daily temperature variations. In addition, vegetation allows fixation of soil by the root system. Furthermore, the ability of the plant to explore the soil with roots is very efficient, especially through the growth of the diameter and strength of the young apical root, as roots may penetrate even very hard soil and the root appears to be a very active tool for deepening the soil.

A common constraint to plant growth in agricultural ecosystems is lack of water. Water is necessary for life at all levels of biological complexity. It acts as both a solvent for life sustaining chemicals and as a structural component to provide form to a living organism. In plants, water is also important to dissolve CO_2 in the mesophyll tissue, transport nutrients from the soil, and cool aboveground leaf tissue. Water also provides turgor pressure that is necessary for leaf and root growth. As a result, lack of water can impact carbon assimilation, affect cell expansion, and result in reduced plant growth as well as impaired nutrient relations.

Because terrestrial plants are largely immobile, they have developed a complex hydraulic system to acquire and transport water to carry out their metabolic functions. This transport system is such an important topic that there are a large number of reviews (Flexas et al., 2006; Hsiao, 1973; Lawlor, 2002) as well as a number of books (Ehlers & Goss, 2004; Wilkinson, 2000) devoted to this subject. The transport system in plants is composed of two major structures called the xylem and phloem. These structures resemble a system of bundles of open channels that facilitate the transport of water and solutes throughout the plant. They also assist in supporting the plant. The xylem bundles primarily carry water and solutes from the roots to the leaves while the phloem bundles transport the products of photosynthesis such as sugars from the leaves and distribute them throughout the plant (Berleth, 2000). The phloem is also involved in transporting phytohormones (Lucas et al., 2013). Osmotic and pressure potentials drive the flow against resistances in the system.

The transport system in plants is a hydraulic network, and thus the system can be mathematically described from an engineering perspective like a physical system of pipes and valves. It is believed that this network develops as a self-organizing structure under the influence of auxin fluxes in the developing plant tissues (Feugier et al., 2005; Fukuda, 2004). There has been significant research into understanding how vascular networks in plants form and operate (Berleth, 2000). These advances have led to the development of fairly sophisticated hydrodynamic models of solute and water flow in plant hydraulic systems (Fukuda, 2004; Pokhilko & Ebenhöh, 2015; Tyree & Ewers, 1991; West et al., 1999; Zhou et al., 2020). The conducting vessels of plants have evolved with specific designs to prevent cavitation in the event of water deficit, prevent water from draining back into the soil under gravitational potentials, and allow water to be uniformly transported to all the branches and leaves of a plant or tree (Tyree & Ewers, 1991). In fact, the allometric scaling laws of the hierarchal branching network of the hydraulic structure of plants are very similar to those of animals (West et al., 1999).

Almost all of the crop models that are currently used to assess growth, development, and yield apply a simple hydrologic model based on Ohm's law to simulate water fluxes in the plant–soil system. These models treat the plant as a simple tube wherein the movement of water and solutes is gradient driven through a system of resistances beginning at the roots and ending up at the leaf surface; stems and branches (if present) are treated only as intermediate pathways. Transport of water and nutrients in the plant vascular system is critical to maintaining photosynthetic processes. Thus, the model focus is the simulation of water uptake and photosynthesis. Biophysical processes involved in photosynthesis and growth within these models may be abstracted to three major processes: diffusion and assimilation of CO_2 into the leaves, transpiration of water and nutrient transport, and conversion of CO_2 into plant matter. The interdependencies of these three processes can be quite complex as they involve both feed-forward and feed-backward relationships.

The scale and scope of crop models ranges from those that contain one or two equations to large system level models that incorporate multiple components. The level of detail describing the soil and plant photosynthesis does vary among the different models. In most crop models, the movement of solutes from the leaves is not explicitly addressed. Models such as the Decision Support System for Agrotechnology Transfer (DSSAT) suite (Jones et al., 2003), the EPIC growth model (Williams et al., 1989), the Agricultural Production Systems Simulator (APSIM) (McCown et al., 1996), and the RZWQM model (Hanson et al., 1998) utilize a Radiation Use Efficiency (RUE) model to simulate daily carbon assimilation as a function of daily light interception. Transpiration is simulated based on canopy level evaporation equations such as the Penman–Montieth (Allen et al., 1998) and Hargreaves equations (Hargreaves & Samani, 1985).

Other models such as SPUDSIM (Fleisher et al., 2007) and MAIZSIM (Kim et al., 2012) use a leaf energy balance to simulate evaporation from the leaf surface and a biochemical model to simulate carbon assimilation (Farquhar et al., 2001; von Caemmerer, 2000).

Dynamic carbon processes were introduced into land surface models (LSMs) in the early 2000s as modeling centers started to focus on the response of climate to the carbon cycle. Understanding the link between photosynthesis and transpiration (i.e., water use efficiency) is a priority for LSM response to climate (Blyth et al., 2021; Delire et al., 2020). A new approach based on economic stomatal optimization theory (Medlyn et al., 2011), where stomata maximize carbon gain while minimizing water loss, has recently been introduced in climate models (De Kauwe et al., 2015; Franks et al., 2017; Lawrence et al., 2019). Despite these advances, a model that accounts for the stomatal conductance response to root zone soil moisture by explicitly modeling root water uptake and water transfer in the plant, in response to climatic demand, has not yet been fully explored in LSMs.

Historical Overview of Water Fluxes in Vegetation Canopy

Water has always been a vital compound, sought after by animals and humans since the dawn of time. The last glaciation, about 18,000 years BC, was followed by a warming with melting of the ice, and this was favorable to the development of populations from the end of the Paleolithic period (15–12 thousand years BC) despite strong secular variations. These changes facilitated the development of cities and introduced scientific developments (writings), technical developments (the wheel), and societal developments (hierarchy, religions, army, crafts, agriculture).

Thus, the first knowledge about water (digging wells) appeared, as did an organization of life often related to rivers, lakes, and marshes. Two conquests emerged: (a) that of gravity, so useful for transporting water on a slight slope; (b) and that of energy, the driving force of water to make the wheels turn. This mastery of water input enabled the formation of urban centers, which were very quickly equipped with sewage disposal systems and the development of techniques to improve agriculture production or human services. All this has favored the transformation of agriculture by improving tillage and integrating irrigation and fertilization (mainly manure).

The first writings on the link between the environment and water supply are found in Hesiod (800 BC) and the first study of the water cycle was presented by Hippocrates (500 BC) and explained shortly afterwards by Aristotle (400 BC), who highlighted the various physical processes of the water cycle.

The Roman world, which took off around 150 BC, was very technical; Seneca (20 BC) and Vitruvius (100 BC) developed the treatises on collection of water for human consumption and agriculture. Transport and regulation of water for towns, agriculture, and industries (mills), used a variety of techniques, aqueducts, and siphons. Pliny (1938) in his "Natural History (37 volumes)" proposed the technique of drainage, useful to evacuate stagnant water and accumulated salt.

The Middle Ages (500–1500 CE) began with the decline of the Roman Empire, which led to a rapid deterioration of knowledge and arguably even a gradual loss of much common knowledge. This period corresponds to the small climatic warming (800–1200 years ago) which allowed a slow but real recovery of agricultural wealth. Overall, agronomic practices remained traditional, but the use of fallow land became more widespread, as did the use of manure to fertilize the soil. More legumes were being grown to help fertilize the soil.

The European Enlightenment (1600–1900 CE) represents the period of renewal in all fields; it was the prelude to the pre-industrial age. The water cycle described by Aristotle with all its physical processes and climatic driving factors still remained without any known quantification or estimation of evaporation from wetlands. It was E. Mariotte who, through his work on gases (Boyle–Mariotte law-1692), became interested in the circulation of sap and water in the plant and its gas exchange with the air. At the same time, Sedileau (1730) set up an experiment to measure a reduced water balance of a plant cover (rain and evaporation). The results showing that rainfall was lower than the evaporation of water (520 mm of rain and 880 mm of evaporation per year) generated new research to analyze the laws of water transfer in plants, leaf evaporation and plant growth through physiological studies. Duhamel De Monceau, quoted by Dupont de Dinechin (1999), described in 1758 the transfer of water from the soil to the roots and then to all the foliage elements through the wood vessels and the return of enriched sap through the phloem. In regard to photosynthesis, which was still a mystery, it was Priestley who, in 1772, showed in an experiment on plants under a bell that there was a gas exchange of oxygen and CO_2 in addition to water vapor.

In the natural environment, the underground transfer of water was still poorly known, although it was essential to understand the fate of rainwater, storage volumes, and their time constant. Also unclear was directionality and temporal variability of the underground transfer. Darcy's law (1856) was very useful for all major public works and brought comprehensive solutions for underground flow analyses. At the beginning of the 20th century, micrometeorology appeared and developed (Bouchet, 1963; Bowen, 1936). Bowen's ratio, defined as the ratio of latent to sensible heat flux from soil or plant surfaces, combined with their energy balance, provided a measure of evaporation. New physical approaches emerged, such as Richards' equation in 1931, which allows us to specify water transfers in soils, and to establish a complete water balance (Thornthwaite & Mather, 1955, 1957) in the natural soil explored by the root system of any plant cover.

Cummings (1929) proposed an approximate energy balance equation which Penman (1948) combined with a mass-transfer equation based on Dalton's work to develop the so-called "Penman equation." This equation gives the evaporation from a water saturated surface from standard climatological records of sunshine, temperature, humidity, and wind speed; it is a potential evaporation. This combination method was further developed by many researchers and extended to cropped surfaces by introducing the effects of plant canopies through resistance factors. Monteith (1965) modified Penman's equation for a single leaf to deal with a canopy which led to the Penman–Monteith model.

For many decades, transpiration models based on the energy balance have been mostly derived from the Penman–Monteith equation. Priestley and Taylor (1972) have shown that transpiration is a rather conservative variable, which can be determined primarily by the available energy. Combined with temperature and vapor pressure deficit (VPD), they obtained good results for well-watered vegetation. Many scientists then developed physical sciences around this microclimatology of the biosphere. Geiger was among the first to analyze the climate near the soil and more specifically evaporation in the atmosphere (Brutsaert, 1982).

However, the use of simplistic approaches often fails to account for several interactions such as between the energy balance and soil water availability. More sophisticated approaches have been proposed to account for these interactive effects and to view the water transfer system in the three domains of soil, plant, and atmosphere as a whole. This need for a soil–plant–atmosphere continuum (SPAC) approach was first recognized by Philip (1966), Cowan (1977), and Perrier (1982), and by many other authors later. In these models, root water uptake is determined by potential transpiration calculated from atmospheric conditions and a reducing function, which depends on the mean soil water content and hydraulic conductivity of soil and roots.

Main Components of Soil–Plant–Atmosphere Continuum Models

Coupling of Energy Balance and Water Balance
Climatic demand, potential evaporation, and evapotranspiration

For a thin interface where the energy capacities per unit area and time and the water storage capacities (vapor or liquid) are negligible compared to the flows of radiative and convective exchanges, the energy balance is written as:

$$R_n = H + LE \qquad (4.1)$$

where R_n is the net radiation (W m^{-2}); and H and LE are the sensible and latent heat fluxes to and from the air, respectively. The net radiation (Figure 4.1) corresponding to the energy input from radiative sources is expressed as follows:

$$R_n = (1-a)R_g + \varepsilon\left(R_a - \sigma T_s^4\right) \qquad (4.2)$$

where R_g, global radiation, and R_a, longwave atmospheric radiation, are the two terms of incident radiation (W m^{-2}); a and ε define the radiative properties of the surface, the albedo and emissivity; T_S is the radiative surface temperature (°C); and σ is

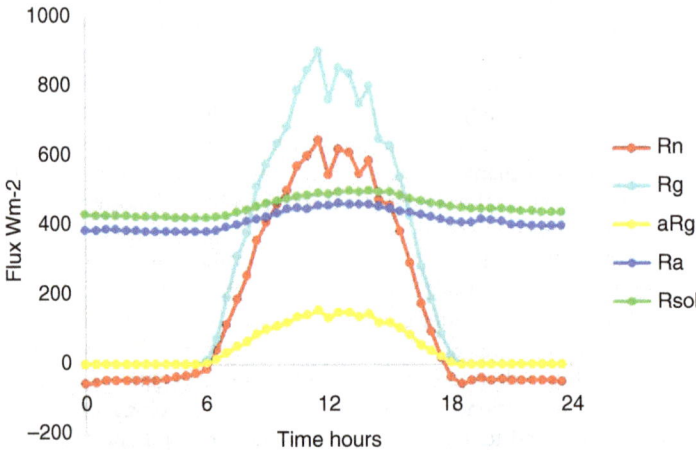

Figure 4.1 Diurnal pattern of the surface radiation balance components (see Equation 4.2). Global solar radiation R$_g$, net radiation R$_n$, reflected solar radiation aR$_g$, longwave atmospheric radiation R$_a$ and surface emitted longwave radiation, Rsol = $\varepsilon\sigma T_s^4$. From Tuzet et al. (1995).

the Stefan–Boltzmann constant ($\sigma = 5.67 \ 10^{-8}$ W m^{-2} K^{-4}). Then, εR_a is the absorbed downwelling longwave radiation, and $\varepsilon \sigma T_s^4$ is the emitted longwave radiation.

Net radiation R_n may be written as the sum of the climatic net radiation and a correction term:

$$R_n = R_n^* + f(T_a, T_S) \tag{4.3}$$

with

$$R_n^* = (1 - a)R_G + \varepsilon \left(R_a - \sigma T_a^4\right) \tag{4.4}$$

and

$$f(T_a, T_S) = h_r(T_a - T_S) \text{ where } h_r = 4\varepsilon\sigma T_a^3 \tag{4.5}$$

h_r is the radiative exchange coefficient (W m^{-2} K^{-1}); it is of the same dimension as the convective exchange coefficient h_u used to express the sensible heat flow (H) and the latent heat flow (LE):

$$H = \rho c_p h_u (T_S - T_a) \tag{4.6}$$

$$LE = \frac{LM}{RT} h_u (P(T_{dS}) - P(T_d)) \tag{4.7}$$

where ρ is the volumetric mass of air; c_p is the heat capacity of air; T_a is the air temperature at the reference height z_R; L is the latent heat of water; M is the molar weight of water; R is the constant of perfect gas; $P(T_{ds})$ is the saturation water vapor pressure at dew point surface temperature T_{ds}; and $P(T_d)$ is the saturation water vapor pressure at dew point temperature T_d at the reference height z_R.

Then, the energy balance (Perrier, 1975a, 1975b, 1975c) can be written as:

$$R_n^* = h_H(T_S - T_a) + h_{LE}(T_{dS} - T_d) \text{ with } h_H = \rho c_p h_u + h_r \text{ and } h_{LE} = \frac{LM}{RT} h_u P' \tag{4.8}$$

where P' is the slope of the saturation water vapor pressure. Another expression of evapotranspiration LE is given by:

$$LE = \frac{h_{LE}}{h_H + h_{LE}} R_{n*} + \frac{h_H}{h_H + h_{LE}} (E_a - E_S) \tag{4.9}$$

The derivation of Equation 4.9 is based on combination of Equations 4.1 and 4.8 and on the definition of E_a, the evaporative power of air and E_S the water retention capacity of the interface:

$$E_a = h_{LE}(T_a - T_d) \text{ and } E_S = h_{LE}(T_S - T_{dS}) \tag{4.10}$$

E_a is always positive and represents the maximum amount of water vapor of the air that can be carried away at any time. E_S is zero as long as free water remains at the surface (thermodynamic water potential equal zero) but quickly decreases with evaporation, this brake causing the vapor pressure at the surface to drop from $P(T_s)$ toward $P(T_d)$.

The terms of Equation 4.9 suggest an interpretation that may serve as an aid in understanding the interaction between the surface and the atmosphere (Perrier & Tuzet, 1991). Depending on the value of forcing variables, five different cases may occur.

1. $E_S = 0$ defines the potential evaporation (EP). It is the evaporation from any large uniform surface where the air in contact with it, is saturated with moisture. Such conditions prevail usually after the occurrence of precipitation or dew; in these prevailing climatic conditions, surface temperature takes its minimum value (Figure 4.2).

$$EP = \frac{h_{LE}}{h_H + h_{LE}} (R_{n*} + h_H(T_a - T_d))$$ (4.11)

$$T_{s0} = T_{ds0} = \frac{R_{n*} + h_H T_a + h_{LE} T_d}{h_H + h_{LE}}$$ (4.12)

2. $E_S = E_a$ there is an equilibrium between air water deficit at reference level and at surface level. This value of evapotranspiration, which is called climatic

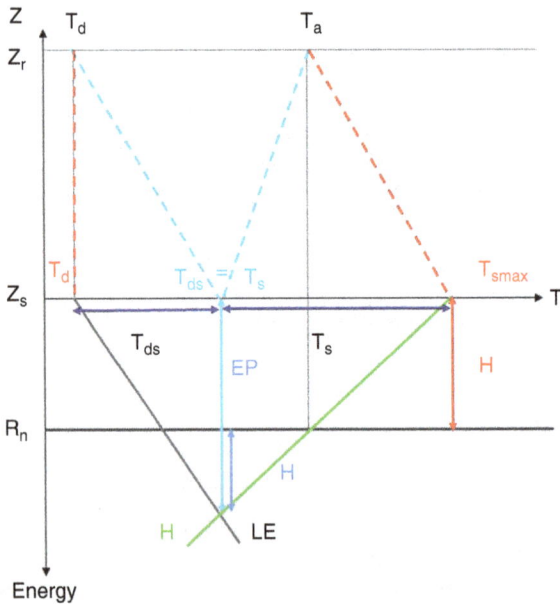

Figure 4.2 The upper graph shows the surface layer (between the heights z_s and z_r) and two schematic profiles. The blue dotted lines correspond to the surface saturation ($T_s = T_{ds}$) and the red ones correspond to a completely dry surface with a maximum temperature T_{smax}. The lower graph represents energy balance partition according to the value of T_s and T_{ds}. The horizontal black line corresponds to net radiation R_n. The gray line LE represents the relationship between evapotranspiration and surface dewpoint temperature (surface saturation LE = EP; dry surface LE = 0). The green line H represents the relationship between sensible heat flux and surface temperature with the origin taken at the point defined by R_n and T_a (H is positive for $T_s > T_a$ and negative for $T_s < T_a$). Thus, on this figure, we can see for given climatic conditions the ranges of T_{ds} and T_s variations between a completely dry surface and a saturated surface. From Perrier and Tuzet (1991).

evapotranspiration, is also an equilibrium value toward which the regional evapotranspiration must naturally tend (Wilson & Rouse, 1972).

$$LE = \frac{h_{LE}}{h_H + h_{LE}} R_{n*} \tag{4.13}$$

$$T_s = T_a + \frac{R_{n*}}{h_H + h_{LE}} \tag{4.14}$$

3. $E_{S\ max}$ corresponds to a dry surface without evapotranspiration (LE = 0) where the surface temperature is maximum.

$$T_{smax} = T_a + \frac{R_{n*}}{h_H} \tag{4.15}$$

4. $E_S < E_a$ the air is drier at the reference level than that at the surface level. This case is referred to an "oasis effect."

5. $E_S > E_a$ the air is wetter at the reference level than that at the surface level. This case is referred to an "island effect."

For a surface having a temperature T_S and a water vapor pressure $P(T_{dS})$, the difference between the saturation water vapor pressure at surface temperature $P(T_S)$ and $P(T_{dS})$ is proportional to the evaporation rate, and the constant of proportionality is described as a "surface resistance," r_S, such that:

$$P(T_S) - P(T_{dS}) = \frac{R\overline{T}}{LM} r_S LE \tag{4.16}$$

The most common example of this case is transpiration through the stomatal pore of a leaf. Extension of this analysis to crop canopies has been the subject of many experimental and theoretical works (Allen et al., 2006; Monteith, 1965, 1981; Perrier, 1976). From Equation 4.16, an expression that relates the water retention capacity of the surface E_S to the surface resistance through the evaporation rate can be written as:

$$E_S = h_u r_S LE \tag{4.17}$$

and the substitution of this expression of E_S in Equation 4.9 gives an equation with the same general form as the original Penman–Monteith formula:

$$LE = \frac{EP}{1 + \dfrac{h_H}{h_H + h_{LE}} h_u r_S} \tag{4.18}$$

The two most frequently encountered uses of the Penman–Monteith equation are, on the one hand, the evaporation of bare soil leading to the formation of a dry layer, called a "mulch," which limits evaporation from the soil and, on the other hand, the transpiration of plants whose stomatal pores regulate transpiration.

Bare Soil Evaporation

After a rainy sequence, soil evaporation is equal to potential evaporation. As evaporation proceeds, a dry surface layer (mulch layer) develops because potential evaporation is far greater than most vertical water diffusion flux in the soil (Gardner & Hillel, 1962; Sifaoui & Perrier, 1978). Soil surface temperature increases and evaporation that takes place at the base of the dry layer is reduced with respect to the climatic demand. The extent of this reduction depends on the thickness of the mulch layer expressed as a mulch layer resistance, r_m given as:

$$r_m = \frac{\xi \, d_m}{f \, D_v} \tag{4.19}$$

where d_m is the thickness of the mulch layer; D_v is the water vapor diffusivity in air; f is the dry layer porosity; and ξ is the tortuosity factor. The thickness of the mulch layer changes according to:

$$\Delta d_m = \frac{E\Delta t}{\rho_w(\theta_{FC} - \theta_R)} \tag{4.20}$$

where E is the total soil evaporation; Δt is the model time step; θ_{FC} is the volumetric soil water content at field capacity ($m^3 \, m^{-3}$); θ_R is the residual soil water content; and ρ_w is the volumetric mass density of water ($kg \, m^{-3}$).

The latent heat fluxes from the soil surface LE_s is given as:

$$LE_S = \frac{LM}{R\overline{T}} \frac{1}{r_m}(P(T_m) - P(T_S)) \tag{4.21}$$

where T_m is soil temperature at the wet-dry interface which is assumed to be saturated; and T_S is soil surface temperature.

In agrometeorological models, a cumulative value of soil evaporation, $\sum_0^t(LE_S.\Delta t)$, between two rainy events is often expressed as a function of soil potential evaporation, $\sum_0^t(EP_S.\Delta t)$:

$$\sum_0^t(LE_S.\Delta t) = \sqrt{A^2 + 2A\sum_0^t(EP_S.\Delta t)} - A \tag{4.22}$$

$$with \; A = L\frac{h_H + h_{LE}}{h_H} \rho_w(\theta_{FC} - \theta_R)\frac{D_v}{h_u}\frac{f}{\xi} \tag{4.23}$$

where A is function of soil characteristics and aerodynamic exchanges (Perrier, 1973; Tuzet et al., 1992).

Plant Transpiration

Transpiration occurs along with carbon assimilation and is a critical component of a plant's physiological processes. The flow of water in the plant and internal pressures due to cell wall structures and osmotic gradients are more important for biological processes than the total amount of water in the plant. The maintenance of a functional hydraulic system in the plant growing in a soil media with varying water availability is essential for survival and growth (Maseda & Fernandez, 2006). The evaporation of water from the leaf surface provides temperature regulation. Atmospheric CO_2 must be dissolved in water in the leaf mesophyll tissue for it to be transported to the photosynthetic organelles of the cells. Evaporation from the leaf surface provides a constant flow of water from the roots in the soil and through the plant to the leaf surfaces. Water movement in the plant vascular system drives the mass flow of nutrients from the soil and transports them to various organs in the plant.

Most existing crop or climate models simulate evapotranspiration by using the physically based Penman–Monteith equation which includes aerodynamic and surface resistance (Equation 4.19). The detailed procedure of crop evapotranspiration is thoroughly explained in the FAO publication paper by Allen et al. (1998). The Penman–Monteith equation is a relatively simple approximation to a nonlinear system of equations in which heterogeneous vegetation and ground surface interact by radiative and diffusive transport. This equation is regarded as one of the most accurate equations to estimate evapotranspiration. Surface resistance in the Penman–Monteith equation is a land cover resistance term which is a measure of available water from the soil surface and the plant canopy.

However, energy budget models which only account for physical and environmental controls of stomatal conductance at the leaf level are incomplete, because, while they describe stomatal response to atmospheric demand for water vapor, they do not consider supply of water to the leaves by transport of water within the soil and plant. Many authors introduced an extra function to account for changes in soil water content (see the section on stomatal conductance below). However, their functions are unable to account for more dynamic responses of stomata to the competing demand and supply of water to leaves on a sub-diurnal timescale. To overcome this deficiency, Feddes et al. (1978) and Tuzet et al. (1992) have, for example, proposed a model of plant transpiration, depending on climatic demand and soil water content. These authors assumed a linear reduction of water uptake or of the ratio, LE/EP, with soil water availability (Figure 4.3). Plant transpiration, limited by the physiological resistance of the plant, evolves according to an optimum value and the soil water content.

To summarize, we can argue that the energy balance, which is the basis of all surface exchanges, only defines these exchanges in terms of surface water availability. Knowledge of this surface water availability necessarily requires good management of the soil water balance and therefore an analysis of the energy balance—water balance coupling. We need to develop models which describe the dynamics of stomatal conductance, transpiration, water transport through the plant and soil, that fully couples stomatal conductance, CO_2 assimilation, and the leaf energy balance. It is this coupled

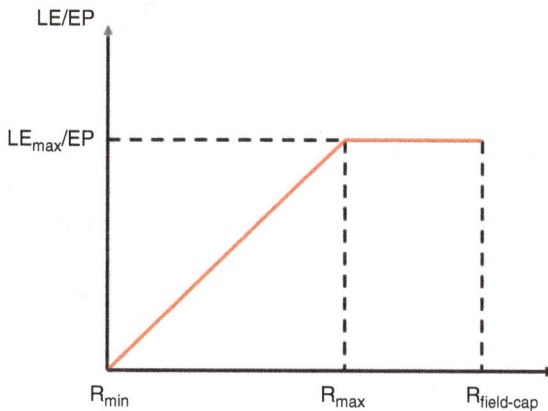

Figure 4.3 General shape of the dimensionless ratio, LE/EP as a function of the soil water supply. $R_{field-cap}$, soil water reserve at field capacity, R_{max}, soil water reserve below which transpiration becomes lower than LE_{max} (due to soil regulation). R_{min}, soil water reserve below which transpiration becomes zero (close to wilting point). LE, evapotranspiration. EP, climatic demand. From Perrier et al. (2003).

management of the energy balance—water balance that enables us to better manage water transfers in the plant and their regulation.

Plant Hydraulic System

The widely accepted model for water movement in the soil–plant–atmosphere system is based on the concept that soil, plant, and atmosphere form a dynamic system like links in a continuum (Tuzet et al., 2017). Water flows through the hydraulic system in response to physical forces. There is a decreasing water potential from the soil to the roots, up through the xylem and leaves, and out into the atmosphere. Therefore, water moves by bulk flow in response to water potential gradients, pulling water from regions of high water potential toward areas of lower water potential. The plant's hydraulic system is the key component in the water transfer from the soil to the atmosphere and transpiration is the dominant factor. Due to climatic demand, evaporation in the stomatal cavities produces a drop in water potential that causes movement of water into and through plants (Figure 4.4). Soil water availability, root water uptake, plant hydraulic resistances and canopy capacitance determine the rate of water supply.

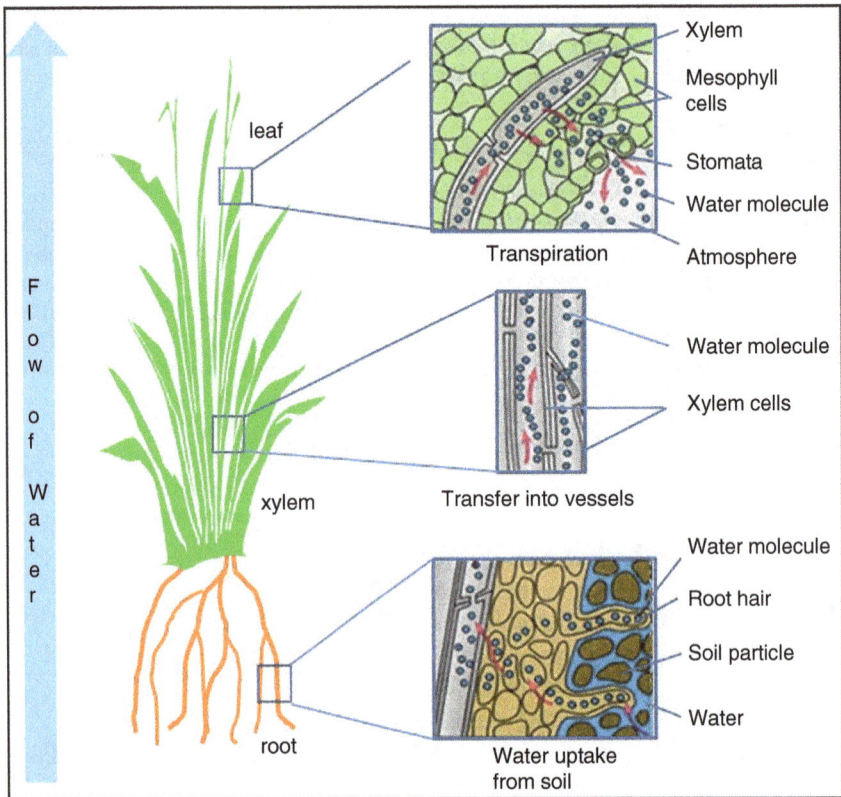

Figure 4.4 Schematic diagram showing path of water from the soil to the atmosphere via the plant. Top panel on right shows a leaf with pathways for water evaporated at walls of mesophyll cells in the substomatal cavity. Middle panel on right shows the transfer of water molecules into the xylem vessels. Lower panel on right shows pathway for transport of water from the bulk soil to the root surfaces and to the root xylem. From personal unpublished work.

Plant structure: leaves, branches, stems, and roots repartition

A mechanistic understanding of the effects of soil–plant hydraulics on transpiration should account for the whole plant's architecture and the various processes of allometry and branching. But, in such a complex architectural system, modeling the movement of water becomes almost impossible even if it is possible to derive local equations for flow within one organ element (Bohrer et al., 2005). Furthermore, measurements conducted with the intent to model water flow through all the plant organs are scarce. Moreover, one can always question the coherence and reliability of such a complex arrangement. And, as did Bohrer et al. (2005), we can wonder whether the architecture of the plant is beneficial for diagnostic and prognostic hydrologic models. This is the reason why plants with very complicated branching patterns are often reduced, by analogy of Ohm's law, to just one or a few resistance elements, which is quite adequate for modeling of stand-level water use over days. However, recently a new functional-structural model called HydroShoot has been developed that simulates the hydro-structure of a plant and water movement (Albasha et al., 2019).

Water flux (E, kg m^{-2} s^{-1}) through the various parts of the SPAC is then modeled as an analog to a simple electrical circuit composed of a series of resistances (or conductance, as the inverse of resistance). Following van den Honert's equation (van den Honert, 1948), the water flux can be written as:

$$E = \frac{\Psi_s - \Psi_l - \rho_w g h}{R_{plant}} \tag{4.24}$$

where Ψ_s and Ψ_l are soil and leaf water potentials (MPa), respectively; and $\rho_w g h$ is the gravitational pull on water column of height h and density ρ_w. For agricultural crops about 1 m tall, gravitational potential is relatively low (i.e., 0.0098 MPa for each increase in height) and therefore not taken into account. In this approach, we include only pressure potential component for flow through xylem conduits. The hydraulic resistance R_{plant} is seen as the resultant resistance of the root, stem, and leaf in series, and water uptake by the roots is assumed equal to the flow of water through the plant canopy.

This simple hydraulic model, used by Tuzet et al. (2003), provides a first-order estimate of leaf water potential diurnal variations, which is helpful to understand the diurnal response of stomata to the competing demand of water vapor from the atmosphere and supply of water to leaves by the soil via the roots and xylem.

In addition, the tissues in a plant may be considered as a number of alternative sources of water linked in parallel with each other and the soil. Thus, the total flux from the plant as transpiration (E) is made up of a number of water exchanges between plant storage compartments. Water storage compartments are grafted onto the main flow of the hydraulic system (Tuzet et al., 2017)

Storage and capacitance

Components of the plant water transport pathway—roots, stems, and leaves—can also serve as water storage compartments and therefore act transiently as intermediate sources of water for transpiring leaves. Water storage capacity per unit area of soil (m^3 m^{-2}) is an extensive property of the plant canopy and consequently it is a measure of the amount of water stored in plants or in particular tissues (roots, stems, foliage).

Figure 4.5 provides a schematic representation of the general approach used to represent the movement of water to and from water storages within a plant canopy

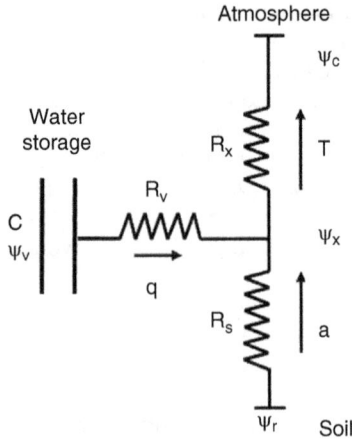

Figure 4.5 Schematic representation of the resistance-capacitance network used to estimate water transfer from storage pool. T, transpiration rate; a, root water uptake; q, storage flux; C, canopy capacitance; ψ_v, storage water potential; ψ_r, root water potential; ψ_x, xylem water potential; ψ_c, leaf water potential; R_v, storage hydraulic resistance; R_s, xylem hydraulic resistance in the lower section; R_x, xylem hydraulic resistance in the upper section. With our sign convention, the direction of arrows corresponds to positive values of flux (i.e., a storage discharge flux for the reservoir). From Tuzet et al. (2017) / with permission of Elsevier.

(Tuzet & Perrier, 2008). The water storage and its associated set of hydraulic resistances (R_x, R_s, R_v) are characterized by a direct link between the flux of water stored into or removed from the reservoir per unit area of soil, q, and the storage water potential, ψ_v. The driving force for liquid water flow through the hydraulic system is the water potential difference between compartments. Therefore, a water potential gradient develops between the storage pool and the stem compartment (ψ_x), causing depletion of the tissue storage. Water potential gradients also develop between the stem compartment and the root (ψ_r) and leaf (ψ_c) sections, causing water flow through the system (transpiration rate, T, and root water uptake, a). The change in storage water potential, ψ_v is defined as:

$$\frac{d\psi_v}{dt} + \frac{1}{C(R_s + R_v)}\psi_v - \frac{1}{C(R_s + R_v)}(\psi_r - R_s T) = 0 \tag{4.25}$$

where $C\,(\mathrm{m^3\,m^{-2}\,MPa^{-1}})$ is the storage capacitance, defined as the ratio of the change in storage water content to the change in storage water potential; it reflects intrinsic differences in the biophysical properties of the storage tissues and their potential ability to release water to the xylem in response to variations in water status of different plant organs. The time dependence of ψ_v after a step change in t is given as:

$$\psi_v = \psi_{v0}e^{-\frac{t}{\tau}} + (\psi_r - R_s T)\left(1 - e^{-\frac{t}{\tau}}\right) \text{ with } \tau = C(R_s + R_v) \tag{4.26}$$

In this equation, ψ_{v0} is the value of ψ_v before the step change in t and τ is the time constant.

In summary, we show that water storage compartments in plants act transiently as intermediate sources of water for transpiring leaves. The water depletion in plants delays the onset of stomatal regulation in the morning and then limits and regulates its closure in the afternoon when climatic demand is high. Furthermore, plant water storages are useful for buffering daily fluctuations in xylem tension, thus reducing the risks of xylem embolism and hydraulic failure under dynamic conditions.

Hydraulic conductance

In plant water transfer, transpiration is the driving force that pulls water from the soil to the leaves. When it occurs, the water movement is a passive process along a very complex network of very fine capillaries (vessels and tracheids), which form the xylem conducting system (Cruiziat et al., 2002). As noted above, the use of the electrical analogy and the associated hydraulic resistances are the main formalism used to deal with water transport in the soil–plant–atmosphere continuum.

Quantitative studies on whole-plant hydraulic resistance and its relation to evaporative flux have been conducted by measuring resistances of excised shoots (Melcher et al., 2012; Schubert et al., 1999; Tyree et al., 1994); a difficulty with this approach is that the complexity of the hydraulic structure raises parameter requirements. Then, many authors prefer to determine whole-plant hydraulic resistance with the aid of models (Smallman et al., 2013; Smallman & Williams, 2019; Sperry et al., 2003; Williams et al., 2001). With this approach, under conditions of ample water supply, a comparison of simulated leaf water potential values with measured values allows the determination of hydraulic resistance values to be refined; the xylem hydraulic resistance determines the amplitude of diurnal variation of the water potentials (Tuzet et al., 2017). Furthermore, most of the authors focus on a modeling approach based on the strong observed coordination between the stomatal and hydraulic properties of plants (Meinzer, 2002). This link is intuitive if stomata operate to maximize carbon assimilation, while maintaining leaf water potential at or above a critical threshold that avoids cavitation.

Stomatal Conductance

An elaborate system of control has evolved to allow the plant to maintain water vapor and CO_2 gas exchange without damaging its hydraulic structure (Sperry, 2004) as soil water availability and transpiration demand varies. Stomata control the movement of both water and CO_2 in and out of the internal areas on the surfaces of plant leaves and hence control gas exchange rates. This control helps maintain leaf water potentials, prevents cavitation and embolisms in plant circulatory organs, and may optimize carbon gain per unit of water uptake (Cowan & Troughton, 1971).

An important parameter used to mechanistically model stomatal behavior is stomatal conductance (g_s, mol H_2O m^{-2} s^{-1}). Stomatal conductance is intimately tied to many environmental and biological processes in the plant and involves feed forward and feed backward control mechanisms. Therefore, simulation of stomatal conductance is necessary for a comprehensive model of photosynthesis, leaf energy balance and water flow from soil through the plants to the atmosphere (Tuzet et al., 2003).

Plant transpiration is mainly controlled by the atmospheric demand and limited by stomatal conductance. Stomatal conductance is regulated to a certain extent by the soil and plant water status, under a range of environmental conditions. Responses of stomata to environmental factors (soil water status, irradiance, atmospheric water VPD, air temperature, and CO_2 concentration) have been extensively studied over the past 50 years and important advances have been achieved in understanding and mathematically modeling the physiology of stomatal movement. Damour et al. (2010) and Kim et al. (2020) provide reviews, which are by no means exhaustive but offer recent summaries of the state of the art in empirical stomatal studies. Models of stomatal conductance can be broadly grouped into four different categories: multiplicative or limiting-factors models, semi-empirical models based on the stomatal conductance-photosynthesis relationship, optimal stomatal behavior models and water stress response models based on abscisic acid (ABA) signaling, leaf water potential or hydraulic control.

The early phenomenological model described by Jarvis (1976) and taken up by many other authors (Mutiibwa & Suat, 2013; Wang et al., 2020) provides a practical approach for interpreting field measurements of stomatal conductance in relation to environmental variables. The stomatal resistance is expressed as the product of independent response functions of environmental variables. Each environmental variable is represented either by a more or less linear function or by an optimum response curve. This approach has often been adapted to the variables measured in various field experiments and for different types of vegetation. Although successfully tested several times multiplicative or limiting factor-based models are essentially empirical and require new parameterization for each new environmental condition. This is their main drawback likely resulting from the assumption that environmental factors have independent effects.

Numerous empirical models have been built on the relationship existing between stomatal conductance and photosynthesis rate. The model proposed by Ball et al. (1987) and later extended by Leuning et al. (1995) received wide attention and applications. This model expresses the dependence of stomatal conductance on net carbon assimilation, relative humidity or VPD, and leaf CO_2 concentration. The great interest in this approach relies on the opportunity to integrate it in general circulation models by coupling stomatal conductance, transpiration, and CO_2 concentration for predicting greenhouse gas induced climate change. It is also advantageous because the dependence of stomatal conductance on various factors can be achieved indirectly through their effects on photosynthesis since the model operates as a function of carbon assimilation. The model is relatively easy to use and to parameterize, but it does not capture responses to soil water status.

An alternative approach to parameterize stomatal regulation is to assume that stomatal aperture is regulated to maximize carbon gain, while minimizing water loss (Cowan, 1977, 1982; Katul et al., 2000; Mäkelä et al., 1996). Medlyn et al. (2011) have shown that this theory of optimal stomatal conductance can be coupled to a photosynthesis model (Farquhar et al., 1980) to derive a model of stomatal conductance that is closely analogous to the empirical Ball-Berry model and it provides a theoretical interpretation for model parameter values. This approach has recently been considered for climatic models (Franks et al., 2017). However, the approach has not yet gained widespread acceptance, and there are accompanying constructive criticisms and calls for extending this framework (Buckley, 2017; Miner et al., 2017). Furthermore, evidence from laboratory and field measurements, which shows that the sensitivity of transpiration and photosynthesis to changes in stomatal conductance is not constant, but changes throughout the day with increasing water stress, does not support this hypothesis (Fites & Teskey, 1988).

Under conditions of water stress, an equally supported hypothesis for stomatal closure is based on the notion that ABA, generated in droughted roots is transported through the transpiration stream to stomata, where it induces stomatal closure. Models based on ABA signaling have been developed. Correct parameterization of these models provides close agreement between simulated and observed stomatal conductance values (Davies & Zhang, 1991; Tardieu et al., 2015; Tardieu & Simonneau, 1998). However, the empirical nature of these models makes it difficult to extrapolate it into different environmental conditions. The increase of ABA concentration in the xylem sap after prolonged water stress is the result of long-distance ABA cycling after ABA synthesis in leaves and phloem-to-xylem exchange along the transport to the roots (Malcheska et al., 2017). Then, continuing progress toward a useful process model requires advances in several areas, including the kinetics of both ABA synthesis and catabolism (Buckley, 2017), the upstream intra- and intercellular signaling cascades that lead to ABA accumulation in guard cells (Batool et al., 2018). Also, it may be necessary to analyze the impact of ABA on the activity of NADPH oxidases for the production of reactive oxygen species (ROS), important regulators also of stomatal closure (Tee, 2018).

Current evidence strongly suggests that stomatal responses to changes in both evaporative demand and soil moisture involve an actively mediated feedback response to leaf water status (Anderegg et al., 2017; Buckley, 2016, 2017). Then, it is to a change in transpiration rate and not to a change in air vapor pressure to which stomata respond (Mott & Parkhurst, 1991). Therefore, any such control mechanism must incorporate the distribution of water potential along the flow path from soil to roots to leaves and be sensitive to changes that affect both liquid and vapor phases of the transpiration stream. Tuzet et al. (2003) were among the first to develop a physiological whole-plant-model that integrated most factors to predict stomatal regulation (Duursma & Medlyn, 2012; Landsberg & Waring, 2017). They propose a key modification of the Ball–Berry–Leuning model where the vapor pressure dependence of stomatal conductance is replaced by a sigmoidal response (Brodribb & Holbrook, 2003) to leaf water potential (ψ_{leaf}):

$$f_\psi = \frac{1 + exp\left(s_f \psi_{ref}\right)}{1 + exp\left(s_f\left(\psi_{ref} - \psi_{leaf}\right)\right)} \tag{4.27}$$

where ψ_{ref} is a reference potential, and s_f is a sensitivity parameter. The value of the parameters s_f and ψ_{ref} depends on morphological adaptations in different species and with environmental conditions experienced during vegetation period. The stomatal conductance is then calculated as:

$$g_{CO2} = g_0 + \frac{a_g A}{c_i - \Gamma} f_\psi \tag{4.28}$$

where g_0 is the residual conductance (limiting value of g_{CO2} at the light compensation point); a_g is an empirical coefficient; A is the assimilation rate; c_i is the CO_2 concentration in the intercellular spaces; and Γ is the c_i compensation point (Tuzet et al., 2017). This stomatal feedback control mechanism is activated by the level of leaf water potential (Klein, 2014). It is expressed in terms of stomatal sensing of the soil water uptake, via changes in the gradient of water potential between the leaves and roots. Furthermore, linking stomatal conductance to leaf water potential allows capturing stomatal closure due to increasing humidity deficit at the leaf surface (Ds) even if a feed-forward response of stomata to Ds is not invoked in the model (Tuzet et al., 2003). In this approach, water uptake by the roots is calculated dynamically to get the detailed distribution of water near the roots. This precise description is particularly important as the soil dries out because of the strongly non-linear relationship between moisture content and hydraulic conductivity, which affects the water transport from the bulk soil to the root surfaces.

Root Water Uptake

Water uptake by plant roots is important to the hydrogeologic water cycle. It is also an important contribution to the CO_2 and energy balance in soils (Bonan, 2015). Plants have an internal vascular system for moving water and dissolved substances that has some similarities to the system in animals. Plants, however, depend on water potential and osmotic gradients rather than a pump to move fluids through the vascular system. The negative pressures needed to drive water from the soil to the atmosphere are generated by evaporation of water from the plant leaves to the atmosphere that create gradients in leaf water potential. Turgor and internal positive water potentials are maintained by osmotic gradients. Because a large amount of water may be necessary for plants to maintain evaporation rates (up to $20 \, l \, m^{-2} \, d^{-1}$ by corn plants) an extensive root system is necessary to provide a large area of contact with capillary water stored in the soil.

Not all of the water in soil is available for uptake by plants. The water content at the low end of availability (called wilting point [WP] water content) has generally been accepted to be the water content in the soil at a water potential of $-1,500$ kPa. See Garg et al. (2020) for a recent review of the concept of WP. Because of the small change in water with change in matric potential at this potential, models are not very sensitive to changes in it. The upper level water content, sometimes called field capacity (FC) or drained upper limit (DUL) can be much more variable and there is less agreement about the best matric potential to use. It is generally accepted that the portion of rainfall that drains out of a soil layer is not available for plants. The amount of drainage, however, greatly depends on the time scale of plant water uptake and drainage as well as soil hydraulic conductivity. Water from rainfall will not significantly contribute to water availability if it drains completely from the root zone within an hour or two. The values for FC and DUL are generally agreed to correspond to the water content of a soil 24 hours after rainfall ceases. One should note that, for the same texture, the values of the drained water content can vary depending on the depth of the measurement. The value can also vary with depth to a water table. Generally, a value of water content at -33 kPa matric potential for fine textured soils and -20 kPa for coarser textured soils is accepted for FC where measurements are not available (Richards, 1944). Soil models that use a bucket or capacitance model analog (Manabe, 1969) for infiltration do not have capillary influence and are more sensitive to the values for DUL and WP than those that use the Richards equation (Richards, 1931).

There are a range of methods to simulate water uptake (see review by Camargo & Kemanian, 2016). The simplest is to use a distribution of root density in the soil and allocate water uptake from a particular location in the soil as a function of root length density and water content in that location. Models such as the Agricultural Production Systems Simulator (APSIM) (Keating et al., 2003) and Decision Support System for Agrotechnology Transfer (DSSAT) (Jones et al., 2003) use this method. Another approach is a modification of Campbell (1991) used in CropSyst (Stöckle et al., 2003) which uses leaf and soil water potentials. The average soil water potential is weighted by root density. In a comparison of different uptake models Camargo and Kemanian (2016) reported that the different models had a range of responses to variations in soil texture, layering and water distribution. CropSyst and DSSAT had similar drying fronts, but CropSyst and similar models tended to be more conservative of water deeper in the profile. These differences will result in variations in response to water stress.

The Feddes model (Feddes et al., 1976) defines water uptake as a sink term in the Richards equation (Richards, 1931). This model is used in HYDRUS2D (Šimůnek et al., 2008). The model describes a shape function where the sink term increases linearly from the WP matric potential to a maximum rate at about −50 kPa of matric potential. The sink term then stays constant and equal to the maximum rate until the matric potential is near the air entry point of the soil ($\Psi \approx -5.0$ kPa). Then it falls to zero as the soil becomes more wet, simulating the negative effect of near saturated soils on water uptake. The exact matric potentials where the slopes change can be varied as input parameters. When the matric potential is less than −50 kPa in this example, the sink will be less than that required to support potential transpiration. At matric potentials between −50 and −5.0 kPa the sink terms will satisfy potential transpiration. The soil is typically divided into layers to solve the water flow equation and total transpiration is calculated as the sum of the sinks over the layers. The rooting depth can vary over time but root density is not considered in the sink calculations. The Feddes model (Feddes et al., 1976) does not take root axial resistances into account and thus does not simulate the variation of water uptake with depth. It only models a reduction coefficient that is dependent on soil water potential. This function does describe water uptake that increases with depth as the surface soil dries because the surface sink value will decrease as the matric potential decreases and the larger sinks will compensate in the wetter layers. Šimůnek and Hopmans (2009) extended the Feddes model to include compensated water and solute uptake. This modification allowed the water uptake to be compensated by uptake in wetter layers as a layer dried. They also added passive and active nutrient uptake. Peters et al. (2017) improved the model by making the coefficients of the relationship between matric potential and the sink functions of soil texture and improving the calculation of matric potential for the point where soil wetness will affect water uptake.

A more mechanistic model of water uptake by roots (J_w, cm^3 water cm^{-2} d^{-1}) first proposed by van den Honert in van den Honert, 1948 is a resistance model similar to Ohm's law for electric current. A schematic of the resistances and driving forces is given in Figure 4.6. Gardner (1960) formalized this relationship to a complete set of equations to calculate water uptake by plants. The equations for water uptake are given as:

$$J_w = \frac{(\Psi_s - \Psi_l)}{R_p + R_{so}} \tag{4.29}$$

$$R_p = R_{rr} + L_r R_{ra} + R_s \tag{4.30}$$

where Ψ_s is the matric potential of the soil (-kPa); Ψ_l is the leaf water potential (-kPa); R_p is the plant resistance (kPa cm^2 g^{-1} H$_2$0 d^{-1}); and R_{so} is the soil resistance (kPa cm^2 g^{-1} d^{-1}). Generally, Ψ_s is the average over the area of root influence. For a layered, one-dimensional model, this is the average matric potential over the layer or area occupied by the root. The total resistance in the plant, R_p (kPa d^{-1} cm^{-2} g^{-1} H$_2$0) is the sum of the root radial resistance (R_{rr}), the root axial resistance ($L_r R_{ra}$), and resistance of the stem/leaves (R_s). The soil resistance is the inverse of the hydraulic conductivity at the current matric potential. Axial root resistance (kPa cm^2 g^{-1} d^{-1} cm^{-1} root) is a function of the length of roots as measured from the distance to the stem (cm). The radial resistance is the resistance across the root membrane and the soil interface. Axial resistance

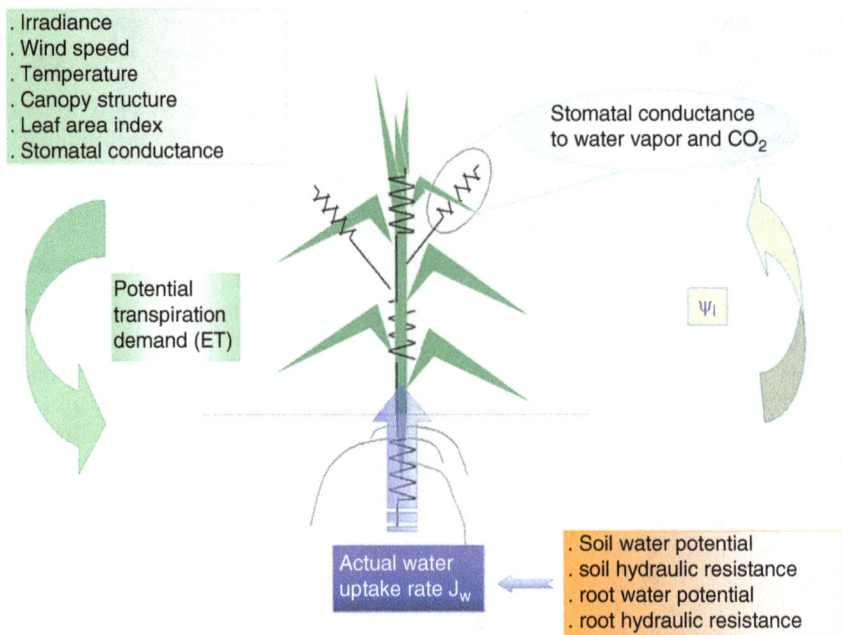

Figure 4.6 Schematic of resistances and driving forces for simulating water uptake according to Equations 4.29 and 4.30. The items listed above the soil are the driving forces and factors that define potential uptake. The items listed below the soil define the potential factors. From Yang et al. (2009) / with permission of American Society of Agricultural and Biological Engineers (ASABE).

increases with root length while radial resistance is a function of the diameter of the root. Radial resistance controls water entry into the root and the gradients can osmotic or hydraulic (Steudle, 2000) although in most models, osmotic gradients are ignored. Osmotic gradients, however, are important for nutrient uptake (Steudle, 2000). An important outcome of von den Honert's paper was that it was shown that stomatal and vascular resistances in plants can be substantial. Until then it was believed that there was little resistance to water flow in plants.

The flux (J_w) can be determined based solely on soil properties and static plant resistances (Campbell, 1991). A realistic response to drying soil can also be simulated. Experiments with water uptake show that water uptake in a uniformly wet soil begins near the plant crown and then extends to deeper in the soil as the soil dries (Doussan et al., 2006). Yet, this flux does not take the state of the plant into account. The plant will also have a demand based on VPD, (CO_2), light and leaf water potential. The maize model MAIZSIM (Yang et al., 2009) will use the minimum of the value of flux determined by requirements for leaf photosynthesis or what is available from the soil. Thus, under shaded conditions, low VPD or high CO_2, the plant will take up less than the capacity of the soil to supply that water. In the case of soil drying or very high demand, the water available from the soil will be less than the demand by the plant. In this case, the resulting low leaf water potential will result in stomatal closure and reduced water uptake. Over a period of time, the reduced leaf water potential and reduced carbon assimilation from stomatal closure will reduce plant growth. An example of water uptake simulations in MAIZSIM for both well watered and water stress conditions is shown in Figure 4.7 (from Yang et al., 2009).

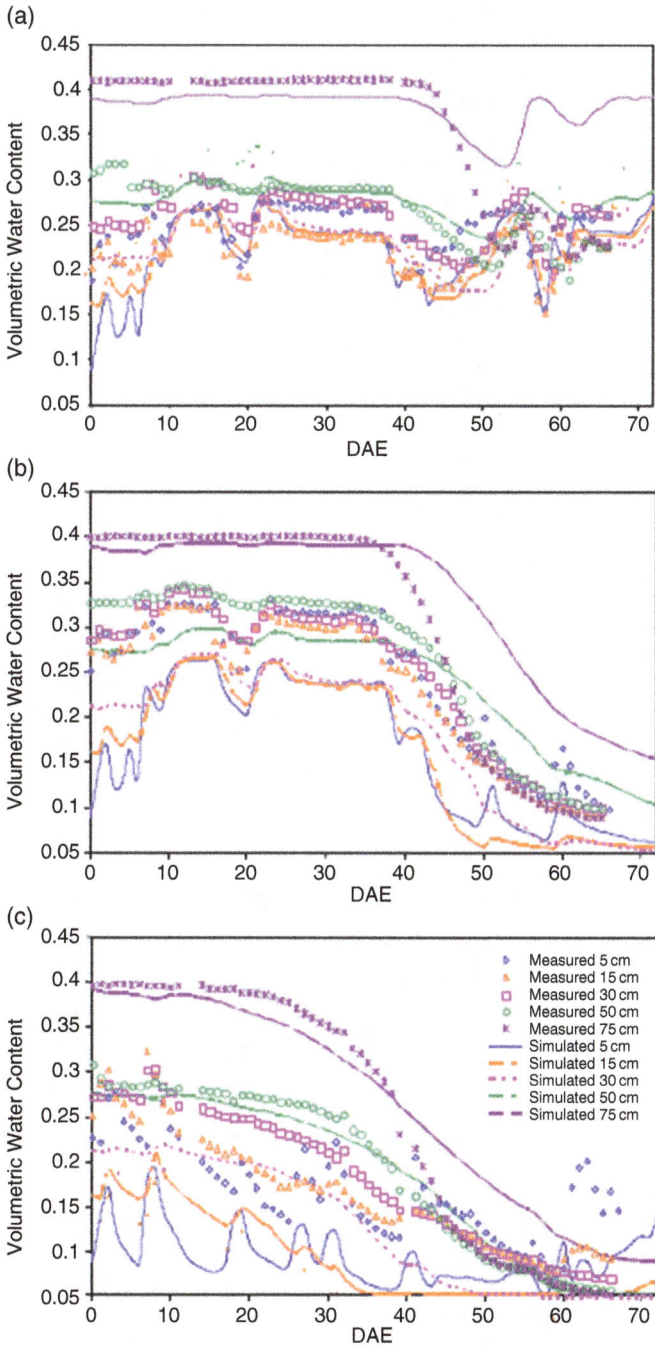

Figure 4.7 Water uptake simulated by MAIZSIM using an Ohm's law analog. The data are from Yang et al. (2009) and represent three irrigation regimes: (a) well watered, (b) medium stress, and (c) high water stress.

Cylindrical geometry and the concept of effective soil volume per unit root length have been used to model uptake of water by roots (Gardner, 1960; Philip, 1957). This approach is also used by Tuzet et al. (2003, 2017); where water transport from soil to roots is simulated in the radial direction through solution of the Richards' diffusion equation.

It is still difficult to measure root dynamics in soil because any direct observation of roots will damage them. However recent advances in the use of two-dimensional light transmission and x-ray imaging (Garrigues et al., 2006) has allowed visualization of actual water uptake in soil cylinders (Doussan et al., 2006). This research suggests that improvement in simulating water uptake will come from more detailed and realistic three-dimensional models of root growth. Such models will also provide a better representation of the interface between the soil and the root and allow the introduction of new factors that affect water uptake such as interaction with fungi and other soil biota.

Plant Regulation of Water Use

Leaf Water Potential Regulation

Leaf water potential regulation is a key process in whole plant and ecosystem functioning (Ratzmann et al., 2019). It depends on the balance between loss of water through transpiration and supply of water to the leaf from the soil. To investigate the dynamics of leaf water potential, we need a process-based model of the SPAC to ensure the links between energy budget and water balance, and to emphasize the key processes involved in the control of transpiration and water status of plant.

A typical temporal change in leaf water potential, predicted by a SPAC model (Tuzet et al., 2017), during a soil drying cycle is shown in Figure 4.8. From these results, it is interesting to discuss and analyze the role of different model parameters on the

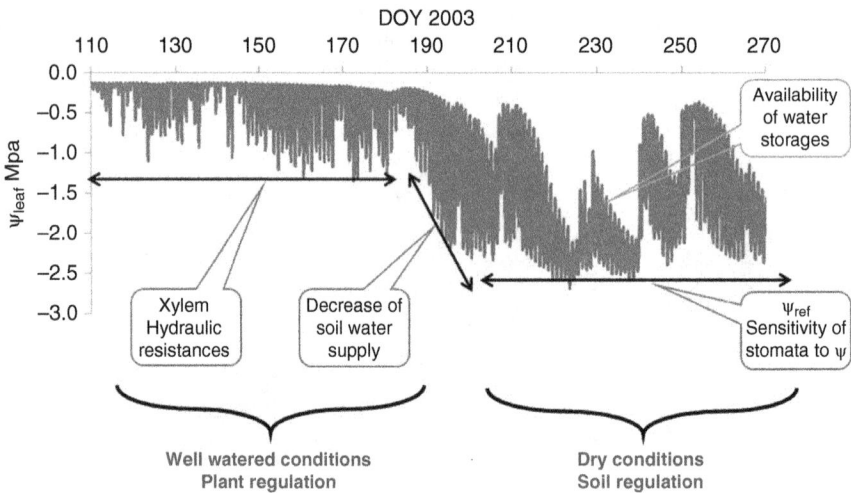

Figure 4.8 Typical temporal change in leaf water potential, ψ_{leaf}, predicted by a SPAC model, during a soil drying cycle; schematic illustration of the key processes involved in the control of leaf water potential, transpiration, and plant water status. From Tuzet et al., (2017) / with permission of Elsevier.

shape of the curve. In this figure, we see two very distinct periods, the first corresponding to a good water supply for the plants and the second to an onset of drought.

Early in the season when soil moisture is not limiting, simulations indicate that the minimum daytime value of leaf water potential $\psi_{leaf\ min}$ depends strongly on xylem hydraulic resistance R_{xylem}. For a given climatic demand, lower hydraulic resistances result in less negative leaf water potentials and greater transpiration. Thus, we can say that in conditions of good water supply, the plant mainly ensures the regulation of water use, that is, of transpiration.

When the soil begins to dry out near the root, it makes it more difficult to transfer water and the rate of soil water supply to the roots starts to decrease, thus limiting transpiration (around day 190 in Figure 4.8). There is also a progressive decrease of both minimum and maximum (i.e., predawn) daytime leaf water potentials. A similar trend would be observed for greater values of xylem hydraulic resistances, except that the onset of soil limitation to transpiration will occur progressively later.

At this stage of the drying cycle, the minimum daytime value of leaf water potential mainly depends on the sensitivity of stomatal conductance to leaf water status. The effect of soil drought, which results in a progressive drop in leaf water potential, leads to a closure of stomata. This closure is a feedback response to water loss from the leaf tissue. In the Tuzet SPAC model, this response is approximated using an empirical logistic function $f_{\psi v}$ (Equation 4.4) where the parameter ψ_{ref} is a "threshold" potential value at which stomatal conductance is reduced to half its maximum value. The smaller ψ_{ref}, the more the plant can survive very low leaf water potential values, reducing stomatal conductance as needed to regulate transpiration and prevent hydraulic failure.

There is also a progressive decrease in the values of the water potential of the leaves before dawn as the soil dries out; this decrease mainly depends on the storage water availability and therefore on the hydraulic resistances connecting water storages to the main flow. If the resistances are low, the contribution of water storage to the transpiration will be greater and root uptake will continue at night to refill water reservoirs leading to lower values of predawn leaf water potential.

These results show that there are two key plant characteristics involved in the control of transpiration and plant water status. The first is the ability of the plant to transfer water from the roots to the leaves, which corresponds to the xylem hydraulic resistance. The second is the sensitivity of stomata to leaf water status, which is approximated by a logistic function of leaf water potential, defining a minimum value of leaf water potential below which the stomata are closed.

Isohydric and Anisohydric Plants

Plants fall into two categories across the continuum of stomatal regulation of water status, labeled isohydric and anisohydric regulation (Tardieu & Simonneau, 1998; Villalobos-González et al., 2019). Isohydric plants close stomata rapidly as soil water potential decreases or as atmospheric conditions dry, thereby restricting excessive water loss and maintaining a relatively constant midday leaf water potential regardless of drought conditions. Anisohydric plants, in contrast, maintain open stomata and allow midday leaf water potential to decline as soil water potential decreases with drought (McDowell et al., 2008; Woodruff et al., 2015). In other words, isohydry is dehydration avoidance, while anisohydry is dehydration tolerance (Lavoie-Lamoureux et al., 2017).

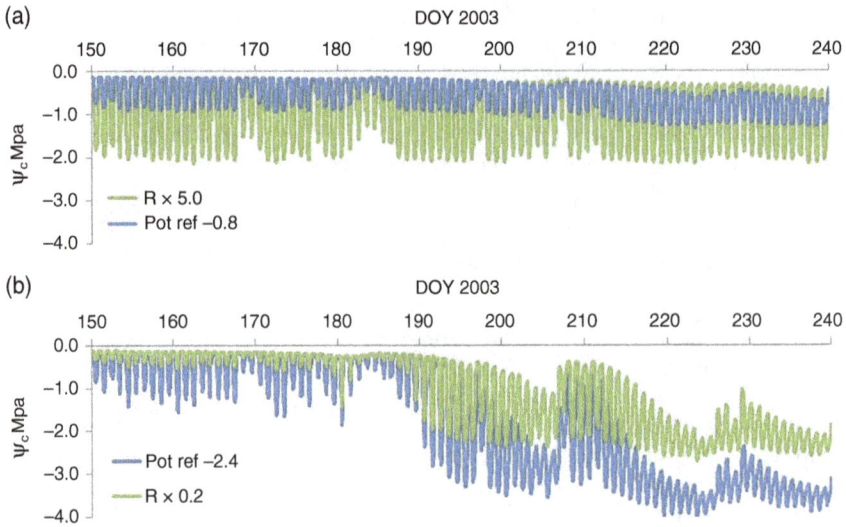

Figure 4.9 Typical temporal change in leaf water potential, ψ_c, predicted by a SPAC model, illustrating the contrasting behavior of isohydric (a) and anisohydric (b) plants in terms of differing plant hydraulic resistances, R, and stomatal sensitivities of stomata to leaf water potential, Pot ref Based on Tuzet et al. 2017.

The Tuzet SPAC model can explain the contrasting behavior of anisohydric and iso-hydric plants in terms of differing plant hydraulic resistances and stomatal sensitivities of stomata to leaf water potential, without explicitly invoking other controls. Figure 4.9a shows that "isohydric" plants, which maintain constant minimum leaf water potentials, are expected to have high xylem hydraulic resistance and/or high stomatal sensitivity to leaf water potential. "Anisohydric" plants (Figure 4.9b), which have variable minimum leaf water potentials, are expected to have lower hydraulic resistances and/or lower stomatal sensitivity to leaf water potential.

Hydraulic Failure

The hydraulic failure theory states that reduced soil water supply coupled with high evaporative demand can cause high tension in the xylem vessels leading to the formation of air bubbles and to a column breakage, called xylem embolism.

Once emboli have formed, flow continues through adjacent functional xylem vessels, bypassing embolized vessels at the cost of increased hydraulic resistance. Different experimental methods are used to describe how much the hydraulic conductivity of a given stem segment decreases when it is exposed to a gradual decrease in xylem water potential. Effects of the vessel embolism on xylem conductance are approximated using a sigmoid function of xylem water potential (Cochard et al., 2013; Sperry & Tyree, 1988). But, the lack of universal xylem refilling seen with whole plant physiology, process models assume that xylem hydraulic conductance fully recovers as soon as drought in the soil is relieved (Mackay et al., 2015).

Furthermore, most of this research on cavitation and embolism have been conducted on woody plants and less attention has been paid to herbaceous plants and to the physiological significance of embolism for the understanding of water relations in crop canopy (Cochard, 2002; de Campos Siega et al., 2018). According to the cohesion-tension theory of sap ascent, the intermolecular bonds between water molecules or between water and solids in xylem vessels are often sufficient to maintain the continuity of water columns in xylem. Then the xylem embolism in the leaf veins always remained low, even when the plants exhibited symptoms of clear water stress. This suggests that the stomata respond to the water status in the leaf as determined by the transpiration rate and that leaf water potential may be the physiological parameter regulating the closure of stomata during water stress (Brodribb & Holbrook, 2003), which would have the effect of preventing extensive developments of embolism. Then, most of the SPA models assume a constant xylem hydraulic resistance. This is of course a simplification but simulation results (Tuzet et al., 2017) show that the soil-to-root resistance increases considerably as the soil dries; it can reach values one to two orders of magnitude higher than the resistance of the xylem. This may account for the variability in observed soil-to-leaf resistance (Williams et al., 2001; Zeppel et al., 2008).

Photosynthesis-Transpiration Relationship

Photosynthesis includes the processes of CO_2 assimilation and water vapor release by plant leaves. It is one of the more important processes for life on Earth as it provides carbohydrate for food and oxygen for respiration. Because plant growth depends on photosynthesis, it is an essential building block of plant simulation models. As noted above, the biophysical processes involved in photosynthesis and growth can be abstracted to three major processes: (a) diffusion and assimilation of CO_2 into the leaves, (b) transpiration of water, and (c) conversion of CO_2 and water into carbohydrate and O_2. The interdependencies of these three processes can be quite complex as they also involve feed-forward and feed-backward relations. Assimilation of carbon as CO_2 is linked to transpiration because the CO_2 must be absorbed onto water films inside the plant mesophyll. Thus, the diffusion of CO_2 into the leaf is accompanied by evaporation of water from the moist surfaces within the leaf mesophyll. To prevent desiccation, an elaborate control mechanism evolved to maintain homeostatic leaf water contents. The stomata which are adjustable openings in the leaf, control the flow of CO_2 and water vapor.

The ability to estimate photosynthesis as a function of light, water and temperature has been an interest of plant physiologists for as long as plants have been studied. An excellent summary on the history of photosynthesis research and modeling can be found in Stirbet et al. (2019). One of the earliest attempts to quantify photosynthesis was by Bay and Pearlstein (1963) who proposed a theory of energy transfer within the photosynthetic organelles. Their goal was to describe the energy transfer of light reactions. There are some very intricate and complex photosynthetic models that simulate parts of the photosynthetic processes in complete detail (Stirbet et al., 2019). Because of the need for detailed parameterization, they are not yet useful in crop models.

The study of photosynthesis in the field and attempts to quantify capture of light and assimilation of CO_2 at the canopy level were carried out in micrometeorological studies by E. R. Lemon and his colleagues in the 1960s (e.g., Lemon & Wright, 1969; Wright & Lemon, 1966) along with Monteith and Szeicz (1960). This work became the basis for the first canopy level photosynthesis models (de Wit, 1965 and Duncan et al., 1967). These early models were simple hyperbolic equations used to model the light response of CO_2 assimilation in a leaf. The leaf models were then integrated

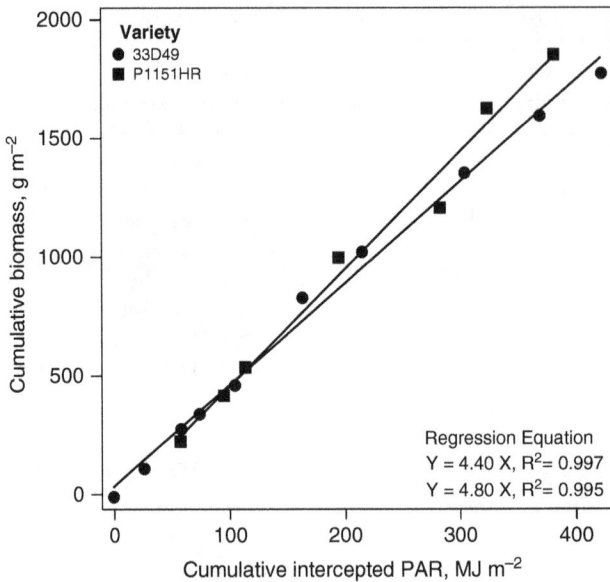

Figure 4.10 Relationship between intercepted solar radiation and biomass increase for two maize varieties. The figure shows the linear relationship between intercepted radiation and biomass increase. Radiation Use Efficiency (RUE) is the slope of the line. Note how there is a small difference between varieties. Hao et al., 2016 / with permission of John Wiley & Sons

over the canopy depending on light interception (Baldocchi & Amthor, 2001). About this time research by Warren-Wilson (1967) established the concept of RUE, an empirical linear relationship between intercepted solar radiance and carbon assimilation (Figure 4.10 from Hao et al., 2016). This relationship became the basis of biomass accumulation as a function of intercepted radiation in many crop models (Kiniry et al., 1989). Until recently these two approaches were used in almost all crop models. A modification of RUE that is also used is the Transpiration Use Efficiency (TUE) approach where biomass accumulation is related to crop transpiration and VPD but modified by radiation (Stöckle et al., 2003; Tanner & Sinclair, 1983). This method addresses the VPD problem identified with RUE.

The RUE concept has been popular in crop models because of its ease of use and the purpose for which models have been used. The problem with RUE is that it is not easily generalized to predict the effects of elevated (CO_2) on transpiration nor to predict energy balances in crops. It also does not directly account for the effects of temperature, VPD or other stresses on carbon assimilation, and transpiration must be modeled independently. If crop models are to expand beyond simple yield calculation and be used for carbon sequestration assessments or to assess crop yields and growth in a higher temperature/elevated (CO_2) climate, models will need to have more comprehensive models that address coupled carbon assimilation and transpiration. Recent model comparison studies by the AgMip (Agricultural Model Improvement) group have shown that models that have an explicit dependence of stomatal conductance on (CO_2) had smaller errors for yield and water use and simulated lower water use at elevated and ambient (CO_2) (Durand et al., 2018).

Probably the most important recent advance in modeling photosynthesis is related to the availability of a fairly simple mechanistically based biochemical model of C_3 (Farquhar et al., 1980; Farquhar et al., 2001) and C_4 (von Caemmerer & Furbank, 1999) photosynthesis (FvCB). These models describe photosynthesis mechanistically based on its key biochemical and anatomical characteristics. Although the model appears to require many parameters, almost all of them can be obtained from measurements with gas exchange systems (e.g., LiCor Biosciences, Lincoln, NE, USA) and many of the parameters are physically meaningful (Wang et al., 2017). The FvCB model allows researchers to vary important metabolic and physical parameters to assess environmental effects on carbon assimilation and transpiration. The model has been used in simulations of trees (Medlyn et al., 1999), horticultural crops (Kim & Lieth, 2003) and row crops (Collatz et al., 1991; Kim et al., 2012, 2013; Tuzet et al., 2003; Yang et al., 2009; Yu et al., 2003). More models are using this and related approaches (e.g., Katul et al., 2000; Yu & Flerchinger, 2008) as a useful alternative to RUE and hyperbolic equations. In addition to uses in crop models, the FvCB model has been combined with hyperspectral data to examine photosynthetic efficiencies of different tree species in a landscape (Yu et al., 2021) and used to estimate terrestrial productivity from remotely sensed data (Stocker et al., 2020). The Farquhar model provides a meaningful basis for qualitatively understanding the photosynthetic process to better quantify water stress. However, it is a steady-state model and thus cannot handle transient changes in leaf biochemical properties. Since it is a leaf level model, however, simulation becomes more complicated because it is necessary to quantify light interception in different parts of the canopy (see Kim & Lieth, 2003; Lizaso et al., 2005). The most common approach is to divide the canopy into sunlit and shaded components that vary over the day (de Pury & Farquhar, 1997).

The Farquhar (FvCB) and similar models abstract the biochemical processes in the Calvin cycle that use light to split CO_2 into O_2 and H_2O and store energy in phosphorous compounds. This energy is then used to assemble sugars from C, H, and O. The FvCB model describes the steady-state CO_2 assimilation rates by plant leaves. The model considers the CO_2 assimilation rate to be a function of three potential rates of carboxylation (Von Caemmerer, 2013). In other words, the FvCB model calculates a CO_2 assimilation rate as the minimum of one of the following, (a) carboxylation limited by Rubisco and CO_2 availability (A_c), or (b) carboxylation limited by light and electron transport rate (A_j) or (c) a triosphoshate limited carboxylation rate (A_p, C3 plants)—limited by accumulating starch at the chloroplast. These three processes occur simultaneously within the Calvin cycle and so the most limiting of the three processes determines the net photosynthetic rate. That is:

$$A = min\left(A_c, A_j, A_p\right) \tag{4.31}$$

The model can be described by the following equations from Kim et al. (2020).

$$A_c = V_{cmax}\left(\frac{C_c - \Gamma^*}{C_c + K_c\left(1 + O/K_o\right)}\right) - R_D \tag{4.32}$$

V_{cmax} is the maximum Rubisco carboxylation rate, K_c and K_o are the temperature dependent Michaelis–Menten kinetic constants for Rubisco for CO_2 and O_2 respectively. The Michaelis–Menten constants are similar for many C_3 plants (Von Caemmerer, 2013). The variables O and C_c are the O_2 (assumed to be the same as in air)

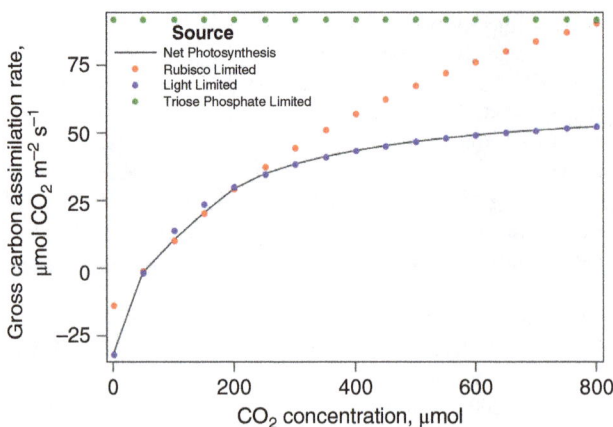

Figure 4.11 Net photosynthesis as a function of Rubisco and CO_2 availability, light and electron transport and Triose Phosphate reactions (sink limitation) calculated from the Farquhar photosynthesis module. For these data $J_{max} = 370\,\mu mol\,m^{-2}\,s^{-1}$, $V_{max} = 13\,\mu mol\,m^{-2}\,s^{-1}$, PAR = $1100\,\mu mol$ quanta $m^{-2}\,s^{-1}$ and temperature = 30 °C. Note that below ~100 µmol (CO_2), the reactions are light limited but above this value, they are rubisco limited. Notice there is no sink limitation, the is because the temperature is near optimum and the sink limitation is mainly a function of temperature. If the temperature were lower, sink limitations are more likely. The equations were adapted from the Farquhar Photosynthesis Model (colostate.edu, checked 2021).

and CO_2 partial pressures at the chloroplast and Γ is the CO_2 compensation point, the partial pressure of CO_2 where $A = R_D$. Finally, R_D is the CO_2 released in respiration and does not include photorespiration.

$$A_j = J\left(\frac{C_c - \Gamma^*}{4C_c - 8\Gamma}\right) - R_D \tag{4.33}$$

$$J = \frac{\alpha_{LL}I_{abs} + J_{max} - \sqrt{(\alpha_{LL}I_{abs} + J_{max})^2 - 4\theta J_{max}\alpha_{LL}I_{abs}}}{2\theta} \tag{4.34}$$

where J is the electron transport rate ($\mu mol\,m^{-2}\,s^{-1}$); J_{max} is the maximum capacity (of J) to absorb electrons; α_{LL} is the quantum yield of electron transport under light limiting conditions (quanta^{-1}); nI_{abs} is the amount of light absorbed or irradiance (µmol quanta $m^{-2}\,s^{-1}$); and θ is an empirical parameter that represents the convexity of the J to I_{abs} response curve.

Figure 4.11 shows how the calculation of net photosynthesis by the three functions vary over internal CO_2 concentration and how the minimum of the equation selects the resulting value.

Plant Response to Water Stress in Extreme Climates

Plants experience water stress either when the water supply to their roots becomes limiting or when the evaporative demand of the atmosphere becomes intense. Climate change and associated increase in extreme climatic events (drought episodes, heat waves, etc.) will lead to a greater occurrence of water stress.

Water stress not only affects plant water relations through the reduction of water content, plant-cell's water potential and turgor, but it also affects stomatal closure, limits gaseous exchanges, reduces transpiration and carbon assimilation (photosynthesis) and generally disrupts all physiological and biochemical processes of plants. It will differentially affect plant growth and production depending on many variables such as the length of the stress, the vegetative status of the crop, and the occurrence of other environmental stress such as high light irradiance, high temperatures (partly as a result of water stress) and elevated ozone concentration.

Water stress, as an abiotic stress, is multidimensional in nature, and it affects plants at various levels of their organization (Lisar et al., 2012; Xu et al., 2010) and their adaptation mechanisms are mainly composed of five categories: recovery, avoidance, tolerance, resistance, and drought escape. Drought recovery is the ability of plants to continue growth after drought injury, as long as the relevant antioxidant metabolism would be involved to annihilate the damage of ROS (Osakabe et al., 2014; Xu et al., 2010). Drought avoidance is the ability of plants to maintain relatively higher tissue water content despite reduced water content in the soil; it is usually achieved through morphological changes in the plant, such as reduced stomatal conductance, decreased leaf area (with curling or wilting leaves and even leaf fall), development of extensive root systems and increased root to shoot ratios. Drought resistance is the ability of plants via altering metabolic paths to survive under severe stress (e.g., increased antioxidant metabolism). Drought tolerance is the ability of plants to endure severe dehydration via osmotic adjustment ability and increasing cell wall elasticity to maintain tissue turgidity. Finally, drought escape is the ability of a plant with a rapid phenological development, to modulate their vegetative and reproductive growth according to water availability and then to complete their life cycle before the onset of drought. For example, in Crassulacean acid metabolism (CAM) plants, CO_2 fixation occurs during the night and stomata are closed during most of the day making life in water-limited environments possible.

Long before the onset of current climate change, a great number of drought escape strategies, since ancient times and particularly in arid areas were experimented and developed. Farmers have implemented various strategies to cope with climatic conditions that are unsuitable to plant development, and in particular to overcome the lack of water. Three main types of design are always used: (a) the first is the simplest and most natural, it consists in collecting and withdrawing water (from the river by a reach or in flat areas, from the water table by a water wheel) and using it for irrigation on the surfaces according to the slope; (b) the second is the increase of surfaces allowing to collect rainfall and run-off and to force water to infiltrate on the cultivated soil; it is used everywhere in sloping areas (terrace cultivation); (c) the third is the stratification ecosystem used in equatorial and humid tropical areas, it is also adapted to drier areas and based on complex interactions between the different species of the agroecosystem. The most effective one adopted worldwide is called agroforestry and its variants such as associated crops and oasis. However, in arid regions, the oasis effect is rather defined as a vegetation cooling effect due to extensive evapotranspiration in oasis compared to the hot and dry surroundings.

Agroforestry is the practice of deliberately integrating woody vegetation (trees or shrubs) with crop systems to benefit from the resulting climate and ecological interactions. The canopy trees act as a radiative and convective screen against the hot and dry environment, the middle vegetation stratum also acts as a radiative screen (Mediterranean fruit trees) and keeps the air cool and moist through evaporation from

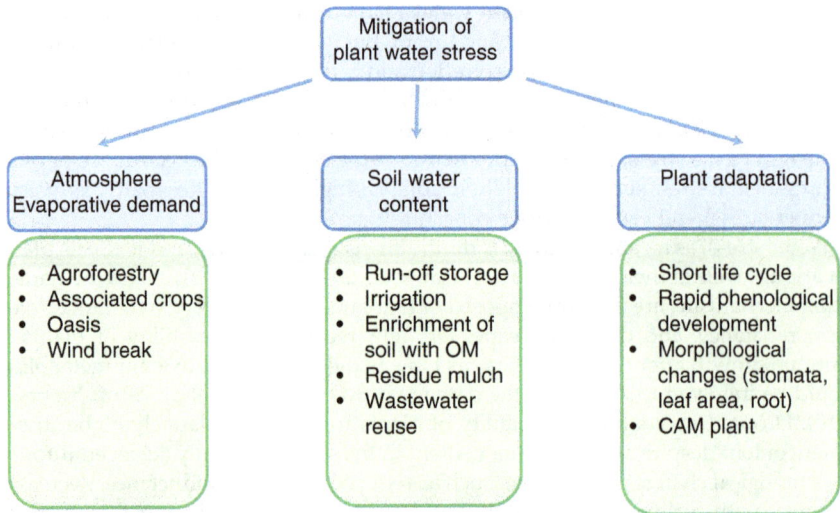

Figure 4.12 Major pathways of water stress mitigation in plants. The description of the different pathways is not meant to be exhaustive.

the crop surface (vegetables, cereals, and fodder). Then, through microclimate regulation, the canopy trees in agroforestry systems can help to control ecosystem water balance by influencing the radiative and convective exchanges of the system that will affect soil evaporation and leaf transpiration. In addition, trees and intercrops can extract water at different soil depths. Thus, there is a better use of soil water and intercrops could improve the water use efficiency and productivity of trees via species interactions. Moreover, crop and tree residues will contribute to enrich the soil with organic matter; higher organic matter content improves the physical properties of soil, such as porosity, water holding capacity, and fertility. Residue mulches are also frequently used in vegetable production to reduce evaporation losses from the soil surface.

In response to climate change, it is clear that more investment is needed in measures that will improve adaptation at the regional, catchment and farm level, such as surface runoff storage structures, conjunctive use of ground and surface water, and wastewater capture and reuse. A simplified representation of the main action levers for mitigating water stress in plants is represented in Figure 4.12. For example, West African tiger bush vegetation is one of the most striking examples of self-organized, banded vegetation pattern (Figure 4.13). It consists of regularly arranged woody stripes alternating with areas free of vegetation, whereby the distance between the vegetation stripes increases with decreasing precipitation. Greater attention will also need to be paid to protecting and supporting mountainous regions where much of the water supply for agricultural areas comes from.

Broad Environmental Feedbacks

Vegetation influences climate and weather through the release of water vapor from stomata during the day. Irrigation on very large surfaces can sustain a better water cycle by increasing annual mean rainfall (Guimberteau et al., 2012). The release of vapor into

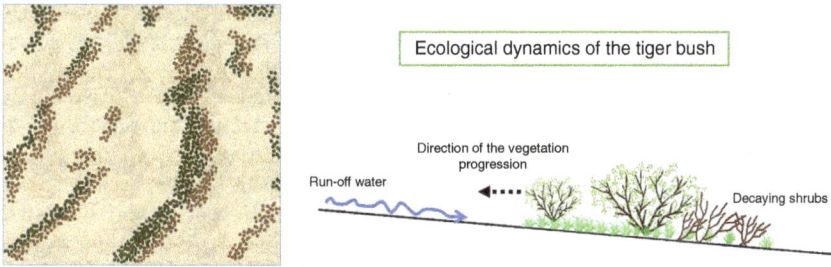

Figure 4.13 **Tiger bush in Niger with indication of the three zones found in each strip. It is in the pioneer zone and the heart of the thicket that regeneration is most vigorous. Photo by F. Mougenot, http://www. secheresse.info/GF/bellefontaine2_p02gf.jpg.**

the air by plant foliage alters energy flows, cools surfaces, and increases humidity, which can lead to cloud formation. Clouds alter the amount of sunlight that can reach the surface, affecting the Earth's energy balance, and in some areas can lead to precipitation. These strong biosphere-radiation feedbacks are often present in several moderately wet regions, for instance in the Eastern United States and in the Mediterranean, where precipitation and radiation increase vegetation growth.

In addition, vegetation can play a role in sedimentation by trapping and retaining some of the sediment eroded within a catchment. There is an effect of flow filtration through vegetation. As we have seen, in the case of the tiger bush, deposits created by vegetation barriers favor colonization by vegetation and the plants of vegetation barriers can also develop on the deposited sediment, and then colonize trapped sediment. Consequently, large quantities of eroded sediment inside catchments are trapped and do not reach catchment exits.

We can see, then, that many random conditions can combine, accentuated by climate change which necessarily led to more energy and moisture in the atmosphere. This will result in more powerful and more frequent thunderstorms or tornadoes, i.e., more rain on average but correspondingly more time to saturate the water vapor-depleted volumes of the atmosphere and therefore sometimes longer droughts. These erratic phenomena with suddenly too much water often cause stagnation and anaerobiosis in flat areas (source of physiological deterioration partially blocking the growth of the plant and especially the root systems) and create lethal conditions (with fungal attacks). In response to waterlogging, a lot of wetland plants (like rice) develop a lacunar parenchyma with useful space for gas transfer and exchange called "aerial parenchyma," allowing oxygen to be transported more easily from leaves down to watered parts of the plant and roots. Aerial parenchyma thus play an important role in the survival of plants under waterlogged conditions. Similarly, droughts and high air temperatures due to low regional and local evaporation cause strong physiological disturbances with slowing down and even stopping production processes, especially with annual agroecosystems. It is the same also in periods of frost with strong radiative cooling and negative air temperatures, especially at sunrise when temperatures have reached their minimum values with a negative radiative balance. The young buds that open, the new leaves, sometimes the flowers and even the young fruits are destroyed by the frost but sometimes, saved by the farmers' spraying of water; the release of energy from the freezing of the water droplets protects the vines from freezing.

It is necessary to be aware that with annual ecosystems, the bare soil regularly suffers the assaults of strong radiation, inducing drying temperatures which, in spite of the plants' water regulation systems, are often quickly lethal (if low evaporation). On the other hand, with heavy rains on a partially uncovered and partly dried surface (following low evaporation) it is the erosion of organic matter and fine particles of the soil that accentuates the loss of biological and physical potential benefits of the soil, and this accelerates the degradation of the vegetation canopy. This is why perennial plants resist drought better than annual plants, as they maintain a root system and lateral buds ready to start again.

Desertification is therefore a phenomenon with multi successive natural feedbacks that are always negative, even those in links with climate; degraded areas strongly accentuate warming at least partly over the year, reducing possible rainfall and accentuating the dryness of the air. Surfaces easily become radiators reaching very high temperatures (from 50 to 80 °C depending on surfaces and local latitudes). This is the case for many surfaces in cities and for all surfaces that have been made impermeable by our societies.

The goal of research is to revitalize degraded agricultural lands by integrating complex systems of plants adapted to very diverse climates. The retained processes, resulting from natural selection, must be analyzed and tested to respond to the implementation of suitable food production systems, the progressive restoration of soil quality, and a water cycle allowing local functioning of the agroecosystem toward water self-sufficiency. Depending on the location, these plants must be adapted to what nature has created, that is, different types of forests, savannahs, dry steppes with trees, and deserts that must be partly revived. Of course, agricultural production of organic food and industrial products must be located in balance with the regional environment; this is the case for agroforestry and all forms of environmentally friendly agriculture. The selection and integration of natural plants adapted to the conditions of the environment must accompany these developments to favor, as much as possible, plants already established locally.

These efforts will often require more water and thus rainfall than the restoration of large-scale ecosystems will produce through its additional evaporation across large scales. Indeed, the average lifetime on the Earth of an evaporated water molecule is only 12 days, so increased evaporation will result in more rainfall return. This is, of course, the best way to reduce global warming, since carbon fixed through restoration (wood elements and soils) reduces the greenhouse effect. This will increase the carbon stock in the soil and promote its restoration (organic matter fixed in humic form). At the same time, this approach is also a way to restore biodiversity.

To conclude, we can say that at all scales, a water balance must be established to verify that all the water coming from precipitation must be used for the development of life or stored (buffer effect) for use, as quickly as possible, for plant evaporation in order to accelerate the renewal of continental precipitation. This is even more true today, with more frequent extreme events and greater loss of water flow to the oceans, which leads to the following question: how to succeed in saving and sharing water between immediate reintroduction into its cycle or storage for later use. Modeling will therefore be increasingly useful to achieve these objectives with new vegetation, new crops, new management of space for the water saved. All this knowledge must therefore be put into perspective in order to move forward into the future.

Concluding Remarks and Future Recommendations

Predictive models, such as those developed to analyze mass and energy-exchange processes within the soil–plant–atmosphere system through stomatal conductance, photosynthesis and transpiration are generally based on knowledge and data of phenomena from the past. They rely on mathematical analyses of this information to increase our understanding of cause-and-effect relationships; they require classifying the processes involved according to importance, which results in the creation of a simplified reality.

Transpiration processes depend on biologically mediated responses to aerial and soil environments that can, in turn, exercise feedback effects via the plant on the energy balance, turbulent exchanges, and air characteristics (CO_2 concentration, temperature and humidity), thereby determining the state of the environment itself. Therefore, the prediction of transpiration fluxes under changing climates involves suitable models describing the complex interaction of plant and environmental variables.

As we have seen, the energy balance, which is the basis of all surface exchanges, only defines these exchanges in terms of surface water availability. Knowledge of this surface water availability necessarily requires good management of the soil water balance. The term "soil water balance" relates the moisture added through precipitation and/or irrigation to that lost through surface evaporation, root water uptake, runoff, and drainage. Future evapotranspiration modeling efforts should take advantage of advances in computer technology to develop field scale soil water balance models which require fewer parameters and less computer time/memory than the commonly used soil water dynamic models. This could also better enable us to model water competition processes at the root level in the case of crop mixtures recommended to better cope with global changes.

Similarly, to refine the transpiration-photosynthesis coupling and their consequences on crop development and yield, a simultaneous coupling of xylem and phloem flows would be physiologically interesting. Both the flow rate of water and that of sugar to and from the leaves set requirements on the stomatal regulation of the transpiration stream.

References

Albasha, R., Fournier, C., Pradal, C., Chelle, M., Prieto, J., Louarn, G., Simonneau, T., & Lebon, E. (2019). Hydro-Shoot: A functional-structural plant model for simulating hydraulic structure, gas and energy exchange dynamics of complex plant canopies under water deficit—application to grapevine (*Vitis vinifera* L.). *bioRxiv*, 542803. http://dx.doi.org/10.1101/542803

Allen, R. G., Pereira, L. S., Raes, D., & Smith, M. (1998). *Crop evapotranspiration: Guidelines for computing crop water requirements*. FAO Irrigation and Drainage Paper 56. FAO.

Allen, R. G., Pruitt, W. O., Wright, J. L., Howell, T. A., Ventura, F., Snyder, R., Itenfisu, D., Steduto, P., Berengena, J., Yrisarry, J. B., Smith, M., Pereira, L. S., Raes, D., Perrier, A., Alves, I., Walter, I., & Elliott, R. (2006). A recommendation on standardized surface resistance for hourly calculation of reference ETO by the FAO56 Penman–Monteith method. *Agricultural Water Management*, 81(1/2), 1–22. http://dx.doi.org/10.1016/j.agwat.2005.03.007

Anderegg, W. R. L., Wolf, A., Arango-Velez, A., Choat, B., Chmura, D. J., Jansen, S., Kolb, T., Li, S., Meinzer, F., Pita, P., Resco de Dios, V., Sperry, J. S., Wolfe, B. T., & Pacala, S. (2017). Plant water potential improves prediction of empirical stomatal models. *PLoS One*, 12(10), 1–17.

Baldocchi, D. D., & Amthor, J. S. (2001). Canopy photosynthesis: History, measurements and models. In J. Roy, B. Saugier, & H. A. Mooney (Eds.), *Terrestrial global productivity* (pp. 9–31). Academic.

Ball, J. T., Woodrow, I. E., & Berry, J. A. (1987). A model predicting stomatal conductance and its contribution to the con trol of photosynthesis under different environmental conditions. In J. Biggins (Ed.), *Progress in photosynthesis research* (Vol. IV, pp. 221–224). Martinus Nijhoff.

Batool, S., Vural, U. V., Rajab, H., Ahmad, N., Waadt, R., Geiger, D., Malagoli, M., Xiang, C., Hedrich, R., Rennenberg, H., Herschbach, C., Hell, R., & Wirtza, M. (2018). Sulfate is incorporated into cysteine to trigger ABA production and stomatal closure. *Plant, Cell and Environment*, 30, 2973–2987.

Bay, Z., & Pearlstein, R. M. (1963). A theory of energy transfer in the photosynthetic unit. *Proceedings of the National Academy of Sciences of the United States of America*, 50, 1071–1078. http://dx.doi.org/10.1073/pnas.50.6.1071

Berleth, T. (2000). Plant development: Hidden networks. *Current Biology*, 10, R658–R661. http://dx.doi.org/10.1016/S0960-9822(00)00683-7

Blyth, E. M., Arora, V. K., Clark, D. B., Dadson, S. J., De Kauwe, M. G., Lawrence, D. M., Melton, J. R., Pongratz, J., Turton, R. H., Yoshimura, K., & Yuan, H. (2021). Advances in land surface modelling. *Current Climate Change Reports*, 7, 45–71.

Bohrer, G., Mourad, H., Laursen, T. A., Drewry, D., Avissar, R., Poggi, D., Oren, R., & Katul, G. G. (2005). Finite element tree crown hydrodynamics model (FETCH) using porous media flow within branching elements: A new representation of tree hydrodynamics. *Water Resources Research*, 41, W11404.

Bonan, G. (2015). *Ecological climatology, concepts and applications*. Cambridge University Press. http://dx.doi.org/10.1111/j.1745-5871.2009.00640.x

Bouchet, R. J. (1963). Evapotranspiration réelle, évapotranspiration potentielle, et production agricole. *Annales Agronomiques*, 14, 743–824.

Bowen, I. S. (1936). The ratio of heat losses by conduction and by evaporation from any water surface. *Physical Review*, 27, 779–787.

Bowen, R. (1980). *Ground water*. Halsted Press, Wiley.

Brodribb, T. J., & Holbrook, N. M. (2003). Stomatal closure during leaf dehydration, correlation with other leaf physiological traits. *Plant Physiology*, 132, 2166–2173.

Brutsaert, W. (1982). *Evaporation into the atmosphere: Theory, history and applications* (pp. 299p). Kluwer Academic Publishers.

Buckley, T. N. (2016). Stomatal responses to humidity: Has the "black box" finally been opened? *Plant, Cell & Environment*, 39, 482–484.

Buckley, T. N. (2017). Modeling stomatal conductance. *Plant Physiology*, 174, 572–582.

Camargo, G. G. T., & Kemanian, A. R. (2016). Six crop models differ in their simulation of water uptake. *Agricultural and Forest Meteorology*, 220, 116–129. http://dx.doi.org/10.1016/j.agrformet.2016.01.013

Campbell, G. (1991). Simulation of water uptake by plant roots. In J. R. Hanks (Ed.), *Modeling plant and soil systems* (pp. 273–285). American Society of Agronomy.

Cochard, H. (2002). Xylem embolism and drought-induced stomatal closure in maize. *Planta*, 215, 466–471.

Cochard, H., Badel, E., Herbette, S., Delzon, S., Choat, B., & Jansen, S. (2013). Methods for measuring plant vulnerability to cavitation: A critical review. *Journal of Experimental Botany*, 64(15), 4779–4791.

Collatz, G. J., Ball, J. T., Grivet, C., & Berry, J. A. (1991). Physiological and environmental regulation of stomatal conductance, photosynthesis and transpiration: A model that includes a laminar boundary layer. *Agricultural and Forest Meteorology*, 54, 107–136.

Cowan, I. R. (1977). Stomatal behaviour and environment. *Advances in Botanical Research*, 5, 117–228.

Cowan, I. R. (1982). Regulation of water use in relation to carbon gain in higher plants. In *Physiological Plant Ecology 11, Water Relations and Carbon Assimilation* (pp. 589–614). Springer.

Cowan, I. R., & Troughton, J. H. (1971). The relative role of stomata in transpiration and assimilation. *Planta*, 97, 325–336.

Cruiziat, P., Cochard, H., & Ameglio, T. (2002). Hydraulic architecture of trees: Main concepts and results. *Annals of Forest Science*, 59, 723–752.

Cummings, N. W. (1929). *Relation between evaporation and humidity as deduced quantitatively from rational equations based on thermodynamics and molecular theory*. Transactions of the American Geophysical Union, 68. National Research Council. pp. 47–56.

Damour, G., Simonneau, T., Cochard, H., & Urban, L. (2010). An overview of models of stomatal conductance at the leaf level. *Plant, Cell and Environment*, 33, 1419–1438.

Davies, W. J., & Zhang, J. H. (1991). Root signals and the regulation of growth and development of plants in drying soil. *Annual Review of Plant Physiology and Plant Molecular Biology*, 42, 55–76.

de Campos Siega, T., Bertoldo, E., de Souza Vismara, L., Nicareta, C., & Wagner, J. A. (2018). Cavitation and embolism in plants: Literature review. *Australian Journal of Basic and Applied Sciences*, 12(5), 1–4.

De Kauwe, M. G., Kala, J., Lin, Y.-S., Pitman, A. J., Medlyn, B. E., Duursma, R. A., et al. (2015). A test of an optimal stomatal conductance scheme within the CABLE land surface model. *Geoscientific Model Development*, 8(2), 431–452. http://dx.doi.org/10.5194/gmd-8-431-2015

de Pury, D. G., & Farquhar, G. D. (1997). Simple scaling of photosynthesis from leaves to canopies without the errors of big-leaf models. *Plant, Cell and Environment*, 20, 537–557.

de Wit, C. T. (1965). *Photosynthesis of leaf canopies*. Agricultural Research Report 663, pp. 1–57. Wageningen, the Netherlands.

Delire, C., Séférian, R., Decharme, B., Alkama, R., Calvet, J.-C., Carrer, D., et al. (2020). The global land carbon cycle simulated with ISBA-CTRIP: Improvements over the last decade. *Journal of Advances in Modeling Earth Systems*, 12, e2019MS001886. http://dx.doi.org/10.1029/2019MS001886

Doussan, C., Pierret, A., Garrigues, E., & Pagès, L. (2006). Water uptake by plant roots: II—Modelling of water transfer in the soil root-system with explicit account of flow within the root system—Comparison with experiments. *Plant Soil*, 283, 99–117. http://dx.doi.org/10.1007/s11104-004-7904-z

Duncan, W. G., Loomis, R. S., Williams, W. A., & Hanau, R. (1967). A model for simulating photosynthesis in plant communities. *Hilgardia*, 38, 181–205.

Dupont de Dinechin, B. (1999). *"DUHAMEL DU MONCEAU" Un savant exemplaire au siècle des lumières-ed. CME, connaissance et mémoires européennes*. 446 p. (Imprimerie E.PAILLART [BP-324] 80103 Abbeville: dépôt légal 4ième trimestre 1998).

Durand, J.-L., Delusca, K., Boote, K., Lizaso, J., Manderscheid, R., Weigel, H. J., Ruane, A. C., Rosenzweig, C., Jones, J., Ahuja, L., Anapalli, S., Basso, B., Baron, C., Bertuzzi, P., Biernath, C., Deryng, D., Ewert, F., Gaiser, T., Gayler, S., … Zhao, Z. (2018). How accurately do maize crop models simulate the interactions of atmospheric CO_2 concentration levels with limited water supply on water use and yield? *European Journal of Agronomy*, 100, 67–75. http://dx.doi.org/10.1016/j.eja.2017.01.002

Duursma, A., & Medlyn, B. E. (2012). MAESPA: A model to study interactions between water limitation, environmental drivers and vegetation function at tree and stand levels, with an example application to $(CO_2)\times$ drought interactions. *Geoscientific Model Development*, 5, 919–940.

Ehlers, W., & Goss, M. (Eds.) (2004). *Water dynamics in plant production*. CABI Publishing.

Farquhar, G. D., von Caemmerer, S., & Berry, J. A. (1980). A biochemical model of photosynthetic CO_2 assimilation in leaves of C3 species. *Planta*, 149, 78–90.

Farquhar, G. D., von Caemmerer, S., & Berry, J. A. (2001). Models of photosynthesis. *Plant Physiology*, 125, 42–45.

Feddes, R. A., Kowalik, P., Kolinska-Malinka, K., & Zaradny, H. (1976). Simulation of field water uptake by plants using a soil water dependent root extraction function. *Journal of Hydrology*, 31(1), 13–26. http://dx.doi.org/10.1016/0022-1694(76)90017-2

Feddes, R.A., Kowalik, P. J., & Zaradny, H. (1978). *Simulation of field water use and crop yield*. Simulation Monographs. Pudoc. Wageningen, the Netherlands.

Feugier, F. G., Mochizuki, A., & Iwasa, Y. (2005). Self-organization of the vascular system in plant leaves: Interdependent dynamics of auxin flux and carrier proteins. *Journal of Theoretical Biology*, 236, 366–375.

Fites, J. A., & Teskey, R. O. (1988). CO_2 and water vapor exchange of Pinus taeda in relation to stomatal behavior: Test of an optimization hypothesis. *Canadian Journal of Forest Research*, 18(2), 150–157.

Fleisher, H. D., Timlin, J. D., Yang, Y., & Reddy, V.R. (2007). *Simulation of potato gas exchange using SPUDSIM*. 2007 ASAE Annual Meeting. St. Joseph, MI: ASABE.

Flexas, J., Bota, J., Galmes, J., Medrano, H., & Ribas-Carbo, M. (2006). Keeping a positive carbon balance under adverse conditions: Responses of photosynthesis and respiration to water stress. *Physiologia Plantarum*, 127, 343–352.

Franks, P. J., Berry, J. A., Lombardozzi, D. L., & Bonan, G. B. (2017). Stomatal function across temporal and spatial scales: Deeptime trends, land-atmosphere coupling and global models. *Plant Physiology*, 174, 583–602.

Fukuda, H. (2004). Signals that control plant vascular cell differentiation. *Nature Reviews Molecular Cell Biology*, 5, 379–391. http://dx.doi.org/10.1038/nrm1364

Gardner, W. R. (1960). Dynamic aspects of water availability to plants. *Soil Science*, 89, 63–73.

Gardner, W. R., & Hillel, D. I. (1962). The relation of external evaporative conditions to the drying of soils. *Journal of Gephysical Research*, 67, 4319–4325.

Garg, A., Bordoloi, S., Ganesan, S. P., Sekharan, S., & Sahoo, L. (2020). A relook into plant wilting: Observational evidence based on unsaturated soil–plant-photosynthesis interaction. *Scientific Reports*, 10, 22064. http://dx.doi.org/10.1038/s41598-020-78893-z

Garrigues, E., Doussan, C., & Pierret, A. (2006). Water uptake by plant roots: I—formation and propagation of a water extraction front in mature root systems as evidenced by 2D light transmission imaging. *Plant and Soil*, 283, 83. http://dx.doi.org/10.1007/s11104-004-7903-0

Guimberteau, M., Laval, K., Perrier, A., & Polcher, J. (2012). Global effect of irrigation and its impact on the onset of the Indian summer monsoon. *Climate Dynamics*, 39(6), 1329–1348.

Hanson, J. D., Ahuja, L. R., Shaffer, M. D., Rojas, K. W., DeCoursey, D. G., Farahani, H., & Johnson, K. (1998). RZWQM: Simulating the effects of management on water quality and crop production. *Agricultural Systems*, 57, 161–195. http://dx.doi.org/10.1016/S0308-521X(98)00002-X

Hao, B., Xue, Q., Marek, T. H., Jessup, K. E., Hou, X., Xu, W., Bynum, E. D., & Bean, B. W. (2016). Radiation-use efficiency, biomass production, and grain yield in two maize hybrids differing in drought tolerance. *Journal of Agronomy and Crop Science*, 202, 269–280. http://dx.doi.org/10.1111/jac.12154

Hargreaves, G. H., & Samani, Z. A. (1985). Reference crop evapotranspiration from temperature. *Applied Engineering in Agriculture*, 1, 96–99.

Hsiao, T. C. (1973). Plant responses to water stress. *Annual Review of Plant Physiology*, 24, 519–570.

Jarvis, P. G. (1976). The interpretation of the variations in leaf water potential and stomatal conductance found in canopies in the field. *Philosophical Transactions of the Royal Society of London B*, 273, 593–610.

Jones, J. W., Hoogenboom, G., Porter, C. H., Boote, K. J., Batchelor, W. D., Hunt, L. A., Wilkens, P. W., Singh, U., Gijsman, A. J., & Ritchie, J. T. (2003). The DSSAT cropping system model. *European Journal of Agronomy*, 18, 235–265. http://dx.doi.org/10.1016/S1161-0301(02)00107-7

Katul, G. G., Ellsworth, D. S., & Lai, C. T. (2000). Modelling assimilation and intercellular CO_2 from measured conductance: A synthesis of approaches. *Plant, Cell & Environment*, 23, 1313–1328.

Keating, B. A., Carberry, P. S., Hammer, G. L., Probert, M. E., Robertson, M. J., Holzworth, D., Huth, N. I., Hargreaves, J. N., Meinke, H., & Hochman, Z. (2003). An overview of APSIM, a model designed for farming systems simulation. *European Journal of Agronomy*, 18, 267–288.

Kim, S.-H., Hsiao, J., & Kinmonth-Schultz, H. (2020). Advances and improvements in modeling plant processes. In K. Boote (Ed.), *Advances in crop modelling for a sustainable agriculture*. Burleigh Dodds Science Publishing.

Kim, S.-H., Jeong, J. H., & Nackley, L. L. (2013). Photosynthetic and transpiration responses to light, CO_2, temperature, and leaf senescence in garlic: Analysis and modeling. *Journal of the American Society for Horticultural Science*, 138, 149.

Kim, S.-H., & Lieth, J. H. (2003). A coupled model of photosynthesis, stomatal conductance and transpiration for a rose leaf (*Rosa hybrida* L.). *Annals of Botany*, 91, 771–781.

Kim, S.-H., Yang, Y., Timlin, D. J., Fleisher, D., Dathe, A., & Reddy, V. R. (2012). Modeling nonlinear temperature responses of leaf growth, development, and biomass in MAIZSIM. *Agronomy Journal*, 104, 1523–1537.

Kiniry, J. R., Jones, C. A., O'Toole, J. C., Blanchet, R., Cabelguenne, M., & Spanel, D. A. (1989). Radiation-use efficiency in biomass accumulation prior to grain-filling for five grain-crop species. *Field Crops Research*, 20, 51–64. http://dx.doi.org/10.1016/0378-4290(89)90023-3

Klein, T. (2014). The variability of stomatal sensitivity to leaf water potential across tree species indicates a continuum between isohydric and anisohydric behaviours. *Functional Ecology*, 28, 1313–1320.

Landsberg, J., & Waring, R. (2017). Water relations in tree physiology: Where to from here? *Tree Physiology*, 37, 18–32.

Lavoie-Lamoureux, A., Sacco, D., Risse, P. A., & Lovisolo, C. (2017). Factors influencing stomatal conductance in response to water availability in grapevine: A meta-analysis. *Physiologia Plantarum*, 159(4), 468–482.

Lawlor, D. W. (2002). Limitation to photosynthesis in water-stressed leaves: Stomata vs. metabolism and the role of ATP. *Annals of Botany*, 89, 871–885.

Lawrence, D. M., Fisher, R. A., Koven, C. D., Oleson, K. W., Swenson, S. C., Bonan, G., Collier, N., Ghimire, B., van Kampenhout, L., Kennedy, D., Kluzek, E., Lawrence, P. J., Li, F., Li, H., Lombardozzi, D., Riley, W. J., Sacks, W. J., Shi, M., Vertenstein, M., … Zeng, X. (2019). The community land model version 5: Description of new features, benchmarking, and impact of forcing uncertainty. *Journal of Advances in Modeling Earth Systems*, 11(12), 4245–4287. http://dx.doi.org/10.1029/2018MS001583

Lemon, E. R., & Wright, J. L. (1969). Photosynthesis under field conditions. XA. Assessing sources and sinks of carbon dioxide in a corn (*Zea mays* L.) crop using a momentum balance approach. *Agronomy Journal*, 61, 405–411.

Leuning, R., Kelliher, F. M., DePury, D. G. G., & Schulze, E. D. (1995). Leaf nitrogen, photosynthesis, conductance and transpiration: Scaling from leaves to canopies. *Plant, Cell and Environment*, 18, 1183–1200.

Lisar, S. Y. S., Motafakkerazad, R., Hossain, M. M., & Rahman, I. M. M. (2012). Water stress in plants: Causes, effects and responses. In I. M. M. Rahman & H. Hasegawa (Eds.), *Water stress*, Chapter: 1. InTech. http://dx.doi.org/10.1111/nph.13170

Lizaso, J., Batchelor, W., Boote, K., & Westgate, M. (2005). Development of a leaf-level canopy assimilation model for CERES-maize. *Agronomy Journal*, 97, 722–733.

Lucas, W. J., Groover, A., Lichtenberger, R., Furuta, K., Yadav, S.-R., Helariutta, Y., He, X.-Q., Fukuda, H., Kang, J., Brady, S. M., Patrick, J. W., Sperry, J., Yoshida, A., López-Millán, A.-F., Grusak, M. A., & Kachroo, P. (2013). The plant vascular system: Evolution, development and functions F. *Journal of Integrative Plant Biology*, 55, 294–388. http://dx.doi.org/10.1111/jipb.12041

Mackay, D. S., Roberts, D. E., Ewers, B. E., Sperry, J. S., McDowell, N. G., & Pockman, W. T. (2015). Interdependence of chronic hydraulic dysfunction and canopy processes can improve integrated models of tree response to drought. *Water Resources Research*, 51, 6156–6176.

Mäkelä, A., Berninger, F., & Hari, P. (1996). Optimal control of gas exchange during drought: theoretical analyses. *Ann Bot*, 77, 461–467.

Malcheska, F., Ahmad, A., Batool, S., Müller, H. M., Ludwig-Müller, J., Kreuzwieser, J., Randewig, D., Hänsch, R., Mendel, R. R., Hell, R., Wirtz, M., Geiger, D., Ache, P., Hedrich, R., Herschbach, C., & Rennenberg, H. (2017). Drought-enhanced xylem sap sulfate closes stomata by affecting ALMT12 and guard cell ABA synthesis. *Plant Physiology*, 174, 798–814.

Manabe, S. (1969). The atmospheric circulation and the hydrology of the earth's surface. *Monthly Weather Review*, 97, 739–774.

Maseda, P. H., & Fernandez, R. J. (2006). Stay wet or else: Three ways in which plants can adjust hydraulically to their environment. *Journal of Experimental Botany*, 57, 3963–3977.

McCown, R. L., Hammer, G. L., Hargreaves, J. N. G., Holzworth, D. P., & Freebairn, D. M. (1996). APSIM: A novel software system for model development, model testing and simulation in agricultural systems research. *Agricultural Systems*, 50, 255–271. http://dx.doi.org/10.1016/0308-521X(94)00055-V

McDowell, N., Pockman, W. T., Allen, C. D., Breshears, D. D., Cobb, N., Kolb, T., Plaut, J., Sperry, J., West, A., Williams, D. G., & Yepez, E. A. (2008). Mechanisms of plant survival and mortality during drought: Why some plants survive while others succumb to drought. *The New Phytologist*, 178, 719–739.

Medlyn, B. E., Badeck, F. W., de Pury, D. G. G., Barton, C. V. M., Broadmeadow, M., Ceulemans, R., De Angelis, P., Forstreuter, M., Jach, M. E., Kellomaki, S., et al. (1999). Effects of elevated CO_2 on photosynthesis in European forest species: A meta-analysis of model parameters. *Plant, Cell and Environment*, 22, 1475–1495.

Medlyn, B. E., Duursma, R. A., & Zeppel, M. J. B. (2011). Forest productivity under climate change: A checklist for evaluating model studies. *Wiley Interdisciplinary Reviews: Climate Change*, 2(3), 332–355.

Meinzer, F. C. (2002). Co-ordination of vapour and liquid phase water transport properties in plants. *Plant, Cell and Environment*, 25, 265–274.

Melcher, P. J. N., Holbrook, N. M., Burns, M. J., Zwieniecki, M. A., Cobb, A. R., Brodribb, T. J., Choat, B., & Sack, L. (2012). Measurements of stem xylem hydraulic conductivity in the laboratory and field. *Methods in Ecology and Evolution*, 2012(3), 685–694.

Miner, G. L., Bauerle, W. L. L., & Baldocchi, D. D. (2017). Estimating the sensitivity of stomatal conductance to photosynthesis: a review. *Plant, Cell and Environment*, 40(7), 1214–1238. http://dx.doi.org/10.1111/pce.12871

Monteith, J. L. (1965). Evaporation and environment. In G. E. Fogg (Ed.), *The state and movement of water in living organisms, Symposia of the Society for Experimental Biology* (Vol. 19, pp. 205–234). Academic Press.

Monteith, J. L. (1981). Evaporation and surface temperature. *Quarterly Journal of the Royal Meteorological Society*, 107, 1–27.

Monteith, J. L., & Szeicz, G. (1960). The CO_2 flux over a field of sugar beets. *Quarterly Journal of the Royal Meteorological Society*, 86, 205–214.

Mott, K. A., & Parkhurst, D. F. (1991). Stomatal responses to humidity in air and helox. *Plant, Cell and Environment*, 14, 509–515.

Mutiibwa, D., & Suat, I. (2013). Transferability of jarvis-type models developed and re-parameterized for maize to estimate stomatal resistance of soybean: Analyses on model calibration, validation, performance, sensitivity, and elasticity. *Transactions of the ASABE*, 56, 409–422.

Osakabe, Y., Osakabe, K., Shinozaki, K., & Tran, L. P. (2014). Response of plant to water stress. *Frontiers in Plant Science*, 5, 86. http://dx.doi.org/10.3389/fpls.2014.00086

Penman, H. L. (1948). Natural evaporation from open water, bare soil and grass. *Proceedings of the Royal Society A: Mathematical, Physical and Engineering Sciences*, 193(1032), 120–145. http://dx.doi.org/10.1098/rspa.1948.0037

Perrier, A. (1973). *Bilan hydrique de l'assolement blé-jachère et évaporation d'un sol nu, en région semi-aride*. In Slatyer R.O., Plant response to climatic factors, Réponse des plantes aux facteurs climatiques, Colloque UNESCO, Uppsala (SWE), 1970/09/15-20, UNESCO, Paris, pp. 477–487.

Perrier, A. (1975a). Etude physique de l'évapotranspiration dans les conditions naturelles. I. Evaporation et bilan d'énergie des surfaces naturelles. *Annales Agronomiques*, 26(1), 1–18.

Perrier, A. (1975b). Etude physique de l'évapotranspiration dans les conditions naturelles. II. Expressions et paramètres donnant l'évapotranspiration réelle d'une surface "mince". *Annales Agronomiques, 26*, 2, 105–123.

Perrier, A. (1975c). Etude physique de l'évapotranspiration dans les conditions naturelles. III. Evapotranspiration réelle et potentielle des couverts végétaux. *Annales Agronomiques, 26*, 3, 229–243.

Perrier, A. (1976). Etude et essai de modélisation des échanges de masse et d'énergie au niveau des couverts végétaux: Profils microclimatiques, évapotranspiration et photosynthèse nette. *Thèse Dr. Etat Sciences Physiques, Univ. Pierre et Marie Curie*, Paris VI. 247 p.

Perrier, A. (1982). Land surface processes: Vegetation. In P. S. Eagleson (Ed.), *Land surface processes in atmospheric general circulation models, WMO, World Meteorological Organization, Geneve (CHE); World Climatic Research Program State (USA); International Council of Scientific Unions (USA), Greenbelt (USA), 1981/01/05-10* (pp. 395–448). Cambridge University Press, GBR.

Perrier, A., Golaz, F., Tuzet, A., Personne, E., Jabbour, B., Meridja, S., & Zurfluh, O. (2003). *Projet d'utilisation du modèle "Bilhyna" en Beauce.* Rapport année 2002, Chambre d'Agriculture d'Eure-et-Loire, INRA INA PG UMR Environnement et Grandes Cultures, Grignon, 30 p.

Perrier, A., & Tuzet, A. (1991). Land surface processes: Description, theoretical approaches, and physical laws underlying their measurements. In T. J. Schmugge & J. C. André (Eds.), *Land surface evaporation. Measurement and parameterization* (pp. 145–155). Springer-Verlag.

Peters, A., Durner, W., & Iden, S. C. (2017). Modified Feddes type stress reduction function for modeling root water uptake: Accounting for limited aeration and low water potential. *Agricultural Water Management, 185*, 126–136. http://dx.doi.org/10.1016/j.agwat.2017.02.010

Philip, J. R. (1957). Evaporation, and moisture and heat fields in the soil. *Journal of Meteorology, 14*, 354–366.

Philip, J. R. (1966). Plant water relations: Some physical aspects. *Annual Review of Plant Physiology, 17*, 245–268.

Pliny. (1938). *Naturali History,English translation by K. Rackham, Vol.I (Praefatio, Libri 1,2),W.Heinemann, Ltd., London.* Harvard University Press.

Pokhilko, A., & Ebenhöh, O. (2015). Mathematical modelling of diurnal regulation of carbohydrate allocation by osmo-related processes in plants. *Journal of the Royal Society Interface, 12*, 20141357.

Priestley, C. H. B., & Taylor, R. J. (1972). On the assessment of surface heat flux and evaporation using large-scale parameters. *Monthly Weather Review, 100*, 81–92.

Ratzmann, G., Zakharova, L., & Tietjen, B. (2019). Optimal leaf water status regulation of plants in drylands. *Nature Scientific Reports, 9*(1), 1–9.

Richards, L. (1944). Moisture retention by some irrigated soils as related to soil moisture tension. *Journal of Agricultural Research, 69*, 215–235.

Richards, L. A. (1931). Capillary conduction of liquids through porous mediums. *Physics, 1*, 318–333.

Schubert, A., Lovisolo, C., & Peterlunger, E. (1999). Shoot orientation affects vessel size, shoot hydraulic conductivity and shoot growth rate in *Vitis vinifera* L. *Plant, Cell and Environment, 22*, 197–204.

Sedileau. (1730). Observations de la quantité de l'eau de pluie tombée à Paris durant près de trois années & de la quantité de l'évaporation, 29 fev.1692. *Mémoires de l'Académie Royale des Sciences, 10*, 29–36.

Sifaoui, M. S., & Perrier, A. (1978). Caractérisation de l'évaporation profonde. *International Journal of Heat and Mass Transfer, 21*, 629–637.

Šimůnek, J., & Hopmans, J. W. (2009). Modeling compensated root water and nutrient uptake. *Ecological Modelling, 220*, 505–521. http://dx.doi.org/10.1016/j.ecolmodel.2008.11.004

Šimůnek, J., van Genuchten, M. T., & Šejna, M. (2008). Development and applications of the HYDRUS and STANMOD software packages and related codes. *Vadose Zone Journal, 7*, 587–600. http://dx.doi.org/10.2136/vzj2007.0077

Smallman, T. L., Moncrieff, J. B., & Williams, M. (2013). WRFv3.2- SPAv2: Development and validation of a coupled ecosystem–atmosphere model, scaling from surface fluxes of CO2 and energy to atmospheric profiles. *Geoscience Model Development, 6*, 1079–1093. https://doi.org/10.5194/gmd-6-1079-2013

Smallman, T. L., & Williams, M. (2019). Description and validation of an intermediate complexity model for ecosystem photosynthesis and evapotranspiration: ACM-GPP-ETv1. *Geoscientific Model Development, 12*, 2227–2253.

Sperry, J. S. (2004). Coordinating stomatal and xylem functioning—An evolutionary perspective. *New Phytologist, 162*, 568–570.

Sperry, J. S., Stiller, V., & Hacke, U. G. (2003). Xylem hydraulics and the soil–plant–atmosphere continuum: Opportunities and unresolved issues. *Agronomy Journal, 95*, 1362–1370.

Sperry, J. S., & Tyree, M. T. (1988). Mechanism of water stress-induced xylem embolism. *Plant Physiology, 88*, 581–587.

Steudle, E. (2000). Water uptake by plant roots: An integration of views. *Plant and Soil*, 226, 45–56. http://dx.doi.org/10.1023/A:1026439226716

Stirbet, A., Lazár, D., Guo, Y., & Govindjee, G. (2019). Photosynthesis: Basics, history and modelling. *Annals of Botany*, 126, 511–537. http://dx.doi.org/10.1093/aob/mcz171

Stocker, B. D., Wang, H., Smith, N. G., Harrison, S. P., Keenan, T. F., Sandoval, D., Davis, T., & Prentice, I. C. (2020). P-model v1.0: An optimality-based light use efficiency model for simulating ecosystem gross primary production. *Geoscientific Model Development*, 13, 1545–1581. http://dx.doi.org/10.5194/gmd-13-1545-2020

Stöckle, C. O., Donatelli, M., & Nelson, R. (2003). CropSyst, a cropping systems simulation model: Modelling cropping systems: Science, software and applications. *European Journal of Agronomy*, 18, 289–307.

Tanner, C. B., & Sinclair, R. R. (1983). Efficient water use in crop production: Research or re-search? *Limitations to Efficient Water Use in Crop Production.*, 1–27.

Tardieu, F., & Simonneau, T. (1998). Variability among species of stomatal control under fluctuating soil water status and evaporative demand: Modelling isohydric and anisohydric behaviours. *Journal of Experimental Botany*, 49, 419–432.

Tardieu, F., Simonneau, T., & Parent, B. (2015). Modelling the coordination of the controls of stomatal aperture, transpiration, leaf growth, and abscisic acid: Update and extension of the Tardieu–Davies model. *Journal of Experimental Botany*, 66(8), 2227–2237.

Tee, E. E. (2018). Uncovering the steps before: Sulfate induces ABA biosynthesis and stomatal closure. *The Plant Cell*, 30, 2894–2895.

Thornthwaite, C. W., & Mather, J. R. (1955). The water balance. *Publications in Climatology*, 8(1), 5–86.

Thornthwaite, C. W., & Mather, J. R. (1957). Instructions and tables for computing potential evapotranspiration and the water balance. *Publications in Climatology*, 10(3), 311.

Tuzet, A., Granier, A., Betsch, P., Peiffer, M., & Perrier, A. (2017). Modelling hydraulic functioning of an adult beech stand under non-limiting soil water and severe drought condition. *Ecological Modelling*, 348, 56–77.

Tuzet, A., & Perrier, A. (2008). Modeling the dynamics of water flow through plants: Role of capacitance in stomatal conductance and plant water relations. In L. R. Ahuja, et al. (Eds.), *Response of crops to limited water: Understanding and modeling water stress effects on plant growth processes*, Advances in Agricultural Systems Modeling Series 1, Chapter 5 (pp. 145–164). ASA, CSSA, and SSSA.

Tuzet, A., Perrier, A., Castell, J. F., & Zurfluh, O. (1995). Energy and water vapour transfers in a particular shrub-grass intercrop: the fallow savanna. In H. Sinoquet, P. Cruz (Eds.) *Ecophysiology of tropical intercropping*, Symposium International, INRA, Gosier (Guadeloupe), 06-10/12/1993, INRA Editions, Paris, pp. 137–152.

Tuzet, A., Perrier, A., & Leuning, R. (2003). A coupled model of stomatal conductance, photosynthesis and transpiration. *Plant, Cell and Environment*, 26(7), 1097–1116.

Tuzet, A., Perrier, A., & Masaad, C. 1992 Crop water budget: Estimation of irrigation requirement. In L. S. Pereira, A. Perrier, M. Ait Kadi, P. Kabat (Eds.), Crop-water models, I.C.I.D. Bulletin, No Sp., 41, 2, pp. 1–17.

Tyree, M. T., & Ewers, F. W. (1991). The hydraulic architecture of trees and other woody plants. *New Phytologist*, 119, 345–360. http://dx.doi.org/10.1111/j.1469-8137.1991.tb00035.x

Tyree, M. T., Yang, S., Cruiziat, P., & Sinclair, B. (1994). Novel methods of measuring hydraulic conductivity of tree root systems and interpretation using AMAIZED. *Plant Physiology*, 104, 189–199.

van den Honert, T. H. (1948). Water transport in plants as a catenary process. *Discussions of the Faraday Society*, 3, 146–153.

Villalobos-González, L., Muñoz-Araya, M., Franck, N., & Pastenes, C. (2019). Controversies in midday water potential regulation and stomatal behavior might result from the environment, genotype, and/or rootstock: Evidence from Carménère and Syrah grapevine varieties. *Frontiers in Plant Science*, 10, 1522.

von Caemmerer, S. (2000). *Biochemical models of leaf photosynthesis*. CSIRO Publishing.

Von Caemmerer, S. (2013). Steady-state models of photosynthesis. *Plant, Cell and Environment*, 36, 1617–1630. http://dx.doi.org/10.1111/pce.12098

von Caemmerer, S., & Furbank, R. T. (1999). The modelling of C4 photosynthesis. In R. F. Sage & R. K. Monson (Eds.), *C4 plant biology (pp. 173–211). Academic Press.*

Wang, H., Guan, H., Liu, N., Soulsby, C., Tetzlaff, D., & Zhang, X. (2020). Improving the Jarvis type model with modified temperature and radiation functions for sap flow simulations. *Journal of Hydrology*, 587. http://dx.doi.org/10.1016/j.jhydrol.2020.124981

Wang, Q., Chun, J. A., Fleisher, D., Reddy, V., Timlin, D., & Resop, J. (2017). Parameter estimation of the Farquhar–von Caemmerer–Berry biochemical model from photosynthetic carbon dioxide response curves. *Sustainability*, 9. http://dx.doi.org/10.3390/su9071288

Warren-Wilson, J. W. (1967). Ecological data on dry-matter production by plants and plant communities. In E. F. Bradley & O. T. Denmead (Eds.), *The collection and processing of field data* (pp. 77–123). Interscience Publishers.

West, G. B., Brown, J. H., & Enquist, B. J. (1999). A general model for the structure and allometry of plant vascular systems. *Nature, 400*, 664–667. http://dx.doi.org/10.1038/23251

Wilkinson, R. E. (2000). *Plant-environment interactions.* CRC Press.

Williams, J. R., Jones, C. A., Kiniry, J. R., & Spanel, D. A. (1989). EPIC crop growth model. *Transactions of the American Society of Agricultural Engineers, 32*, 497–511.

Williams, M., Bond, B. J., & Ryan, M. G. (2001). Evaluating different soil and plant hydraulic constraints on tree function using a model and sap flow data from ponderosa pine. *Plant, Cell and Environment, 24*, 679–690.

Wilson, R. G., & Rouse, W. R. (1972). Moisture and temperature limits of the equilibrium evapotranspiration model. *Journal of Applied Meteorology, 11*, 436–442.

Woodruff, D. R., Meinzer, F. C., Marias, D. E., Sevanto, S., Jenkins, M. W., & McDowell, N. G. (2015). Linking nonstructural carbohydrate dynamics to gas exchange and leaf hydraulic behavior in *Pinus edulis* and *Juniperus monosperma. New Phytologist, 206*(1), 411–421. http://dx.doi.org/10.1111/nph.13170.Woodruff

Wright, J. L., & Lemon, E. R. (1966). Photosynthesis under field conditions. VIII. Analysis of windspeed fluctuation data to evaluate turbulent exchange within a corn crop. *Agronomy Journal, 58*, 255–261.

Xu, Z., Zhou, G., & Shimizu, H. (2010). Plant responses to drought and rewatering. *Plant Signaling & Behavior, 5*(6), 649–654.

Yang, Y., Kim, S.-H., Timlin, D., Fleisher, D., Quebedeaux, B., & Reddy, V. (2009). Simulating canopy transpiration and photosynthesis of corn plants under contrasting water regimes using a coupled model. *Transactions of the ASABE, 52*, 1011–1024.

Yu, G.-R., Kobayashi, T., Zhuang, J., Wang, Q.-F., & Qu, L.-Q. (2003). A coupled model of photosynthesis-transpiration based on the stomatal behavior for maize (*Zea mays* L.) grown in the field. *Plant and Soil, 249*(2), 401–415.

Yu, Q., & Flerchinger, G. N. (2008). Extending the simultaneous heat and water (SHAW) model to simulate carbon dioxide and water fluxes over wheat canopy. Chapter 7. Response of crops to limited water: Understanding and modeling water stress effects on plant growth processes. In L. R. Ahuja, V. R. Reddy, S. A. Saseendran, & Q. Yu (Eds.), *Advances in Agricultural Systems Modeling, Ser. 1* (pp. 191–214). ASA, CSSA, SSSA.

Yu, Y., Yang, X., & Fan, W. (2021). Remote sensing inversion of leaf maximum carboxylation rate based on a mechanistic photosynthetic model. *IEEE Transactions on Geoscience and Remote Sensing.* http://dx.doi.org/10.1109/TGRS.2021.3050105

Zeppel, M., Macinnis-Ng, C., Palmer, A., Taylor, D., Whitley, R., Fuentes, S., Yunusa, I., Williams, M., & Eamus, D. (2008). An analysis of the sensitivity of sap flux to soil and plant variables assessed for an Australian woodland using a soil-plant-atmosphere model. *Functional Plant Biology, 35*, 509–520.

Zhou, X.-R., Schnepf, A., Vanderborght, J., Leitner, D., Lacointe, A., Vereecken, H., & Lobet, G. (2020). CPlantBox, a whole-plant modelling framework for the simulation of water- and carbon-related processes. *In Silico Plants, 2.* http://dx.doi.org/10.1093/insilicoplants/diaa001

5

Marvin J. Shaffer and
Lajpat R. Ahuja

Soil Organic Matter and Carbon–Nitrogen Transformations

Abstract

This chapter is an update on the soil organic matter model called OMNI. As with several other models, OMNI handles chemical complexity of the substrate by using the pools concept. It simulates all the major pathways of C–N transformations including mineralization–immobilization of crop residues, manure, and other organic materials; mineralization of the soil humus fractions; inter-pool transfers of C and N; nitrification of NH_4–N to produce NO_3–N and N_2O gas; denitrification (production of N_2 and N_2O); gaseous loss of NH_3 and CH_4 gas; hydrolysis of urea; production and consumption of CO_2; and microbial biomass growth and death. In OMNI, the growth and subsequent death of microorganisms drive most of the processes and are a function of environmental variables such as soil temperature and water content, soil pH, soil O_2 levels, and solution concentrations (or activities) of nutrients.

This chapter is an update of Shaffer et al. (2000). The computer program for Organic Matter/Nitrogen cycling (OMNI) was developed as a major sub-model of the USDA–ARS Root Zone Water Quality Model (RZWQM). OMNI is a state-of-the-art model for C and N cycling in soil systems. An attempt was made to unify the form of the process rate equations governing C and N transformations or cycling in soils and provide a firm theoretical basis for environmental interactions as well as future extensions. Significant use was made of concepts and principles found in existing models such as NTRM (Shaffer & Larson, 1987); Phoenix (Juma & McGill, 1986); CENTURY (Parton et al., 1983); and Frissel's nitrogen model (Frissel & van Veen, 1981). An

Modeling Processes and Their Interactions in Cropping Systems: Challenges for the 21st Century, First Edition.
Edited by Lajpat R. Ahuja, Kurt C. Kersebaum, and Ole Wendroth.
© 2022 American Society of Agronomy, Inc. / Crop Science Society of America, Inc. / Soil Science Society of America, Inc.
Published 2022 by John Wiley & Sons, Inc.

excellent introduction to C and N transformations and processes in soil can be found in the appropriate chapters of Alexander (1977), Hansen et al. (1995), and Shaffer and Ma (2001). OMNI combines many features such as crop residue and soil organic matter (SOM) pools found in existing models and adds basic principles of chemical rate process theory, soil microbial growth, and environmental interactions. For example, OMNI uses transition-state rate equations (Laidler, 1969; Shaffer, 1985; Shaffer & Dutt, 1974) that include Arrhenius temperature response functions and reactive constituent concentrations and simulate microbial responses to soil O_2 levels, pH, water content, and salinity. The model includes direct interactive linkages to related sub-models in RZWQM such as the soil chemistry, plant growth, and solute transport models. Linkages such as NH_4-N, NO_3-N, and H ion concentrations, CO_2 partial pressures, and ionic strength are among the current primary transfer points. Additional information about how OMNI is linked to other modules can be found in other chapters.

General flow diagrams of the soil C–N cycle are shown in Figures 5.1 and 5.2. OMNI is a model for the homogeneous plot to field scale. As with several other models, OMNI handles chemical complexity of the substrate by using the *pools* concept. It simulates all the major pathways illustrated including mineralization–immobilization of crop residues, manure, and other organic materials; mineralization of the soil humus fractions; inter-pool transfers of C and N; nitrification of NH_4^+ to produce NO_3-N and N_2O gas; denitrification (production of N_2 and N_2O); gaseous loss of NH_3 and CH_4 gas; hydrolysis of urea; production and consumption of CO_2, and microbial biomass (BM or MBM) growth and death. In OMNI, the growth and subsequent death of microorganisms drive most of the processes and are a function of environmental variables such as soil

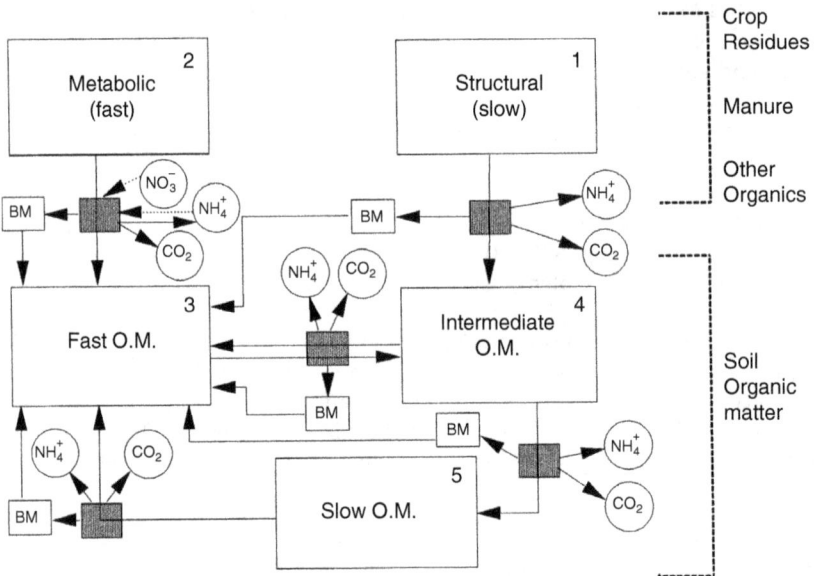

Figure 5.1 Aerobic decay of organic matter (OM) in the OMNI sub-model.

soil organic matter and carbon–nitrogen

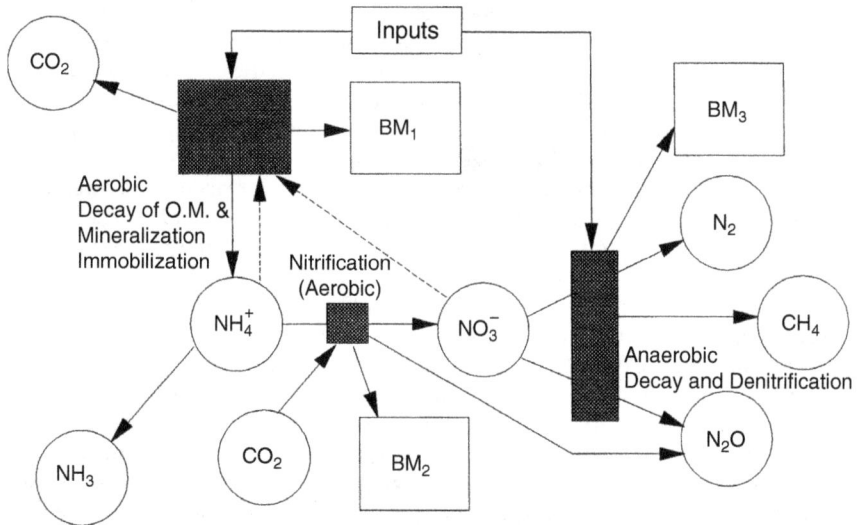

Figure 5.2 Aerobic and anaerobic processes in the OMNI sub-model; BM is microbial biomass and OM is organic matter.

temperature and water content, soil pH, soil O_2 levels, and solution concentrations (or activities) of nutrients.

During the last three decades, several process-based simulation models have been developed for a uniform plot to field scale, besides RZWQM-OMNI in 2000, for understanding the agricultural ecosystem C and N biogeochemical cycle and the response of N_2O emissions to different agricultural management (Chen et al., 2008; Pattey et al., 2007). Some examples of commonly used models are APSIM (McCown et al., 1996), DNDC (Li et al., 1992), DayCent (Del Grosso et al., 2000; Parton et al., 1998), *ecosys* (Grant, 2001a, 2001b; Grant & Pattey, 1999), ExpertN (Engel & Priesack, 1993), FASSET (Olesen et al., 2002), the NASA–Ames version of the Carnegie–Ames–Stanford Approach (CASA) model (Potter et al., 1997), NOE (Hénault et al., 2005), and WNMM (Li et al., 2007). In all these models, the nitrification and denitrification rates are generally estimated as a zero- or first-order kinetics function of the basic substrates (Chen et al., 2008). The rates are also affected by environmental factors such as soil water content, soil temperature, and soil pH. Some models include soil nitrifier or denitrifier microbial populations for estimating nitrification or denitrification (such as DNDC, *ecosys*, and RZWQM), whereas other models do not directly consider microbial involvement (such as NOE, ExpertN, FASSET, and WNMM). Nitrous oxide emissions are assumed to occur during both nitrification and denitrification. No one model has shown the best results in all situations. Mechanistic understanding of soil C/N dynamics, pools that can be related to the complexity of organic matter structure and measured data, and computational approaches that may be widely applicable are continuously evolving (Campbell & Paustian, 2015; Robertson et al., 2018).

Parton et al. (2015) presented the current state-of-the-science principles and approaches for modeling the dynamics of SOM and nutrient cycling in agriculture and ecosystems. They presented the mathematical background for various reaction kinetics: zero- and first-order reactions, Michaelis–Menten kinetics for enzyme reactions, and the most general Monod kinetics for microbial growth. For modeling soil C and N dynamics, particularly N mineralization, based on these principles and mathematical approaches, they described the analytical exponential model for first-order kinetics, which assumes that the role of microbes is not limiting and applies reasonably well in simple cases of relatively homogeneous organic matter. To account for the changing nature of the organic matter during decomposition, the reaction rate constant (k) has been made a function of time or additional components are included. The summation of two or more first-order exponential equations has been shown to fit the experimental data (e.g., Molina et al., 1980; Richter et al., 1982). Agren and Bosatta (1998) proposed that the heterogeneity of the SOM could be described using a time-varying quality variable, representing accessibility of the C atom to microbial decomposers that produce compounds of different quality. On the contrary, Yang and Janssen (2000) proposed a single-pool, first-order model that allows k to change. It does account for the decrease in the mineralization rate with time. To account for the very heterogeneous nature of SOM, it is most commonly divided into multiple compartments or pools, with first-order kinetics in each pool and allowing for inter-pool transfers. Most often, the pools are defined conceptually, but they can also be measurable fractions. Defining the measurable pools is the current challenge. The pools generally are based on the ease of decomposition of the plant constituents of the SOM, such as the water-extractable materials, lignin, and cellulose contents (Paul, 2016). Biochemically, they reflect differences in stabilization mechanisms, bioavailability, and kinetic parameters.

Most current soil-C models simulate decay of the SOM pools using first-order kinetics and assume that the microbes decay the SOM but do not have much effect on the rate of decay. A number of models have included the effect of microbial population (microbial biomass) and physiological activity on decay rates (Ma et al., 2008; Parton et al., 2015). Assumptions in some models are that extracellular enzymes control the depolymerization of soil organic C to dissolved organic C (DOC) and the DOC controls the decomposition. Enzyme production is proportional to microbial biomass. These models typically use Michaelis–Menten type functions with known input parameters. Environmental factors, such as soil moisture, temperature, pH, and N level, can affect the formation of enzymes and their impact on decay. Microbial uptake of DOC for growth and energy are included. The microbial C use efficiency (CUE) is an important factor. The substrate–enzyme–microbial models assume that the soil environmental factors and the SOM pool characteristics affect the CUE, whereas conventional models generally use fixed values of CUE and assume CUE as a C transfer among pools. There is great potential to improve conventional models by including the impact of microbial activity, enzyme production, and environmental variables on CUE. Models of soil N dynamics generally include links to physical processes of water, solute, heat, and gas transport and physiochemical processes such as volatilization, adsorption, and fixation.

Some widely used SOM dynamics models are the Century/DayCent and DNDC models, while the most dynamic and complex of them is the *ecosys* model (Grant, 2001a, 2001b). The Century model (Parton et al., 1987) divides the surface litter and belowground litter based on the lignin/N ratio of structural components and partitions metabolic components into active, slow (intermediate), and passive (long-term decay) pools. Soil silt and clay contents have an important role in stabilizing the slow and passive pools. The first-order decay rates of each pool are a function of monthly values of soil water, temperature, and pH. The DayCent model is a daily time step version of the

Century model. It also simulates N_2, NO, and N_2 gas emissions from nitrification and denitrification (Parton et al., 2001) and CH_4 (Cheng et al., 2013). Denitrification is a function of soil NO_3, available labile C, water content, and an index of gas diffusivity. Redox potential is not represented explicitly. The Century and DayCent models do not simulate microbial growth and activity.

The DNDC (DeNitrification–DeComposition) model has four major sub-models—soil-climate, plant growth, denitrification, and decomposition—and has discrete SOM pools: residue litter, living microbes, humus, and passive humus (Li et al., 1992). Each of the pools has two to three sub-pools, with specific C/N ratios and decomposition rates. The decomposition module provides NH_4^+ and DOC for nitrification and denitrification sub-modules. The nitrification sub-module predicts the growth and death of nitrifiers, the nitrification rate, and N_2O and NO production as regulated by DOC and soil environmental variables. The denitrification module simulates denitrification and changes in the population of denitrifiers as a function of the substrates. The fluxes are based on substrates, soil aeration, and gas diffusion. Chemo-denitrification is considered a source of NO production. The soil anaerobic volume depends on O_2 diffusion and the respiratory activity of microbes and roots. Denitrification is governed by the Nernst and Michaelis–Menten equations. A larger size of the anaerobic zone decreases diffusion of gases and increases N_2 gas emissions.

The *ecosys* model (Grant, 2001a, 2001b) is a three-dimensional model developed in Canada and represents a full range of management practices and soil processes that affect SOM and nutrient dynamics. The processes include solute and gaseous transport, the full range of soil ionic equilibria, oxidation–reduction reactions, microbial activities, and plant growth and nutrient uptake. It contains one of the most detailed and broad-based soil nutrient models currently available.

A comprehensive review and comparison of C–N dynamics in nine U.S. models were made in 2001 (Ma & Shaffer, 2001). Many of the above-mentioned models and processes were discussed in some detail. A similar review of C and N for soil management was also completed for European N dynamics models (McGechan & Wu, 2001).

Based on these reviews of the current science, the RZWQM-OMNI model remains a leading one-dimensional, multicomponent model of SOM and C–N transformations that includes simulation of microbial growth. It was carefully formulated and parameterized based on field data, the knowledge gained from other models, and literature research, as noted above. We have added and tested a section on N_2O gas emissions (Gillette et al., 2017). Important details of OMNI are presented below.

Carbon and Nitrogen Pools

In the model, organic matter (OM) is distributed across five computational pools and is decomposed by three microbial biomass populations (Table 5.1, Figure 5.2). The OM pools consist of slow and fast pools for crop residues and other organic amendments and fast, medium, and slow pools for decaying soil OM, respectively. The fast and medium soil OM pools approximately correspond to the potentially mineralizable N pool (N_0) frequently mentioned in the literature (Stanford & Smith, 1972). For use in OMNI, N_0 input data should be determined on soils that do not contain fresh additions of crop residues, manure, or other forms of organic matter. An even better technique to establish initial values for the SOM and microbial population pools is to run the model using 10–20 years of historical management and climate data. This allows the model to initialize itself with respect to the faster pools and avoids problems with

Table 5.1 OMNI State Variables

State variables	C or N pool	Interpretation
s_1	C	Organic matter Pool 1 (slow decaying structural material), μg C g^{-1} soil
s_2	C	Organic matter Pool 2 (fast decaying metabolic material), μg C g^{-1} soil
s_3	C	Organic matter Pool 3 (fast decaying soil OM), μg C g^{-1} soil
s_4	C	Organic matter Pool 4 (medium decaying soil OM), μg C g^{-1} soil
s_5	C	Organic matter Pool 5 (slow decaying soil OM), μg C g^{-1} soil
s_6	C	CO_2 sink/source, μg C g^{-1} soil
s_7	C	Heterotrophic biomass population 1 (soil decomposers) MBM$_1$, no. of organisms g^{-1} soil
s_8	C	Autotrophic biomass population (nitrifiers) BM$_2$, no. of organisms g^{-1} soil
s_9	C	Heterotrophic biomass population 2 BM$_3$ (facultative anaerobes), no. of organisms g^{-1} soil
s_{10}	C	Urea, mol urea L^{-1} pore water
s_{11}	N	NO_3, mol NO_3 L^{-1} pore water
s_{12}	N	NH_4, mol NH_4 L^{-1} pore water
s_{13}	N	N_2 sink, mol N L^{-1} pore water
s_{14}	N	NH_4 mineralized, mol N L^{-1} pore water
s_{15}	N	NO_3 immobilized, mol N L^{-1} pore water
s_{16}	N	N_2O sink, mol N L^{-1} pore water
s_{17}	N	NH_3 volatilized, mol N L^{-1} pore water
s_{18}	N	NH_4 immobilized, mol N L^{-1} pore water
s_{19}	C	CH_4 source/sink, μg C g^{-1} soil

field or laboratory estimates of pool sizes and their application to the soil horizon structure of a particular simulation. The user should also make use of data on measured total SOM levels. This can be used to help establish values for the slow soil organic pool using relations to climate conditions (Nendel et al., 2019) and to set the pool totals to field measurements. Mineralization rates from the model can be tested against independent laboratory and field methods to see if the expected ranges are being simulated.

The partitioning of OM in the model is based, in part, on the work of Parton et al. (1983), Juma and McGill (1986), and others. The BM populations include two heterotrophic groups (soil fungi and facultative bacteria, Populations 1 and 3) and one autotrophic group (nitrifiers, Population 2). Populations 1 and 2 are strict aerobes, while Population 3 is primarily anaerobic. For computational (mass balance) and conceptual purposes, C sink–source compartments for CO_2 and CH_4 gases are also included. These gases behave as byproducts as well as sources during various parts of the C and N cycle. Nitrogen concentrations are simulated in organic form as organic residues, SOM, and microbial biomass according to specified OM pool and microbial C/N ratios, in mineral form as NO_3_N and NH_4_N, as urea fertilizer, and as general N sinks (N_2 and N_2O). There is a total of $5 + 1 + 3 + 1 + 1$ C pools and $2 + 1 + 2 + 2 + 1$ direct N pools in the model, giving a total of 19 state vectors. Each pool corresponds to a particular C or N species. Most C compartments also indirectly contain N that is maintained via appropriate C/N ratios.

In the model, soil humus is depleted by microbial decay and built up by the addition of dead biomass and inter-pool transfers. Biomass populations are depleted through death and built up by biomass assimilation and growth during decay of OM and nitrification. Growth of heterotrophs occurs primarily by aerobic decay of OM. However, some facultative bacteria grow during denitrifying activity or CH_4 production under anaerobic conditions. The autotrophs grow exclusively as a result of nitrifying activity under aerobic conditions. Urea is converted to NH_4^+ by the enzymatic process of hydrolysis. The C source–sink storage slots (s_6 and s_{19}) increase as a result of MBM respiration, receiving CO_2 from aerobic respiration and CH_4 from anaerobic respiration. Carbon dioxide behaves as a C source during nitrification, and CH_4 is a C source during aerobic OM decay.

Mineral N in solution exists as NO_3^- and NH_4^+. Ammonium is the primary form in the sense that NO_3^- is formed only by microbial oxidation of NH_4^+ through the process of nitrification. Solution NH_4^+ may be adsorbed onto the surfaces of soil clays by the process of cation exchange, making it temporarily unavailable for leaching. Conversion of organic N to mineral N first produces NH_4^+, then NO_3^-. Nitrate is removed by the process of denitrification and immobilization to complex organic forms. Ammonium is formed by the hydrolysis of urea and by transformations of organic N along mineralizing pathways. Conversely, NH_4^+ is removed via NH_3 volatilization and nitrification and by immobilization.

There are many processes operating simultaneously on the various N and C species in OMNI. However, only a subset of those processes is modeled independently by rate equations. For mass balance consistency, the remaining processes are modeled as functions of the specified independent rates. For example, OM decay is calculated by specified first-order rate equations. Aerobic microbial heterotrophic growth, on the other hand, is calculated as a function of the decay based on the OM and MBM C/N ratios and specified assimilation efficiency factors. Similarly, the augmentation of OM pool s_3 is calculated as a function MBM death, which is modeled by an independent set of rate equations. This conceptual model of OM and N cycling in OMNI is relatively straightforward. However, mass balance must be maintained in the system when sources become limited. For example, when the demand for mineral N (NH_4^+ and NO_3^-) during immobilization exceeds the supply (storage plus mineralization), various rates must be adjusted to bring demand into balance with the available supply. Not to do so would result in negative mineral N quantities (physically meaningless) and produce numerical instability. Similar adjustments are required when the total OM decay (via aerobic decay and denitrification-related anaerobic decay) exceeds the available OM supply and when calculated death rates exceed the available biomass population. The algorithms for handling these situations are complex.

This description has intentionally been kept at a general conceptual level. The quantitative specification of the algorithm is bound up with its implementation in a computer program. Therefore, a description of the basic rate equations, the associated bookkeeping operations for calculating dependent process rates, the algorithms for handling special cases, and the physical and chemical assumptions are given in the context of the OMNI computer program.

Fundamental Process Rate Equations

Additional detailed information on OMNI rate process equations can be found in Shaffer and Ma (2001), Ma et al. (2001), and Ma et al. (2008).

Aerobic Decay of Organic Matter

As noted above, OM is divided into five separate pools: plant or other organic structural material (slow residue pool), plant or other organic metabolic material (fast residue pool), and fast-, medium-, and slow-decaying SOM. Concentrations of these OM pools (in $\mu g \, C \, g^{-1}$ soil) comprise the first five state variables. The basic form of the decay rate equations is the same for all pools, the only difference being in the values of the user-supplied rate coefficients. The equations are first order with respect to the C substrate source, and decomposition rates are given by

$$r_{dec,i} = K_{dec,i}s_i \tag{5.1}$$

where i is an OM pool index, $1 \leq i \leq 5$; $r_{dec,i}$ is the decomposition rate (μg C g^{-1} soil d^{-1}); and s_i is the C substrate concentration (μg C g^{-1} soil). The first-order rate coefficient for decay, $K_{dec,i}$ (d^{-1}), is determined as

$$K_{dec,i} = f_{aer}\left(\frac{k_b T}{h_p}A_i\right)\exp\left(-\frac{E_{ai}}{R_g}T\right)\frac{[O_2]}{[H]^{kh}\lambda_1^{kh}}POP_{het} \tag{5.2}$$

where f_{aer} is the factor for the extent of aerobic conditions ($0 \leq f_{aer} \leq 1$); k is the Boltzman constant = 1.383×10^{-23} (J K^{-1}); T is the temperature factor (K) (for soil temperature $\leq T_{maxi}$, T = soil temperature; for soil temperature > T_{maxi}, $T = 2T_{maxi}$ – soil temperature); h_p is the Planck constant = 6.63×10^{-34} (J s); A_i is a pool-specific rate coefficient (s d^{-1}); E_{ai} is the apparent activation energy for the ith pool (kcal mol^{-1}) = $E_{0i} + k_i u$, where E_{0i} is the reference activation energy for the ith pool (kcal mol^{-1}), k_i is a salinity coefficient for the ith pool, and u is ionic strength (mol L^{-1} pore water); R_g is the universal gas constant = 1.99×10^{-3} kcal mol^{-1} K^{-1}; $[O_2]$ is the O_2 concentration in soil water assuming soil air O_2 is not limited (mol O_2 L^{-1} pore water); $[H]$ is the H ion concentration (mol H L^{-1} pore water); λ_1 is the activity coefficient for a monovalent ion; kh is the H ion exponent for decay of organic matter (=0.167 for pH $\leq$ 7.0 and –0.333 for pH > 7.0); and POP_{het} is the population of aerobic heterotrophic microbes (no. organisms g^{-1} soil).

These decay rates are due to aerobic action by heterotrophic fungi and bacteria. Hence, this is termed *aerobic decay*. The rates are a function of the microbial population size, POP_{het}. Even though decay reduces the amount of OM in the system, the decay rates are calculated as positive numbers by convention. Note that the apparent activation energy for the reaction, E_{ai}, is a function of the ionic strength u (salinity) of the soil solution (Shaffer, 1985). The factor f_{aer} for aerobic conditions is included to correct the soil solution O_2 concentrations for depletion of O_2 in the soil atmosphere. The method is based on work by Linn and Doran (1984) and Nommik (1956) and makes use of the soil water content expressed as water-filled pore space percentage. When a suitable direct algorithm for soil atmosphere O_2 gas concentrations is developed, this method of estimation will no longer be needed.

Nitrification

Nitrification is the conversion of NH_4^+ (pool s_{12}) to NO_3^- (pool s_{11}) by nitrifying autotrophic bacteria. Here both zero- and first-order rate equations are used based on current NH_4^+ concentrations (activities):

$$r_{nit,1} = -K1_{nit}ss_{12}\gamma_1 \tag{5.3}$$

and

$$r_{nit,0} = -K0_{nit} \tag{5.4}$$

where $r_{nit,1}$ is the first-order nitrification rate (mol NH_4 L^{-1} pore water d^{-1}); $r_{nit,0}$ is the zero-order nitrification rate (mol NH_4 L^{-1} pore water d^{-1}); s_{12} is the concentration of NH_4^+ ions (mol L^{-1} pore water); $K0_{nit}$ is the zero-order rate coefficient for nitrification (mol L^{-1} pore water d^{-1}), used when NH_4^+ activity is >3.0 $\times$ 10^{-3} mol L^{-1} pore water (the first-order rate coefficient multiplied by 3.0×10^{-3} to become zero order); and $K1_{nit}$ is the first-order rate coefficient for nitrification (d^{-1}), used when NH_4^+ activity is $\leq 3.0 \times 10^{-3}$ mol L^{-1} pore water, determined as

$$K1_{nit} = f_{ear}\left(\frac{k_b T_n}{h_p A_{nit}}\right) \exp\left(-\frac{E_{an}}{R_g} T_n\right) \frac{[O_2]^{0.5}}{[H]^{khn} \gamma_1^{khn}} POP_{aut} \tag{5.5}$$

where A_{nit} is the nitrification rate coefficient (s d^{-1} organism^{-1}); E_{an} is the apparent activation energy for nitrification (kcal mol^{-1}), $= E_{0n} + k_n u$, where E_{0n} is the reference activation energy for nitrification (kcal mol^{-1}) and k_n is the salinity coefficient for nitrification; T_n is the nitrification temperature (K) (for soil temperature $\leq T_{maxn}$, where T_{maxn} is the temperature at which the maximum nitrification rate occurs (K), T_n is the soil temperature; for soil temperature $> T_{maxn}$, $T_n = 2T_{maxn}$ − soil temperature); khn is the H ion exponent for nitrification ($=0.167$ for pH ≤ 7.0 and -0.333 for pH > 7.0); and POP_{aut} is the population of autotrophic microbes (no. organisms g^{-1} soil).

The concentration of NH_4^+ is reduced by nitrification at the rate r_{nit}. Conversely, the concentration of NO_3^- increases at the same rate minus the assimilation rate of NH_4–N by nitrifiers in the production biomass. The rate r_{nit} is taken as a negative number because it reduces the amount of NH_4^+ in the system.

Denitrification

Denitrification is the conversion of NO_3^- to N_2 and N_2O. Thus, denitrification serves to deplete NO_3^- in the system. The first-order rate equation is

$$r_{den} = -K_{den} s_{11} \gamma_1 \tag{5.6}$$

where r_{den} is the denitrification rate (mol NO_3 L^{-1} pore water d^{-1}); s_{11} is the concentration of NO_3^- ions (mol L^{-1} pore water); K_{den} is the first-order rate coefficient for denitrification (d^{-1}), determined as

$$K_{den} = f_{anaer}\left(\frac{k_b T}{h_p} A_{den}\right) \exp\left(-\frac{E_{ad}}{R_g} T_d\right) \frac{CSUBST}{[H]^{khd} \gamma_1^{khd}} POP_{ana} \tag{5.7}$$

where f_{anaer} is a factor for anerobic effect ($0 \leq f_{anaer} \geq 1$); A_{den} is the denitrification rate coefficient (s d^{-1} organism^{-1}); E_{ad} = apparent activation energy for denitrification (kcal mol^{-1}), $= E_{0d} + k_d u$, where E_{0d} is the reference activation energy for denitrification (kcal mol^{-1}) and k_d is the salinity coefficient for denitrification; T_d is the denitrification temperature (K) (for soil temperature $\leq T_{maxd}$, the temperature at which the maximum denitrification rate occurs (K), T_d = soil temperature; for soil temperature $> T_{maxd}$, $T_d = 2T_{maxd}$ − soil temperature); CSUBST is the weighted C substrate concentration (μg C g^{-1} soil); khd is the H ion exponent for denitrification (= 0.167 for pH ≤ 7.0 and -0.333 for pH > 7.0), and POP_{ana} is the population of anaerobic microbes (no. organisms g^{-1} soil).

The rate r_{den} is taken as negative because NO_3^- is diminished by denitrification. Under strongly reducing conditions where O_2 is depleted, the model produces essentially 100% N_2 gas and no N_2O. As the soil water content decreases toward the limit for denitrification of 60% water-filled pore space (WFPS), the system becomes less reducing and a greater percentage of N_2O gas is produced, reaching 100% at 60% WFPS. To accomplish this transition, a linear function is used between 60 and 90%WFPS.

Nitrous Oxide Gas Emission

Nitrous oxide emission occurs during both processes of nitrification and denitrification as a leakage. Fang et al. (2015) obtained four different algorithms from other models for N_2O emission during nitrification and denitrification, linked them to RZWQM, and tested them against the field experimental data. Better results were obtained by incorporating the algorithm for the NOE model (Hénault et al., 2005) for N_2O emission during nitrification and from the DayCent model (Del Grosso et al., 2000; Parton et al., 1998) during denitrification.

Nitrous oxide emissions during nitrification is estimated from the NOE model as

$$N_2O_{nit} = Fr_{N2O\ Nit\ NOE}F_{SW\ Nit\ NOE}R_{nit} \tag{5.8}$$

$$F_{SW\ Nit\ NOE} = \frac{0.4WFPS - 1.04}{WFPS - 1.04} \tag{5.9}$$

where $Fr_{N2O\ Nit\ NOE}$ is the fraction of nitrification for N_2O emissions, and 0.0016 was used as default value (Bessou et al., 2010); and $F_{SW\ Nit\ NOE}$ is the soil water factor for the O_2 availability effect on N_2O emission during nitrification (Khalil et al., 2004).

Nitrous oxide emissions from denitrification (N_2O_{den}) is calculated from the DayCent model as

$$N_2O_{den} = Fr_{N_2O\ Den\ DayCent}R_{den} \tag{5.10}$$

$$Fr_{N2O\ Den\ DayCent} = \frac{1}{1 + R_{NO-N2} + R_{N2-N2O}} \tag{5.11}$$

$$R_{NO-N2O} = 4 + \frac{9\tan^{-1}[0.75\pi(10D - 1.86)]}{\pi} \tag{5.12}$$

$$R_{N2-N2O} = \max\left[0.16k_1, k_1 \exp\left(-\frac{0.8[NO_3]}{[CO_2]}\right)\right] \\ \times \max(0.1, 0.015 \times WFPS \times 100 - 0.32) \tag{5.13}$$

$$k_1 = \max(1.5, 38.4 - 350D) \tag{5.14}$$

where $Fr_{N2O\ Den\ DayCent}$ is the fraction of denitrification for N_2O emissions; R_{NO-N2O} is the ratio of NO to N_2O; R_{N2-N2O} is the ratio of N_2 to N_2O; $[NO_3]$ is the soil NO_3–N content; D is diffusivity in the soil (Davidson & Trumbore, 1995); and WFPS is the water filled pore space (%).

Production of Methane Gas

Methane gas is produced under anaerobic conditions by facultative anaerobes when NO_3^- has been entirely depleted and sufficient C substrate is available. The production of the process rate equation is written

$$r_{meth} = -K_{meth}CSUBST \tag{5.15}$$

where r_{meth} is the rate of CH_4 production (mol CH_4 L^{-1} pore water d^{-1}) and K_{meth} is the first order rate coefficient for CH_4 production (d^{-1}), determined as

$$K_{meth} = \frac{f_{anaer}[(k_b T/h_p)A_{meth}]\exp(-E_{am}/R_g T)1.0}{[H]^{khm}\gamma_1^{khm}POP_{ana}} \tag{5.16}$$

where A_{meth} is a rate coefficient (s d^{-1}); E_{am} is the apparent activation energy for CH$_4$ production (kcal mol^{-1}), $= E_{0m} + k_m u$, where E_{0m} is the reference activation energy for CH$_4$ production (kcal mol^{-1}) and k_m is the salinity coefficient for CH$_4$ production; T_m is the CH$_4$ production temperature (K) (for soil temperature $\leq T_{maxm}$, where T_{maxm} is the temperature at which the maximum CH$_4$ production rate occurs [K], T_m = soil temperature; for soil temperature > T_{maxm}, $T_m = 2T_{maxm}$ − soil temperature); and khm is the H ion exponent for CH$_4$ production (=0.167 for pH ≤ 7.0 and −0.333 for pH > 7.0).

Hydrolysis of Urea

Urea [CO(NH$_2$)$_2$] is converted to NH$_4^+$ by the process of hydrolysis. The enzyme involved is urease. Urease activity in soil does not appear to be a function of microbial population size or soil water content. Again, this process is modeled as a first-order equation:

$$r_u = -K_u s_{10} \tag{5.17}$$

where r_u is the rate of hydrolysis of urea (mol urea L^{-1} pore water d^{-1}), s_{10} is the urea concentration (mol urea L^{-1} pore water), and K_u is the first-order rate coefficient for urea hydrolysis (d^{-1}), determined as

$$K_u = f_{aer} \left(\frac{k_b T_u}{h_p} A_u \right) \exp \left(-\frac{E_{au}}{R_g T_u} \right) \tag{5.18}$$

where A_u is the rate coefficient for urea hydrolysis (s d^{-1}), E_{au} is the apparent activation energy for the reaction, and T_u is the hydrolysis temperature (K) (for soil temperature > T_{maxh}, where T_{maxh} is the temperature at which the maximum hydrolysis rate occurs [K], T_u = soil temperature; for soil temperature > T_{maxh}, $T_u = 2T_{maxh}$ − soil temperature).

The hydrolysis rate r_u is taken as negative because urea is removed from the system by hydrolysis.

Aerobic Growth of Biomass: Heterotrophs

Biomass growth is not modeled directly with state equations specific to microbial development. Rather, growth is calculated from organic matter decay, assuming efficiency factors for conversion of decayed OM to biomass C in a manner such that mass is conserved. Decayed OM-C has three possible ultimate destinations: transfer to another OM pool (Figure 5.1), biomass assimilation, or the CO$_2$ sink via biomass respiration. Assume (a) a fraction FRT$_i$ of gross decay from each biomass pool is lost to inter-pool transfer, (b) an efficiency factor e_{max} for conversion of the remaining OM decay rate to biomass C, and (c) the remaining OM-C goes off as CO$_2$ to the C sink. Then the rate of C assimilation from the ith OM pool to the jth biomass population is

$$r_{Cass,i,j} = FRT_i e_{max} fr_j r_{dec,j} \quad \text{for } i = 1,5 \text{(OM pools)}, \ j = 1 \text{ or } 3 \text{(MBM pools)} \tag{5.19}$$

where fr$_j$ is the fraction of MBM Populations 1 and 3.

The corresponding net assimilation of decayed OM-N by biomass is

$$r_{Nass,i,j} = fr_j \left(r_{dec,j} - FRT_j r_{dec,i} \right) \left(\frac{1}{CN_i} - \frac{e_{max}}{CN_{j+6}} \right) \tag{5.20}$$
$$\text{for } i = 1,5 \text{ (OM pools)}, \ j = 1 \text{ or } 3 \text{ (MBM pools)}$$

where CN$_i$ is the C/N ratio of the ith OM pool (see Table 5.1 for correspondence between state variables, OM pools, and MBM populations) and CN$_{j+6}$ is the C/N ratio of MBM Populations 1 and 3. Note again that some decay and growth pathways are inherently mineralizing ($1/CN_i > e_{max}/CN_j$) while others are inherently immobilizing ($1/CN_i < e_{max}/CN_j$). It is important to remember that aerobic biomass growth is not an independent process; it is functionally dependent on the underlying aerobic OM decay processes.

The net N mineralization(+)–immobilization(–) by the jth heterotrophic biomass population across all OM pools is calculated by summing across the r_{Nass} values. If net immobilization is occurring, then sufficient NH_4^+ plus NO_3^- must be present to meet the demand. Otherwise, microbial growth and OM decay cannot proceed. OMNI tests for sufficient NH_4^+ plus NO_3^- and shuts down the decay process if the supply has been exhausted.

The assimilated C is less than the amount of OM-C decayed, since some of that decayed C goes to other OM pools via inter-pool transformations. Likewise, the assimilated C is also less than the total C uptake by biomass populations. The uptake is partitioned between assimilated C and C ultimately released as CO_2 in respiration.

Growth of Biomass: Autotrophs

Autotrophic organisms are distinguished by their reliance on inorganic sources of energy and CO_2 as a C source. In the OMNI model, the autotrophs are the agents of the nitrification process. They meet their energy requirements by converting NH_4^+ to NO_3^-. In order to relate this rate of decrease in NH_4^+ to an autotrophic growth rate, a conversion efficiency factor e_{nit} must be assumed. This factor will partition the nitrified NH_4–N between uptake by autotrophs and ultimate release as NO_3–N. The growth rate (as the rate of assimilation of C) is given as

$$r_{Cass,2} = -a_c e_{nit} CN_8 r_{nit} \qquad (5.21)$$

where $r_{Cass,2}$ is the autotrophic C assimilation during growth ($\mu g\ C\ g^{-1}$ soil d^{-1}), a_c is a units conversion factor [($\mu g\ N\ g^{-1}$ soil L^{-1})/(mol $NH_4\ L^{-1}$ pore water)], CN_8 is the C/N ratio of autotrophic biomass, and r_{nit} is the nitrification rate (mol $NH_4\ L^{-1}$ pore water).

The negative sign is necessary because r_{nit} is calculated as a negative number. In OMNI, the C value is taken from the CO_2 source/sink (state variable c_6) for purposes of mass balance. The corresponding immobilization of N (<0) by the autotrophic biomass is $r_{Nass,2}$ (mol $NH_4\ L^{-1}$ pore water), calculated as

$$r_{Nass,2} = e_{nit} r_{nit} \qquad (5.22)$$

The remaining fraction ($1 - e_{nit}$) is converted to NO_3^-.

Special Case: Growth under Anaerobic Conditions

Heterotrophs normally flourish only under aerobic conditions. This effect is handled through the f_{aer} factor in the OM decay rate equations. However, the facultative heterotrophs (Pool 3) are unique in that they can also survive under anaerobic conditions. As noted above, the facultative anaerobic bacteria participate in denitrification or CH_4 gas production depending on the supply of NO_3^-. Denitrification occurs exclusively if a sufficient supply of NO_3^- is available, and CH_4 production proceeds exclusively once the NO_3^- is depleted. In both cases, a sufficient supply of OM-C must be available to meet the needs of the microbes. A zero NO_3^- supply coupled with anaerobic conditions results in the production of CH_4 gas.

The total "anaerobic" OM decay is calculated as

$$r_{\text{adec}} = -a_{\text{den}} r_{\text{den}} \tag{5.23}$$

where r_{adec} is the rate of an aerobic OM decay (μg C g^{-1} soil d^{-1}), and a_{den} is the denitrification decay conversion factor [(μg C g^{-1} soil)/(mol NO$_3$ L^{-1} pore water)].

Remember that decay rates are calculated as positive numbers in OMNI. This decayed OM is taken up by the facultative biomass (Population 3, state variable s_9). The fraction e_{den} of the OM-C is assimilated (retained) by the biomass, while the remaining fraction ($1 - e_{\text{den}}$) is ultimately given off in respiration (CO_2 and/or CH_4):

$$r_{\text{CassA}} = e_{\text{den}} r_{\text{adec}} = -e_{\text{den}} a_c r_{\text{den}} \tag{5.24}$$

where r_{CassA} (μg C g^{-1} soil d^{-1}) is the rate of OM-C assimilation into the MBM, and e_{den} is the anaerobic (denitrification) assimilation efficiency factor.

The total decay rate used must now be distributed across the five OM pools. The rule used is that the relative aerobic and anaerobic decay rates are identical for each pool. Total anaerobic decay is given by

$$rr_{\text{dec},i} = \frac{r_{\text{dec},i}}{r_{\text{dec}}} \tag{5.25}$$

The relative (aerobic and anaerobic) decay rates (μg C g^{-1} soil d^{-1}) are then given by

$$r_{\text{dec}} = \sum_{i=1}^{5} r_{\text{dec},i} \tag{5.26}$$

These relative rates are used to scale the total anaerobic rate to give the individual pool anaerobic decay rates (μg C g^{-1} soil d^{-1}):

$$r_{\text{adec},i} = rr_{\text{dec},i} r_{\text{adec}} \quad 1 \le i \le 5 \tag{5.27}$$

Biomass Death

Death rates for the three biomass populations (s_i) are each calculated as first-order equations (no. of organisms g^{-1} soil d^{-1}):

$$r_{\text{det},i} = K_{\text{det},i} s_i, \quad i = 7, 8, 9 \tag{5.28}$$

Heterotrophs-1 (soil decomposers) (d^{-1}):

$$K_{\text{det},7} = f_{\text{aer}}^{-1} \left(\frac{k_b T}{h_p} A_{d,7} \right) \exp \left(\frac{-E_a}{R_g T} \right) \frac{[H] \gamma_1}{[O_2] \text{ CSUBST}} \tag{5.29}$$

where $A_{d,7}$ is a coefficient (d^{-1}).
Autotrophs (nitrifiers) (d^{-1}):

$$K_{\text{det},8} = f_{\text{aer}}^{-1} \left(\frac{k_b T}{h_p} A_{d,8} \right) \exp \left(\frac{-E_a}{R_g T} \right) \frac{[H]}{[O_2] c_{12}} \tag{5.30}$$

where $A_{d,8}$ is a coefficient (d^{-1}).
Heterotrophs-2 (denitrifiers):
aerobic part (d^{-1}):

$$K_{\text{det},9,a} = f_{\text{aer}}^{-1} \left(\frac{k_b T}{h_p} A_{d,8} \right) \exp \left(\frac{-E_a}{R_g T} \right) \frac{[H] \gamma_1}{[O_2] \text{CSUBST}} \tag{5.31}$$

anaerobic part (d^{-1}):

$$K_{det,9,an} = f_{anaer}^{-1} \left(\frac{k_b T}{h_p} A_{d,9} \right) \exp \left(\frac{-E_a}{R_g T} \right) \frac{[H]\gamma_1}{c_1 CSUBST} \tag{5.32}$$

sum of aerobic and anaerobic parts:

$$K_{det,9} = K_{det,9,a} + K_{det,9,an} \tag{5.33}$$

Even though death depletes the biomass populations, the death rates are calculated as positive numbers.

Computation of O_2 Saturation Solubility in Water

Whenever the pore water concentration of O_2 is used in calculating first-order rate coefficients, it is assumed that the pore water is saturated with O_2. Saturation O_2 solubility is tabulated in most standard references; alternatively, it can be calculated by Henry's Law. Calculation of the O_2 concentration would be very difficult if saturation were not assumed.

Saturation O_2 solubility (in mol L^{-1} pore water) is calculated either by Henry's Law or by interpolation using a look-up table, depending on the prevailing temperature. The look-up table for the temperature range $0.0 \leq T$ (°C) ≤ 30) is given in Metcalf and Eddy, Inc. (1978). For temperatures T (°C) > 30, O_2 solubility is calculated by Henry's Law:

$$[O_2] = \frac{a \, pO_2}{1000} H_1 \left[\text{moles LPW}^{-1} \right] \tag{5.34}$$

where $[O_2]$ is the $O_2(g)$ saturation solubility in water (mol L^{-1} pore water), pO_2 is the partial pressure of O_2 (atm), H_1 is Henry's Law coefficient, and a is the mol H_2 L^{-1} pore water = 55.556 mol H_2 L^{-1} pore water. The coefficient H_1 is a function of temperature (Metcalf and Eddy, Inc., 1978). Coefficient values for temperatures between the given values are calculated by linear interpolation.

Correction Factors for Oxygen and Moisture Effects

Correction for the relative effects of aerobic and anaerobic conditions is accomplished by scaling the calculated pseudo-first-order rate coefficients by the factors f_{aer} (aerobic) and f_{anaer} (anaerobic). They vary between 0 and 1. For strongly aerobic conditions, f_{aer} is near 1 and f_{anaer} is near 0. The opposite holds for strongly anaerobic conditions. The actual O_2 content of the soil water is not directly estimated. Rather, the soil water content and O_2 saturation solubility are used as a surrogate for determining aerobicity and anaerobicity (Caskey & Schepers, 1985). When the soil water content is high, conditions are strongly anaerobic. The opposite holds at low water contents, the assumption being that the higher O_2 content of the soil raises the O_2 content of the small amount of water in the soil. The values of the factors f_{aer} and f_{anaer} are summarized in Table 5.2 for several values of WFPS.

The WFPS is a function of soil moisture content and is calculated as

$$\text{WFPS} = \frac{100w}{1.0 - \text{bd/pd}} \tag{5.35}$$

where w is moisture content (cm^3 cm^{-3}), bd is soil bulk density (g cm^{-3}), and pd is soil particle density (g cm^{-3}).

Table 5.2 Correction Factors f_{aer} and f_{anaer} for Aerobic and Anaerobic Effects, Respectively, on Rate Coefficients

PWFPS	f_{aer}	PWFPS	f_{anaer}
≤60	1.00	≤60	0.00
70	0.40	80	0.13
80	0.10	100	1.00
≥87	0.00		

Linear interpolation between tabulated factors is used when the actual percent water filled pore space (PWFPS) WFPS is PWFPS/100 value falls between the tabulated values. It is important to note that even though they are inversely related in a crude manner, they are calculated independently of one another. Note that the values for f_{aer} equal 1.0 for PWFPS ≤60. Reductions in microbial activity at low soil water contents are estimated using the relationships for the effects of ionic strength on apparent activation energy.

Model Testing, Verification, and Calibration

Complex models such as OMNI must be thoroughly tested for mass balance before calibration and use. The OMNI model includes internal algorithms for mass balance with respect to C and N. Both constituents were tested under various scenarios to be sure the model strictly obeys the First Law of Thermodynamics. The method involves tracking all input, output, and storage compartments in the model to be certain that mass is being conserved. To accomplish this, a variety of mineral and organic sources of N were applied to the soil and the model was run under both aerobic and anaerobic conditions. In all cases, the maximum observed mass balance error was no greater than 0.01 and 0.001% of the total initial N and C, respectively, in the system after simulations of 300 d.

Typical results obtained with the model are shown in Figures 5.3–5.5. Figure 5.3 shows the dynamic simulated results obtained for various N species under aerobic conditions after a source of organic residues high in fast-pool constituents (such as alfalfa) was added to the soil along with 20 µg g^{-1} urea fertilizer. Note the rapid growth of the aerobic heterotrophs along with the relatively rapid decay of the fast residue pool. Also note the transient buildup of NH_4–N followed by the appearance of NO_3–N as mineralization progresses, the initial urea-N is hydrolyzed to NH_4–N, and the NH_4–N is nitrified. The gradual increase in NO_3–N after about Day 50 is due to microbial death and the subsequent mineralization and nitrification of the microbial biomass N.

Figure 5.4 shows simulated results obtained under conditions similar to those in Figure 5.3 except that a source of organic residues high in slow-pool constituents (such as wheat straw) was added to the system. In this case, NH_4–N and NO_3–N were rapidly immobilized by the aerobic heterotrophs. Note the relatively large buildup of N associated with the microbes. Once the slow residue pool was exhausted, microbial death and the subsequent mineralization and nitrification of the microbial biomass N resulted in increased NO_3–N after about 180 days.

Figure 5.5 illustrates denitrification of NO_3–N, with accompanying decrease in the population of aerobic heterotrophs. In this anaerobic situation, the typical aerobic processes seen in Figures 5.3 and 5.4 do not operate.

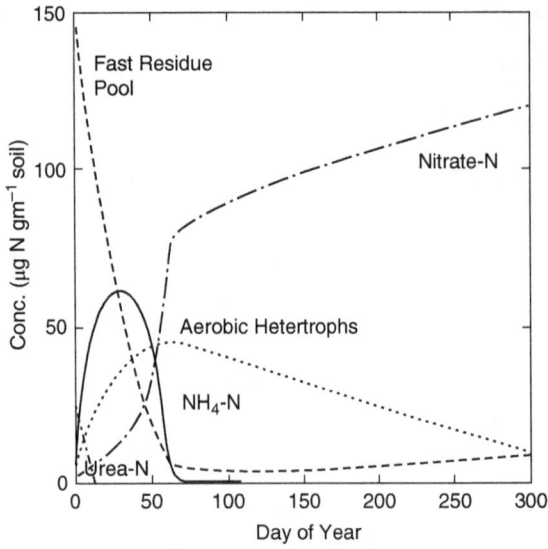

Figure 5.3 Mineralization of the fast residue pool under aerobic conditions.

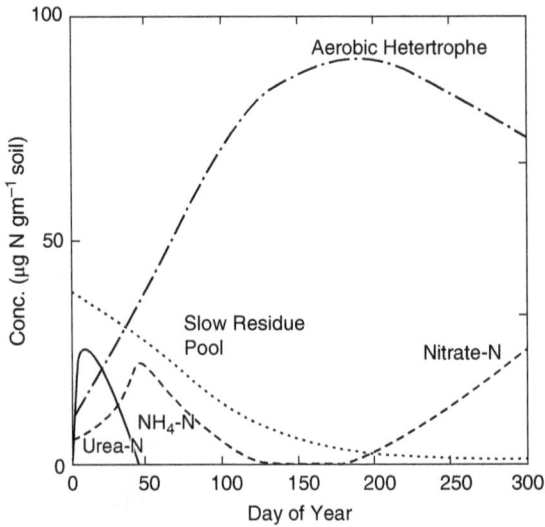

Figure 5.4 Immobilization of mineral N under aerobic conditions.

Figures 5.6–5.8 show comparisons of calibrated results with observed field data published by Broder and Wagner (1988) for decay of wheat, corn, and soybean residues incorporated into Mexico silt loam soil in Columbia, MO. Note that the OMNI two-pool model for residue decay fits the observed decay curves quite well. The C/N ratios

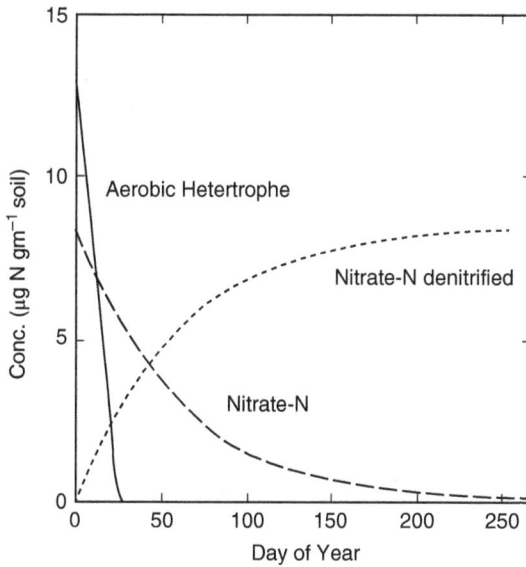

Figure 5.5 Denitrification under anaerobic conditions.

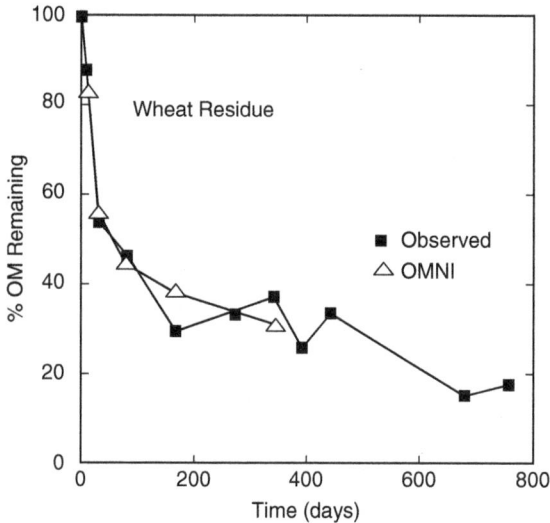

Figure 5.6 Calibrated versus observed decay of wheat residue; OM is organic matter.

for the residues used in the study were 43.5, 39.6, and 18.2 for wheat, corn, and soybean residues, respectively. The rather close values of the C/N ratios for wheat and corn explain why the observed (and predicted) curves for these residues were similar. More typical values for wheat C/N ratios range from 60 to 120.

171

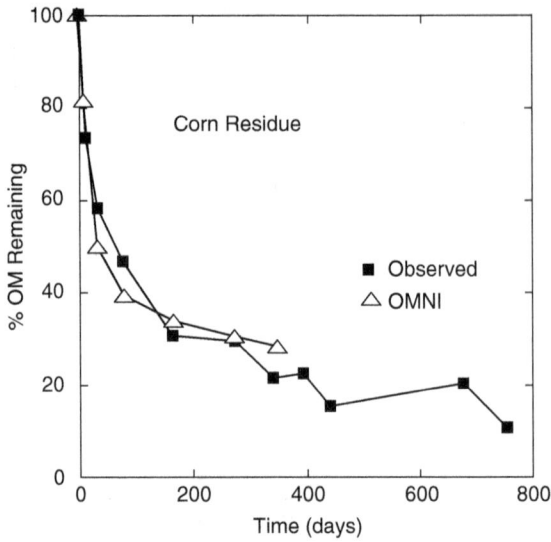

Figure 5.7 Calibrated versus observed decay of corn residue; OM is organic matter.

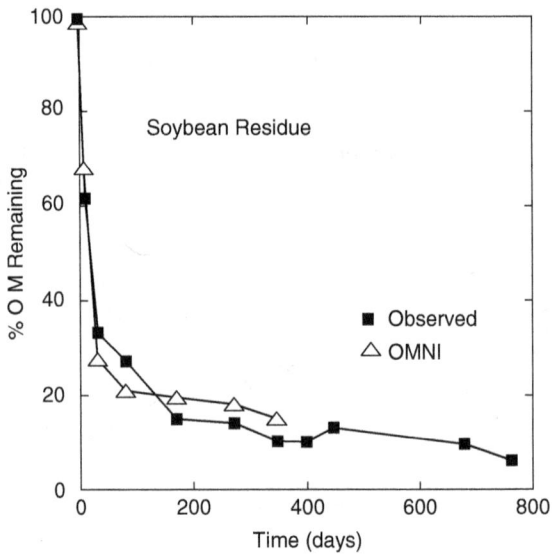

Figure 5.8 Calibrated versus observed decay of soybean residue; OM is organic matter.

Field Tests of OMNI in the RZWQM

Sensitivity analyses (Ma et al., 2000) along with extensive field application, parameterization, and testing of the RZWQM model and the OMNI sub-model have been done across the United States (Bakhsh et al., 2004; Gillette et al., 2017; Hu et al., 2006;

Ma et al., 2001, 2007, 2008, 2012; Malone et al., 2019; Nolan et al., 2010) and with international applications (Cameira et al., 2007; Fang et al., 2013). OMNI was tested with field applications of manure (Ma et al., 1998a; 1998b).

The OMNI sub-model has been used as a template or basis for N equations in related models such as GPFARM (Shaffer et al., 2004), the GPFARM-Range model (Qi et al., 2012a), and the NLEAP model (Ascough & Delgado, 2015; Shaffer et al., 2001).

OMNI has been shown to function well with companion N balance, leaching, and drainage submodels in the RZWQM (Fang et al., 2008, 2013; Ma et al., 2008; Nolan et al., 2010; Qi et al., 2012b; Singer et al., 2011; Sophocleous et al., 2009).

These theoretical verification–calibration tests and field tests have shown that OMNI is a robust model for soil organic matter and C–N transformations. Further enhancements and tests of the model with field experimental data will continue for broader practical applications.

Future Work on C–N Transformations

The most important thing with all soil process models (including OMNI) is to continuously field test them under a variety of soil types and cropping systems and different climates, with comprehensive experimental data. We also need to evaluate and possibly refine the SOM pools based on the complexity of chemical–physical–biological structures and measurable fractions.

We need C–N models that are based on a variety of approaches from related disciplines. For example, the OMNI model is derived from a physical chemistry rate process background as opposed to a more classical laboratory enzyme decay or empirical statistical analysis. Parallel development by related disciplines helps provide alternative comparable results and reduces dependence on single approaches.

Many of the current C–N models have become increasingly complex to the point where only the developers and their students can effectively make use of these tools. Thus, state and federal regulators and resource managers often do not have direct access to proper evaluation tools, or they may even misapply research-level models. Developers of C–N models need to consider downsizing their research simulation models to versions more useful to local managers and regulators. This task may actually be more difficult (and less exciting) than extending theory to the next level. However, these types of tools are badly needed—provided they actually work and have been field tested.

To be useful under field conditions, C–N models must be coupled to soil–crop–solute transport models with associated GIS data systems capable of supplying the soil and climate spatial and temporal input information required. Multi-process coupling has been achieved for about 30 yr, but the introduction of GIS coupling at the field scale is relatively new (Ascough & Delgado, 2015). This allows the application of one-, two-, and three-dimensional models to farm and field conditions. The introduction and now widespread use of precision agriculture techniques provides the opportunity for management applications of GIS-coupled C–N models—again, another case for the use of downsized C–N models, this time on farms.

With a comprehensive review of the literature, Campbell and Paustian (2015) suggested further studies and research model developments in the following areas:

- the role of microbes in SOM decomposition and stabilization in relation to soil aggregation, clay–humus complexes, and less biodegradable litter components.

- SOM saturation, which will limit the soil C increase.

- temperature and moisture controls on C–N transformations—the soil temperature and moisture response functions can be modulated by soil matrix properties and the chemical nature of the substrate; thus the response functions could vary with the pool.

- extension of C–N transformations to deeper soil layers below the top 20–30-cm layer. This will need to consider (a) movement of dissolved or labile C; (b) root growth exudates and turnover; (c) physical deeper movement of matter through bioturbation or physical soil processes; and (d) microbial access.

References

Agren, G. I., & Bosatta, E. (1998). *Theoretical ecosystem ecology: Understanding element cycles.* Cambridge University Press.

Alexander, M. (1977). *Introduction to soil microbiology.* John Wiley and Sons.

Ascough, J. C., II, & Delgado, J. A. (2015). Modeling landscape-scale nitrogen management for conservation. In T. Mueller & G. F. Sassenrath (Eds.), *GIS applications in agriculture. Volume 4, Conservation planning* (pp. 101–118). CRC Press. https://doi.org/10.1201/b18173-8

Bakhsh, A., Hatfield, J. L., Kanwar, R. S., Ma, L., & Ahuja, L. R. (2004). Simulating nitrate drainage losses from a Walnut Creek watershed field. *Journal of Environmental Quality, 33,* 114–123. https://doi.org/10.2134/jeq2004.1140

Bessou, C., Mary, B., Léonard, J., Roussel, M., Gréhan, E., & Gabrielle, B. (2010). Modelling soil compaction impacts on nitrous oxide emissions in arable fields. *European Journal of Soil Science, 61,* 348–363. https://doi.org/10.1111/j.1365-2389.20. https://doi.org/10.01243.x

Broder, M. W., & Wagner, G. H. (1988). Microbial colonization and decomposition of corn, wheat, and soybean residue. *Soil Science Society of America Journal, 52,* 112–117. https://doi.org/10.2136/sssaj1988.03615995005200010020x

Cameira, M. R., Fernando, R. M., Ahuja, L. R., & Ma, L. (2007). Using RZWQM to simulate the fate of nitrogen in field soil–crop environment in the Mediterranean region. *Agricultural Water Management, 90,* 121–136.

Campbell, E. E., & Paustian, K. (2015). Current developments in soil organic matter modeling and the expansion of model applications: A review. *Environmental Research Letters, 10,* 123004.

Caskey, W., & Schepers, J. (1985). Modeling of microbial activity: Mineralization, immobilization, nitrification, and denitrification. In D. G. Decoursey (Ed.), *Proceedings of the Natural Resources Modeling Symposium,* (Publication ARS-30 (pp. 197–201). USDA–ARS.

Chen, D., Li, Y., Grace, P., & Mosier, A. R. (2008). N₂O emissions from agricultural lands: A synthesis of simulation approaches. *Plant and Soil, 309,* 169–189.

Cheng, K., Ogle, S. M., Parton, W. J., & Pan, G. (2013). Predicting methanogenesis from rice paddies using the DayCent ecosystem model. *Ecological Modeling, 261,* 19–31.

Davidson, E. A., & Trumbore, S. E. (1995). Gas diffusivity and production of CO_2 in deep soils of the eastern Amazon. *Tellus B, 47,* 550–565.

Del Grosso, S. J., Parton, W. J., Mosier, A. R., Ojima, D. S., Kulmala, A. E., & Phongpan, S. (2000). General model for N₂O and N₂ gas emissions from soils due to denitrification. *Global Biogeochemical Cycles, 14,* 1045–1060.

Engel, T., & Priesack, E. (1993). Expert-N: A building block system of nitrogen models as resource for advice, research, water management and policy. In H. J. P. Eijsackers & T. Hamers (Eds.), *Integrated soil and sediment research: A basis for proper protection* (pp. 503–507). Kluwer Academic Publishers.

Fang, Q., Ma, L., Yu, Q., Malone, R. W., Saseendran, S. A., & Ahuja, L. R. (2008). Modeling nitrogen and water management effects in a wheat–maize double-cropping system. *Journal of Environmental Quality, 37,* 2232–2242. https://doi.org/10.2134/jeq2007.0601

Fang, Q. X., Ma, L., Halvorson, A. D., Malone, R. W., Ahuja, L. R., Del Grosso, S. J., & Hatfield, J. L. (2015). Evaluating four nitrous oxide emission algorithms in response to N rate on an irrigated corn field. *Environmental Modeling & Software, 72,* 56–70.

Fang, Q. X., Ma, L., Yu, Q., Hu, C. S., Li, X. X., Malone, R. W., & Ahuja, L. R. (2013). Quantifying climate and management effects on regional N leaching and crop yield in the North China Plain. *Journal of Environmental Quality, 42,* 1466–1479. https://doi.org/10.2134/jeq2013.03.0086

Frissel, M. J., & van Veen, J. A. (1981). Simulation model for nitrogen immobilization and mineralization. In J. K. Iskandar (Ed.), *Modeling wastewater renovation: Land treatment* (pp. 359–381). John Wiley & Sons.

Gillette, K., Ma, L., Malone, R. W., Fang, Q. X., Halvorson, A. D., Hatfield, J. L., & Ahuja, L. R. (2017). Simulating N_2O emissions under different tillage systems of irrigated corn using RZ-SHAW model. *Soil & Tillage Research*, 165, 268–278.

Grant, R. F. (2001a). A review of the Canadian ecosystem model—*ecosys*. In M. J. Shaffer, L. Ma, & S. Hansen (Eds.), *Modeling carbon and nitrogen dynamics for soil management* (pp. 173–264). Lewis Publishers.

Grant, R. F. (2001b). Modeling transformations of soil organic carbon and nitrogen at differing scales of complexity. In M. J. Shaffer, L. Ma, & S. Hansen (Eds.), *Modeling carbon and nitrogen dynamics for soil management* (pp. 597–630). Lewis Publishers.

Grant, R. F., & Pattey, E. (1999). Mathematical modelling of nitrous oxide emissions from an agricultural field during spring thaw. *Global Biogeochemical Cycles*, 13, 679–694.

Hansen, S., Shaffer, M. J., & Jensen, H. E. (1995). Developments in modeling nitrogen transformations in soil. In P. E. Bacon (Ed.), *Nitrogen fertilization and the environment* (pp. 83–107). Marcel Dekker.

Hénault, C., Bizouard, F., Laville, P., Gabrielle, B., Nicoullaud, B., Germon, J. C., & Cellier, P. (2005). Predicting in situ soil N_2O emission using NOE algorithm and soil database. *Global Change Biology*, 11, 115–127.

Hu, C., Saseendran, S. A., Green, T. R., Ma, L., Li, X., & Ahuja, L. R. (2006). Evaluating nitrogen and water management in a double-cropping system using RZWQM. *Vadose Zone Journal*, 5, 493–505. https://doi.org/10.2136/vzj2005.0004

Juma, N. G., & McGill, W. B. (1986). Decomposition and nutrient cycling in agro-ecosystems. In M. J. Mitchell & J. P. Nakas (Eds.), *Microfloral and faunal interactions in natural and agro-ecosystems* (pp. 74–136). Martinus Nijhoff/Dr. W. Junk Publishers.

Khalil, K., Mary, B., & Renault, P. (2004). Nitrous oxide production by nitrification and denitrification in soil aggregates as affected by O_2 concentration. *Soil Biology & Biochemistry*, 36, 687–699.

Laidler, K. J. (1969). *Theories of chemical reaction rates*. McGraw-Hill.

Li, C., Frolking, S., & Frolking, T. A. (1992). A model of nitrous oxide evolution from soil driven by rainfall events: I. Model structure and sensitivity. *Journal of Geophysical Research*, 97, 9759–9776.

Li, Y., Chen, D. L., White, R. E., Zhang, J. B., Li, B. G., Zhang, Y. M., Huang, Y. F., & Edis, R. (2007). A spatially referenced water and nitrogen management model (WNMM) for (irrigated) intensive cropping systems in the North China Plain. *Ecological Modelling*, 203, 395–423.

Linn, D. M., & Doran, J. W. (1984). Effect of water-filled pore space on carbon dioxide and nitrous oxide production in tilled and non-tilled soils. *Soil Science Society of America Journal*, 48, 1267–1272. https://doi.org/10.2136/sssaj1984.03615995004800060013x

Ma, L., Ahuja, L. R., Nolan, B. T., Malone, R. W., Trout, T. J., & Qi, Z. (2012). Root Zone Water Quality Model (RZWQM2): Model use, calibration and validation. *Transactions of the ASABE*, 55, 1425–1446.

Ma, L., Ascough, J. C., II, Ahuja, L. R., Shaffer, M. J., Hanson, J. D., & Rojas, K. W. (2000). Root Zone Water Quality Model sensitivity analysis using the Monte Carlo simulation. *Transactions of the ASAE*, 43, 883–895.

Ma, L., Malone, R. W., Heilman, P., Karlen, D. L., Kanwar, R. S., Cambardella, C. A., Saseendran, S. A., & Ahuja, L. R. (2007). RZWQM simulation of long-term crop production, water and nitrogen balances in northeast Iowa. *Geoderma*, 140, 247–259.

Ma, L., Malone, R. W., Jaynes, D. B., Thorp, K., & Ahuja, L. R. (2008). Simulated effects of nitrogen management and soil microbes on soil nitrogen balance and crop production. *Soil Science Society America Journal*, 72, 1594–1603. https://doi.org/10.2136/sssaj2007.0404

Ma, L., Scott, H. D., Shaffer, M. J., & Ahuja, L. R. (1998a). RZWQM simulations of water and nitrate movement in a manured tall fescue field. *Soil Science*, 163, 259–270.

Ma, L., Shaffer, M. J., Boyd, J. K., Waskom, R., Ahuja, L. R., Rojas, K. W., & Xu, C. (1998b). Manure management in an irrigated silage corn field: Experiment and modeling. *Soil Science Society America Journal*, 62, 1006–1017. https://doi.org/10.2136/sssaj1998.03615995006200040023x

Ma, L., Shaffer, M. J., & Ahuja, L. R. (2001). Application of RZWQM for soil nitrogen management. In M. J. Shaffer, L. Ma, & S. Hansen (Eds.), *Modeling carbon and nitrogen dynamics for soil management* (pp. 265–302). Lewis Publishers.

Ma, L., & Shaffer, M. J. (2001). A review of carbon and nitrogen processes in nine U.S. soil nitrogen dynamics models. In M. J. Shaffer, L. Ma, & S. Hansen (Eds.), *Modeling carbon and nitrogen dynamics for soil management* (pp. 55–102). Lewis Publishers.

Malone, R. W., Herbstritt, S., Ma, L., Richard, T. L., Cibin, R., Gassman, P. W., Zhang, H. H., Karlen, D. L., Hatfield, J. L., Obrycki, J. F., Helmers, M., Jaynes, D., Kaspar, T., Parkin, T., & Fang, Q. X. (2019). Corn stover harvest and N losses in central Iowa. *Science of the Total Environment*, 663, 776–792.

McCown, R. L., Hammer, G. L., Hargreaves, J. N. G., Holzworth, D. P., & Freebairn, D. M. (1996). APSIM: A novel software system for model development, model testing, and simulation in agricultural systems research. *Agricultural Systems*, 50, 255–271.

McGechan, M. B., & Wu, L. (2001). A review of carbon and nitrogen processes in European soil nitrogen dynamics models. In M. J. Shaffer, L. Ma, & S. Hansen (Eds.), *Modeling carbon and nitrogen dynamics for soil management* (pp. 103–171). Lewis Publishers.

Metcalf and Eddy, Inc. (1978). *Wastewater engineering: Treament, disposal, reuse* (2nd ed., pp. 275–878). McGraw-Hill.

Molina, J. A. E., Clapp, C. E., & Larson, W. E. (1980). Potentially mineralizable nitrogen in soil: The simple exponential model does not apply for the first 12 weeks of incubation. *Soil Science Society of America Journal*, 44, 442–443. https://doi.org/10.2136/sssaj1980.03615995004400020054x

Nendel, C., Meltzer, D., & Thorburn, P. J. (2019). The nitrogen nutrition potential of arable soils. *Scientific Reports*, 9, 5851. https://doi.org/10.1038/s41598-019-42274-y

Nolan, B. T., Puckett, L. J., Ma, L., Green, C. T., Bayless, E. R., & Malone, R. W. (2010). Predicting unsaturated zone nitrogen mass balances in diverse agricultural settings of the United States. *Journal Environmental Quality*, 39, 1051–1065. https://doi.org/10.2134/jeq2009.0310

Nommik, N. (1956). Investigations of denitrification in soil. *Acta Agriculturae Scandinavica*, 6, 195–228.

Olesen, J. E., Petersen, B. M., Berntsen, J., Hansen, S., Jamieson, P. D., & Thomsen, A. G. (2002). Comparison of methods for simulating effects of nitrogen on green area index and dry matter growth in winter wheat. *Field Crops Research*, 74, 131–149.

Parton, W. J., Anderson, D. W., Cole, C. V., & Stewart, J. W. B. (1983). Simulation of soil organic matter formations and mineralization in semiarid ecosystems. In R. R. Lowrance, R. L. Todd, L. E. Asmussen, & R. A. Leonard (Eds.), *Nutrient cycling in agricultural ecosystems* (Special Publication 23 (pp. 533–550). University of Georgia, College of Agriculture Experiment Station.

Parton, W. J., Del Grosso, S. J., Plante, A. F., & Adair, C. (2015). Modeling the dynamics of soil organic matter and nutrient cycling. In E. A. Paul (Ed.), *Soil microbiology, ecology, and biochemistry* (4th ed., pp. 505–537). Academic Press.

Parton, W. J., Hartman, M., Ojima, D. S., & Schimel, D. W. (1998). DAYCENT and its land surface submodel: Description and testing. *Global and Planetary Change*, 19, 35–48.

Parton, W. J., Holland, E. A., Del Grosso, S. J., Hartman, M. D., Martin, R. E., Mosier, A. R., & Ojima, D. S. (2001). Generalized model for NO_2 and N_2O emissions from soils. *Journal of Geophysical Research*, 106, 17403–17420.

Parton, W. J., Schimel, D. S., Cole, C. V., & Ojima, D. S. (1987). Analysis of factors controlling soil organic matter levels in Great Plains grasslands. *Soil Science Society of America Journal*, 51, 1173–1179. https://doi.org/10.2136/sssaj1987.03615995005100050015x

Pattey, E., Edwards, G. C., Desjardins, R. L., Pennock, D. J., Smith, W., Grant, B., & MacPherson, J. I. (2007). Tools for quantifying N_2O emissions from agroecosystems. *Agricultural and Forest Meteorology*, 142, 103–119.

Paul, E. A. (2016). The nature and dynamics of soil organic matter: Plant inputs, microbial transformations, and organic matter stabilization. *Soil Biology and Biochemistry*, 98, 109–126.

Potter, C. S., Riley, R. H., & Klooster, S. A. (1997). Simulation modeling of nitrogen trace gas emissions along an age gradient of tropical forest soils. *Ecological Modelling*, 97, 179–196.

Qi, Z., Bartling, P. N. S., Ahuja, L. R., Derner, J. D., Dunn, G. H., & Ma, L. (2012a). Development and evaluation of the carbon–nitrogen cycle module for the GPFARM-Range model. *Computers and Electronics in Agriculture*, 83, 1–10. https://doi.org/10.1016/j.compag.2012.01.007

Qi, Z., Ma, L., Helmers, M. J., Ahuja, L. R., & Malone, R. W. (2012b). Simulating NO_3–N concentration in tile drainage at various nitrogen application rates using RZWQM2. *Journal Environmental Quality*, 41, 289–295.

Richter, J., Nuske, A., Habenicht, W., & Bauer, J. (1982). Optimized N-mineralization parameters of loess soils from incubation experiments. *Plant and Soil*, 68, 379–388.

Robertson, A. D., Paustian, K., Ogle, S., Wallenstein, M. D., Lugato, E., & Cotrufo, M. M. (2018). Unifying soil organic matter formation and persistence frameworks: The MEMS model. *Biogeosciences Discussions*. https://doi.org/10.5194/bg-2018-430

Shaffer, M. J. (1985). Simulation model for soil erosion-productivity relationships. *Journal of Environmental Quality*, 14, 144–150. https://doi.org/10.2134/jeq1985.00472425001400010029x

Shaffer, M. J., Bartling, P. N. S., & McMaster, G. S. (2004). GPFARM modeling of corn yield and residual soil nitrate-N. *Computers and Electronics in Agriculture*, 43, 87–107.

Shaffer, M. J. & Dutt, G. R. (1974). Nitrification in soil-water systems: A computerized activated complex model. In *Proceedings of the First World Congress on Water Resources, Chicago, IL* (pp. 284–294). International Water Resources Association.

Shaffer, M. J., & Larson, W. E. (1987). *NTRM, a soil–crop simulation model for nitrogen, tillage, and crop-residue management* (Conservation Research Report 34-1). USDA–ARS.

Shaffer, M. J., Lasnik, K., Ou, X., & Flynn, R. (2001). NLEAP internet tools for estimating NO_3–N leaching and NO_2 emission. In M. J. Shaffer, L. Ma, & S. Hansen (Eds.), *Modeling carbon and nitrogen dynamics for soil management* (pp. 403–425). Lewis Publishers.

Shaffer, M. J., & Ma, L. (2001). Carbon and nitrogen dynamics in upland soils. In M. J. Shaffer, L. Ma, & S. Hansen (Eds.), *Modeling carbon and nitrogen dynamics for soil management* (pp. 11–26). Lewis Publishers.

Shaffer, M. J., Rojas, K. W., Decoursey, D. G., & Hebson, C. S. (2000). Nutrient chemistry processes—OMNI. In L. R. Ahuja, K. W. Rojas, J. D. Hanson, M. J. Shaffer, & L. Ma (Eds.), *Root Zone Water Quality Model: Modelling management effects on water quality and crop production* (pp. 119–136). Water Resources Publications.

Singer, J., Malone, R. W., Jaynes, D. B., & Ma, L. (2011). Cover crop effects on nitrogen load in tile drainage from Walnut Creek Iowa using RZWQM. *Agriculture Water Management*, 98, 1622–1628.

Sophocleous, M., Townsend, M. A., Vocasek, F., Ma, L., & Ashok, K. C. (2009). Soil nitrogen balance under wastewater management: Field measurements and simulation results. *Journal Environmental Quality*, 38, 1286–1301. https://doi.org/10.2134/jeq2008.0318

Stanford, G., & Smith, S. J. (1972). Nitrogen mineralization potentials of soils. *Soil Science Society of America Proceedings*, 36, 465–472. https://doi.org/10.2136/sssaj1972.03615995003600030029x

Yang, H. S., & Janssen, B. H. (2000). A mono-component model of carbon mineralization with a dynamic rate constant. *European Journal of Soil Science*, 51, 517–529. https://doi.org/10.1046/j.1365-2389.2000.00319.x

6

Donald L. Suarez and
Todd H. Skaggs

Equilibrium Soil Chemistry Submodels

Abstract

The inorganic composition of soil water impacts biological, chemical, and physical processes in the soil. Plant growth (biomass) is impacted by the availability of essential nutrients as well as the toxicity of specific elements, ion imbalances, and overall salinity. The soil solution composition also impacts water transport, runoff, and erosion as well as transport of individual ions via interaction with the soil mineral phase. Complex models attempting simulation of soil and plant processes including solute transport, nutrient and solution chemistry impacts on plant growth, and water transport must in turn be able to simulate major chemical processes that impact solution chemistry. Important chemical processes include dissolution–precipitation of minerals, ion exchange, ion adsorption, and oxidation–reduction. In this chapter we review chemical process models used in equilibrium soil chemistry submodels.

Introduction

The inorganic composition of soil water impacts biological, chemical, and physical processes in the soil. Plant growth (biomass) is impacted by the availability of essential nutrients as well as the toxicity of specific elements, ion imbalances, and the overall salinity. The soil solution composition also impacts water transport, runoff, and erosion as well as transport of individual ions via interaction with the soil mineral phase. Chemical and biological processes impact the soil solution composition and are in turn impacted by the solution composition. Models simulating mineral–water interactions and soil chemical processes have existed for about six decades. Among the earliest aqueous computer models are EQBRAT (Detar, 1969), WATCHEM (Barnes & Clarke,

---

Modeling Processes and Their Interactions in Cropping Systems: Challenges for the 21st Century, First Edition.
Edited by Lajpat R. Ahuja, Kurt C. Kersebaum, and Ole Wendroth.
© 2022 American Society of Agronomy, Inc. / Crop Science Society of America, Inc. / Soil Science Society of America, Inc.
Published 2022 by John Wiley & Sons, Inc.

1969), WATEQ Truesdell & Jones, 1974), and SOLMNEQ (Kharaka & Barnes, 1973). Early soil water models include those developed by Dutt (1961) for $CaCO_3$ and water and Dutt et al. (1972) adding cation exchange, the U.S. Bureau of Reclamation Irrigation Return Flow Model (Shaffer et al., 1977), and WATSUIT (Oster & Rhoades, 1975). More advanced models that coupled chemical processes with one-dimensional solute transport include LEACHM (Wagenet & Hutson, 1987), hydrosalinity models presented by Shaffer and Gupta (1981), the Nitrogen–Tillage–Residue Management (NTRM) model (Shaffer & Larson, 1987), UNSATCHEM (Suarez & Šimůnek, 1993, 1997), the Root Zone Water Quality Model (RZWQM) (Ahuja et al., 2000; Shaffer et al., 2000), and models coupled with two-dimensional transport (UNSATCHEM 2-D [Šimůnek & Suarez, 1994] and Hydrus 2D/3D [Šimůnek et al., 2018]).

Complex models attempting simulation of soil, plant processes including water and solute transport, plant growth including nutrient effects, and the impact of solution chemistry on plant growth and water transport must in turn be able to simulate major chemical processes that impact solution chemistry. The important processes include dissolution–precipitation of minerals, ion exchange, ion adsorption, and oxidation–reduction. With modern and continuing advances in computational speed, it is now possible to calculate equilibrium reactions for a huge database of mineral solubilities and couple them to solute transport models. Unfortunately, this is of limited value because classical thermodynamics evaluates only what is possible and not what can occur in relevant time frames. The authors of one of the earliest solution chemical equilibria programs, WATEQ (Truesdell & Jones, 1974), cautioned against using such thermodynamic equilibrium models to predict solution composition. Relevant time frames for soil-water modeling is on the order of hours to tens of years, thus we should consider equilibrium only for mineral phases that equilibrate in those time frames. Most soil minerals are thermodynamically unstable yet persist for thousands to hundreds of thousands of years. In addition, the chemical environment including pH, soil water content, and thus ion concentrations are constantly changing due to wetting and drying events, creating a dynamic environment that makes equilibrium even more problematic.

In arid regions, modeling of mineral precipitation–dissolution reactions are generally restricted to equilibrium with calcite and gypsum. Other chemical process often considered are ion exchange and, very rarely, oxidation–reduction. Prediction of redox equilibrium currently requires input of the redox status of couples (i.e., Fe^{3+}–Fe^{2+}) considered to be controlling other species, often via Eh measurements. Such predictions do not consider changes due to biological processes or production and transport of reactive gases (such as O_2, CO_2, and H_2S), and there is an understanding that redox couples are not generally in equilibrium with one another due to kinetic considerations.

For major ions, the ability to predict ion composition also requires the ability to calculate the soil pH as well as the concentrations of the major cations and anions, such as Ca^{2+}, Mg^{2+}, Na^+, NO_3^-, HCO_3^-, CO_3^{2-}, Cl^-, and SO_4^{2-} and minor elements such as NH_4^+, B, P, Fe, and Mn. Some models can predict B (UNSATCHEM 2.1) or crop NO_3^- and NH_4^+ uptake (the SOLCHEM module of RZWQM), but no model accurately predicts a suite of minor species with strong adsorption.

An important consideration in arid land soils is the maintenance of good soil physical properties for sustained agricultural production. Reclamation of saline and sodic soils and improving or maintaining adequate hydraulic conductivity require information about solution salinity, cation composition, and pH. A useful and unique feature of UNSATCHEM is the prediction of changes in relative hydraulic conductivity as related to these variables.

A chemical equilibrium approach is preferred to a chemical kinetic method primarily due to its simplicity. Highly soluble salts are assumed not present in the soil and also assumed to not form due to high solubility, with consideration of only the reaction or formation of calcite and gypsum. This simplification is generally valid except for highly salinized soils where the solubility of other sulfate salts ($Na_2SO_4 \cdot 10H_2O$) must be considered, especially in the surface crust under dry conditions. In some cases, limited chemical kinetics have been included, such as silicate weathering (UNSATCHEM), but these processes are generally not relevant to short- term predictions. Nonetheless, kinetic approaches are essential to predicting long-term processes related to mineral dissolution and formation.

Chemical Equilibrium Relationships

The basic reactions used in most models are presented below. Application of thermodynamics to natural systems was first popularized by Garrels and Christ (1965) in the geochemical literature and by Lindsay (1979) in soil chemistry. More modern texts that consider solution chemistry, such as Stumm and Morgan (1996) and Appelo and Postma (2005), place more emphasis on kinetic processes. Evaluation of solution status with regard to mineral solubility requires comparison of the solubility to ion activities in solution. The use of an equilibrium constant considers forward and back reactions.

A simple soil chemical reaction is

$$Aa + Bb \rightleftarrows Cc + Dd \tag{6.1}$$

In this reaction, a moles of reactant A combine with b moles of reactant B to form c moles of product C and d moles of product D. In the case where both the forward reaction

$$\text{RATE}_1 = K_{1R} A^a B^b \tag{6.2}$$

and the backward reaction

$$\text{RATE}_2 = K_{2R} C^c D^d \tag{6.3}$$

are at equilibrium, then

$$K_{1R} A^a B^b = K_{2R} C^c D^d \tag{6.4}$$

and

$$K_{eq} = \frac{K_{1R}}{K_{2R}} = \frac{C^c D^d}{A^a B^b} \tag{6.5}$$

where K_{eq} is the equilibrium constant.

The equilibrium constant K_{eq} can be determined experimentally or calculated from the Gibbs free energies of formation for the reactants and products. The Gibbs free energy for many ions, solid phases, and elements as well as equilibrium constants are readily available in various references including Wagman et al. (1982), Woods and Garrels (1987), Nordstrom et al. (1990), and Stumm and Morgan (1996), often with data only at 25 °C and 1 atm pressure but often with thermodynamic data that permit calculation of temperature dependence.

The standard change in free energy for a chemical reaction is, by convention, equal to the sum of the free energies of the products, ΔG_{pr}°, minus the sum of the free energies of the reactants, ΔG_{fre}°. This can be written as

$$\Delta G_r^{\circ} = \sum \Delta G_{fpr}^{\circ} - \sum \Delta G_{fre}^{\circ} \tag{6.6}$$

where ΔG_f° refers to the Gibbs free energies of formation in the standard states. The Gibbs free energy of the reaction, ΔG_r°, is related to the equilibrium constant K_{eq} by the relationship

$$\Delta G_r^{\circ} = -RT \ln K_{eq} \tag{6.7}$$

where R is the universal gas constant and T is temperature (K).

Because the equilibrium constant is temperature dependent and temperatures in natural systems can deviate by more than 25° from 25 °C, the K_{eq} values often need to be corrected for temperature. This can be determined either experimentally or, in the absence of experimental data, using an integrated form of the van't Hoff equation assuming ΔH° is independent of temperature (Stumm & Morgan, 1996):

$$\ln \frac{K_{eq}(T_2)}{K_{eq}(T_1)} = \frac{-\Delta H^{\circ}}{R} \left(\frac{1}{T_2} - \frac{1}{T_1} \right) \tag{6.8}$$

where T_1 is the reference temperature (298.15 K), T_2 is the temperature of interest (K), and ΔH° is the change in heat content. Some models include temperature correction (UNSATCHEM), but most do not (LEACHM, RZWQM). Alternatively, the following temperature-dependent equation for the correction of equilibrium constants is preferred because it is accurate across a wider temperature range, with constants available for a number of mineral solubility and carbonic acid dissociation constants (Truesdell & Jones, 1974):

$$\log K_{eq} = a_1 + \frac{a_2}{T} + a_3 T + a_4 \log T + \frac{a_5}{T^2} \tag{6.9}$$

Ionic interactions in solution cause a departure from "ideal" behavior, with increasing departure as the solution concentration increases. Activity values rather than species concentrations are utilized to account for non-ideal behavior. Individual ion activities are obtained by multiplying the species concentration by an experimentally determined activity coefficient γ_i that varies with ionic strength. Ionic mobility is impacted by interaction with other species, primarily electrostatic effects. For very dilute solutions, the Debye–Huckel equation is used to compute appropriate activity coefficients for the species (Lewis et al., 1961):

$$-\log \gamma_i = 0.509 z_i^2 \frac{\sqrt{I}}{1 + \sqrt{I}} \tag{6.10}$$

where z_i is the charge of the ith constituent and I is the ionic strength computed by the relationship

$$I = 0.5 \sum_{i=1}^{M} z_i^2 c_i \qquad (6.11)$$

where M is the number of species and c_i is the concentration (mol kg^{-1}) of the ith constituent. Equation 6.10 is used in RZWQM but is accurate only at very low ionic strengths.

An extended form of Equation 6.10, sometimes named the Debye–Huckel equation (Garrels & Christ, 1965) and sometimes the extended Debye–Huckel equation (Stumm & Morgan, 1996), is

$$-\log \gamma_i = 0.509 z_i^2 \frac{\sqrt{I}}{1 + Ba\sqrt{I}} \qquad (6.12)$$

where B is a characteristic of water and a is an adjustable parameter related to ion size. Equation 6.12 is considered applicable up to 0.1 M (Stumm & Morgan, 1996).

Many soil chemical models have used the Davies equation, modified from the Debye–Huckel limiting law:

$$-\log \gamma_i = 0.509 z_i^2 \left(\frac{\sqrt{I}}{1 + \sqrt{I}} - 0.2I \right) \qquad (6.13)$$

Stumm and Morgan (1996) and others have indicated that the Davies equation is applicable up to 0.5 M.

Note that the Debye–Huckel limiting law equation (Equation 6.10) and the Davies equation (Equation 6.13) consider only charge. It is well known that individual ions of the same charge have differing activity coefficients thus the use of these equations will lead to substantial errors at moderate ionic strength (Suarez, 1999). If using the ion association model with individual activity coefficients, it is much preferable to use the extended form of the Debye–Huckel equation proposed by Truesdell and Jones (1974):

$$\ln \gamma_i = \frac{A z_i^2 \sqrt{I}}{1 + Ba_i \sqrt{I}} + b_i I \qquad (6.14)$$

where A and B are constants depending on the dielectric constant of water, density, and temperature, and a_i and b_i are adjustable parameters provided for each species. The a_i and b_i parameters are available for a large set of individual ions (Truesdell & Jones, 1974). These parameters were developed from mean molal salt activity coefficients based on experimental data (where molal concentration is moles per kilogram of water). Their single ion activity coefficients were obtained from the assumption that $\gamma_{K^+} = \gamma_{Cl^-}$ and utilize mean salt data for KCl up to 4.0 M to obtain these two calculated γ_i values as a function of ionic strength. Next, the ion activity coefficients for other ions as a function of ionic strength were calculated from other mean salt data (e.g., $CaCl_2$), with excellent fits up to 4.0 M. However, as there are no specific ion–ion interactions considered other than of the individual pairs examined, the fit to mixed salt solutions is less accurate, probably up to 0.3–0.5 m (molal) depending on the solution composition. Equation 6.14 is utilized in UNSATCHEM. Some chemical speciation models make Equation 6.14 available as an option but utilize the Davies equation for ions where the a_i and b_i parameters are not available (MINTEQA2; Allison et al., 1991).

Utilizing a single ion association model such as Equation 6.14 requires consideration of complexes and ion pairs. Complexes are ion associations where the ligand is sufficiently close to partially or completely displacing the hydration shell surrounding the cation and forming bonds observable via spectroscopic analysis, with corresponding reactions such as

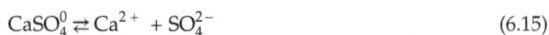

$$CaSO_4^0 \rightleftarrows Ca^{2+} + SO_4^{2-} \tag{6.15}$$

The corresponding equilibrium expression is

$$K_{eq} = \frac{[Ca^{2+}][SO_4^{2-}]}{CaSO_4^0} \tag{6.16}$$

where square brackets indicate activity. Note that this is a standard convention in the chemical and geochemical literature, but in the soil chemistry literature brackets often denote concentration.

Ion pairs can be viewed as weak ion associations in which the individual hydration shells are maintained, thus there is no chemical evidence of their existence as individual entities. An example is the species $CaHCO_3^+$. Thus they are numerical constructs necessary to predict the activity of individual species in mixed electrolytes when using the ion association model. At higher ionic strength (>0.3 M) the solution is sufficiently concentrated so that all ion–ion interactions must be considered for calculation of activity coefficients, thus ion pairs as separate chemical entities are not utilized. In this instance, the Pitzer expressions (Pitzer, 1979) are utilized. The activity coefficients are expressed in a virial-type expansion having the form

$$\ln \gamma_i = \ln \gamma_i^{DH} + \sum_j B_{ij}(I) m_j + \sum_j \sum_k C_{ijk} m_j m_k + \dots \tag{6.17}$$

where γ_i^{DH} is a modified Debye–Huckel activity coefficient, B_{ij} and C_{ij} are coefficients specific to each ion interaction, and m is the molal concentration of each species. The Pitzer approach considers ion–ion interactions for every species in solution and thus it does not consider the individual ion pairs and complexes such as $NaSO_4^-$ described above as species. The model is considered suitable for prediction of species activity in solutions up to $20 \, mol \, kg^{-1}$, a concentration well above the intended use of the models relevant to agriculture. In UNSATCHEM the user can select activities calculated from the Pitzer equations or by the extended Debye–Huckel expressions.

For most solutions of low-salinity neutral molecules can be assumed to have activity coefficients equal to 1, as used in RZWQM. At higher ionic strength, neutral species activity coefficients can be calculated using Equation 6.14, available for a number of species (Truesdell & Jones, 1974) and used in UNSATCHEM. The actual use of activity coefficients is illustrated in

$$K_{eq} = \frac{C^c \gamma^c D^d \gamma^d}{A^a \gamma^a B^b \gamma^b} \tag{6.18}$$

Soil Chemistry Equations

The full soil water system can be represented as a system of simultaneous nonlinear equations (in the form of Equation 6.18) together with mass and charge balance equations for the system. The system is generally solved numerically using Newton's method. Processes typically included are chemical dissolution–precipitation involving calcium carbonate (calcite), calcium sulfate di hydrate (gypsum), and gibbsite (aluminum hydroxide) or the more soluble amorphous aluminum oxyhydroxide [$Al(OH)_3$]; the bicarbonate buffer system including interactions with CO_2 gas, ion pairing, and ion exchange involving Ca^{2+}, Mg^{2+}, Na^+, K^+, NH^{4+}, and Al^{3+}. Most models consider pH as a fixed model input; this is suitable for the examination of saturation status of an analyzed solution but is not suitable for coupling with water and solute transport, as discussed below.

Ion Pairs and Complexes

Consideration of the complexes of calcium sulfate and magnesium sulfate, and the ion pairs $CaCl^+$, $CaHCO_3^+$, $CaCO_3^0$, $MgCl^+$, $MgHCO^{3+}$, $MgCO_3^0$, $NaSO_4^-$, and KSO_4^- are important in all but very dilute systems. In acid systems, Al ion species such as $AlOH_2^+$ and $AlSO_4^+$ are important components. Significant amounts of Ca^{2+}, Mg^{2+}, Al^{3+}, and SO_4^{2-} can be in ion pairs and complexes thereby either decreasing the activities of the uncomplexed ions if there is no solid phase interaction or, in the presence of a solid phase, the complexes and ion pairs result in increased total element concentrations.

Calcium sulfate ion pair ($CaSO_4^0$):

$$\frac{[Ca^{2+}][SO_4^{2-}]}{[CaSO_4^0]} = K_{CaSO_4} \tag{6.19}$$

Magnesium sulfate ion pair ($MgSO_4^0$):

$$\frac{[Mg^{2+}][SO_4^{2-}]}{[MgSO_4^0]} = K_{MgSO_4} \tag{6.20}$$

Sodium sulfate ion pair ($NaSO_4^-$):

$$\frac{[Na^+][SO_4^{2-}]}{[NaSO_4^-]} = K_{NaSO_4} \tag{6.21}$$

Aluminum hydroxide ion pair ($AlOH^{2+}$):

$$\frac{[Al^{3+}]}{[H^+][AlOH^{2+}]} = K_{AlOH} \tag{6.22}$$

Aluminum sulfate ion pair ($AlSO_4^+$):

$$\frac{[Al^{3+}][SO_4^{2-}]}{[AlSO_4^-]} = K_{AlSO_4} \tag{6.23}$$

Bicarbonate Buffer System

The pH in soil–water systems under acidic conditions is buffered by Al species and H as well as organic species and soil organic matter; under neutral to alkaline conditions, it is buffered by carbonate species and to a lesser extent organic species and organic matter. In the carbonate system, the pH is part of the combined pH–P_{CO_2}–HCO_3^-–alkalinity system—specifying any two entities fixes the third (where P_{CO_2} is the partial pressure of CO_2).

Carbon dioxide in the soil air reacts with water to form carbonic acid (H_2CO_3). The reaction,

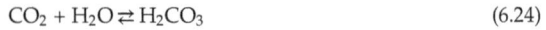

$$CO_2 + H_2O \rightleftarrows H_2CO_3 \tag{6.24}$$

is generally written as a form of Henry's law.

The total aqueous-phase CO_2 concentration, c_w, is defined as the sum of $CO_2(aq)$ and H_2CO_3, and is related to the gas-phase CO_2 concentration, c_a, by (Stumm & Morgan, 1996)

$$c_w = K_H R T c_a \tag{6.25}$$

where K_H is the Henry's law constant, R is the universal gas constant, and T is the absolute temperature. An expression for the value of K_H as a function of temperature was given by Harned and Davis Jr. (1943). Equation 6.25 is commonly expressed as

$$K_{CO_2} = \frac{[H_2CO_3^*]}{P_{CO_2}[H_2O]} \tag{6.26}$$

where P_{CO_2} is the partial pressure of CO_2 (atm), the activity of water is 1.0, and $H_2CO_3^* \equiv c_w$ is the preferred notation for the total aqueous-phase CO_2 concentration. The weak acid H_2CO_3 dissociates to form bicarbonate, HCO_3^-, and a hydrogen ion:

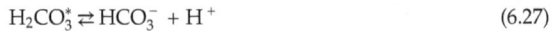

$$H_2CO_3^* \rightleftarrows HCO_3^- + H^+ \tag{6.27}$$

with the equilibrium expression

$$K_1 = \frac{[HCO_3^-][H^+]}{[H_2CO_3^*]} \tag{6.28}$$

Bicarbonate dissociates to form a carbonate ion, CO_3^{2-}, and another hydrogen ion

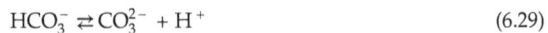

$$HCO_3^- \rightleftarrows CO_3^{2-} + H^+ \tag{6.29}$$

with the corresponding equilibrium constant

$$K_2 = \frac{[CO_3^{2-}][H^+]}{[HCO_3^-]} \tag{6.30}$$

These three equations (6.26, 6.28, and 6.30) define the carbonate species. Often chemical speciation models coupled with a transport model consider pH as a fixed input that, along with the measured alkalinity, defines the system. This approach is valid for speciation of an analyzed solution where P_{CO_2} is determined from total inorganic C along with the other carbonate species (closed system) but is not valid for predicting solution composition in the soil. In the soil system, the P_{CO_2} is best regarded as an independent variable, determined from production and transport of the gas, independent of the inorganic chemical reactions (open system) (Šimůnek & Suarez, 1993). This approach is utilized by UNSATCHEM and is justified because the production and consumption of CO_2 from inorganic mineral reactions is generally orders of magnitude smaller than the production from biological processes in soil. The pH is then calculated from the alkalinity and P_{CO_2}.

The UNSATCHEM model considers CO_2 production from plant and microbial respiration. Because these processes are very dependent on temperature and water and to a lesser extent salt stress, these relations are included as well. The model also assumes that the nutritional status is ideal. This assumption will, in many instances, overestimate CO_2 production and thus the concentration, especially in non-agricultural soils. The complexity of the numerous CO_2 production processes and their dependence on temperature and nutrient status means that the prediction is difficult and the model parameters may need to be optimized differently for different environments. In some cases it may be preferable to input a fixed soil CO_2 concentration, which can also be varied for the environment and season to be predicted. This assumption is much preferable to that of a fixed pH as used in most models because it allows prediction of the impacts of water composition and additions of gypsum, calcite, and acid on soil pH, plus subsequent changes in solution composition and soil hydraulic conductivity. The closed system approach is exclusively used in subsurface transport modeling (e.g., NETPATH; Plummer et al., 1991) and is justified by the consideration that there is no change in total C in or out of the system (neglecting gas diffusion or biological production), a generally reasonable assumption for groundwater but not for soil.

The chemical speciation model in UNSATCHEM includes consideration of nine major aqueous components: Ca, Mg, Na, K, SO_4, Cl, alkalinity, NO_3, and B. Alkalinity is defined as

$$
\begin{aligned}
\text{alkalinity} = &\left(HCO_3^-\right) + 2\left(CO_3^{2-}\right) + 2\left(CaCO_3^0\right) + \left(CaHCO_3^+\right) + 2\left(MgCO_3^0\right) \\
&+ \left(MgHCO_3^+\right) + 2\left(NaCO_3^-\right) + \left(NaHCO_3^0\right) + \left(B(OH)_4^-\right) - \left(H^+\right) + \left(OH^-\right)
\end{aligned}
\tag{6.31}
$$

where parentheses represent concentrations (mol kg^{-1}). From these components we obtain 11 species: Ca^{2+}, Mg^{2+}, Na^+, K^+, SO_4^{2-}, Cl^-, HCO_3^-, CO_3^{2-}, NO_3^-, $B(OH)_4^-$, and H_3BO_3. It is reasonable to assume that all aqueous species are in equilibrium as defined by the ion association expressions and constants, as these reactions are essentially instantaneous. Alkalinity as defined in Equation 6.31 is a conservative species with respect to changes in CO_2, in this instance affected only by dissolution or precipitation of a carbonate phase (such as calcite). However, the combined species $H_2CO_3^*$ does have a slower reaction process as it consists of both dissolved CO_2 and the H_2CO_3 aqueous species. Nonetheless, for soil systems where changes in CO_2 are relatively slow (minutes to days), it is enough to also consider $H_2CO_3^*$ to be in equilibrium with soil CO_2 gas.

Solid Phases

Soil–water systems are not in equilibrium with the numerous solid phases in the soil, and the use of a thermodynamic equilibrium approach to predict solution composition will lead to very large errors, specifically underestimating the solubility of almost all elements. Thus geochemical models such as PHREEQE (Parkhurst et al., 1980),

PHREEQE C (Parkhurst & Appelo, 2013), and MINTEEQA2 (Allison et al., 1991; USEPA, 1999) contain a large list of solid phases for a large number of elements. The model developers indicate that these are assessment tools rather than predictive tools. The models are of more value to subsurface hydrochemical systems than for predicting soil solution compositions. They are useful to evaluate the extent of nonequilibrium and the possible occurrence of particular reactions in soil. However, even that approach is limited as there is nonequilibrium not just with respect to solid phases but also with respect to solution-redox sensitive species, probably due to the effects of microbial processes. Thus reactions that should not occur within an overall thermodynamic equilibrium system can occur, such as the presence of unstable reduced species within an aerobic system.

The predominant silicate minerals in soils are very slow to both dissolve and precipitate. In almost all instances, primary minerals are thermodynamically unstable (feldspars, micas, hornblendes, etc.) in soil environments yet persist for more than tens of thousands of years. Most soil oxides are also thermodynamically unstable. Slow dissolution of primary minerals results in the formation of unstable silicate and oxide phases (such as illite, smectites, and poorly crystalline amorphous phases) rather than the less soluble phases predicted by thermodynamics. The soil solid phase is thus a combination of mostly unstable primary minerals and unstable secondary minerals slowly transforming to more stable forms.

It is generally considered that soil is in chemical equilibrium with certain relatively soluble minerals such as calcite, gypsum, and gibbsite. These mineral phases form the basis for the soil chemical submodels used in agronomic applications such as coupled water solute transport and plant models. If these minerals are present, we assume that the solution phase is in equilibrium; if the solution is at or above saturation, we assume that the mineral will precipitate and the solution will be in equilibrium. Solubility equations are thus coupled to the speciation equations to allow for dissolution and precipitation of these minerals:

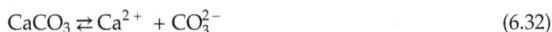

$$CaCO_3 \rightleftarrows Ca^{2+} + CO_3^{2-} \tag{6.32}$$

The equilibrium expression is

$$[Ca^{2+}][CO_3^{2-}] = K_{SP} \tag{6.33}$$

where brackets denote activities and $K_{SP} = 10^{-8.47}$ at 25 °C. Because CO_3^{2-} is a generally minor carbonate species in solution, it is greatly preferable to solve for equilibrium using an expression considering bicarbonate rather than carbonate. Thus the equilibrium condition of a solution with calcite in the presence of CO_2 can be described by

$$[Ca^{2+}][HCO_3^-]^2 = \frac{K_{SP}K_{CO_2}K_{a1}}{K_{a2}} P_{CO_2}[H_2O] = K_{SP}K_T \tag{6.34}$$

where K_{CO_2} is the Henry's law constant for the solubility of CO_2 in water, K_{a1} and K_{a2} are the first and second dissociation constants of carbonic acid in water, and K_{SP} is the solubility product for calcite. The saturation status of a solution is determined by calculation of the ion activity product (IAP):

$$[Ca^{2+}][CO_3^{2-}] = IAP \tag{6.35}$$

equilibrium soil chemistry submodels

where in this case the solution ion activities are calculated from the concentration and the IAP calculated. To attain equilibrium, that is, when the IAP is equal to the solubility product K_{SP}, a quantity x of Ca^{2+} and HCO_3^- must be added or removed from the solution to satisfy the equilibrium condition. The quantity x is obtained by solving the following third-order equation:

$$\left(Ca^{2+} + x\right)\left(HCO_3^- + 2x\right)^2 = \frac{K_{SP}K_T}{\gamma_{Ca^{2+}} + \gamma_{HCO_3^-}^2} \tag{6.36}$$

It has been shown that waters below irrigated lands in arid zone regions are supersaturated with respect to calcite (Suarez, 1977a; Suarez et al., 1992), thus the equilibrium condition underestimates the Ca solubility in soil water. The cause of supersaturation has been shown to be due to poisoning of crystal surfaces by dissolved organic matter (Inskeep & Bloom, 1986; Lebron & Suarez, 1996), thus effectively preventing crystal growth. This is consistent with the observation that we do not see calcite crystals in the soil. Nonetheless, the solid phase of calcium carbonate found in soils is almost always calcite and not less thermodynamically stable calcium carbonates such as aragonite and vaterite. There is thus no solid phase that can accurately represent solution Ca and alkalinity.

Calcite crystal growth models are not applicable to soil systems because the concentrations of dissolved organic C in near-surface natural environments (Lebron & Suarez, 1996) are usually comparable to the levels found by Inskeep and Bloom (1986) to completely inhibit calcite crystal growth. Lebron and Suarez (1996) developed a precipitation rate model that considers the effects of dissolved organic C on both crystal growth and heterogeneous nucleation. The combined rate expression is

$$R_T = R_{CG} + R_{HN} \tag{6.37}$$

where R_T is the total precipitation rate (mmol L^{-1} s^{-1}), R_{CG} is the precipitation rate related to crystal growth, and R_{HN} is the precipitation rate due to heterogeneous nucleation. Because for soil systems the crystal growth rate can be neglected, only nucleation is an important process. The R_{HN} term is given by

$$R_{HN} = K_{HN}f(SA)\left[\log\left(\Omega - 2.5\right)\right]\left(3.37 \times 10^4 \, DOC^{-1.14}\right) \tag{6.38}$$

where K_{HN} is the precipitation rate constant due to heterogeneous nucleation, $f(SA)$ is a function of the surface area of the particles (e.g., clay) upon which heterogeneous nucleation occurs (=1.0 if no solid phase is present), Ω is the calcite saturation value, and 2.5 is the Ω value above which heterogeneous nucleation can occur. This equation leads to calcite precipitation rates that are independent of the calcite surface area, consistent with the experimental data of Lebron and Suarez (1996).

The precipitation model predicts that calcite nucleation is not initiated until the solution phase is approximately 2.5-fold supersaturated with respect to calcite. This value corresponds reasonably well to the observed average of threefold calcite supersaturation found in waters beneath irrigated lands (Suarez, 1977a; Suarez et al., 1992). Thus in this instance an empirical constant or apparent K_{sp} of $10^{-8.0}$ gives a similar result to the more complex kinetic approach. This constant is best utilized instead of the calcite equilibrium constant ($10^{-8.47}$) for prediction purposes, although thermodynamically meaningless. Numerically one could also assume that the solid phase is not equal to 1 and adjust by this approximate factor of 3; however, this is chemically not justified because there is no evidence that the calcite formed in the soil is more soluble than well-formed crystalline calcite (Suarez & Rhoades, 1982).

In contrast to the calcite case, solid-phase activity coefficients could be used to predict Al solubility, using a solid-phase coefficient <1 for Al predictions based on gibbsite solubility, because the formed solid is less stable (more soluble than gibbsite). Alternatively, the solution could be equilibrated with amorphous $Al(OH)_3$, which has a larger equilibrium constant. Pedogenic calcites in the arid, irrigated, southwestern United States often contain 1–7% Mg substituted for Ca in the structure (Suarez & Rhoades, 1982), which may if anything result in more stable (less soluble) calcite according to the data of Busenberg and Plummer (1989), thus not explaining calcite supersaturation in soils.

When gypsum is present in the soil, it can generally be assumed that the soil solution will be in equilibrium with the solid. Crystal growth is possible despite the presence of organic matter, as evidenced by the formation of relatively large assicular (needle like) formed gypsum crystals in arid zone soils. Although the assumption of gypsum equilibrium is generally reasonable, it is known that supersaturation when precipitating and undersaturation when dissolving does occur in soil. The biggest error will probably occur when applying gypsum to a sodic soil for reclamation. In this instance a kinetic approach may be preferred if the objective is to predict the dynamics of the system during infiltration of water during reclamation; however, the equilibrium assumption has been found suitable for predicting the final exchangeable ion composition after reclamation (Suarez, 2001).

As with calcite, predictive models require input as to the presence or absence of the mineral phase. If the mineral is present, the models force equilibrium of the solution with the solid phase; if the solution is supersaturated, the model forces precipitation and equilibrium regardless of whether or not the mineral was initially present.

Sometimes only the presence or absence of the mineral is specified. However, many models (including UNSATCHEM) can track the concentrations of the solid phase, enabling prediction of the extent of dissolution—important during leaching for sodic soil reclamation. The reaction for dissolution and precipitation of gypsum is

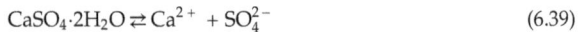

$$CaSO_4 \cdot 2H_2O \rightleftarrows Ca^{2+} + SO_4^{2-} \tag{6.39}$$

with the equilibrium expression given by

$$[Ca^{2+}][SO_4^{2-}] = \frac{K_{SP}}{[H_2O]^2} \tag{6.40}$$

where K_{SP} is the solubility product in solution, taken as $10^{-4.61}$. To obtain equilibrium, that is, when the IAP is equal to the solubility product K_{SP}, a quantity of gypsum, x, must be added or removed from the Ca^{2+} and SO_4^{2-} concentrations in solution, obtained by solving the quadratic equation

$$(Ca^{2+} - x)(SO_4^{2-} - x) = \frac{K_{SP}}{\gamma_{Ca^{2+}} \gamma_{SO_4^{2-}} [H_2O]^2} \tag{6.41}$$

In addition to calcite and gypsum, other phases are sometimes considered. Under high-pH conditions, magnesium carbonate precipitation should be considered. The UNSATCHEM model includes provision for the precipitation–dissolution of a Mg phase.

The Mg carbonate regarded as most thermodynamically stable under Earth surface conditions is magnesite ($MgCO_3$), with a solubility of about three times that of calcite. The phase is thus considered in some (MINTEEQA2) but not all thermodynamic geochemical assessment models (PHREEQE, PHREEQE C). However, it is not suitable for predictive models such as coupled water and solute transport and plant models. The soil predictive models should consider only phases that either precipitate under Earth surface conditions or occur frequently and are reactive under Earth surface conditions; these need not be the thermodynamically the most stable. With this consideration, magnesite, the thermodynamically more stable Mg carbonate, should be neglected because it apparently does not form under Earth surface temperatures, is extremely rare, and its dissolution rate is exceedingly slow, such that its solubility has not been satisfactorily determined from dissolution studies at or near 25 °C. Including magnesite in predictive models will result in large underestimation of Mg concentrations in arid environments.

Magnesium precipitation can occur as a carbonate (either nesquehonite or hydromagnesite) or as a silicate (sepiolite) under very saline and high-pH conditions. All three phases readily dissolve and precipitate in time frames suitable for soil water transport and are considered in the UNSATCHEM model.

At 25 °C and at CO_2 partial pressures above $10^{-3.27}$ kPa, nesquehonite ($MgCO_3 \cdot 3H_2O$) is stable relative to hydromagnesite, thus this is the Mg carbonate phase of potential relevance to soils in hypersaline environments. The precipitation (if saturation is achieved) or dissolution of nesquehonite (if specified as a solid phase) in the presence of CO_2 can be described by

$$MgCO_3 \cdot 3H_2O + CO_2(g) \rightleftarrows Mg^{2+} + 2HCO_3^- + 2H_2O \tag{6.42}$$

with the solubility product K_{SP} defined by

$$K_{SP} = [Mg^{2+}][CO_3^{2-}][H_2O]^3 \tag{6.43}$$

where $K_{SP} = 10^{-5.62}$. Substituting the equation for Henry's law for the solubility of CO_2 in water and the equations for the dissociation of carbonic acid in water into the solubility product, we obtain

$$[Mg^{2+}][HCO_3^-]^2 = \frac{K_{SP}K_{CO_2}K_{a1}}{K_{a2}}\frac{P_{CO_2}}{[H_2O]^2} = \frac{K_{SP}K_T}{[H_2O]^3} \tag{6.44}$$

This relation is solved for equilibrium in a manner similar to that used for calcite, with a third order equation and substitution of Mg^{2+} concentration and $\gamma_{Mg^{2+}}$ for activity of Mg^{2+} and substitution of HCO_3^- concentration and $\gamma_{HCO_3^-}$ for activity of HCO_3^- in Equation 6.44.

Sepiolite will readily precipitate into a solid but with a K_{SP} greater than that of well crystallized sepiolite. Formation of this mineral requires high pH, high Mg concentrations, and low CO_2 partial pressure.

The precipitation or dissolution of sepiolite in the presence of CO_2 can be described by the reaction

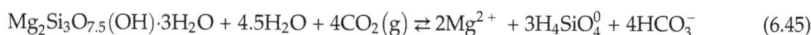

$$Mg_2Si_3O_{7.5}(OH) \cdot 3H_2O + 4.5H_2O + 4CO_2(g) \rightleftarrows 2Mg^{2+} + 3H_4SiO_4^0 + 4HCO_3^- \tag{6.45}$$

with the solubility product K_{SP} defined by

$$K_{SP} = \frac{\left[Mg^{2+}\right]^2 \left[H_4SiO_4^0\right]^3 \left[OH^-\right]^4}{\left[H_2O\right]^{4.5}} \tag{6.46}$$

In this instance, UNSATCHEM utilizes the precipitated sepiolite solubility value given by Wollast et al. (1968) rather than the well crystallized equilibrium value. Freshly precipitated sepiolite has been prepared in the laboratory at IAP values of 10^{-35} comparable to the K_{SP}^S listed by Wollast et al. (1968), but the K_{SP} for a precipitated phase has been more recently listed as $10^{-37.2}$ (USEPA, 1999). These differences are probably related to the extent of reaction or aging of the solid. A kinetic expression for precipitation would be preferred but is not available for the prediction of unsaturated zone solution composition. The equilibrium condition is expressed as

$$\left[Mg^{2+}\right]^2 \left[HCO_3^-\right]^4 = \frac{K_{SP}K_{CO_2}^4 K_{a1}^4}{K_w^4} \frac{P_{CO_2}^4 (H_2O)^{4.5}}{\left(H_4SiO_4^0\right)^3} = K_{SP}K_T^+ \tag{6.47}$$

Solution of Equation 6.47 requires knowledge of Si concentrations. Relatively little information exists on the controls on Si concentrations in soil waters, especially in arid zones. In soil systems, Si concentrations are not fixed by quartz solubility but rather by the dissolution and precipitation of aluminosilicates and Si adsorption onto oxides and aluminosilicates. As a result of these reactions, Si concentrations in soil solution follow a U-shaped curve with pH, similar to Al oxide solubility, with a Si minimum around pH 8.5 (Suarez, 1977b) rather than following the solubility curve of quartz or amorphous Si, which give a constant total Si in solution until higher pH when total Si increases as the weak acid H_4SiO_4 dissociates to H^+ and $H_4SiO_4^-$. Data from eight arid land soils reacted at various pH values for two weeks by Suarez (1977b) were fitted to a second-order relationship as

$$\sum SiO_2 = 0.001\left(6.34 - 1.34pH + 0.0819pH^2\right) \tag{6.48}$$

where SiO_2 is the sum of all silica species (mol L^{-1}). This U-shaped relationship with pH probably provides only a rough estimate of Si concentrations, but we consider it acceptable because it is used only to restrain Mg concentrations at high levels of evapotranspiration, when Mg concentrations become very high at low CO_2 and elevated pH. Formation of sepiolite in soils has been documented by numerous researchers (Diaz-Hernandez et al., 2018).

In the presence of calcite or whenever Ca in solution equals or exceeds the values in equilibrium with calcite, the most thermodynamically stable phase containing Mg is dolomite, not a pure Mg carbonate. Dolomite [CaMg(CO$_3$)$_2$] has an equilibrium expression of

$$K_{SP} = \left[Ca^{2+}\right]\left[Mg^{2+}\right]\left[CO_3\right]^2 \tag{6.49}$$

with a K_{SP} of $10^{-17.09}$. Thus if calcite and gypsum were in equilibrium, the ion activity product of [Mg][CO$_3$] would be $10^{-17.09}/10^{-8.48}$ or $10^{-8.61}$ and Mg would be below Ca and thus always be undersaturated with respect to magnesite, hydromagnesite, and nesquehonite. However, true dolomite, a well-ordered mineral with layers of Mg, Ca, and CO$_3$, appears to rarely form in soil environments. As a result, dolomite precipitation and equilibrium from a supersaturated solution is not considered in UNSATCHEM. When dolomite is present in the soil (from parent geological material), predictions of solution composition could use the kinetic model of Busenberg and Plummer (1982) to represent the dissolution process. The dissolution rate of dolomite is slow, especially as the solution IAP values approach within two to three orders of magnitude of the solubility product. In arid zone soils, when dolomite is present calcite is also present. In this instance in the presence of calcite and evapotranspiration, Ca concentrations in solution typically exceed Ca solubility and, given typical Mg concentrations, solutions are generally supersaturated with respect to well-ordered dolomite. Thus dolomite dissolution needs to be considered, primarily in the case where calcite is not present (such as liming an acid soil with dolomitic material). In this case a kinetic expression for dolomite dissolution is necessary.

Under elevated Mg concentrations, Mg is incorporated into calcite, forming high-Mg calcites, which may over time transform into Ca-enriched poorly ordered dolomite, with solutions remaining very supersaturated with respect to dolomite. The Mg carbonate precipitated (nesquehonite), combined with precipitation of calcite, will probably represent the mixed Ca–Mg precipitate that is observed in hypersaline environments, sometimes called protodolomite. However, the resulting solution composition is much different than that produced by simply forcing equilibrium with respect to dolomite, as the model forms this mixed precipitate (calcite + magnesium carbonate) under conditions of approximately 200–500 times supersaturation with respect to dolomite. This result is consistent with the high levels of dolomite supersaturation maintained in high-Mg waters (Suarez, 1977a; Suarez et al., 1992). Recently Diaz-Hernandez et al. (2018) reported on the presence and pedogenic formation of calcian (nonstoichiometric) dolomite, calcite, sepiolite, and palygorskite (Mg-rich, poorly ordered clay) in soil formed in an arid region from weathering of volcanic tephra. Solution compositions were not provided, so supersaturation indices cannot be calculated. However, based on the precipitation of calcite and sepiolite, it is evident that solutions would be many orders of magnitude supersaturated with respect to dolomite. The representation of this system by threefold supersaturation with calcite and equilibrium with a poorly crystallized sepiolite might be adequate to predict the solution composition.

Acid soils are often in equilibrium with the mineral gibbsite, $Al(OH)_3$, or an amorphous form (non-crystalline) called an amorphous aluminum oxyhydroxide phase, and its solubility can be the chemical reaction expressed as

$$Al(OH)_3 + 3H^+ \rightleftarrows Al^{3+} + 3H_2O \qquad (6.50)$$

with the equilibrium equation

$$\frac{[Al^{3+}]}{[H^+]^3} = K_{SP} \qquad (6.51)$$

Ion Exchange

Soil clays have negative surface charge that is neutralized by solution cations that are loosely bound to the surface (exchangeable cations). Cation exchange reactions are extremely important in soil–water systems. Exchange ions can be readily exchanged for other ions but are not available for leaching, thus maintaining electrical neutrality. They also are not directly available to participate in reactions occurring in the solution phase. Ion exchange equations generally included are Ca^{2+}, Mg^{2+}, Na^+, and, to a lesser extent, K^+, Al^{3+}, and NH_4^+. These represent the major cations found in many soil systems. Anion exchange is generally less important and often omitted. Three forms of the ion exchange relationships are commonly utilized: the Gapon (1933), the Vanselow (Vanselow, 1932), and the Gaines–Thomas (1953) equations. All three forms have been successfully used to fit exchange data. Assuming ideal exchange phase behavior (concentration = activity), they all result in deviations from measured exchange data across exchanger phase composition.

The exchanger selectivity coefficients change with exchanger phase composition for a number of reasons, including non-ideal behavior of individual minerals as well as the consideration that in soils there is usually a number of exchange sites related to different minerals, exchange associated with organic matter, and even different affinities on different sites within a specific mineral.

The Vanselow convention has been considered preferable based on theoretical considerations, but it is not clear if this results in any less variability in the coefficient across exchanger composition. For homovalent exchange, for example, the reaction can be written as

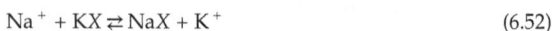

$$Na^+ + KX \rightleftarrows NaX + K^+ \tag{6.52}$$

All conventions are similar with the following equilibrium expression:

$$\frac{[K^+]}{[Na^+]}\frac{[NaX]}{[KX]} = k_{Na/K} \tag{6.53}$$

where X is either the mole fraction of the exchange ion or the charge fraction of the exchange ion. In the case of heterovalent exchange, the approaches differ. For Na–Ca exchange, using the Gapon convention, the reaction is written as (where $Ca_{0.5}$ represents ½ mol of Ca):

$$Na^+ + Ca_{0.5}X \rightleftarrows NaX + 0.5Ca^{2+} \tag{6.54}$$

with the following equilibrium expression:

$$\frac{\sqrt{[Ca^{2+}]}}{[Na^+]}\frac{[NaX]}{[Ca_{0.5}X]} = k_{Na/Ca} \tag{6.55}$$

where X is the fraction of charged sites occupied by the ion (moles of charge).

With the Vanselow convention, the Na–Ca exchange reaction is written as

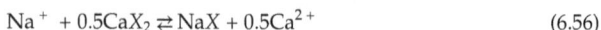

$$Na^+ + 0.5CaX_2 \rightleftarrows NaX + 0.5Ca^{2+} \tag{6.56}$$

and the corresponding equilibrium expression is

$$\frac{\sqrt{[Ca^{2+}]}}{[Na^+]}\frac{[NaX]}{\sqrt{[CaX_2]}} = k_{Na/Ca} \tag{6.57}$$

where X represents the mole fraction of the ion on the exchange complex, X_2 is two times the exchange mole fraction, and k is the selectivity constant. In cation exchange application, the term *selectivity* rather than *equilibrium* constant is used because these relationships do not meet the criteria for thermodynamic relations, as discussed below.

equilibrium soil chemistry submodels

Because the selectivity coefficients are dependent on the convention chosen, users must be careful when selecting values to insert into the various chemical models. UNSATCHEM uses the Gapon expressions, shown to provide good prediction of exchangeable Na across a wide range of arid zone soils and exchanger composition (U.S. Salinity Laboratory Staff, 1954). These constants should ideally consider activities of the solution species as well as exchanger phase activity. The use of activity coefficients for only the solute species will lead to overestimation of exchangeable Na in saline soils (Appelo & Postma, 2005). Hence, either the same activity coefficients are used for the exchanger phase as the solute species, as in PHREEQC, or only concentrations are used, as in UNSATCHEM.

The RZWQM models Ca–Mg ion exchange using the expression

$$\frac{[Ca^{2+}]}{[Mg^{2+}]} \frac{\left[X_{Mg^{2+}}\right]}{[X_{Ca^{2+}}]} = k \tag{6.58}$$

where X is the mole fraction of the ion on the exchanger phase, with a selectivity coefficient $k = 0.67$.

UNSATCHEM uses the expression

$$\frac{(Ca^{2+})}{(Mg^{2+})} \frac{\sqrt{X_{Mg^{2+}}}}{\sqrt{X_{Ca^{2+}}}} = k \tag{6.59}$$

where X is the charge fraction on the exchanger and $k = 0.9$.

The RZWQM models Na–NH_4 ion exchange using

$$\frac{[Na^+]}{[NH_4^+]} \frac{\left[X_{NH_4^+}\right]}{[X_{Na^+}]} = k \tag{6.60}$$

where $k = 0.22$. UNSATCHEM does not consider NH_4^+.

The RZWQM simulates Ca–Al ion exchange using

$$\frac{[Ca^{2+}]^3}{[Al^{3+}]^2} \frac{\left[X_{Al^{3+}}\right]^2}{[X_{Ca^{2+}}]^3} = k \tag{6.61}$$

where $k = 0.67$.

The newer versions of UNSATCHEM and SWS version (Suarez, 2012) use a multi-site mixing model containing separate consideration of Ca–Na selectivity on organic matter versus clays, accounting for the much greater preference of organic matter for Ca and enabling improved prediction of Ca^{2+}–Na^+ cation exchange for soils of varying organic matter. UNSATCHEM also considers K exchange, which RZWQM does not. Predictions of cation exchange are adequate with all the approaches; the limiting factor is that generalized cation selectivity coefficients, even with consideration of organic matter content, may not be adequate in many instances.

Aluminum and Sodium Complexes with Soil Organic Matter

Aluminum and Na are known to form complexes with charged particles of soil organic matter (SOM). In acid systems, the Al–organic matter complex in RZWQM is given as

$$\frac{[Al^{3+}][SOM^-]}{[AlSOM^{2+}]} = 1 \times 10^6 \tag{6.62}$$

where SOM^- is soil organic matter and $AlSOM^{2+}$ is the divalent Al–organic matter complex.

In alkaline systems in the RZWQM, the Na–organic matter complex is written as

$$\frac{[Na^+][SOM^-]}{[NaSOM^0]} = 1.06 \tag{6.63}$$

Adsorption–Desorption

Adsorption, or what is often deemed *specific adsorption*, refers to the attachment of solution species that are not exchanged. They may be either charged or noncharged species and thus may impact the surface charge of the mineral. The Langmuir isotherm is the most utilized adsorption equation in soil science and has been used in several models to describe non-competitive adsorption. The difficulty in using the Langmuir isotherm is that it is specific to the soil and chemical conditions being evaluated while many adsorption reactions are pH dependent; hence changing the pH will adversely affect the prediction, for example, of B transport during reclamation of a sodic alkaline soil. The oxyanions of B, As, and Mo, with potential plant and animal toxicity, are of interest in arid land soils. In this instance, a surface speciation model is preferred rather than the more utilized Langmuir isotherm. A relatively simple adsorption model, the constant capacitance model (CCM) has been successfully applied in soils (Goldberg, 1992). This model considers the pH dependence of the adsorption using pH-dependent apparent equilibrium constants obtained from experimental data. The practical application of the CCM to modeling was limited by the need to obtain adsorption envelopes for each soil of interest, requiring extensive adsorption experiments, each with a large range of solutions with different pH values. However, it has been established that for B (Goldberg et al., 2000) and Mo (Goldberg et al., 2002) the apparent equilibrium constants for the CCM can be predicted from several known soil properties: clay surface area, organic matter content, and Fe and Al oxide content. This application is incorporated into the more recent versions of UNSATCHEM and SWS (Suarez, 2012). Alternatively, the Langmuir isotherm approach could be used with the addition of pH-dependent relationships or the more detailed CD-MUSIC adsorption model (Hiemstra & van Riemsdijk, 1996), but in both instances detailed soil-specific data would need to be obtained.

The B adsorption reaction is

$$SOH + H_3BO_3 \rightleftarrows SH_3BO_4^- + H^+ \tag{6.64}$$

The intrinsic equilibrium constants are given as

$$K_+ = \frac{(SOH_2^+)}{(SOH)(H^+)} \exp\left(\frac{F\psi}{RT}\right) \tag{6.65}$$

$$K_- = \frac{(SO^-)(H^+)}{(SOH)} \exp\left(-\frac{F\psi}{RT}\right) \tag{6.66}$$

$$K_{B^-} = \frac{(SH_3BO_4^-)(H^+)}{(SOH)(H_3BO_3)} \exp\left(\frac{F\psi}{RT}\right) \tag{6.67}$$

(Goldberg et al., 2000) where F is the Faraday constant (C mol$_c$ L^{-1}), ψ is the surface potential (V), R is the molar gas constant, T is the absolute temperature (K), S denotes the surface site, and parentheses indicate concentrations (mol L^{-1}). Goldberg et al. (2000) developed regression equations for the prediction of the three CCM surface complexation constants based on generally available soil properties. They developed the following equations based on batch equilibrium adsorption experiments conducted with 32 arid land soils:

$$\log K_{B^-} = -9.14 - 0.375 \ln(SA) + 0.167 \ln(OC) + 0.11 \ln(IOC) + 0.466 \ln(Al) \quad (6.68)$$

$$\log K_+ = 7.85 - 0.102 \ln(OC) - 0.198 \ln(IOC) - 0.622 \ln(Al) \quad (6.69)$$

$$\log K_- = -11.97 + 0.302 \ln(OC) + 0.0584 \ln(IOC) + 0.302 \ln(Al) \quad (6.70)$$

where SA is the surface area measured by ethylene glycol monoethylether (EGME), OC is the organic C, IOC is the inorganic C, and Al is the extractable Al (absorbed, hydroxides, and oxides). Using these relationships, Goldberg et al. (2000) satisfactorily predicted the adsorption envelopes (adsorption as a function of pH) for a series of arid land soils. They concluded that the fits using the CCM with the constants determined from the above predictive equations were acceptable for use in modeling adsorption. Of these properties, only the surface area term is not readily available with descriptions of soil properties. Surface area can be directly determined or estimated from the clay content and mineralogy of a soil. UNSATCHEM application of these equations provided good predictions for both B adsorption and desorption in soil columns packed for three different soils and with a range of solution pH values without the need to conduct B adsorption experiments (Suarez et al., 2012).

Other adsorption models have been included in solution transport models including CD MUSIC (in PREEEQEC), a comprehensive model requiring extensive input information incorporating the triple-layer model in addition to a diffuse layer mode. These models are well suited to specific critical applications, such as U transport from a proposed containment facility, but in the absence of generalized input parameters are less practical for general application in models that consider variably saturated water transport, plant water uptake, etc., applicable to soil–water systems.

Model Testing and Validation

Models need to be tested for internal consistency and to ensure that the equations utilized are accurately solved. Models also need to be evaluated by comparison with sets of observed data, often called *validation*. Surprisingly, some models do not converge on the equations provided, most commonly due to expressing solubility with minor species (i.e., solving for calcite solubility from the primary expression using Ca and CO$_3$ species) and insufficient iteration. Most models have the necessary mass balance tests to be certain that the model conserves both mass and charge. True validation is probably impossible for a number of reasons, such as insufficient change in the variables examined, range of concentrations evaluated, and scale of measurement (in both time and space). A distinction also needs to be made between prediction and simulation. In many models, a number of "calibration" parameters serve, in essence, as curve fitting. Using the same or similar data sets to calibrate the model (inverse modeling) provides evidence of good *simulation* without indicating predictive capability in a general sense.

There are a very large number of publications where models have provided good representation (simulation) of experimental data but far fewer where actual predictions were successfully made. Successful UNSATCHEM predictions include soil CO_2 and water content predictions (Suarez & Šimůnek, 1993), prediction of electrical conductivity and changes in the sodium adsorption ratio (SAR, where SAR = $Na^+/(Ca^{2+} = Mg^{2+}$, in units of mmol L^{-1}) with reclamation (Suarez, 2001), and prediction of B transport for various soils and pH conditions (Suarez et al., 2012). The HYDRUS model includes a module with the earlier version of UNSATCHEM and has been successfully utilized in a number of studies including gypsum reclamation of sodic soil (Reading et al., 2012; Shaygan et al., 2018; Wang et al., 2016) and prediction of the effects of solution composition on hydraulic conductivity (Reading et al., 2012; Wang et al., 2016). Successful applications of RZWQM include water content and Br predictions (Ahuja et al., 1996) and water content and NO_3 (Cameira et al., 1998).

Questions relevant to model application that need to be addressed in future research:

1. Under what circumstances might equilibrium models not be suitable to predict soil solution Ca concentrations?

2. If using a kinetic model for mineral dissolution, what soil information would be required that is not generally available?

3. In addition to the carbonate system, what other pH buffer systems are present in arid zone soils? In acid soils?

Notation

R	universal gas constant
T	temperature (K)
T_1	reference temperature for van't Hoff equation (K)
T_2	temperature of interest for van't Hoff equation (K)
γ_i	Single ion activity coefficient (unitless)
ΔH°	change in heat content
ΔG_f°	Gibbs free energies of formation in the standard states
I	ionic strength (mol per kg of solvent)
z_i	charge of ith ion in solution
c_i	solution concentration of the ith constituent
K_{eq}	equilibrium constant
k	selectivity constant for exchange
P_{CO_2}	partial pressure of carbon dioxide in soil air (atm)
m_i	molal concentration of ion in solution (moles per kg water). For dilute solutions this is equal to M (moles per liter of water)
X	Exchange phase, mole fraction of an exchangeable ion in Vanselow convention (unitless)
X	Exchange phase, charge fraction of exchangeable ion in Gapon convention (unitless)
X_2	Exchange phase, two times mole fraction of exchangeable ions in Vanselow convention (unitless)
SOM^-	concentration of charged soil organic matter in solution (mol L^{-1}),
$AlSOM^{2+}$	concentration of aluminum-organic matter complex in solution (mol L^{-1})
$NaSOM^0$	concentration of sodium-organic matter complex in solution (mol L^{-1})
S	concentration of surface sites in the constant capacitance adsorption model
$[\cdot]$	species activity (dimensionless, =1 for solid phases by convention)

References

Ahuja, L. R., Ma, Q. L., Rojas, K. W., Boesten, J. J., & Farahani, H. J. (1996). A field test of the Root Zone Water Quality Model: Pesticide and bromide behavior. *Pesticide Science*, 48, 101–108.

Ahuja, L. R., Rojas, K. W., Hanson, J. D., Schaffer, M. J., & Ma, L. (Eds.) (2000). *Root Zone Water Quality Model: Modelling management effects on water quality and crop production*. Water Resources Publications.

Allison, J. D., Brown, D. S., & Novo-Gradic, K. J. (1991). *MINTEQA2/PRODEFA2: A geochemical assessment model for environmental systems: Version 3.0 users manual* (EPA/600/3-9/021). USEPA Office of Research and Development.

Appelo, C. A., & Postma, D. (2005). *Geochemistry, groundwater and pollution* (2nd ed.). A.A. Balkema.

Barnes, I., & Clarke, F. E. (1969). *Chemical properties of ground water and their corrosion and encrustation effects on wells* (USGS Professional Paper 498-D). U.S. Government Printing Office. https://doi.org/10.3133/pp498D

Busenberg, E., & Plummer, L. N. (1982). The kinetics of dissolution of dolomite in CO_2–H_2O systems at 1.5 to 65 °C and 0 to 1 atm PCO_2. *American Journal of Science*, 282, 45–78.

Busenberg, E., & Plummer, L. N. (1989). Thermodynamics of magnesium calcite solid solutions at 25 °C and I atm total pressure. *Geochimica et Cosmochimica Acta*, 53, 1189–1208.

Cameira, M. R., Sousa, P. L., Farahani, H. J., Ahuja, L. R., & Pereira, L. S. (1998). Evaluation of the RZWQM for the simulation of water and nitrate movement in level-basin, fertigated maize. *Journal of Agricultural Engineering Research*, 69, 331–341.

Detar, D. F. (1969). *Computer programs for chemistry*. W.A. Benjamin.

Diaz-Hernandez, J. L., Sanchez-Navas, A., Delgado, A., Yepes, J., & Garcia-Casco, A. (2018). Textural and isotopic evidence for Ca–Mg carbonate pedogenesis. *Geochimica et Cosmochimica Acta*, 222, 485–507.

Dutt, G. R. (1961). *Quality of percolating water: 1. Development of a computer program for calculating the ionic composition of percolating waters* (Contribution no. 50). Water Resources Center, Department of Irrigation, University of California.

Dutt, G. R., Shaffer, M. J., & Moore, W. J. (1972). *Computer simulation model of dynamic bio- physicochemical processes in soils* (Technical Bulletin 196). Agricultural Experiment Station, University of Arizona.

Gaines, G. I., & Thomas, H. C. (1953). Adsorption studies on clay minerals: II. A formulation of the thermodynamics of exchange adsorption. *The Journal of Chemical Physics*, 21, 714–718.

Gapon, E. N. (1933). On the theory of exchange adsorption in soil. *Journal of General Chemistry of the USSR*, 3, 144–152.

Garrels, R. M., & Christ, C. L. (1965). *Solutions, minerals, and equilibria*. Harper & Rowe.

Goldberg, S. (1992). Use of surface complexation models in soil chemical systems. *Advances in Agronomy*, 7, 233–329.

Goldberg, S., Lesch, S. M., & Suarez, D. L. (2000). Predicting boron adsorption by soils using soil chemical parameters in the constant capacitance model. *Soil Science Society of America Journal*, 64, 1356–1363.

Goldberg, S., Lesch, S. M., & Suarez, D. L. (2002). Predicting molybdenum adsorption by soils using soil chemical parameters in the constant capacitance model. *Soil Science Society of America Journal*, 66, 1836–1842.

Harned, H. S., & Davis, R., Jr. (1943). The ionization constant of carbonic acid and the solubility of carbon dioxide in water and aqueous salt solutions from 0 to 50 °C. *Journal of the American Chemical Society*, 653, 2030–2037.

Hiemstra, T., & van Riemsdijk, W. H. (1996). A surface structural approach to ion adsorption: The charge distribution (CD) model. *Journal of Colloid and Interface Science*, 179, 488–508.

Inskeep, W. P., & Bloom, P. R. (1986). Kinetics of calcite precipitation in the presence of water soluble organic ligands. *Soil Science Society of America Journal*, 50, 1167–1172.

Kharaka, Y. K., & Barnes, I. (1973). *SOLMNEQ: Solution–mineral equilibrium computation*. Technical Report PB214-899. USGS. https://doi.org/10.3133/25339

Lebron, I., & Suarez, D. L. (1996). Calcite nucleation and precipitation kinetics as affected by dissolved organic matter at 25 °C and pH >7.5. *Geochimica et Cosmochimica Acta*, 60, 2767–2776.

Lewis, G. N., Randall, M., Pitzer, K. S., & Brewer, L. (1961). *Thermodynamics* (2nd ed.). McGraw-Hill.

Lindsay, W. L. (1979). *Chemical equilibria in soils*. John Wiley & Sons.

Nordstrom, D. K., Plummer, L. N., Langmuir, D., Busenberg, E., May, H., Jones, B. F., & Parkhurst, D. L. (1990). Revised chemical equilibrium data for major water–mineral reactions and their limitations. In D. C. Melchoir & R. L. Bassets (Eds.), *Chemical modeling in aqueous systems II*, Symposium Series 416 (pp. 389–413). American Chemical Society. https://doi.org/10.1021/bk-1990-0416.ch031

Oster, J. D., & Rhoades, J. D. (1975). Calculated drainage water compositions and salt burdens resulting from irrigation with river waters in the western United States. *Journal of Environmental Quality, 4*, 73–79. https://doi.org/10.2134/jeq1975.00472425000400010017x

Parkhurst, D. L., & Appelo, C. A. (2013). Description of input and examples for PHREEQC v. 3: A computer program for speciation, batch reaction, one dimensional transport and inverse geochemical calculations. In *Techniques and methods* (Book 6: Modeling techniques, Section A: Groundwater, Chap. 43). USGS. https://doi.org/10.3133/tm6A43

Parkhurst, D. L., Thorstenson, D. C., & Plummer, L. N. (1980). *PHREEQE: A computer program for geochemical calculations* (Water Resources Investigations Rep. 80-96). USGS.

Pitzer, K. S. (1979). Theory: Ion interaction mode. In K. S. Pitzer (Ed.), *Activity coefficients in electrolyte solutions* (pp. 158–208). CRC Press.

Plummer, L. N., Prestemon, E. C., & Parkhurst, D. L. (1991). *An interactive code (NETPATH) for modeling NET geochemical reactions along a flow PATH.* (Water Resources Investigations Rep 91-4078). USGS.

Reading, L. P., Baumgarti, T., Bristow, K. L., & Lockington, D. A. (2012). Applying HYDRUS to flow in a sodic clay soil with solution composition-dependent hydraulic conductivity. *Vadose Zone Journal, 11*(2). https://doi.org/10.2136/vzj2011.0137

Shaffer, M. J., & Gupta, S. C. (1981). Hydrosalinity models and field validation. In I. K. Iskandar (Ed.), *Modeling wastewater renovation: Land treatment* (pp. 136–181). John Wiley & Sons.

Shaffer, M. J., & Larson, W. E. (Eds.) (1987). *NTRM, a soil–crop simulation model for nitrogen, tillage, and crop-residue management* (Conservation Research Rep. 34-1). USDA–ARS.

Shaffer, M. J., Ribbens, R. W., & Huntley, C. W. (1977). *Prediction of mineral quality of irrigation return flow. Vol. V: Detailed return flow salinity and nutrient simulation model* (EPA-6001, 2-77- 179e). USEPA.

Shaffer, M. J., Rojas, K. W., & DeCoursey, D. G. (2000). The equilibrium soil chemistry process—SOLCHEM. In L. R. Ahuja, K. W. Rojas, J. D. Hanson, M. J. Shaffer, & L. Ma (Eds.), *Root Zone Water Quality Model: Modelling management effects on water quality and crop production* (pp. 145–161). Water Resources Publications.

Shaygan, M., Baumgartl, T., Arnold, S., & Reading, L. P. (2018). The effect of soil physical amendments on reclamation of a saline–sodic soil: Simulation of salt leaching using HYDRUS-1D. *Soil Research, 56*, 829–845.

Šimůnek, J., & Suarez, D. L. (1993). Modeling of carbon dioxide transport and production in soil: 1. Model development. *Water Resources Research, 29*, 487–497.

Šimůnek, J., & Suarez, D. L. (1994). Two-dimensional transport model for variably saturated porous media with major ion chemistry. *Water Resources Research, 30*, 1115–1133.

Šimůnek, J., van Genuchten, M. T., & Šejna, M. (2018). *The HYDRUS software package for simulating two- and three-dimensional movement of water, heat, and multiple solutes in variably-saturated media. Technical manual, Version 3.0.* PC Progress.

Stumm, W., & Morgan, J. J. (1996). *Aquatic chemistry: Chemical equilibria and rates in natural waters* (3rd ed.). John Wiley & Sons.

Suarez, D. L. (1977a). Ion activity products of calcium carbonate in waters below the root zone. *Soil Science Society of America Journal, 41*, 310–315.

Suarez, D. L. (1977b). *Magnesium, carbonate, and silica interactions in soils* (U. S. Salinity Laboratory Annual Report). U.S. Salinity Laboratory.

Suarez, D. L. (1999). Thermodynamics of the soil solution. In D. L. Sparks (Ed.), *Soil physical chemistry* (2nd ed., pp. 97–134). CRC Press.

Suarez, D. L. (2001). Sodic soil reclamation: Model and field study. *Australian Journal of Soil Research, 39*, 1225–1246.

Suarez, D. L. (2012). Modeling transient rootzone salinity (SWS model). In W. W. Wallender & K. K. Tanji (Eds.), *Agricultural salinity assessment and management* (ASCE manuals and reports on engineering practice no. 71 (2nd ed., pp. 855–897). American Society of Civil Engineers.

Suarez, D. L., & Rhoades, J. D. (1982). The apparent solubility of calcium carbonate in soils. *Soil Science Society of America Journal, 46*, 716–722.

Suarez, D. L., & Šimůnek, J. (1993). Modeling of carbon dioxide transport and production in soil: 2. Parameter selection, sensitivity analysis and comparison of model predictions to field data. *Water Resources Research, 29*, 499–513.

Suarez, D. L., & Šimůnek, J. (1997). UNSATCHEM: Unsaturated water and solute transport model with equilibrium and kinetic chemistry. *Soil Science Society of America Journal, 61*, 1633–1646.

Suarez, D. L., Wood, J. W., & Ibrabim, I. (1992). Reevaluation of calcite supersaturation in soils. *Soil Science Society of America Journal, 56*, 1776–1784.

Suarez, D. L., Wood, J. W., & Taber, P. E., Jr. (2012). Adsorption and desorption of boron in column studies as related to pH: Results and model predictions. *Vadose Zone Journal*, 11(2). https://doi.org/10.2136/vzj2011.0073

Truesdell, A. H., & Jones, B. F. (1974). WATEQ, a computer program for calculating chemical equilibria of natural waters. *Journal of Research United States Geological Survey*, 2, 234–248.

USEPA. (1999). *MINTEQA2/PRODEFA2: A geochemical assessment model for environmental systems. User manual supplement for Version 4.0* (EPA/600/3-9/021). USEPA Office of Research and Development.

U.S. Salinity Laboratory Staff. (1954). *Diagnosis and improvement of saline and alkali soils* (Agriculture Handbook no. 60). U.S. Government Printing Office.

Vanselow, A. P. (1932). Equilibria of the base-exchange reactions of bentonites, permutates, soil colloids and zeolites. *Soil Science*, 33, 95–113.

Wagenet, R. J., & Hutson, J. L. (1987). *LEACHM: Leaching estimation and chemical model, a process based model of water and solute movement, transformations, plant uptake and chemical reactions in the unsaturated zone* (Continuum 2). Department of Agronomy, Cornell University.

Wagman, D. D., Evans, W. H., Parker, V. B., Schumm, R. H., Halow, I., Bailey, S. M., Churney, K. L., & Nuttall, R. L. (1982). The NBS tables of chemical thermodynamic properties: Selected values for inorganic and C_1 and C_2 organic substances in SI units. *Journal of Physical and Chemical Reference Data*, 11(Supplement 2).

Wang, J., Bai, Z., & Yang, P. (2016). Using HYDRUS to simulate the dynamic changes of Ca^{2+} and Na^+ in sodic soils reclaimed by gypsum. *Soil and Water Research*, 11, 1–10. https://doi.org/10.17221/14/2015-SWR

Wollast, R., Mackenzie, F. T., & Bricker, O. P. (1968). Experimental precipitation and genesis of sepiolite at Earth-surface conditions. *American Mineralogist*, 53, 1645–1662.

Woods, T. L., & Garrels, R. M. (1987). *Thermodynamic values at low temperature for natural inorganic materials: An uncritical summary*. Oxford University Press.

7

R. Don Wauchope

Pesticide Dissipation and Fate in Agricultural Settings: A Review of Research, 2000–2020

Abstract

Progress from 2000 to the present in our understanding of the fate of pesticides after application in agricultural settings is reviewed in the context of predicted environmental concentrations (PEC) simulation modeling. Formulation secrecy remains an impediment to the study of nearly all processes. Spray drift from aerial and ground equipment has been characterized and is predictable by probabilistic empirical rules. Volatilization of pesticide active ingredients from application deposits on plants and soil have been physically modeled and are part of the mass balance of a few PEC models. Measurement and modeling of plant uptake, translocation, and degradation of pesticides has considerably advanced recently. Soil sorption understanding has benefited from some revealing isotope-exchange studies, and the coupling of sorption and biotic degradation in soils has become a standard part of advanced PEC modeling. Spatial variability of sorption and degradation from point to field scale is much better understood and has been incorporated into modeling to a limited extent. Understanding pesticide leaching via preferential flow has been aided by modeling studies. Effects and limitations of conservation tillage practices on pesticide runoff have been quantified by rainfall simulation studies. Several specific research needs remain.

Introduction

When an agricultural pesticide is applied to a crop or soil, numerous chemical and physical processes affect the pesticide's efficacy, dispersion, persistence, and ultimate environmental impact. In 2000, the USDA released the Root Zone Water Quality Model (RZWQM) (Ahuja et al., 2000), which was envisioned to provide a simulation accounting for all such processes and their interactions at a single point in a field. Input data

Modeling Processes and Their Interactions in Cropping Systems: Challenges for the 21st Century, First Edition.
Edited by Lajpat R. Ahuja, Kurt C. Kersebaum, and Ole Wendroth.

describing a given soil, crop, pesticide, and weather scenario led to a broad array of predicted environmental concentrations (PECs) as a function of time. Practical constraints intervened and the model was incomplete—especially, pesticide spray drift and soil erosion were not included and volatilization was not adequately considered. Still, the list of processes and parameters included in the simulation was the most extensive at the time:

- application target
- pesticide formulation
- soil sorption kinetics
- pesticide acid–base properties
- root uptake
- wash-off from foliage and crop residue
- increased persistence in the vadose zone
- upward leaching due to wicking
- half-lives for each compartment
- aerobic and anaerobic soils
- a new mixing model for runoff
- dissipation from plant surfaces
- binding (aging) of pesticides in soil

The transformation and transport processes considered and the agricultural environment compartments where and between which they operate are diagrammed in Figure 7.1.

The level of knowledge of these processes from experiment, and the level of detail incorporated into the model, varied widely. In some cases (e.g., foliar dissipation) RZWQM simply allowed for the process and provided a parameter (i.e., a "lumped" dissipation half-life). In others (e.g., soil sorption), a rich and immense literature existed and complex algorithms were incorporated.

The purpose of this chapter is to survey progress in understanding of these process details since 2000 and, secondarily, progress in the incorporation of this process detail into PEC models. This is a large area; almost 1,000 studies were considered as a result of literature searches. The result is a survey of the literature, not a critical review. I will not exhaust any of the topics and necessarily have limited this report to studies that have provided significant new process insight. Occasionally an interesting or unusual experiment is included.

In 2000, RZWQM was only one model in a very intense and competitive arena (Cohen et al., 1995; Klein, 2011; Šimůnek et al., 2003). Regulatory needs for PECs in ground and surface waters for registrations of new pesticides drove much of this development. Sarmah et al. (2004) provided an interesting "outsider" review of the field and the international transferability of models at the time. The PEARL model (Leistra et al., 2001) is currently the most comprehensive PEC model for pesticides; it has incorporated most of the pesticide processes in RZWQM and added an advanced, mechanistic volatilization model.

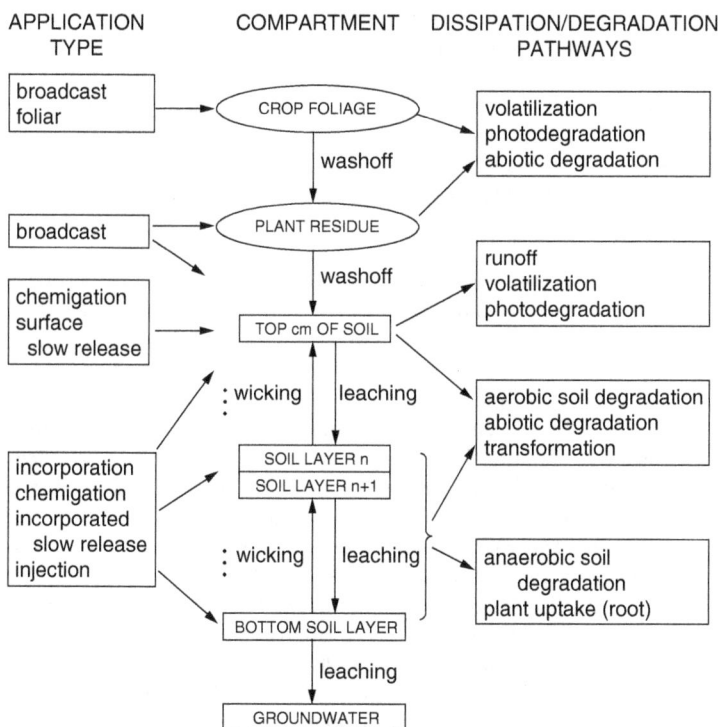

Figure 7.1 Location of applied pesticide deposits, dissipation or degradation pathways in the environmental compartments, and transfer processes between compartments of the RZWQM model. Modified from Wauchope et al. (2000, 2004).

Pesticide Application, Drift, Efficiency, and Retention

Spray Drift

In North America, pesticide spray drift experimentation received a significant boost in the 1990s with USEPA demands for data from the industry. Industry responded with the Spray Drift Task Force, a complex and intensive project by 32 companies to provide a generic empirical model of drift for aerial and ground applications. The results for aerial application, with its greater potential drift, were reviewed by Bird et al. (1996) and compared with 45 previously published trials: "...Both sets of data show median values of pesticide deposition dropping from ~5% of the nominal application rate at 30 m downwind to ~0.5% at 150 m during low-flight applications" (p. 1095).

These trials demonstrated that differences in equipment, wind speed, and practice lead to a wide range of drift; they proposed that specification of equipment and operating conditions on pesticide use labels, in addition to the already mandated wind maxima, was needed. The Spray Drift Task Force results were reported by Hewitt et al. (2002) and Hewitt et al. (2001) for aerial and ground equipment applications, respectively. Aerial deposition ranged from 20 to 70% 10 m from the spray path to

0.002–0.01% at 1,000 m, with a relatively constant variation of about 3–5× at all distances. These data have been incorporated into the proprietary model AgDRIFT (Teske et al., 2002).

Felsot et al. (2011) published a global review of spray drift measurement, modeling, and regulation and spray mitigation approaches. Overall, they stated that drift to surface waters causes about a tenth of the pollution by runoff (i.e., about 0.2%)—probably a very simplified generalization. Uses and improvements in AgDRIFT were reported and the more widely used agricultural pesticide spray application drift model (AgDISP) (Nsibande et al., 2015; Woodward et al., 2008) and other models. Many countries have established spray buffer zones and equipment operating rules for minimizing drift. Aerial applications have been banned in some countries (Paul Hendley, personal communication, 2020), and within countries, mandated spray buffer zones might vary by state, province, or other administrative boundaries (Annemieke Farenhorst, personal communication, 2020).

Hilz and Vermeer (2013) reviewed the technology and practice for risk reduction of drift and discussed the relationships among spray solution viscosity, surface tension, and inhomogeneities (solid particles and emulsions). They discussed the potential effects of commercial formulations on risk, but generalizations were difficult.

In the EU, after the drift risk assessment of an active ingredient has been accepted "in principle" based on drift data on a single formulation, individual countries may evaluate and accept additional formulations. For example, in the Netherlands, the IDEFICS drift model (Holterman et al., 1997; Holterman et al., 2017) is used to evaluate drift (J.B.H.J. Linders, personal communication, 2020). The EU BROWSE (Bystanders, Residents, Operators and WorkerS Exposure models for plant protection products) project uses Netherlands and U.K. drift data to develop new models including exposure from orchard sprayers (Butler Ellis et al., 2017; Butler Ellis et al., 2010; van de Zande et al., 2014).

A few specific studies of interest: Koch et al. (2003) reported a study of drift to neighboring hedgerows from applications to turf and the effects of drift-reducing nozzles, comparing spray traps versus foliar deposition. The canopy structure of the receiving vegetation is important. Application losses including drift reduction by adjuvants was reviewed by de Ruiter et al. (2003). Winchell et al. (2018), in an unusual exercise with a watershed hydrologic model, were able to estimate spray drift to aquatic water bodies. De Cock et al. (2017) used deposition and retention models to determine the optimum droplet size for both efficacy and drift minimization.

In the context of spray drift, a database exists of 9,141 military aerial applications of herbicides in Vietnam between 1961 and 1971, including flight paths, chemical used, and doses applied. Because details of U.S. military missions in the area were also recorded, these data have been used to model exposures of U.S. personnel (Stellman & Stellman, 2004).

Application Efficiency and Retention

Most pesticide applications in agriculture involve broadcast pressurized tank mixes of chemical(s) and water applied through hydraulic spray nozzles onto crops or soils. This results in a complex sequence of possible processes: (a) spray droplet formation; (b) drift of spray away from the target area; (c) volatilization of the spray prior to impact; (d) impact and deposit or deflection of spray drops on plants, crop residues, and soil surfaces; and (e) volatilization and sorption of spray deposits on plant, crop residue, and soil surfaces. Other modes of application such as injection into the soil, additions to irrigation water, or seed treatment are assumed to be much less uncertain as to the initial deposition. For our purposes, we define *application efficiency* as the fraction of the total pesticide spray that deposits on plants, plant residue, and soil surfaces within the target area. This will be decreased by processes (a), (b), and (c) above. We define *foliar retention* as the fraction of the total spray that impacts and remains on plant surfaces if present. This will depend on the fraction of the spray area subtended by plant surfaces (leaf area index) and the amount of spray that sticks to those surfaces. In the RZWQM, pesticide application efficiency was an input value defaulting to 60%. This fraction was then partitioned between soil and plants (if present) using the leaf area index calculated by the plant module of the model. Leaf area index is commonly used in this way in models.

Jensen and Olesen (2014) reviewed 89 ground equipment application studies, mostly using dye tracers. Although near-complete efficiencies can occur, losses were typically 10–20% in field crops and as much as 50% from orchard sprayers. This suggests that the 60% default efficiency in the RZWQM is simplistic.

For retention, the European Food Safety Authority (2014) published a table of crop interception values; however, these can be very different from U.S. planting densities (Paul Hendley, personal communication, 2020).

Differences in retention among plant species can be dramatic. Dorr et al. (2014, 2016) presented a foliar spray retention model that accounts for plant wettability and spray chemistry and compared the results to experimental data. For cotton, which is quite wettable, typical retention was 40%; for less wettable wheat, typical retention was only 10–20%.

A few notable experiments: Wolters et al. (2004), in a unique near-total mass balance accounting, used wind tunnel data and modeling to estimate the fractions of broadcast radiolabeled methyl parathion deposited on the soil (18.0%), in or on winter wheat plants (22.0%), volatilized (32.5% including metabolites), mineralized to CO_2 (6.4%), and attached to airborne particles (8.2%), accounting for 87% of the amount applied. Dang et al. (2016), in an innovative experiment, determined the spray fraction of six herbicides deposited on sugarcane residues and found 43–80%.

Recently, studies using an aerial drone sprayer flying close to the crop canopy gave similar efficiencies as ground spray equipment (Gao et al., 2019; Wang et al., 2019). These drones have potential for precision application of pesticides and even insect predators, as shown by Lan et al. (2017) and Lost Filho et al. (2019).

Progress Summary—Processes During Application

A large amount of spray drift data has been collected and a range of empirical models developed, mainly to satisfy regulatory risk assessment screening needs. As of 2020, aerial drift has been much better characterized. As suggested by the Wolters et al. (2004) experiment, drift is not necessarily a small part of the pesticide mass balance at an agricultural site, and PEC models should allow for this possibility.

Application efficiency and plant foliage retention determine the initial total pesticide concentrations per unit field area and then the distribution of the total among plant cover, soil, and plant residues. This is critical for PEC modeling, and some good experimental work has been done, but it would appear that predictability in this area, whether empirical or theoretical, has not improved significantly since 2000.

Pesticide Uptake by and Dissipation in and on Plants

Pesticide Penetration and Photodegradation on Foliage

Zabkiewicz (2007), in a review, defined (foliar) *spray efficacy* as determined in four steps:

- **deposition**: the fraction of spray reaching the target area (80–90%)

- **retention**: the fraction of deposit captured by the plant (10–100%)

- **uptake**: the fraction of retained chemical absorbed by foliage (30–80%)

- **translocation**: the fraction of absorbed material translocated (10–50)

It is a simple calculation to determine the worst and best cases: efficacy just to the translocation stage can range from 0.2 to 40%. Zabkiewicz (2007) described the beginnings of efforts to model these processes.

Forster et al. (2006) reviewed foliar uptake, emphasizing its complexity; Forster and Kimberley (2015) stated "Differences [in retention] among plant species remain the most difficult to explain, and this is thought to provide the largest challenge to models for foliar uptake" (p. 1332). Massinon et al. (2017), among others, used high-speed photographs of droplets impacting on plants with widely different leaf surfaces, revealing a complex set of droplet behaviors and plant species differences. Taylor (2011), in a broad review, covered leaf wax studies and droplet behavior, super-hydrophobicity of leaf wax surfaces, and agrochemical deposit structures and mentioned the uncertainties of unknown commercial spray ingredients. The beginnings of a physical (as opposed to empirical) model for spray retention was described.

Photodegradation has received relatively little attention since 2000. Katagi (2004) reviewed photodegradation of pesticides on plant and soil surfaces. Lavieille et al. (2009) reported a half-life of less than an hour in bright sunlight for the herbicide mesotrione on foliar surfaces, and the degradation was greatly accelerated by surfactant spreading. Eyheraguibel et al. (2009) reported on the photolysis of bentazon, clopyralid, and triclopyr with simulated sunlight on cuticular wax film. Photodegradation on the film was faster than in water and in some cases significantly affected by the formulation.

Pesticide Root Uptake and Plant Metabolism

The RZWQM used the empirical relationship observed by Briggs et al. (1982) between the octanol–water partition coefficient of a nonpolar pesticide (K_{OW}) and the ratio of its concentration in soil water to that in the root translocation stream of contacting roots, called the transpiration stream concentration factor (TSCF). This was combined with plant water uptake in the hydrology module of the model to calculate uptake. Half-lives for pesticides in plants (as a whole) and on foliar surfaces were provided for, but only if the user had an estimate for the scenario. Storage was not accounted for. This part of the RZWQM has never been tested.

The TSCF has remained a centerpiece of root uptake studies of chemicals, although the octanol–water relation does not always apply. Examples of studies include Hwang et al. (2017), Felizeter et al. (2014), and Gredelj et al. (2020). Fantke et al. (2016) proposed a protocol for plant uptake studies to provide more useful information for modeling. Juraske et al. (2008) reported a correlation between pesticide metabolism half-lives in soils and plants for 42 pesticides. Fantke and Juraske (2013) reviewed a vast collection of published pesticide half-lives in and on plants, including edible portions of plants. Almost all of the half-lives are in within the range of 1–10 days. Thorbek and Hyder (2006) looked for correlations between regulatory food tolerances, or maximum residue levels (MRLs) sorted by use (herbicide, fungicide, insecticide) and crop type (e.g., fruits, leafy vegetables, seeds) and pesticide physical and chemical properties. Regression models using crop type and other properties often accounted for some 50% of variation—in log–log-transformed data.

Brunetti et al. (2019) recently reviewed soil and root uptake models since Briggs et al. (1982), referred to generally as dynamic plant uptake or DPU process models, and their combination with soil hydrologic models. Based on an analysis of the shortcomings of previous efforts, they coupled a modified version of the well-developed DPU model by Trapp (2000, 2007, 2009) with the widely used and hydrologically sophisticated (Richards equation) HYDRUS-1D soil hydrology model (Šimůnek et al., 2003). The resulting model has unique capabilities, tracking root and foliar uptake, transport, and metabolism of pesticides in the growing roots, stems, leaves, and fruits of plants, and includes volatilization from foliage. Brunetti et al. (2019) made it clear that the model is a work in progress (many input parameters required) and that much experimental work, particularly in the plant metabolism area, needs to be done. M. Silburn (personal communication, 2020) has suggested that this Brunetti–Trapp model would be an innovative way to analyze the uptake and efficacy of herbicides.

For a recent important experimental methodology study and review of plant uptake factors (a modified TSCF) by German industry, see Lamshöft et al. (2018), (P. Hendley, personal communication, 2020).

Progress Summary—Pesticide Uptake by and Dissipation in and on Plants

Foliar uptake is affected by spray solution chemistry, foliar surface structure, foliar wax properties, plant morphology, leaf surface weathering, and plant growth stage. Foliar cuticle penetration and photodegradation of the deposit have received significant experimental attention, but there appears to be little predictive capability at this point. Clearly, plant root uptake and metabolism by plants have had significant advances since the postulation of the TSCF. Brunetti et al. (2019) has provided both a recent summary of progress and a report of a significant modeling advance. Lamshöft et al. (2018) offered a proposed protocol for root uptake experiments.

Pesticide Volatilization from Plants and Soil

Early Reviews: The State of the Art in the Early 2000s

Majewski (1999) reviewed methodology of field flux methods of volatilization measurement and estimated typical uncertainty to be 50%. Bedos et al. (2002) reviewed field data and noted effects of pesticide vapor pressure, water solubility and soil sorption. In general pesticide volatilization from foliage was much higher than from soil because of higher air turbulence, more rapid spray deposit drying, and movement into the soil subsurface. Losses can be 90% in the first 24 hours. Initial volatilization is strongly dependent on vapor pressure, especially from foliage, and vapor pressure is very sensitive to temperature. They called for the development of a mechanistic model for volatilization. De Ruiter et al. (2003) in a thorough review of volatilization losses indicated that significant fractions can be lost from soil and nearly 100% losses from plant foliage are not uncommon. They also discussed the potential role of adjuvants in both drift reduction and post-application volatilization. This was a good overall survey of understanding at the time but no quantitative generalizations of adjuvant effects were able to be made.

Key Experiments and Foundations of PEC Modeling of Volatilization

Pesticide volatilization was a latecomer to modeling investigations, following water pollution processes, because of the early regulatory emphasis on water quality (Foreman et al., 2000). By the time atmospheric pollution began to receive more attention, the use of systems computer modeling had become standard and provided an approach to problem formulation and the sorting out of the many variables controlling volatilization.

Yates et al. (2002) modeled methyl bromide volatilization from under a plastic cover under irrigated California agricultural conditions: losses were 24–74% of applied amounts. Their model assumed Henry's law, a stagnant air layer, and first-order degradation in soil, water, and air. They obtained an excellent prediction of losses totaling 55%, versus measured values of 58–70%, and a detailed prediction of observed diurnal fluctuations in loss. They showed the critical influence of atmospheric conditions in predicting instantaneous emissions to the atmosphere. Subsequently, Yates (2006a, 2006b) measured volatilization losses of triallate for 6.5 days, then used the model to calculate long-term total emissions of 80%, which could be cut to 5% with incorporation. This was pioneering work.

At this time much new experimental data began appearing on post-application volatilization of pesticides from plants and soil. For example, Gish et al. (1995) demonstrated that the presence of crop residue (mulch) promoted volatilization of applied pesticides, affirming diminished losses from soil. A 20-days monitoring of five heavy-use (at the time) chemicals by Rice et al. (2002) under rainy conditions indicated revealing differences in volatilization losses: atrazine 8%, metolachlor 12%, β-endosulfan 15%, chlorpyrifos 23%, trifluralin 27%, and α-endosulfan 35%.

Guth et al. (2004) published a large 24-hours volatilization versus vapor pressure data set for 80 active ingredients generated by industry laboratory and wind-tunnel studies (e.g., Rüdel, 1997; Wolters et al., 2004). The very impressive experimental setup is shown in Figure 7.2. These studies followed German regulatory protocols; many used the same soil and crop. The identities of the active ingredients were kept secret. Formulation effects were not discussed even though 10 "formulation types" were listed without explanation: SC, EC, WG, SL, WP, Acetone, Methanol, Acetonitrile, FW, and ME. Some correlation of volatilization with vapor pressure and, in the case of soil applications, the soil organic C (SOC) sorption coefficient was noted.

In an important advance, Leistra (2005) reviewed volatilization from plants and defined processes and parameters needed for modeling and the research needs for each. Deposition and retention data and knowledge of deposit formation was needed. Vapor pressure is crucial, and critically evaluated data were needed. Temperature dependence is roughly similar for many pesticides. Emission was to be calculated by a boundary layer diffusion model. Three post-deposition processes can also strongly mediate volatilization: wash-off of deposits by rain (for which the Willis equation was proposed, see below), penetration of the active ingredient into foliage, and degradation of deposits, principally by photolysis. The latter two processes had little discernable pattern. In a series of masterly experimental and modeling studies

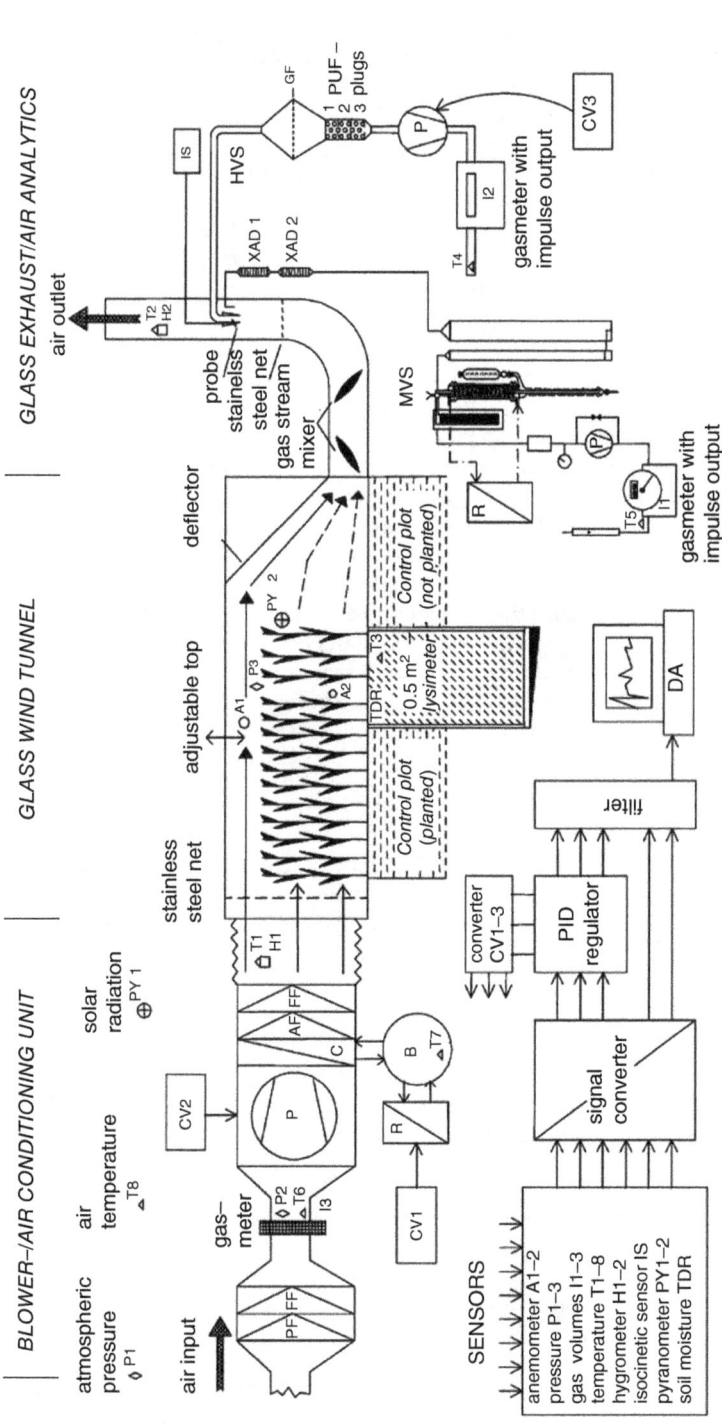

Figure 7.2 Wind tunnel experiment for simulating field volatilization of pesticides from soil and plant surfaces. AF = active charcoal filter, B = brine tank, C = cooler, CV = converter, DA = data acquisition, FF = line filter, GF = glass fiber filter, HVS = high volume sampler, IS = isocinetic sensor, MVS = medium volume sampler ($^{14}CO_2$), P = pump and blower, PF = prefilter, PUF = polyurethane foam, R = refrigeration, TDR = time domain reflectometry, XAD = adsorbing resins (AmberliteR XAD-4). Wolters et al. (2004).

exploring these processes, Leistra and colleagues developed a volatilization module for the PEARL PEC model (Leistra et al., 2001) that overall incorporated most of the nuances of foliar dissipation (Leistra et al., 2008; Leistra & van den Berg, 2007).

Modeling of volatilization has since been incorporated into several other models. Magri et al. (2009) developed the TURFP model for simultaneous pesticide losses from turf by leaching, runoff, and volatilization. Bedos et al. (2010) used the aerodynamic gradient method and inverse modeling to measure the volatilization of chlorothalonil and fenpropidin. Fenpropidin, with a much higher vapor pressure, did not exhibit as big a difference in volatilization as would be predicted by the vapor pressure models; they had postulated a rapid foliar penetration for fenpropidin. Wang et al. (2011) accurately simulated shank injection of chloropicrin. Nelson et al. (2013) combined the HYDRUS model (Šimůnek et al., 2003) with a volatilization model to simulate the distribution in soil of methyl isothiocyanate after drip irrigation application. Garcia et al. (2014) modeled gas-phase sorption and volatilization of trifluralin from bare soil under dry soil conditions; the model indicated that sorption by dry soil could be important for trifluralin, which could then be released when the soil was moistened, affirming a hypothesis made 38 years earlier by Harper et al. (1976).

Houbraken et al. (2015) measured volatilization of 10 pesticides after application; a dramatic graph of the results showed almost no correlation with vapor pressure P_v except for pesticides with $P_v > 10^{-5}$ Pa. For four chemicals (trifloxystrobin, pyrimethanil, propyzamide, and fluazinam) the correlation was calculated to not be significant: processes not related to vapor pressure determined the volatilization rate. They urged more knowledge of formulation effects. Recently, Bedos et al. (2017) measured the volatilization of S-metolachlor and benoxacur after spraying on soil with and without crop residues. Deposits intercepted by crop residues were more volatile and degraded more quickly.

Volatilization Prediction with Pesticide Environmental Parameters

In 1982–1984, in a series of influential studies by Jury, Spencer, and Farmer (Jury et al., 1983, 1984a, 1984b, 1984c; Spencer et al., 1982) demonstrated the ability to estimate an agricultural pesticide's potential to pollute soil, air, and water based on two equilibrium constants: the Henry's law constant K_h, and soil organic carbon (OC) sorption constant K_{oc}, plus a kinetic constant k for field or soil degradation or chemical half-life. This model was valuable for its ability to screen active ingredients for their potential for atmospheric pollution. Those studies reviewed two decades of work on the use of equilibrium coefficients to develop environmental "partition models" for organic pollutants and on soil sorption effects on pesticide volatilization. Woodrow et al. (1997) showed that the logarithm of initial pesticide volatilization or "emission" rate E from applications to inert surfaces, water, and soils correlated very well ($r^2 = .97$–.99) with the logarithms of combinations of the vapor pressure, soil OC sorption coefficient, and water solubility. Woodrow et al. (2011) extended these relationships to subsurface-applied fumigants with large vapor pressure by simply adding the depth of application.

van Wesenbeeck et al. (2008) showed that the logarithm of vapor pressure (VP) versus the log(E) data from (a) Woodrow et al. (1997), (b) American Society for Testing and Materials (1987) test data, and (c) the data of Guth et al. (2004)—89 compounds in all, covering about eight orders of magnitude in VP—were all statistically in the same data set with an r^2 of .98 and a predictive error of about one order of magnitude. Mackay and van Wesenbeeck (2014) demonstrated that, for the same set of compounds used by van

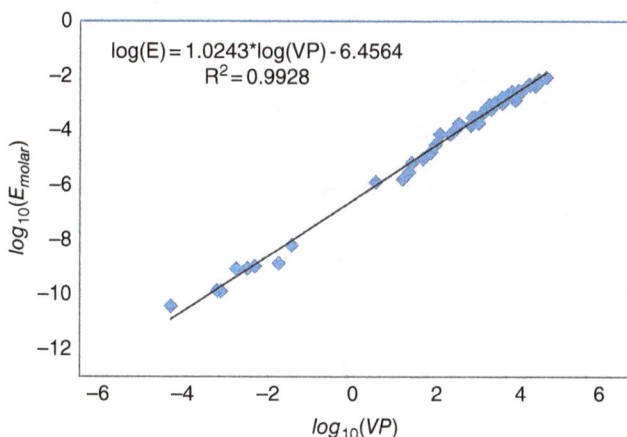

Figure 7.3 Base 10 log–log correlation between vapor pressure (VP) and molar emission (volatilization) rates of molecules (E_{molar}). Adapted from Mackay and van Wesenbeeck (2014) / with permission of American Chemical Society.

Wesenbeeck et al. (2008) (except for those that could not be identified) on inert surfaces, an even better correlation ($r^2 = .9928$) is between log(VP) and log(E_{molar}), that is, the latter in units of moles/area·time because of the greater intermolecular attraction of larger molecules (Figure 7.3). In fact, with this more physically correct version, a non-logarithmic plot of the data passes close to zero and thus a single-parameter relationship results if forced through zero:

$$E_{molar}\left(mol\,m^{-2}\,s^{-1}\right) = 4.07 \times 10^{-7} VP(Pa) \tag{7.1}$$

Although the constant is certainly complex dimensionally, this is an extraordinarily simple and powerful relationship. Mackay and van Wesenbeeck (2014) stressed that, in agricultural situations, the effects of soil sorption and water solubility are better expressed by the correlations of Woodrow, which account for those processes.

It should be emphasized that these correlations all involve the most intensive, earliest volatilization losses—typically from hours to a day—from pesticides and other chemicals deposited on surfaces. For pesticides, such deposits form within minutes after application as the carrier evaporates, and the excellent correlation indicates that essentially pure active ingredient is exposed. Long-term, multiyear volatilization studies, in contrast, typically show very large variations in total losses, year on year: Gish et al. (2011) in an eight-years study of metolachlor and atrazine observed total losses of 5–63% for metolachlor and 2–12% for atrazine, depending mainly on the weather. Prueger et al. (2017) observed 5–63% losses of metolachlor during 17 years; total volatilization increased exponentially with soil moisture.

Effects of Formulation, Deposit Site, and Foliar Penetration on Volatilization

Houbraken et al. (2015) conducted a unique, illuminating study in which they compared the volatilization of commercially formulated fenpropimorph, pyrimethanil, chlorpyrifos-ethyl, and lindane with the pure compounds and also formulated with three different well-characterized surfactants. In general, the initial volatility of the pure active ingredients was well correlated with vapor pressure but was (usually) strongly curbed by the addition of an adjuvant. Lichiheb et al. (2014) and Lichiheb et al. (2016) modeled the effect of foliar sorption and photodegradation of pesticides on leaves. They reviewed the state of modeling and offered a model compromise between theory and the availability of test data. By accounting for these processes, their model was able to well predict volatilization at the field scale. They used the octanol–water coefficient K_{ow} as a measure of foliar penetration but also observed that formulation had a big effect, especially for systemic pesticides.

Recent Progress

Mueller and Steckel (2019) showed that, in a tank mix of glyphosate and dicamba, the strong acid glyphosate decreased the mix pH enough to shift the weaker acid dicamba in the direction of the nonionized form, significantly increasing its volatility. Mao et al. (2018), coupling physical models for soil and foliar chlorpyrifos losses, demonstrated the great differences between the two: foliar losses were large and immediate (hours) while soil losses were much smaller but long lasting (weeks). Taylor et al. (2020) demonstrated a multipartition approach to both soil and foliar volatilization including soil–air partitioning; the model predicted the observed higher volatilization from foliage relative to soil and also reflected differences between plant growth stage and plant species.

Progress Summary—Pesticide Volatilization from Plants and Soil

Vapor pressure has been shown to be an important predictor of early volatilization and provides a simple way to factor volatilization into the mass balance of pesticide fate. In the individual field scenario, however, there are a myriad of possible mediations. Excellent, physically based volatilization models have been developed and validated. This is one of the most significant accomplishments of 2000–2020.

Pesticide Wash-Off from Foliage and Crop Residues: The Willis Equation

Wash-off of pesticide residues from plant foliage or crop mulch by rain or irrigation is an important transfer process, especially if it occurs shortly after application. The classic studies by Willis and colleagues (Willis et al., 1980) resulted in an empirical logarithmic pesticide loss function:

$$C_{ps} = C_{ps}^0 F_{wo} \exp_e(-P_{wo}I_r \Delta t) \tag{7.2}$$

where C_{ps}^0 and C_{ps} (mass/area) are area concentrations of the pesticide before and after the time period Δt, F_{wo} is the fraction of pesticide deposit available for wash-off, P_{wo} is a power constant, and I_r (mm/time) is the average rainfall intensity during Δt; P_{wo} and F_{wo} are approximately correlated with the water solubility of the pesticide, and a table of values was provided for RZWQM users. Equation 7.2 will be referred to as the Willis equation. Again, formulation is typically a highly significant and not well understood factor, especially for rainfall shortly after application.

Since 2000, Fife and Nokes (2002) confirmed that rainfall total (as opposed to intensity) is the dominant parameter. Silburn and Kennedy (2007) described the utility of rainfall simulators for wash-off studies. Matocha et al. (2006) demonstrated the dynamics of wash-off contributions to runoff in an important study of trifloxysulfuron in cotton.

Potter et al. (2003) studied three cotton defoliants in strip-till and conventional till-age: rainfall was applied one hour after application. Interestingly, dimethipin, with a higher solubility and lower soil sorption than the other two defoliants, gave lower run-off losses (2–5 vs. 12–15%), probably because it was washed off prior to the beginning of runoff and had time to be leached below the soil surface before runoff began. In later studies, herbicide runoff was shown to be affected by wash-off of residues intercepted by crop residues (Potter et al., 2006). Fomesafen, a moderately strong acid formulated as the sodium salt and extremely water soluble, was quickly washed off crop residues but still fit the Willis equation (Potter et al., 2011). Because Formesafen is a nonvolatile salt, they got 95% spray efficiency from spray traps.

Joyce et al. (2008) and Joyce et al. (2010) modeled data for runoff of diazinon with and without cover vegetation and looked at the timing of wash-off and infiltration rel-ative to runoff. Modeling the lag time between wash-off and runoff allowed recom-mendations to be made for "raining in" the insecticide with irrigation, lowering the potential for runoff in subsequent rainfall. Melland et al., 2016 demonstrated that spot spraying reduces herbicide runoff in both water and the sediment phase and also saw pre-runoff infiltration effects. Dang et al. (2016) reported wash-off of six herbicides and potassium bromide from sugarcane residues, which are used as an erosion-control mulch. They placed the cane residues in metal trays under two-hours, 100-mm rains. Their wash-off data fitted the Willis equation well, and they compared their fitted para-meters to RZWQM recommendations, generally confirming them. They also measured the persistence of residues and increasing rainfastness of the pesticide deposits with time. This study provided good evidence that the Willis equation, which was based originally mostly on extremely insoluble (early) insecticides, also can describe much more soluble compounds.

Progress Summary—Pesticide Wash-Off from Foliage and Crop Residues: The Willis Equation

The empirical Willis equation has been shown to be able to describe pesticide wash-off from crop foliage or residues well and has been extended beyond the hydro-phobic compounds originally studied. Given the lack of available formulation information for pesticides, the prospects of a physical prediction model for wash-off remains poor.

Pesticide Sorption by Soil

In the 1990s, simulation modeling sensitivity analysis demonstrated that, even given the great dependence of pesticide fate and transport on many weather, site, and crop parameters, detailed knowledge of a pesticide's intensity of soil sorption and speed of soil degradation, and the interactions of these two processes, is critical for realistic PEC simulation and prediction. Dann et al. (2006) suggested that site-specific sorption data are more important than the choice of model.

The RZWQM soil sorption model described the complexities as understood at the time (Figure 7.4). Figure 7.4 suggests that the soil phase is homogeneous. Much of the soil sorption research since the very beginning has demonstrated it is quite the con-trary. Some significant experiments since 2000 have teased out some of the physical complexities of the linkages between soil persistence and mobility and the nature of the sorption process.

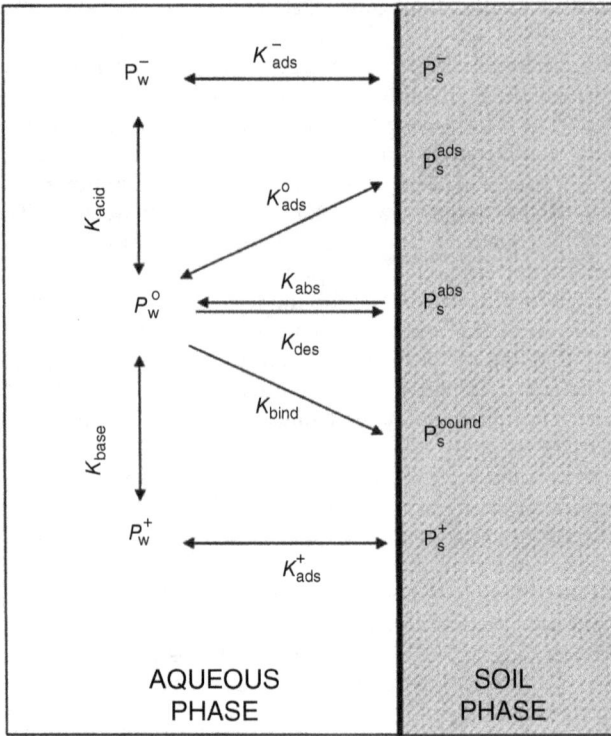

Figure 7.4 Equilibrium (K) and kinetic (k) processes within and between the soil solid and aqueous phases. P is the pesticide molecule; superscripts indicate neutral and (if relevant) anionic and cationic, bound, absorbed, and adsorbed species.

Reviews

Cleveland (1996) reviewed the methodology and regulatory use of soil sorption data, warned of the limitations of applying laboratory-derived data to field conditions, and suggested a more realistic and revealing approach for registrants conducting soil mobility studies. Senesi and Loffredo (1999) offered an exhaustive review of soil organic matter (SOM) chemistry. Delle Site (2001) reviewed the results of 681 organic compound soil sorption studies, mainly reporting sorption isotherm data including ionizable compounds, some discussion of sorption kinetics, and 86 pages of pesticide sorption experiment results. Wauchope et al. (2002) reviewed the mounting evidence for the nonequilibrium and partially irreversible nature of soil sorption and the limitations on the perception of its being a "constant," as embodied in the conventional assumption that a 24-h mixing experiment represents equilibrium, and then proposed general rules for sorption constant variability between soils and between chemicals. Huang et al. (2003) reviewed soil heterogeneity effects on sorption. Sabljic and Nakagawa (2014) reviewed quantitative structure–activity relationship (QSAR) predictions of sorption coefficients.

The Nature of Soil Organic Matter

Early pesticide soil sorption studies, often using pesticides that were much more hydrophobic than typical today, showed a strong correlation with SOM content, so much so that it became conventional to report pesticide soil OC sorption coefficients, K_{oc}, calculated by dividing a measured sorption constant K by the OC fraction of the soil. This "normalized" coefficient provided, in many cases, a reasonable estimate for other soils of known soil SOM fraction. Soil organic matter continues to receive study as usually the most important pesticide sorbent in soils. Leboeuf and Weber Jr. (2000) modeled SOM as a complex of rubbery and glassy macromolecules into which hydrophobic solutes in soil water will have rapid and slow sorption and desorption kinetics, respectively. The model describes sorption–desorption experiments well even with extremely long equilibration times. Leistra and Matser (2004) showed that two fungicides competed for the same sorption sites. Kleber et al. (2007) assumed that SOM is a self-assembling "membrane-like bilayer" that attaches itself through polar attraction to soil mineral surfaces and then forms a hydrophobic surface or even a thin liquid-like phase, providing a liquid–liquid phase exchange for solutes. Similarly, Phillips et al. (2011), using model humic acid and fulvic acid molecular structures from the literature, modeled SOM as an organic solvent composed of humic acids and fulvic acids. Sorption was predicted as an organic solvent–water partition coefficient using the quantum mechanics-based model COSMO-SAC. The logarithm of K_{oc} values for a set of 440 diverse, environmentally relevant chemicals were predicted with a root mean square error of 0.84–1.08, depending on which model humic or fulvic acid was used.

In 2000 it was well understood that any arable soil provides a wide range of sorption mechanisms able to attract a very diverse range of pesticide molecules, from the extremely hydrophobic molecule dichlorodiphenyltrichloroethane (DDT) to the extremely water-soluble multi-ion glyphosate. It has also become clear that as the SOM content in soils decreases from high in organic soils, which are by definition mostly decomposed plant and microorganism matter, to very low in subsurface clay soils, which may have 1% SOM or less, the mineral and metal hydrous oxide surfaces in the soil become uncovered and become important for sorption. Since 2000, Gebremariam et al. (2010), in an exhaustive review of chlorpyrifos soil sorption experiments, showed that sorption correlated poorly with SOM at low levels, indicating other sorption mechanisms. Ghafoor et al. (2013) demonstrated that glyphosate, bentazone, and isoproturon sorption to soil begins to correlate with Al and Fe oxides, soil pH, and clay content as opposed to SOM, especially in subsurface soils or other geosorbents lower than about 2% organic matter. Kookana et al. (2014) reviewed the shortcomings of a simple SOM sorption assumption and proposed soil spectroscopic (infrared) measurements as an integrated measure of sorptivity. Jarvis (2016), using 785 data points gathered from 34 different published studies on 21 different compounds in subsoils, showed that the apparent K_{oc} on average doubled for each 10× decrease in SOM, the difference due to clay playing an increasing role in sorption as the SOM fraction decreases. Tureli et al. (2015) showed that soils with similar SOM content (2.4–3.5%) but different specific surface areas (measured by N_2 sorption, 19–84 $m^2\,g^{-1}$) had sorption equilibria for chlorpyrifos and 3,5,6-trichloro-2-pyridinol that depended on their specific surface area rather than SOM content.

This brings up a study that appears to have significance but has received little attention, at least from researchers on soil sorption of organic compounds. Dexter et al. (2008) observed, based on four soil property databases, that when SOM (expressed as SOC, a standard soils analysis metric) exceeds 1/10 of the clay content (mass/mass for both) the "excess" SOC becomes less significant in correlations with soil properties. For example, soils with OC <1/10 of the clay content have more dispersible clay. These results appear to suggest a general relationship that will allow soils that will deviate from "normal" SOM sorption to be identified. In other words (recalculating in terms of SOM), 1 g of SOM is needed to complex about 5.8 g of clay. A soil with SOM/clay content below this ratio will exhibit additional sorption by mineral surfaces. Thus, Soares et al. (2015) measured the sorption of phenanthrene by soil samples taken from a 15- by 15-m grid in two fields, one with 1–2% OC and the other with 2–9% OC, then used the 1:10 Dexter ratio to estimate "noncomplexed" SOM.

Sorption of Ionizable Pesticides

The RZWQM (Wauchope et al., 2004) models, and the PELMO (Klein, 2011) and PEARL (Leistra et al., 2001) models use similar sorption equations for those acidic and basic pesticides that can be partially dissociated within the normal range of soil pH. Assuming that total sorption is a weighted combination of a sorption constant for neutral and ionic species, a method of calculating a combined sorption equilibrium constant, which is strongly dependent on soil pH, is given. A question that all the modelers agreed needed further research was the surface excess acidity or pH shift typically observed when sorption was plotted for these compounds versus soil pH: the observed shift in sorption occurs about ½ to 1½ pH units above the pK_a of the solute. Leistra et al. (2001) discussed this succinctly (where HA, A$^-$, and K_{oc} are undissociated and dissociated acid and SOC sorption constants, respectively; a few minor edits are made for clarity):

A question is whether the pH of the inflection point in the [sorption]–pH relationship can deviate from the pK_a–value of the compound. Near the negatively-charged surfaces in soil, [the concentration of] H$^+$ is higher than in soil solution. This tends to increase the concentration of HA near the surfaces. However, this effect is counteracted by the lower concentration in A$^-$ due to repulsion near the negatively charged surfaces.

The negative charge at the organic matter surfaces (with weak-acid groups) is dependent on the pH in solution. As the pH decreases, the association of H$^+$ at the organic matter surfaces is increased. This may be expected to enhance the hydrophobic interactions. Because of this effect, the inflection point in the [sorption]–pH relation may be at a higher pH value than that corresponding to the pK_a of the compound.

Another complication is that the value of soil pH obtained in an experiment is dependent on the way it is measured. The pH can be measured in solution above the soil layer and within the soil slurry. Further, pH-values are measured in different ways as pH(H₂O), pH(KCl) or pH(CaCl₂), with the salts at different concentrations. It is likely that the concentration of exchangeable cations and the way in which the pH is measured affect the pH-value obtained. The value of the pH-shift under different conditions and with different experimental procedures requires further investigation (p. 110).

There has been some significant work on this question. Spadotto and Hornsby (2003) measured and modeled the sorption of 2,4-D and showed that for this relatively strong acid, SOM was the principal sorbent of the anion and, importantly, the K_{oc} of the anion was a function of pH effects on SOM sorptivity. Sassman and Lee (2005) measured the sorption of three tetracycline drugs, which have three dissociation constants between pH 3 and 9 and four species present (+ 0 , + – 0 , + – –, and 0 – –, where 0 indicates an undissociated labile hydrogen location). The cationic form dominated sorption, and sorption could be modeled with combinations of sorption coefficients after normalizing to the cation exchange sorption constants of the soils. Kah and Brown (2007a) measured the sorption of six acidic and four basic pesticides in nine soils; normalizing for SOM left a strong dependence on pH for the acids, but the bases were more complicated. Similarly, Franco and Trapp (2008) and Franco et al. (2009) developed regression equations assuming ionized and un-ionized species ratios from pK_a values, octanol–water regression relations to calculate un-ionized species K_{oc}, and an empirical surface acidity factor averaging 0.6 pH units (but it was very variable between compounds). This work also suggested that for basic (cationic) compounds, assuming SOM-only sorption is problematic. Van der Linden et al. (2009) showed the significant impact of allowing for pH effects on both degradation and sorption in modeling leaching (see below).

In more recent work, Singh et al. (2014), in a study of 140 soil profiles from two fields, showed that SOM was important for predicting 2,4-D and atrazine sorption variance, but glyphosate was complicated by subsurface clay and metal oxide sorption. Vitoratos et al. (2016) were able to estimate well (within about a factor of 2) sorption of weak-base pesticides using neutral and cationic species. Spadotto et al. (2020) got excellent fits for the PEARL/RZWQM procedure and also suggested that the neutral/anion sorption ratio may be nearly a constant for monovalent acidic pesticides.

In contrast to most pesticides, glyphosate and its soil metabolite aminomethylphosphonate, both multi-anionic compounds at higher pH, show little dependence on SOM, according to work by Paradelo et al. (2015) and Sidoli et al. (2016), who showed that an effective sorption coefficient can best be calculated using soil pH ($CaCl_2$, not water or KCl, according to Sidoli et al., 2016), Al and Fe oxides, and an Olsen phosphate availability index.

Nonequilibrium "Kinetic" Soil Sorption

A 24-hours batch mixing experiment serves very well, by convention, to estimate a pesticide's relative soil sorption tendency, especially on the many orders of magnitude scale covered by all pesticides. However, by 2000 it was clear that the "equilibrium" thus calculated is an experimental artifact: the sorption process is probably never at equilibrium—soil is much too complex chemically and physically. When longer mixing times are used, or when field observations are made, a more complex model is needed. Indeed, batch mixing tends to "overexpose" the surfaces in undisturbed soils. Sur et al. (2014) looked at both time-dependent and nonlinear (Freundlich) isotherm sorption effects on groundwater PECs using FOCUS-PRZM (a FOCUS version that allows for both modifications) and found predicted leaching decreased by 2/3 and 4/5, respectively.

The two-site model

In a batch experiment, much of the sorption can occur in minutes but slow sorption is still observable for many days. A two-site model at a minimum, assuming "fast" and "slow" sorption sites, is required to describe such experiments (Boesten & van der Pas, 1988; Wauchope & Myers, 1985). A two-site option is now available for most PEC models.

Kookana et al. (1992) compared sorption kinetics as observed in soil column leaching experiments with batch measurements; they showed that the "instantaneous" part of sorption is overrepresented in batch studies because of thorough mixing. However, batch studies, which remove much of the diffusional hindrance, can be better at observing "true" sorption kinetics. They analyzed this using a two-site model. Similarly, Smith et al. (2003) used a thin-disk soil sorption experiment that displayed the rapid adsorption–desorption more characteristic of the first moments of batch sorption. Kah and Brown (2007b) used a centrifugation equilibration technique that allowed more realistic soil/water ratios; interestingly, they observed lower sorption than a typical batch mixture for high-sorbing soils but higher sorption for lower sorbing soils. Cáceres et al. (2010) fit a variety of models to the sorption kinetics of metsulfuron-methyl and found that the two-site model worked best. Vereecken et al. (2011), in a review of batch versus column and batch versus field sorption studies, observed that batch measurements overestimated sorption, more so as sorption increased.

Two-site models that invoke slow diffusion kinetics into soil pores or particles (which may be indistinguishable experimentally from "fast/slow" sorption sites) give similarly significant improvements over equilibrium models (Beulke et al., 2004; Gamble et al., 2012; van Beinum et al., 2005).

"Aging" and "Time-Dependent" sorption: A three-site model adequate for describing field studies

Beyond batch or column experiments, with their time frame of hours to days, when field observations are made of available pesticides in soils there is an "aging" effect that takes place over months, during which soil water concentrations become increasingly small. Pedit and Miller (1994, 1995), in classic studies, analyzed long-term (4 mo) slurry sorption and fit the data to models with increasing degrees of complexity, including instantaneous and diffusion-limited sorption, and diffusion as controlled by soil particle size distribution. Although two-site instant/kinetic models fit the data well, allowing for additional slower particle size distribution effects on diffusion improved fits.

Aging thus has come to refer to this third, slow process (or processes) required to adequately describe field dissipation studies (Ahmad et al., 2004; Chen et al., 2014a; Cryer, 2014; Negley et al., 2014; Suddaby et al., 2014; Sur et al., 2014) and can significantly affect PECs (Boesten, 2017). In the RZWQM, a simple "binding" half-life is used. In the Riparian Ecosystem Management Model (REMM), which uses the same sorption model, a default binding half-life of one year is used and the bound pesticide becomes lost (Lowrance et al., 2000; Potter et al., 2012). This slow "binding," however, can be reversible (Suddaby et al., 2013; see below). Kan et al. (2000) modeled desorption of chlorohydrocarbons from contaminated sediments with two steps, one with half-lives of two to seven days and a second with half-lives from four months to nine years.

It is difficult if not impossible to experimentally distinguish between an extremely slowly desorbing sorption site and a permanently bound site. The typical soil probably has both (Celis et al., 1999; Beulke et al., 2004; van Beinum et al., 2005; Katayama et al. (2010); Gamble et al., 2012).

Ter Horst et al. (2013) proposed a pattern of acceptability of (second- and third-site) kinetic parameters obtained from inverse modeling of field studies: "…we found a relationship between the accuracy of a parameter estimate and its CV (coefficient of variation) provided by the inverse modelling technique. Using this relationship, we calculated the likelihood of rightly or wrongly accepting or rejecting a parameter estimate as a function of this CV…" (p. 260).

The Question of Reversibility: Bound Residues, Non-extractable Residues, and Bioavailability

Even 24-hours sorption data reveal that a diversity of sorption sites exists in soil, revealed by the need, in most cases, to use a Freundlich isotherm to fit the data even at quite low concentration ranges. Still, some relatively simple combination of equilibria and kinetic expressions can be found to fit almost any soil sorption data over any time frame. But the fit says nothing much about the actual mechanisms of sorption except to clearly show that in most cases that there are more than one, and indeed there may be many, fast and slow processes. This was well reviewed by Suddaby et al. (2014), on whom this section heavily relies.

Northcott and Jones (2000) reviewed the terminology and approaches for determining bound residues and argued that SOM as sorbent needed much more study. Gevao et al. (2000) emphasized that *bound* in terms of experimental extractability and *bound* in terms of long-term bioavailability are two different things.

In the EU, studies of soil persistence for registration require that "non-extractable residues" (NER) be determined by exhaustive extraction at the end of the studies, and the amount of NER allowable is limited. But what is the relationship between NER and slowly desorbed and permanently bound residues? Barriuso et al. (2008) reviewed "the coherent" data from EU NER determinations, much of it from registration studies. Figure 7.5 shows the NER found for different chemical classes of pesticides after (typically) 100 days. The strong dependence on chemical class suggests that chemical reaction plays a prominent role in NER formation.

Generally, NER increases with application rates and frequency and is decreased by competition with degradation (which, it appears, in general leads to products that are more extractable). Barriuso et al. (2008) ended by saying,

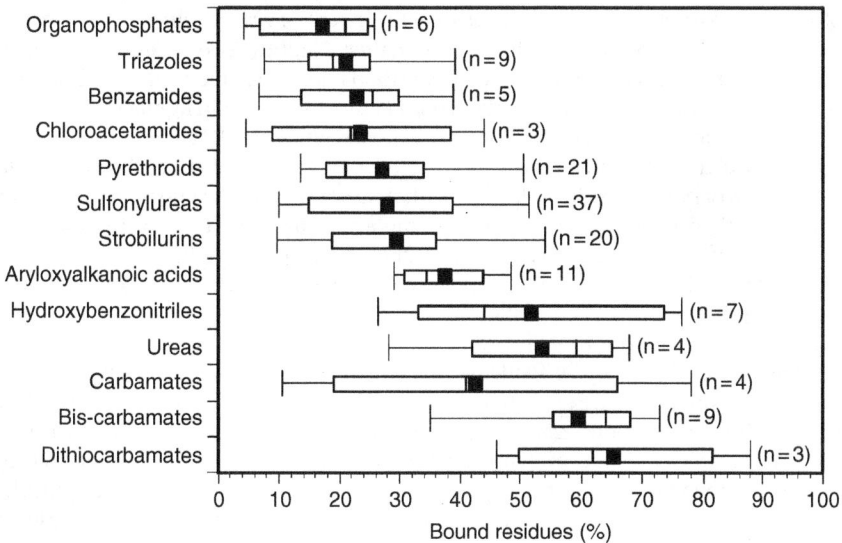

Figure 7.5 Ranges of fractions of bound (unextractable) residues of pesticides found after ~100 days in soil for classes of pesticides. Barriuso et al. (2008).

...In the end, the important matter is not so much how the residue is defined, [i.e., by an extraction procedure] but the question of the reversibility between unavailable and available forms of the residues and their biological availability. Accounting for potential biological effects may lead to improved risk assessments..." (p. 1851).

It would be interesting, considering really long-term "protected" residues, to compare these data with comparable older organochlorine pesticide studies if such were available.

Katayama et al. (2010) published a very wide-ranging review of bioavailability and soil and sediment sorption for pesticides, suggesting that exoenzymes may affect bioavailability to a greater extent than would be expected from only soil–water equilibrium. They stated, "...In contrast to metals, and despite attempts, no similar adequate methods yet exist to accurately estimate bioavailability of organic chemicals in soil and sediments..." (p. 5). They described chemical uptake mechanisms by soil microorganisms and higher plants, effects of soil and solute properties, and procedures for describing, measuring, and predicting bioavailability. The role of soil fauna, especially as processors of SOM, is described. There are many processes here that have the potential for more detailed modeling of bioavailability.

The most significant advance in the mechanistic understanding of soil sorption in 2000–2020 began with innovative experiments by Celis and Koskinen (1999a, 1999b) using isotopic substitution. Batch sorption experiments were done with and without ^{14}C-labeled solute, then centrifuged, the supernatants exchanged, and desorption and mixing of the isotopes into the near-equilibrium solution phase followed. This experiment avoids most of the artifacts of typical desorption "hysteresis" experiments and unambiguously showed that about 10% of the pesticides adsorbed in 24 hours could not be re-equilibrated by desorption in several days. Suddaby et al. (2013) reexamined this experiment and showed that the time frame used was too short: equilibrium had not been obtained in the initial 24-hours mixing, and a slow irreversible step was required to fit the data. Sander and Pignatello (2005, 2009) used similar batch and isotope exchange experiments to determine the extent of reversibility of naphthalene and *p*-dichlorobenzene sorption to organic matter. Naphthalene on lignite appeared to be completely reversible at low concentrations and hysteretic but non-bound at higher concentrations. *Para*-dichlorobenzene behaved similarly in lignite but exhibited an extremely slow-desorbing fraction in a Pahokee peat soil (46% C).

Suddaby et al. (2014), in a penetrating review of the terminology and experimental investigation of bound residues, suggested that the term *irreversible* be reserved for permanently sorbed molecules, *bound* refer to molecules that are resistant to desorption but capable of release, and *non-extractable* should mean precisely that and nothing more. Given that the conventional definition (Roberts et al., 1984) of a bound residue relates to the fraction of sorbed pesticide resistant to extraction, in fact a non-extractable residue is not necessarily a bound residue. The isotope exchange experiments show that "bound" residues may be available for many months at ever-decreasing amounts. Conceptual models for such sorption include physical entrapment, interphase diffusion, and mobile–immobile water regions. In her studies, Suddaby (2012) performed long-term "forced" isotope exchange experiments with ^{12}C- and ^{14}C-chlorotoluron to probe the reversibility of very slow sorption in three soils, with revealing results. An example of her results is shown in Figure 7.6. In this experiment, after almost two months of mixing, an average of 94% of added chlorotoluron was recovered from the mixtures. Of this 70, 57, and 84% of the recovered chlorotoluron was in the soil

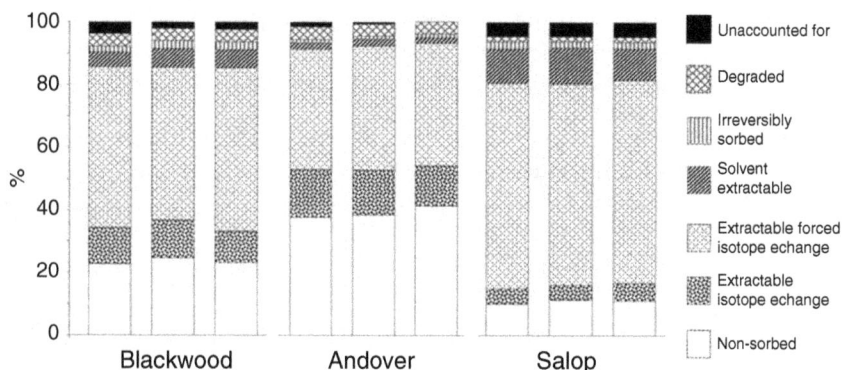

Figure 7.6 Mass balance of ^{14}C-chlorotoluron in nine soil samples adsorbed for 56 days. Three replicate samples per soil type are shown. The non-sorbed fraction reflects ^{14}C-chlorotoluron present in the aqueous phase after the adsorption period, while that which partitioned to the soil is collectively represented by the three extractable fractions and irreversibly sorbed fraction. Suddaby (2012) and Suddaby et al. (2014).

phase of the Blackwood, Andover, and Salop soils, respectively, and *89, 95, and 85%, respectively, of the absorbed chlorotoluron was still exchangeable with the aqueous phase.* Similar results were observed with prometryn and hexaconazole in the same soils with mixing times up to almost six months. In all cases, the total amounts of non-exchangeable sorbed chemical were on the order of 3% or less. This is a significant finding.

In important reviews, Kästner et al. (2014) and Schäffer et al. (2018) characterized NER, emphasizing the role soil microorganisms play in their formation. The NERs of xenobiotics, whether physically entrapped or absorbed, are to be distinguished from covalently bound species and from degraded chemicals (usually biodegraded), which become part of microbial biomass. They cited Barriuso et al. (2008), who said that about 1/3 of pesticides wind up as NER. They quoted the IUPAC definition (Roberts et al., 1984). When NER is determined by destructive ^{14}C analysis (by combustion), no identification of chemical species is possible. They classified NER in three types: Type I, sequestered (and reversible, but slowly), citing Katayama et al. (2010) that covalent bonding of sequestered compounds can occur; Type II, covalently bound and irreversible; and Type III, biotically converted residues that become part of the SOM. Later, Schäffer et al. (2018) described an extraction methodology for determining NER fractions.

Sorption Spatial Variability

Since the review on variability in K_d (soil sorption constant unadjusted for SOM) and K_{oc} by Wauchope et al. (2002), some valuable studies have been done. A catchment-scale (187-ha) study by Coquet (2003) reported variation in 51 samples of about 30%: "…Values of K_d ranged from 0.47 to 1.70 L kg^{-1} for atrazine, 0.47 to 1.81 for isoproturon, and 0.55 to 2.21 for metamitron. A clustering method was used to reduce the number of samples on which to measure sorption isotherms to 14 … more than 97% of variation could be ascribed to OC [SOM] content" (p. 69).

Farenhorst et al. (2009) and Farenhorst et al. (2010) demonstrated how local variability in K_d can affect PRZM leaching predictions and showed that K_{oc} for 2,4-D varied

4× (76–315) across a single field due to differences in the aromaticity of the SOM, an effect shown earlier by Ahmad et al. (2001). In a laboratory study, Motoki et al. (2014) also found order-of-magnitude ranges in K_{oc}, in part due to large differences in SOM aromatic carbon and black carbon content.

Farenhorst et al. (2014), in a review of K_{oc} variability, gave a table of the spatial variability of K_d versus K_{oc}; it was common for the CVs to be similar to each other. They showed that a lognormal distribution best describes K_d and K_{oc} values in a field area, and normalizing for SOM does not give much improvement in CV. Glyphosate, with its three anionic molecular sites, obeys different sorption rules. There were big differences between horizons. They discussed the use of probability distribution functions in K_{oc} modeling and suggested near-infrared soil surface spectra as a quick and inexpensive method for measuring K_{oc} (except for glyphosate), similar to an earlier proposal by Janik and Skjemstad (1995).

Progress Summary—Pesticide Sorption by Soil

This critical component of PEC modeling is still a quite active research area. Isotope exchange experiments have offered new insight into soil sorption reversibility, meaning here the extent to which a sorbed chemical can be recovered by desorption, as opposed to the classical chemical usage meaning non-hysteretic. Terminology and the limits to experiment have been clarified by some excellent reviews. *Bound* has three meanings in the literature: unextractable to the soil analyst, (almost) permanently trapped in geosolids to the environmental fate scientist, and non-bioavailable to the toxicologist. The question of the ultimate reversibility of NER or bound residues remains unclear, but the case of DDT comes to mind: decades after the last applications, parent and metabolite are still to be found in soil and water.

Some valuable work on soil sorption variability across the landscape has also been done.

Pesticide Degradation in Soils

Reviews

A guidance document and review and recommendations for determining pesticide persistence, with much discussion of error estimation, was provided by FOCUS (2006). Ye et al. (2010) reviewed soil degradation of enantiomeric chemicals, with many examples of enantiomers having different degradation rates. Bansal (2012) has an extensive compendium of abiotic, photolytic, and microbial degradation studies, mostly of older pesticides. Fenner et al. (2013) reviewed gene-based identification of degrader proteins. Hussain et al. (2015) gave a review of degradation and other aspects of fate of phenylurea herbicides in soils; accelerated degradation is observed, and biodegradation is the major source of dissipation. A report by the European Food Safety Authority (2014) reviewed and recommended procedures for interpretation of field and laboratory experiments and calculation of dissipation half-lives and their uncertainty.

Some notable studies: Lavy et al. (1996) performed a classic in situ subsurface soil experiment that separated degradation and leaching as loss processes. The work indicated some very slow subsoil degradation for atrazine, metribuzin, metolachlor, picloram, and alachlor, with 2,4-D being much more labile. Mills et al. (2001) found that in situ cores gave similar degradation rates to laboratory culture studies; acetochlor was actively degraded even 3 m deep. However, Gaston et al. (2003) suggested that intact cores exhibit degradation more representative of field conditions than homogenized soil samples. Fang et al. (2014) performed a metagenomic analysis of aquatic sediments, allowing a delineation of degradation pathways and biota for DDT and atrazine. Munira et al. (2018), in a groundwater survey, found that MCPA and 2,4-D detections predominantly occurred in summer and fall samples, respectively, even though the two very similar chemicals were typically applied at the same time. The results demonstrate that pesticides with similar chemistries can still have distinct persistence and mobility behaviors.

Temperature, Soil Moisture, and Tropical Environments

Racke et al. (1997) reviewed pesticide fate in tropical soils: pesticides are dissipated by chemical and biological processes and volatilize more rapidly in tropical soils due to higher soil moisture and temperature. Sanchez-Bayo and Hyne (2011) reached similar conclusions using modeling. In Brazil, 10 pesticides exhibited half-life, DT(50), values in the field that were a fraction of those seen in temperate climates (Laabs et al., 2002). Römbke et al. (2008) published a thoughtful and wide-ranging study of the applicability of EU risk assessment tools to tropical conditions, going beyond exposure to the full eco-risk approach. Some extraordinary generalizations were proposed—with proper caveats: for example, observed tropical soil DT(50) values are in general half of temperate zone values, and in fact the difference has only a minor effect on risks. They offered many useful proposals for needed research. Lewis et al. (2016) discussed the difficulty of extrapolating temperate results to the tropics, as did Thorburn et al. (2013) in a study of water quality impact on the Great Barrier Reef. Sur (2014) compared 50 and 90% degradation times for many compounds across North American and EU field sites. After "normalizing" field data for short-term temperature and moisture effects to 20 °C and soil water capacity (using the Walker and Arrhenius equations, respectively), regional and even continental differences were *eliminated* except for an exceptionally cold and dry area and an area with weathered low-pH soils in the United States. It appears that these statistical non-differences are mainly due to very large variances in the data. In Europe, even with a narrower variation in conditions, there still was no significant difference between north and south; in fact wetter and cooler balanced warmer and drier. In general, even North America–EU comparisons showed few significant differences, again because confidence limits are so wide. Laboratory studies in the tropics have shown that soil activity per se is not different from temperate conditions. Field data are fewer in the tropics, but in general volatilization losses are higher for a majority of compounds. Some low-pH soils may retard degradation. This seems to be generally confirmed by laboratory studies. There is a very loose association between soil properties and degradation between chemicals.

Kookana et al. (2010) published a unique field study showing a strong climate influence on atrazine and metabolite persistence in forest soils. Soil cores were analyzed for two years at five forestry sites including subtropical, temperate, and Mediterranean conditions. Bromide added as a tracer leached rapidly, but the triazines mostly remained in the top 45 cm of soils. Cooler conditions led to much slower degradation of atrazine: half-lives of about two weeks were observed in the subtropical and Mediterranean conditions, but in cooler Tasmania a rate constant increasing in time led to a DT(90) of 375 days.

Daam and van den Brink (2010) suggested that risk in tropical freshwater ecosystems should be little different except that pesticide usage can be higher in tropical environments. Mangala et al. (2010) showed that plant residue decomposition is much faster in the tropics and is slowed by chlorpyrifos impacts on earthworms and termites. Dores et al. (2016) conducted laboratory soil sorption studies and terrestrial field dissipation of chlorpyrifos and endosulfan in central Brazil and monitored runoff and leaching. Their results for runoff suggest that intense tropical rainfall can be offset by rapid dissipation (including volatilization) after application. Chai et al. (2009) found similar results for leaching. In an interesting recent study, Bonfleur et al. (2016) found that high amounts of Al and Fe in a group of tropical and subtropical Oxisols appeared to block SOM sorption sites by sequestering SOM in soil aggregates.

Accelerated Degradation

When a pesticide is used repeatedly on a soil, microbial degrader microorganisms may adapt and the pesticide will exhibit an increasingly short lifetime year after year. This phenomenon is called *accelerated degradation*, and it has been observed particularly with herbicides. Houot et al. (2000) found accelerated degradation of atrazine after as few as two applications. However, a pH < 6.5 inhibited it and formed large amounts of bound residue. Krutz et al. (2010a) reported atrazine and metabolites half-lives of a fraction of conventional times in s-triazine-adapted soils and even subsoils. In an exhaustive review of global soil microbial adaptation to atrazine, Krutz et al. (2010b) found that typical aerobic half-lives in soils reduced from 60 to 6 days. Their emphasis, however, was on surficial soils; the widespread detections of atrazine in groundwaters is almost certainly due to leaching to subsoils, where it is often much more persistent.

Spatial Variability in Soil Degradation

Smith et al. (1987) reported an early study estimating sample requirements for field evaluations of pesticide leaching, using known variability in soil properties to design soil sampling recommendations. In their comprehensive review discussed below, Dechesne et al. (2014) said that these guidelines have been little used. Perhaps small-scale (centimeters) degradation variability begins with small-scale variability of application rates: Suszter and Ambrus (2017) collected 5-cm diameter by 15-cm depth cores taken at random in a soil after surface sprays, and the amounts found had a CV of 100%. Ghafoor et al. (2011) studied degradation variability in a 13-km^2 watershed: soil samples gave CVs for three herbicides of 32–64%. Using a microbial activity predictor combined with a soil sorption predictor could account for 50% of the variability; isoproturon degradation showed a strong pH dependence.

A comprehensive and critical review by Dechesne et al. (2014) is the most significant of the period discussed here in this area. Within-field degradation activity CVs often exceed 50% "...and frequently display[s] non-random spatial patterns (p. 1)." They reviewed experimental techniques and horizontal vs. vertical variation (they vary independently). Almost universally, degradation rates decrease with depth, somewhat varying with microbial populations but often decreasing faster than microbial populations. Nutrient flux and pesticide application history mediate these populations. No generalization was able to be made for correlation with other soil properties:

> ...*Often, several of the measured [soil] biological and physico-chemical parameters are cross-correlated, making the identification of "true" controlling factor(s) elusive, and—often—the true controlling factor might not have been measured. This limitation is for example obvious for biodegradation variability across soil depth as many of the commonly measured soil parameters vary with depth and their correlation with pesticide biodegradation is likely to be spurious [i.e., devoid of causal basis]... (p. 6).*

Variation across fields can be small for particular pesticides but overall is quite high and tends to increase the smaller the scale of sampling. Studies have covered scales from centimeters to 100s of meters. Across-field variability increases with depth, sub-soils having more "hot spots." References and definitions for spatial geostatistics are given for discerning patterns at different scales. There is some correlation with soil pH for 2.4-D and ureas and for soil moisture and OC/clay—most likely sorption effects. Generally correlation is poor with most measures of microbial population or activity because degraders are typically a small part of the total population and also because of lag time effects. Other variables include previous application(s) and soil structure heterogeneity, especially macropores.

> ...*Overall, studies of the spatial distribution of pesticide degraders have so far contributed only modestly to improving our knowledge of their ecology. One of the biggest achievements may have been the demonstration that rapid phenylurea pesticide degradation in several agricultural soils is associated to* Sphingomonas *populations that have a narrow pH optimum (Bending et al., 2003; Shi & Bending, 2007)... (p. 8).*

Dechesne et al. (2014) concluded that currently very few models fully incorporate realistic within-field spatial variability. One notable exception is a recent model that includes observed three-dimensional centimeter-scale heterogeneity of MCPA degradation potential (Rosenbom et al., 2014). The researchers concluded that, due to the generally high degradation potential in the topsoil, the actual distribution of this potential does not impact MCPA leaching unless MCPA is transported mainly in biopores with low degradation potential. Their work marks an important step in the evaluation of the impact of small-scale variability on pesticide fate.

There is evidence that incorporating such variance could be useful in PEC modeling. Leterme et al. (2007a, 2007b) performed Monte Carlo simulation of atrazine leaching in a catchment using uncertainty distributions in both SOM and half-life. The results pinpointed field areas where leaching above 0.1 ppb occurred that did not appear in a deterministic simulation. They later explored the effects of interpolating such input data (which can have nonlinear responses) in a model as opposed to blunt-force calculations across a distribution, also reviewing previous work. There is a complex issue of the costs of a full computation to get a full uncertainty analysis.

Coupling Soil Sorption and Degradation

It is evident that sorption mediates degradation in soils simply by isolating solutes from degrading organisms as described above. By 2000 there was a growing realization that sorption is complex, however, and PEC modeling of degradation based on an instantaneous sorption equilibrium constant was unrealistic. Bosma et al. (1997) stated, "Mass transfer—and not the intrinsic microbial activity—is in most cases the critical factor in bioremediation" (p. 248). Van der Linden et al. (2009), in an elegant and beautifully depicted simulation using a geographically distributed version of PEARL (GEOPEARL), the spatially distributed version of PEARL, accounted for the dependence of both degradation and sorption of mesotrione—and two metabolites—on pH as measured by Dyson et al. (2002). The relationship of mesotrione and its first metabolite's sorption coefficients and half-lives with pH were such that the metabolite was more mobile but degraded faster near the surface, so leaching of the parent was more important than leaching of the metabolite in production of the metabolite, the main contaminant in groundwater. Shaner et al. (2012) also studied mesotrione data in an experiment tracking the parent in plant-available (centrifuged) soil water. On the other hand, Gaultier et al. (2008) incubated and tested soil samples from across a Canadian province for 2,4-D sorption and degradation rate and found little correlation between the two.

In 2014, a notable collection of studies, mostly by industry, on "non-first-order degradation and time-dependent sorption" of pesticides was published with the goal of incorporating this knowledge into regulatory PEC models (Chen et al., 2014b). Some chapters are cited above; in addition, Chen et al. (2014c) postulated that it is slow diffusion of solutes into soil spaces that are inaccessible (because too small) to microbes and which restrict rediffusion back into larger soil spaces that leads to both apparent hysteresis in sorption–desorption and non-first-order biotic degradation. The resultant model accurately describes data from soil degradation studies as affected by soil moisture and temperature and is explicit in terms of soil, biotic, and chemical parameters. Beulke and van Beinum (2014) proposed a procedure for using laboratory aged-sorption study for higher tier modeling. They implemented a two-site first-order model; early apparent-equilibrium sites are determined using 24-hours batch sorption, and slower sorption is by a final Freundlich isotherm. They presented a comprehensive discussion of regulatory modeling issues, especially on the relevance of laboratory studies to field processes. Tang et al. (2014) presented a comparison of kinetics analysis procedures between EU and North American guidance. Jackson (2014) discussed statistical procedures for fitting dissipation curves. Purdy and Cheplick (2014) used the term *bio-accessible* for a two-stage degradation model: "The physical interpretation of this model was found to be unrelated to soil physical properties but associated with the movement of residues between a compartment in which the degradation processes occur, and a compartment in which they do not"(p. 167).

Progress Summary—Pesticide Degradation in Soils

The methodology for field and laboratory persistence measurement has become more standardized. The effects of tropical climate have been characterized, with the apparent generalization that degrader microbe populations and mechanisms are similar everywhere—persistence differences are due to moisture and temperature differences. This may be useful when considering potential climate change effects. The review by Dechesne et al. (2014) is a major contribution to understanding the spatial variability of persistence and the difficulties of incorporating spatial variability into PEC models. A significant advance is that "higher tier" PEC modeling for registration of new pesticides now routinely provides a coupling of more complex sorption with degradation.

Pesticide Losses in Water: Runoff and Leaching

Pesticides in Leachate: Preferential Flow

Some significant work has been done, particularly on the impacts of soil conservation practices on *preferential flow*, the rapid leaching through larger pathways in the soil (e.g., cracks and wormholes) that can bypass soil filtering and chemical adsorption and degradation. Kladivko et al. (2001) reviewed agricultural subsurface tile drain studies in North America and summarized that drains overall decrease runoff and thereby decrease pesticide nonpoint pollution of waters. Even though preferential flow can even allow highly sorbed pesticide movement into subsurface flow, the concentrations are small compared with runoff. Kitchen et al. (1998) and Elliott et al. (2000) reported that in some situations management can have little effect on pesticide leaching. Kung et al. (2000) monitored tile drainage under 18- by 80-m plots under no-till and conventional tillage with simulated rainfall followed by injecting tracers (Cl⁻, Br⁻, and rhodamine WT) into irrigation water. The results showed that column experiments cannot replicate field preferential flow conditions. The sorptivity of tracers did not delay the appearance of their peaks but did cause tailing.

In a series of experiments combining simulated rainfall, the legendary (in agricultural hydrology circles) USDA 30- by 30-cm weighing lysimeters at Coshocton, OH, and RZWQM modeling, Malone and others (Malone et al., 2003, 2004a, 2004b) showed that when macropores and preferential flow are present, rainfall intensity fluctuations during events can produce large differences in pesticide concentrations in the percolate compared with what is found in a constant-intensity experiment. They showed that macroporosity alone could not explain the variation in herbicide percolate loads, but soil hydraulic properties interacted with macroporosity. The time to breakthrough had a dramatic effect on herbicide concentrations. Gish et al. (2004) and Kung et al. (2000) showed that slower flow rates could give a classical breakthrough curve even for a non-sorbing tracer, indicating that no preferential flow is observed at lower water flux rates. Even glyphosate can leach in preferential flow (Vereecken, 2005). Köhne et al. (2009) gave a good overview of progress in modeling chemical transport in preferential flow. Katagi (2013) published a review of soil column leaching of pesticides but which includes lysimeter and field studies and a detailed review of sorption modeling.

Measuring and Modeling Pesticide Runoff

Silburn (2003) and Silburn and Kennedy (2007) compared pesticide runoff experimental scales and the role of rain simulators. In the absence of rill formation, no difference was found in runoff of several pesticides from small and large plots with low slopes. Significantly, they measured pesticide concentrations in the soil from 0 to 25 mm and essentially confirmed Leonard et al.'s (1979) empirical "extraction ratio" for pesticides dissolved in runoff. Silburn (2003) developed a "framework" for extrapolating plot to field to catchment. They observed that partitioning to sediment can be higher than in batch sorption studies, suggesting that erosion control may remediate pesticide runoff more than might be expected based on batch sorption. Capel and Larson (2001) showed that atrazine small-plot runoff results could be extrapolated to very large catchments; they stated that this is probably because of atrazine's combination of "physical/chemical properties, formulation, and application method" (p. 1252).

Ma et al. (2004), using a RZWQM simulation that included experimental soil half-life and sorption data for fenamiphos and its oxidation daughter product, were able to adequately predict runoff losses of both, but they needed to account for crust or seal formation (rainfall-induced surface seals that occurred in their soils) to predict the hydrology well. Muller et al. (2004) simulated extreme runoff on small plots: herbicides were applied to moist soil (25% volumetric water content) on 20 and 30% slopes with rainfall intensities up to 111 mm/hour applied 24 hours later. Five herbicides showed an initial infiltration effect, where infiltration prior to runoff moved solute below the runoff extraction zone, decreasing chemical runoff. Potassium bromide was applied at the same time. Losses up to 65% of applied chemical were observed at the highest rainfall intensity and slope. Joyce et al. (2008, 2010) extended a model developed by Watanabe and Grismer (2003) for vegetative buffer strip simulation and modeled transfer of diazinon spray deposits on foliage and soil to the soil and runoff using rainfall simulation. Their results indicated that, for application to bare soil, the kinetics of soil desorption to soil water can be rate limiting for runoff losses, a rare reference to the role of soil desorption kinetics and how it might be related to pesticide extraction by runoff. Davis et al. (2012) showed that critical, first rains after application contribute the majority of a large sugarcane catchment's transport of herbicides to the Great Barrier Reef.

DeMars et al. (2018) added erosion modeling to the RZWQM based on the USDA Universal Soil Loss Equation (USLE) from the GLEAMS PEC model (Leonard et al., 1987) with some promising but erratic results for predicted pesticide loss in sediment, depending on input scenario.

Pesticide Runoff and Leaching Mediation by Soil Conservation Practices

Practices for remediation of pesticides in runoff and leaching, especially as related to soil and water conservation practices, continue to be studied—because the system can be complicated by the many processes involved and is still inadequately understood (Alletto et al., 2010). Wauchope (2006) reviewed the literature on each of the U.S. Soil Conservation Service's official practices list for observed effects on pesticide losses to ground and surface water.

Tillage practices

Potter et al. (2011) conducted rainfall simulator experiments on the effects of tillage practice and irrigation incorporation on losses of Fomesafen, interesting because it is formulated as a soluble sodium salt of a relatively strong acid with a pK_a of 2.8. Losses under a 55 mm during a two-hours storm one day after application without incorporation and in conventional tillage were about 5%. Irrigation incorporation or conservation tillage reduced runoff by more than 50%. Benoit et al. (2008) measured the sorption of pesticides from runoff onto plant residues.

Elias et al. (2018) reviewed the literature from 1985 to 2016 on pesticide losses under no-till and conventional tillage, focused mostly on field studies in the United States. In general, pesticides with low soil sorption or high water solubility had higher loads and concentrations from no-till fields. Fulton et al. (1999) demonstrated the value of integrated nonpoint-source insecticide runoff control in watersheds draining to estuaries, including in situ toxicity testing.

Buffer zones and water catchments

Braskerud and Haarstad (2003) and Rose et al. (2006) showed that a small constructed wetland can be effective at removing pesticides from runoff. Potter et al. (2012) used the REMM model (Lowrance et al., 2000), designed to model riparian buffer strip interception of pesticides and nutrients to compare tillage effects:

> ...[strip-tillage]-system inputs were attenuated at a lower rate during transport through the buffer. Findings were explainable by examining model processes and agreed with published studies that herbicides transported primarily in the dissolved form and delivered in subsurface flow have lower rates of retention in vegetated buffers... (p. 259).

The model includes soil binding and plant uptake processes.

Maillard and Imfeld (2014) conducted a mass balance study indicating competing degradation versus sorption as effected by season and vegetation for a wetland stormwater interceptor. A thorough analysis included 12 pesticides in water, suspended solids, sediments, and organisms. The wetland was effective for a diversity of pesticide attenuations, in part because of differing conditions occurring during the year.

Williams et al. (2016), coupling REMM to the Agricultural Policy/Environmental eXtender (APEX) runoff model (Williams et al., 2000), showed that movement of atrazine into an estuary through a buffer zone was sensitive to atrazine degradation (which was accelerated in the source field) and that the buffer could be overwhelmed by a tropical storm:

> REMM simulations indicated that the buffer system reduced atrazine transport by as much as 77% (18-day DT(50)) in surface runoff, 100% in subsurface flow, and by as much as 50% in sediment transport for the 3 years of the simulations (p. 759).

In a thorough review that might be the last word on edge-of-field buffer zones, Arora et al. (2010) showed that prediction of mediation of pesticide losses by buffer strips is very probabilistic in nature, highly dependent on weather sequence, site conditions, pesticide properties, experimental approach, and even experimental reporting. Buffer strips typically retained 75% of eroded soil, and pesticides that are strongly sorbed by soil and sediment ($K_{oc} > 1,000$) exhibited 53–100% retention. Retention of less strongly sorbed pesticides ranged from 20 to 100% (Figure 7.7).

Other studies of mediating factors have included the impact of persistence on transport in sediment (Muñoz-Carpena et al., 2015) and precision application and furrow irrigation (Davis & Pradolin, 2016; Melland et al., 2016). The latter showed that remediation with precision application can be higher than the simple reduction in application area fraction.

Progress Summary—Pesticide Losses in Water: Runoff and Leaching

Preferential flow effects on pesticide transport into groundwater and the removal of pesticides from runoff water by buffer strips and catchments have both been much better characterized by field work since 2000, and both processes have found their way

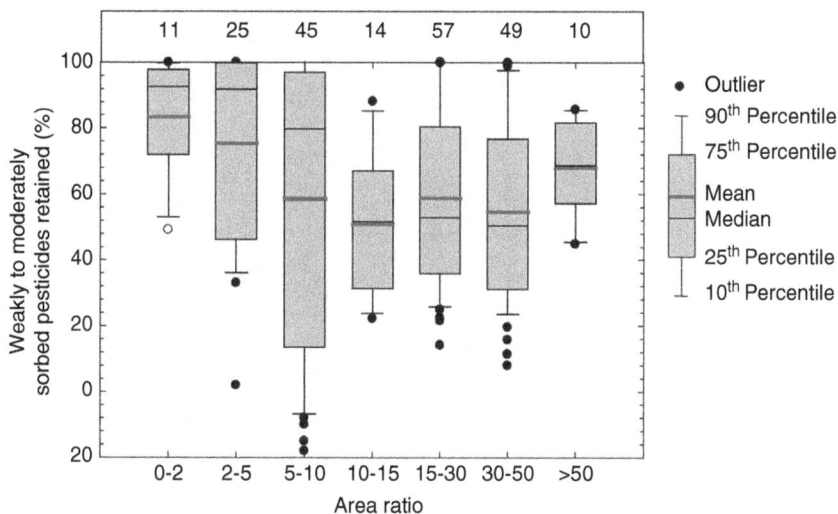

Figure 7.7 Percentage of weakly to moderately sorbed pesticides ($K_{oc} < 1,000$ L kg^{-1}) mass retentions as a function of the ratio of field area to buffer area. Percentage reduction numbers (used here) are on an event basis. Dots are outliers. Numbers above the box plots indicate sample size. Arora et al. (2010).

into PEC modeling, particularly in the RZWQM and REMM models. In general, the uses—and limitations—of soil conservation practices for removing pesticides from runoff are much better understood.

Selected Other Pesticide Studies and Trends, 2000–2020

Here a short mention and representative citations of important work and trends beyond the scope of this chapter are given.

The FOCUS and FOOTPRINT EU projects produced major developments in the use of models for PEC for registration purposes, particularly in the area of standard scenario development, probabilistic modeling, experimental guidance, and transparency of the uncertainty of predictions (Dubus et al., 2009; FOCUS, 2006; Jarvis & Dubus, 2006; Leistra et al., 2001).

Photodegradation of pesticides in aquatic systems has been covered in two comprehensive reviews: Sandin-España and Sevilla-Morán (2012) covered most aspects of the process. Remucal (2014) emphasized "the importance of dissolved organic matter (DOM) as a sensitizer in indirect photodegradation within aquatic systems (p. 628)."

Aquatic pesticide exposure was treated by Nfon et al. (2011), who presented an aquatic food web pesticide uptake and transfer model for small catchments which includes dealing with the issue of transient pesticide exposures. It includes a figure that summarizes trophic levels, in a "picture worth a 1,000 words," instructive to the exposure community (Figure 7.8).

QSAR, "nontarget analysis," and the NORMAN Network were covered by Mamy et al. (2015), who provided a grand tour of the world of QSAR as applied to pesticide fate predictions. "A total of 790 equations involving 686 structural molecular descriptors are reported to estimate 90 environmental parameters..." (p. 1277), with nearly 400 references. An excellent and perhaps more available QSAR source is the OECD website and toolbox: https://www.oecd.org/chemicalsafety/risk-assessment/oecd-qsar-toolbox.htm.

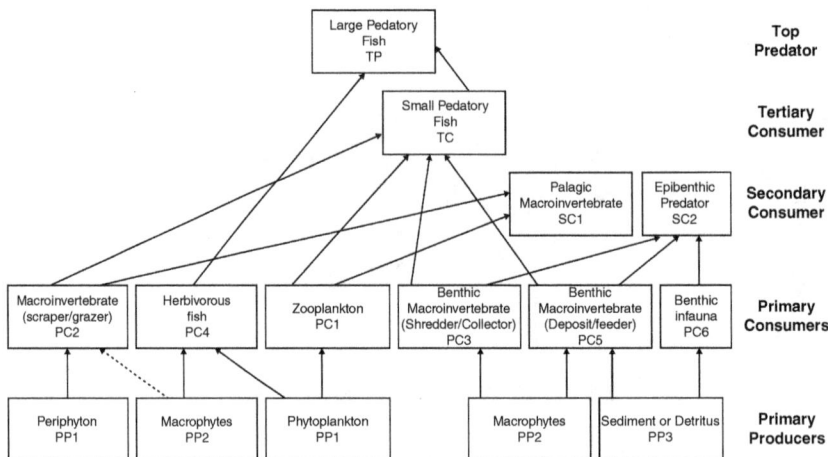

Figure 7.8 Pond food web structure showing representative species in the different trophic levels. The arrows indicate feeding relationships. Nfon et al. (2011) / with permission of Elsevier.

This toolbox provides a way to use the QSARs directly and is kept current (J.B.H.J. Linders, personal communication, 2020).

An important new application of QSAR is its ability to provide estimates for the physical properties of the thousands of emerging environmental contaminants (EECs) being found in nontarget environmental sampling. Nontarget environmental sampling is a result of wondrous advances in high-resolution mass spectral detection and identification in the chromatographic analysis of environmental samples (D.C.G. Muir, personal communication, 2020; Schymanski et al., 2015). This coupled with advanced data analysis and storage and communication has led to international cooperation in examining environmental samples for thousands of anthropogenic chemicals and for predicting their pollution potential (Muir & Howard, 2006; Muir et al., 2019; Escher et al., 2020; NORMAN Network, 2020; Dulio et al., 2018; Williams et al., 2020).

The U.S. Geologic Survey National Water Quality Assessment (NAWQA) Pesticide National Synthesis Project was not discussed here due to space constraints and a field-scale focus. This was a continental-scale air, surface water, and soil pesticide sampling program covering two decades and is arguably the most bold, strategic, well-coordinated, and fruitful single program in the history of pesticide environmental research (Capel et al., 2008). It was preceded by exhaustive preparatory reviews (Larson et al., 1999; Majewski & Capel, 1995; Nowell et al., 1999) and supported by unparalleled laboratory capabilities and Good Laboratory Practice protocols and sampling innovations. A few of many important publications are: Coupe et al. (2000), Majewski et al. (2000), Foreman et al. (2000), Vogel et al. (2008), Majewski et al. (2007), Capel and Larson (2001), Capel et al. (2001), Battaglin et al. (2014).

Conclusions and Research Needs

Communication with toxicologists. With the enormous generation of new anthropogenic chemical contaminant environmental concentration data discussed above, Felsot (personal communication, 2020) has suggested that there is a continuing, urgent need for the exposure scientist community to interpret these data to the toxicity scientist community and to define the boundaries of realistic environmental exposures.

Rainfall interception modeling. Kozak et al. (2007) modeled interception of rainfall by the crop canopy and crop residues and its effects on leaching and runoff. Experimental work indicates that up to 50% of seasonal rainfall may be kept from the soil. Their simulations covering a range of scenarios showed that "…interception was shown to decrease infiltration, runoff, evapotranspiration from soil, deep seepage of water and chemical transport, macropore flow, leaf area index, and crop/grain yield." Interception appears to be understudied relative to its possible influence on rainfall patterns on pesticide PECs.

Modeling pesticide volatilization and other processes. The Leistra and Yates pesticide volatilization models need to be incorporated into the USDA–ARS OM3 Object Modeling System, a modular agricultural modeling system (Ahuja et al., 2005; David et al., 2012), and into the RZWQM. The Leistra model is already part of the PEARL model, and the RZWQM is lacking in this respect. There is still no model that combines volatilization with runoff and erosion. Indeed—is it possible to imagine a single, international PEC model that combines all the best algorithms from anywhere on the planet? Was this not one goal of the Object Modeling System?

Applying contemporary data mining and machine learning model-building. There exist two huge, complex, and under-characterized agricultural datasets.

There is a myriad of data on pesticide residues in the edible portion of food crops, determined in order to set MRLs for registration of hundreds of active ingredients on hundreds of crops. This is probably several billion dollars' worth of data that—beyond its original purpose—has never to any significant extent been examined for patterns. For example, could these data not be combined with the Brunetti et al. (2019) model to explore the controlling parameters of plant uptake and food residues? Could a similar approach be used to attempt to summarize or analyze and model the singularly parameter-laden data available on spray drift? Current language-generating artificial intelligence (AI) models use billions of parameters (The Economist, 2020).

Pesticide dynamics in the top millimeter (or less) of the soil and in soil dust. Pesticide runoff and volatilization from soil are likely to be most dependent on pesticide concentrations at the field surface boundary. But pesticide deposition and dissipation in the top millimeter (or less) of soil are not considered in PEC modeling, where typically the top centimeter or more is assumed to be the "soil surface" compartment. This makes practical sense in terms of soil sampling, but it is obvious that conditions at the field surface boundary are different from the interior of the soil profile. The new PEARL model documentation (van den Berg et al., 2016) discussed the need to document pesticide behavior in the top millimeter of soil.

Forty-five years ago, Spencer et al. (1975) vacuumed up soil surface dust in the dripline of parathion-sprayed citrus trees and found much higher concentrations of parathion and its oxidation product paraoxon in the dust than in the soil below—and methyl parathion is a prominent contaminant of USGS atmospheric air and water samples (Coupe et al., 2000). Chang et al. (2011) found glyphosate in wind-eroded soil in precipitation samples. DeSutter et al. (1996) showed that wind erosion can move as much soil as water erosion, and they detected alachlor, atrazine, and atrazine metabolites in most of the wind sediments adjacent to fields where the same residues were detected.

Some experiments are in order to measure the dissipation and degradation kinetics of pesticides at the field surface, perhaps by placing a screen a millimeter down to isolate this thin layer. Might this also be a useful technique to study formulation effects? Finally, could a wind erosion model like WEPS (Hagen, 2004) be combined with initial deposition amounts from sprays to estimate the upper limits of pesticide loss to atmospheric particulates, thereby determining the potential contribution of wind erosion to the rain and air pesticide contamination so well documented by the USGS? WEPS has been combined with the atmospheric transport model CMACS 3/CMAQ for simulating urban particulate air pollution (Chen et al., 2014d).

The "Dexter ratio" needs more study. The work of Dexter et al. (2008) discussed above suggests that clays can complex only a fixed amount of SOM. If there is less SOM present, then Fe and Al oxides and clay surfaces are exposed. This clay/SOM complexation ratio limitation concept might provide aid in a general approach to the prediction of pesticide sorption in low-SOM soils.

Another USGS water and air sampling expedition is needed. It would be very valuable if one of those extraordinary regional or paired-region comparison USGS NAWQA studies, so revealing of the relations between agricultural pesticide use and water and atmospheric pesticide transport, could be combined with today's non-target environmental sampling methodology. A significantly different spectrum of pesticides and concentrations would probably be observed, as well as a host of new contaminants.

Personal Notes and Acknowledgments

I have not drilled down into the primary scientific literature in about 15 years. A few observations:

- Among all the excellent work that has been done on modeling pesticides in agricultural settings, one name seems to stand out as having had a uniquely long, productive, and creative career: Minze Leistra (Figure 7.9).

- This field now has the good fortune to have a greatly increased number of outstanding, innovative women scientists (e.g., Figure 7.10).

- The People's Republic of China, especially their engineers, are doing world-class work. Examples in the field of application engineering research are Gao et al. (2019), Li et al. (2018), Wang et al. (2018, 2019), and Xin et al. (2018).

Reviewing the extensive literature for this project could not have been possible without the excellent help of Susan Chambers, Reference and Interlibrary Loan Librarian of the Transylvania County Public Library. Tim Strickland, Director of the USDA–ARS Southeast Watershed Research Laboratory, generously provided remote access to the enormous resources and helpful staff of the USDA National Agricultural Library.

My deepest appreciation is expressed to the following eminent scientists, who have reviewed and commented on the manuscript: WenLin Chen, Syngenta Corporation; Annemieke Farenhorst, University of Manitoba, Winnipeg, Canada; Allan Felsot, Washington State University; David Gustafson, formerly of Monsanto Corporation; Paul Hendley, Phasera Ltd.; Rai Kookana, CSIRO Australia; Jan Linders, formerly of the National Institute for Public Health and Environment, the Netherlands; Derek Muir, Environment and Climate Change Canada; and Mark Silburn, Queensland Department of Natural Resources, Mines and Energy, Australia. I am also deeply grateful to Laj Ahuja for the invitation to write this chapter.

Figure 7.9 Minze Leistra of Wageningen University sampling groundwater in a field.

Figure 7.10 Annemieke Farenhorst and students at the University of Manitoba: (L–R) Alison Murata, Lindsey Andronak, Dr. Farenhorst, Jenna Rapai, Geethani Amarawansha.

Acronyms and Abbreviations

AgDISP	agricultural pesticide spray application drift model
APEX	Agricultural Policy/Environmental eXtender model
BROWSE	Bystanders, Residents, Operators and WorkerS Exposure drift models
COSMO-SAC	a quantum mechanics-based model for molecular properties used for QSAR
CREAMS	Chemicals, Runoff, and Erosion from Agricultural Management Systems
DT(50)	Half-life (usually in days) of a pesticide in soil
DT(90)	Time (d) for 90% dissipation of a pesticide in soil
EEC(s)	emerging environmental contaminant(s)
FOCUS	FOrum for the Co-ordination of pesticide fate models and their USe
FOOTPRINT	Functional Tools for Pesticide Risk Assessment and Management
GEOPEARL	geographically distributed version of PEARL
GLEAMS	Groundwater Loading from Agricultural Management Systems
NER	non-extractable (soil) residues
OC	organic carbon
PEC	predicted environmental concentration(s)
PELMO	Pesticide Leaching Model

QSAR	quantitative structure–activity relationship
REMM	Riparian Ecosystem Management Model
RZWQM	Root Zone Water Quality Model
SOC	soil organic carbon
SOM	soil organic matter
TSCF	transpiration stream concentration factor
TURFP	a pesticide PEC model for turf
USLE	Universal Soil Loss Equation
WEPS	Wind Erosion Prediction System (model)
VP	vapor pressure

References

Ahmad, R., Kookana, R. S., Alston, A. M., & Skjemstad, J. O. (2001). The nature of soil organic matter affects sorption of pesticides: 1. Relationships with carbon chemistry as determined by ^{13}C CPMAS NMR spectroscopy. *Environmental Science & Technology*, 35, 878–884.

Ahmad, R., Kookana, R. S., Megharaj, M., & Alston, A. M. (2004). Aging reduces the bioavailability of even a weakly sorbed pesticide (carbaryl) in soil. *Environmental Toxicology and Chemistry*, 23, 2084–2089.

Ahuja, L. R., Ascough, J. C., & David, O. (2005). Developing natural resource models using the object modeling system: Feasibility and challenges. *Advances in Geosciences*, 4, 29–36.

Ahuja, L. R., Rojas, K. W., Hanson, J. D., Shaffer, M. J., & Ma, L. (Eds.) (2000). *Root Zone Water Quality Model: Modeling management effects on water quality and crop production*. Water Resources Publications.

Alletto, L., Coquet, Y., Benoit, P., Heddadj, D., & Barriuso, E. (2010). Tillage management effects on pesticide fate in soils: A review. *Agronomy for Sustainable Development*, 30, 367–400.

American Society for Testing and Materials. (1987). *Method D3539-87: Standard test methods for evaporation rates of volatile liquids by shell thin-film evaporometer*. Author. (According to the ASTM website this standard has been withdrawn.)

Arora, K., Mickelson, S. K., Helmers, M. J., & Baker, J. L. (2010). Review of pesticide retention processes occurring in buffer strips receiving agricultural runoff. *Journal of the American Water Resources Association*, 46, 618–647.

Bansal, O. P. (2012). Degradation of pesticides. In H. S. Rathore & L. M. L. Nollet (Eds.), *Pesticides: Evaluation of environmental pollution* (pp. 47–78). CRC Press.

Barriuso, E., Benoit, P., & Dubus, I. (2008). Formation of pesticide nonextractable (bound) residues in soil: Magnitude, controlling factors and reversibility. *Environmental Science & Technology*, 42, 1845–1854.

Battaglin, W. A., Meyer, M. T., Kuivila, K. M., & Dietze, J. E. (2014). Glyphosate and its degradation product AMPA occur frequently and widely in U.S. soils, surface water, groundwater, and precipitation. *Journal of the American Water Resources Association*, 50, 275–290.

Bedos, C., Alletto, L., Durand, B., Fanucci, O., Brut, A., Bourdat-Deschamps, M., Giuliano, S., Loubet, B., Ceschia, E., & Benoit, P. (2017). Observed volatilization fluxes of S-metolachlor and benoxacor applied on soil with and without crop residues. *Environmental Science and Pollution Research*, 24, 3985–3996.

Bedos, C., Cellier, P., Calvet, R., Barriuso, E., & Gabrille, B. (2002). Mass transfer of pesticides into the atmosphere by volatilization from soils and plants: Overview. *Agronomie*, 22, 21–33.

Bedos, C., Rousseau-Djabri, M. F., Loubet, B., Durand, B., Flura, D., Briand, O., & Barriuso, E. (2010). Fungicide volatilization measurements: Inverse modeling, role of vapor pressure, and state of foliar residue. *Environmental Science & Technology*, 44, 2522–2528.

Bending, G. D., Lincoln, S. D., Sørensen, S. R., Morgan, J. A. W., Aamand, J., & Walker, A. (2003). In-field spatial variability in the degradation of the phenyl-urea herbicide isoproturon is the result of interactions between degradative *Sphingomonas* spp. and soil pH. *Applied and Environmental Microbiology*, 69, 27–834.

Benoit, P., Madrigal, I., Preston, C. M., Chenu, C., & Barriuso, E. (2008). Sorption and desorption of non-ionic herbicides onto particulate organic matter from surface soils under different land uses. *European Journal of Soil Science*, 59, 178–189.

Beulke, S., Brown, C. D., Fryer, C. J., & van Beinum, W. (2004). Influence of kinetic sorption and diffusion on pesticide movement through aggregated soils. *Chemosphere*, 57, 481–490.

Beulke, S., & van Beinum, W. (2014). Principles of the use of aged sorption studies in EU regulatory exposure assessments. *ACS Symposium Series*, 1174, 133–146. https://doi.org/10.1021/bk-2014-1174.ch007

Bird, S. L., Esterly, D. M., & Perry, S. G. (1996). Off-target deposition of pesticides from agricultural aerial spray applications. *Journal of Environmental Quality*, 25, 1095–1104. https://doi.org/10.2134/jeq1996.00472425002500050024x

Boesten, J. J. T. I. (2017). Effects of aged sorption on pesticide leaching to groundwater simulated with PEARL. *Science of the Total Environment*, 576, 498–507. https://doi.org/10.1016/j.scitotenv.2016. https://doi.org/10.099

Boesten, J. J. T. I., & van der Pas, L. J. T. (1988). Modeling adsorption/desorption kinetics of pesticides in a soil suspension. *Soil Science*, 146, 221–231.

Bonfleur, E. J., Kookana, R. S., Tornisielo, V. L., & Regitano, J. R. (2016). Organomineral interactions and herbicide sorption in Brazilian tropical and subtropical Oxisols under no-tillage. *Journal of Agricultural and Food Chemistry*, 64, 3925–3934. https://doi.org/10.1021/acs.jafc.5b04616

Bosma, T. N. P., Middeldorp, P. J. M., Schraa, G., & Zehnder, A. J. B. (1997). Mass transfer limitation of biotransformation: Quantifying bioavailability. *Environmental Science & Technology*, 31, 248–252. https://doi.org/10.1021/es960383u

Braskerud, B. C., & Haarstad, K. (2003). Screening the retention of thirteen pesticides in a small constructed wetland. *Water Science and Technology*, 48, 267–274. https://doi.org/10.2166/wst.2003.0332

Briggs, G. G., Bromilow, R. H., & Evans, A. A. (1982). Relationships between lipophilicity and root uptake and translocation of non-ionised chemicals by barley. *Pesticide Science*, 13, 495–504.

Brunetti, G., Kodešová, R., & Šimůnek, J. (2019). Modeling the translocation and transformation of chemicals in the soil–plant continuum: A dynamic plant uptake module for the HYDRUS model. *Water Resources Research*, 55, 8967–8989. https://doi.org/10.1029/2019WR025432

Butler Ellis, M. C., Underwood, B., Peirce, M. J., Walker, C. T., & Miller, P. C. H. (2010). Modelling the dispersion of volatilised pesticides in air after application for the assessment of resident and bystander exposure. *Biosystems Engineering*, 107, 149–154.

Butler Ellis, M. C., van den Berg, F., van de Zande, J. C., Kennedy, M. C., Charistou, A. N., Arapaki, N. S., Butler, A. H., Machera, K. A., & Jacobs, C. M. (2017). The BROWSE model for predicting exposures of residents and bystanders to agricultural use of pesticides: Comparison with experimental data and other exposure models. *Biosystems Engineering*, 154, 122–136. https://doi.org/10.1016/j.biosystemseng.20. https://doi.org/10.08.002

Cáceres, L., Escudey, M., Fuentes, E., & Báez, M. E. (2010). Modeling the sorption kinetic of metsulfuron-methyl on Andisols and Ultisols volcanic ash-derived soils: Kinetics parameters and solute transport mechanisms. *Journal of Hazardous Materials*, 179, 795–803.

Capel, P. D., & Larson, S. J. (2001). Effect of scale on the behavior of atrazine in surface waters. *Environmental Science & Technology*, 35, 648–657.

Capel, P. D., Larson, S. J., & Winterstein, T. A. (2001). The behavior of 39 pesticides in surface waters as a function of scale. *Hydrological Processes*, 15, 1251–1269.

Capel, P. D., McCarthy, K. A., & Barbash, J. E. (2008). National holistic watershed-scale approach for studying agricultural chemicals. *Journal of Environmental Quality*, 37, 983–993. https://doi.org/10.2134/jeq2007.0226

Celis, R., Hermosin, M. C., Cox, L., & Cornejo, J. (1999). Sorption of 2,4-dichlorophenoxyacetic acid by model particles simulating naturally occurring soil colloids. *Environmental Science & Technology*, 33, 1200–1206.

Celis, R., & Koskinen, W. C. (1999a). An isotopic exchange method for the characterization of the irreversibility of pesticide sorption–desorption in soil. *Journal of Agricultural and Food Chemistry*, 47, 782–790.

Celis, R., & Koskinen, W. C. (1999b). Characterization of pesticide desorption from soil by the isotopic exchange technique. *Soil Science Society of America Journal*, 63, 1659–1666. https://doi.org/10.2136/sssaj1999.6361659x

Chai, L. K., Mohd-Tahir, N., Hansen, S., & Hansen, H. C. B. (2009). Dissipation and leaching of acephate, chlorpyrifos, and their main metabolites in field soils of Malaysia. *Journal of Environmental Quality*, 38, 1160–1169. https://doi.org/10.2134/jeq2007.0644

Chang, F., Simcik, M. F., & Capel, P. D. (2011). Occurrence and fate of the herbicide glyphosate and its degradate aminomethylphosphonic acid in the atmosphere. *Environmental Toxicology and Chemistry*, 30, 548–555.

Chen, L., Han, B., Bai, Z., & Zhao, H. (2014a). Combined use of WEPS and Models-3/CMAQ for simulating wind erosion source emission and its environmental impact. *Science of the Total Environment*, 466–467, 762–769.

Chen, W., Cheplick, M., Reinken, G., & Jones, R. (2014b). Implementation of sorption kinetics coupled with differential degradation in the soil pore water system for FOCUS-PRZM. *ACS Symposium Series*, 1174, 275–297. https://doi.org/10.1021/bk-2014-1174.ch015

Chen, W., Laabs, V., Kookana, R. S., & Koskinen, W. C. (2014c). Coupled sorption and degradation kinetics and non-first order behavior. *ACS Symposium Series*, 1174, 39–56. https://doi.org/10.1021/bk-2014-1174.ch003

Chen, W., Sabljic, A., Cryer, S. A., & Kookana, R. S. (Eds.) (2014d). Non-first order degradation and time-dependent sorption of organic chemicals in soil. *ACS Symposion Series*, 1174. https://doi.org/10.1021/bk-2014-1174

Cleveland, C. B. (1996). Mobility assessment of agrichemicals: Current laboratory methodology and suggestions for future directions. *Weed Technology*, 10, 157–168.

Cohen, S. Z., Wauchope, R. D., Klein, A. W., Eadsforth, C. V., Graney, R., Lin, J., & Jones, R. D. (1995). Offsite transport of pesticides in water: Mathematical models of pesticide leaching and runoff. *Pure and Applied Chemistry*, 67, 2109–2148.

Coquet, Y. (2003). Variation of pesticide sorption isotherm in soil at the catchment scale. *Pest Management Science*, 58, 69–78.

Coupe, R. H., Manning, M. A., Foreman, W. T., Goolsby, D. A., & Majewski, M. S. (2000). Occurrence of pesticides in rain and air in urban and agricultural areas of Mississippi, April–September 1995. *Science of the Total Environment*, 248, 227–240.

Cryer, S. A. (2014). Quantifying transient sorption behavior of agrochemicals using simple experiments and modeling. *ACS Symposium Series*, 1174, 241–254. https://doi.org/10.1021/bk-2014-1174.ch013

Daam, M. A., & van den Brink, P. J. (2010). Implications of differences between temperate and tropical freshwater ecosystems for the ecological risk assessment of pesticides. *Ecotoxicology*, 19, 24–37.

Dang, A., Silburn, M., Craig, I., Shaw, M., & Foley, J. (2016). Washoff of residual photosystem II herbicides from sugar cane trash under a rainfall simulator. *Journal of Agricultural and Food Chemistry*, 64, 3967–3974. https://doi.org/10.1021/acs.jafc.5b04717

Dann, R. L., Close, M. E., Lee, R., & Pang, L. J. (2006). Impact of data quality and model complexity on prediction of pesticide leaching. *Journal of Environmental Quality*, 35, 628–640. https://doi.org/10.2134/jeq2005.0257

David, O., Ascough, J. C., Lloyd, W., Green, T. R., Rojas, K. W., Leavesley, G. H., & Ahuja, L. R. (2012). A software engineering perspective on environmental modeling framework design: The object modeling system. *Environmental Modelling and Software*, 30, 1–13.

Davis, A. M., Lewis, S. E., Bainbridge, Z. T., Glendenning, L., Turner, R. D. R., & Brodie, J. E. (2012). Dynamics of herbicide transport and partitioning under event flow conditions in the lower Burdekin region, Australia. *Marine Pollution Bulletin*, 65, 182–193. https://dx.doi.org/10.1016/j.marpolbul.2011.08.025

Davis, A. M., & Pradolin, J. (2016). Precision herbicide application technologies to decrease herbicide losses in furrow irrigation outflows in a northeastern Australian cropping system. *Journal of Agricultural and Food Chemistry*, 64, 4021–4028. https://dx.doi.org/10.1021/acs.jafc.5b04987

De Cock, N., Massinon, M., Salaha, O. S. T., & Lebeau, F. (2017). Investigation on optimal spray properties for ground based agricultural applications using deposition and retention models. *Biosystems Engineering*, 162, 99–111.

de Ruiter, H., Holterman, H. J., Kempenaar, C., Mol, H. G. J., de Vlieger, J. J., & van de Zande, J. C. (2003). *Influence of adjuvants and formulations on the emission of pesticides to the atmosphere* (Report 59). Plant Research International.

Dechesne, A., Badawi, N., Aamand, J., & Smets, B. F. (2014). Fine scale spatial variability of microbial pesticide degradation in soil: Scales, controlling factors, and implications. *Frontiers in Microbiology*, 5, 667. https://dx.doi.org/10.3389/fmicb.2014.00667

Delle Site, A. (2001). Factors affecting sorption of organic compounds in natural sorbent/water systems and sorption coefficients for selected pollutants: A review. *Journal of Physical and Chemical Reference Data*, 30, 187–439. https://doi.org/10.1063/1.1347984

DeMars, C., Zhan, Y., Chen, H., Heilman, P., & Zhang, X. (2018). Integrating GLEAMS sedimentation into RZWQM for pesticide sorbed sediment runoff modeling. *Environmental Modelling Software*, 109, 390–401.

DeSutter, T. M., Clay, S. M., & Clay, D. E. (1996). Transport of agrichemicals by wind eroded sediments to non-target areas. *Proceedings of the South Dakota Academy of Science*, 75, 147.

Dexter, A. R., Richard, G., Arrouays, D., Czyż, E. A., Jolivet, C., & Duval, O. (2008). Complexed organic matter controls soil physical properties. *Geoderma*, 144, 620–627.

Dores, E. F. G. C., Spadotto, C. A., Weber, O. L. S., Dalla Villa, R., Vecchiato, A. B., & Pinto, A. A. (2016). Environmental behavior of chlorpyrifos and endosulfan in a tropical soil in central Brazil. *Journal of Agricultural and Food Chemistry*, 64, 3942–3948.

Dorr, G. J., Forster, W. A., Mayo, L. C., McCue, S. W., Kempthorne, D. M., Hanan, J., Turner, I. W., Belward, J. A., Young, J., & Zabkiewicze, J. A. (2016). Spray retention on whole plants: Modelling, simulations and experiments. *Crop Protection*, 88, 118–130.

Dorr, G. J., Kempthorne, D. M., Mayo, L. C., Forster, W. A., Zabkiewicz, J. A., McCue, S. W., Belward, J., Turner, I., & Hanan, J. (2014). Towards a model of spray–canopy interactions: Interception, shatter, bounce and retention of droplets on horizontal leaves. *Ecological Modelling*, 290, 94–101.

Dubus, I.G., Reichenberger S., Allier D., Azimonti G., Bach M., Barriuso E., Giovanni Bidoglio, Stephen Blenkinsop, F Boulahya, Faycal Bouraoui, A Burton, Tiziana Centofanti, Olivier Cerdan, Y Coquet, B Feisel, Wieslaw Fialkiewicz, Hayley J. Fowler, Francesco Galimberti, Andy Green, ..., Wurm M. (2009). *FOOT-PRINT: Functional tools for pesticide risk assessment and management* [Final report of the EU project FOOT-PRINT (SSPI-CT-2005-022704)]. www.eu-footprint.org

Dulio, V., van Bavel, B., Brorström-Lundén, E., Harmsen, J., Hollender, J., Schlabach, M., Slobodnik, J., Thomas, K., & Koschorreck, J. (2018). Emerging pollutants in the EU: 10 years of NORMAN in support of environmental policies and regulations. *Environmental Sciences Europe*, 30, 5–18. https://dx.doi.org/10.1186/s12302-018-0135-3

Dyson, J. S., Beulke, S., Brown, C. D., & Lane, M. C. G. (2002). Adsorption and degradation of the weak acid mesotrione in soil and environmental fate implications. *Journal of Environmental Quality*, 31, 613–618.

Elias, D., Wang, L., & Jacinthe, P.-A. (2018). A meta-analysis of pesticide loss in runoff under conventional tillage and no-till management. *Environmental Monitoring and Assessment*, 190, 79. https://dx.doi.org/10.1007/s10661-017-6441-1

Elliott, J. A., Cessna, A. J., Nicholaichuk, W., & Tollefson, L. C. (2000). Leaching rates and preferential flow of selected herbicides through tilled and untilled soil. *Journal of Environmental Quality*, 29, 1650–1656. https://doi.org/10.2134/jeq2000.00472425002900050036x

Escher, B. I., Stapleton, H. M., & Schymanski, E. L. (2020). Tracking complex mixtures of chemicals in our changing environment. *Science*, 367, 388–392.

European Food Safety Authority. (2014). EFSA guidance document for evaluating laboratory and field dissipation studies to obtain DegT50 values of active substances of plant protection products and transformation products of these active substances in soil. *EFSA Journal*, 12, 3662–3699. https://dx.doi.org/10.2903/j.efsa.2014.3662

Eyheraguibel, B., ter Halle, A., & Richard, C. (2009). Photodegradation of bentazon, clopyralid, and triclopyr on model leaves: Importance of a systematic evaluation of pesticide photostability on crops. *Journal of Agricultural and Food Chemistry*, 57, 1960–1966. https://doi.org/10.1021/jf803282f

Fang, H., Cai, L., Yang, Y., Ju, F., Li, X., & Yu, Y. (2014). Metagenomic analysis reveals potential biodegradation pathways of persistent pesticides in freshwater and marine sediments. *Science of the Total Environment*, 470–471, 983–992.

Fantke, P., Arnot, J. A., & Doucette, W. J. (2016). Improving plant bioaccumulation science through consistent reporting of experimental data. *Journal of Environmental Management*, 181, 374–384.

Fantke, P., & Juraske, R. (2013). Variability of pesticide dissipation half-lives in plants. *Environmental Science & Technology*, 47, 3548–3562.

Farenhorst, A., McQueen, D. A. R., Saiyed, I., Hilderbrand, C., Li, S., Lobb, D. A., Messing, P., Schumacher, T. E., Papiernik, S. K., & Lindstrom, M. J. (2009). Variations in soil properties and herbicide sorption coefficients with depth in relation to PRZM (Pesticide Root Zone Model) calculations. *Geoderma*, 150, 267–277.

Farenhorst, A., McQueen, R., Kookana, R. S., Singh, B., & Malley, D. (2014). Spatial variability of pesticide sorption: Measurements and integration to pesticide fate models. *ACS Symposium Series*, 1174, 255–274. https://dx.doi.org/10.1021/bk-2014-1174.ch014

Farenhorst, A., Saiyed, I. M., Goh, T. B., & McQueen, P. (2010). The important characteristics of soil organic matter affecting 2,4-dichlorophenoxyacetic acid sorption along a catenary sequence. *Journal of Environmental Science and Health, Part B*, 45, 204–213. https://doi.org/10.1080/03601231003613542

Felizeter, S., McLachlan, M. S., & De Voogt, P. (2014). Root uptake and translocation of perfluorinated alkyl acids by three hydroponically grown crops. *Journal of Agricultural and Food Chemistry*, 62, 3334–3342. https://doi.org/10.1021/jf500674j

Felsot, A. S., Unsworth, J. B., Linders, J. B. H. J., Roberts, G., Rautman, D., Harris, C., & Carazo, E. (2011). Agrochemical spray drift; assessment and mitigation—A review. *Journal of Environmental Science and Health, Part B*, 46, 1–23.

Fenner, K., Canonica, S., Wackett, L. P., & Elsner, M. (2013). Evaluating pesticide degradation in the environment: Blind spots and emerging opportunities. *Science*, 341, 752–758.

Fife, J. P., & Nokes, S. B. (2002). Evaluation of the effect of rainfall intensity and duration on the persistence of chlorothalonil on processing tomato foliage. *Crop Protection*, 21, 733–740.

FOCUS. (2006). *Guidance document on estimating persistence and degradation kinetics from environmental fate studies on pesticides in EU registration* (EC Document Reference Sanco/10058/2005 version 2.0). https://esdac.jrc.ec.europa.eu/public_path/projects_data/focus/dk/docs/finalreportFOCDegKinetics.pdf

Foreman, W. T., Majewski, M. S., Goolsby, D. A., Wiebe, F. W., & Coupe, R. H. (2000). Pesticides in the atmosphere of the Mississippi River Valley: II. Air. *Science of the Total Environment*, 248, 213–216.

Forster, W. A., & Kimberley, M. O. (2015). The contribution of spray formulation component variables to foliar uptake of agrichemicals. *Pest Management Science*, 71, 1324–1334.

Forster, W. A., Zabkiewicz, J. A., & Liu, Z. Q. (2006). Cuticular uptake of xenobiotics into living plants: 2. Influence of the xenobiotic dose on the uptake of bentazone, epoxiconazole and pyraclostrobin, applied in the presence of various surfactants, into *Chenopodium album, Sinapis alba and Triticum aestivum* leaves. *Pest Management Science*, 62, 664–672.

Franco, A., Fu, F., & Trapp, S. (2009). Influence of soil pH on the sorption of ionizable chemicals: Modeling advances. *Environmental Toxicology and Chemistry*, 28, 458–464.

Franco, A., & Trapp, S. (2008). Estimation of soil–water partition coefficient normalized to organic carbon for ionizable organic chemicals. *Environmental Toxicology and Chemistry*, 27, 1995–2004.

Fulton, M. H., Moore, D. W., Wirth, E. F., Chandler, G. T., Key, P. B., Daugomah, J. W., Strozier, E. D., Devane, J., Clark, J. R., Lewis, M. A., Finley, D. B., Ellenberg, W., Karnaky, K. J., Scott, G. I., & Geoffrey, G. I. (1999). Assessment of risk reduction strategies for the management of agricultural nonpoint source pesticide runoff in estuarine ecosystems. *Toxicology and Industrial Health*, 15, 201–214.

Gamble, D. S., Webster, G. R. B., & Lamoureux, M. (2012). Propanil in a Manitoba soil: An interactive spreadsheet model based on conventional chemical kinetics. *Journal of Environmental Monitoring*, 14, 1167–1173.

Gao, S., Wang, G., Zhou, Y., Wang, M., Yang, D., Yuan, H., & Yan, X. (2019). Water-soluble food dye of Allura red as a tracer to determine the spray deposition of pesticide on target crops. *Pest Management Science*, 75, 2592–2597.

Garcia, L., Bedos, C., Genermont, S., Benoit, P., Barriuso, E., & Cellier, P. (2014). Modeling pesticide volatilization: Testing the additional effect of gaseous adsorption on soil solid surfaces. *Environmental Science & Technology*, 48, 4991–4998.

Gaston, L. A., Boquet, D. J., & Bosch, M. A. (2003). Fluometuron sorption and degradation in cores of silt loam soil from different tillage and cover crop systems. *Soil Science Society of America Journal*, 67, 747–755. https://doi.org/10.2136/sssaj2003.7470

Gaultier, J., Farenhorst, A., Cathcart, J., & Goddard, T. (2008). Regional assessment of herbicide sorption and degradation in two sampling years. *Journal of Environmental Quality*, 37, 1825–1836.

Gebremariam, S. Y., Beutel, M. W., Yonge, D. R., Flury, M., & Harsh, J. B. (2010). Adsorption and desorption of chlorpyrifos to soils and sediments. *Reviews of Environmental Contamination and Toxicology*, 215, 123–175. https://doi.org/10.1007/978-1-4614-1463-6_3

Gevao, B., Semple, K. T., & Jones, K. C. (2000). Bound pesticide residues in soils: A review. *Environmental Pollution*, 108, 3–14.

Ghafoor, A., Jarvis, N. J., & Stenström, J. (2013). Modelling pesticide sorption in the surface and subsurface soils of an agricultural catchment. *Pest Management Science*, 69, 919–929. https://dx.doi.org/10.1002/ps.3453

Ghafoor, A., Jarvis, N. J., Thierfelder, T., & Stenstrom, J. (2011). Measurements and modeling of pesticide persistence in soil at the catchment scale. *Science of the Total Environment*, 409, 1900–1908.

Gish, T. J., Kung, K.-J. S., Perry, D. C., Posner, J., Bubenzer, G., Helling, C. S., Kladivko, E. J., & Steenhuis, T. S. (2004). Impact of preferential flow at varying irrigation rates by quantifying mass fluxes. *Journal of Environmental Quality*, 33, 1033–1040. https://doi.org/10.2134/jeq2004.1033

Gish, T. J., Prueger, J. H., Daughtry, C. T., Kustas, W. P., McKee, L. G., Russ, A. L., & Hatfield, J. L. (2011). Comparison of field-scale herbicide runoff and volatilization losses: An eight-year field investigation. *Journal of Environmental Quality*, 40, 1432–1144. https://doi.org/10.2134/jeq20. https://doi.org/10.0092

Gish, T. J., Sadeghi, A., & Wienhold, B. J. (1995). Volatilization of alachlor and atrazine as influenced by surface litter. *Chemosphere*, 31, 2971–2982.

Gredelj, A., Polesel, F., & Trapp, S. (2020). Model-based analysis of the uptake of perfluoroalkyl acids (PFAAs) from soil into plants. *Chemosphere*, 244, 125534.

Guth, J. A., Reischmann, F. J., Allen, R., Arnold, D., Hassink, J., Leake, C. R., Skidmore, M. W., & Reeves, G. L. (2004). Volatilisation of crop protection chemicals from crop and soil surfaces under controlled conditions: Prediction of volatile losses from physico-chemical properties. *Chemosphere*, 57, 871–887.

Hagen, L. J. (2004). Evaluation of the Wind Erosion Prediction System (WEPS) erosion submodel on cropland fields. *Environmental Modelling and Software*, 19, 171–176.

Harper, L. A., White, A. W., Jr., Bruce, R. R., Thomas, A. W., & Leonard, R. A. (1976). Soil and microclimate effects on trifluralin volatilization. *Journal of Environmental Quality*, 5, 236–242. https://doi.org/10.2134/jeq1976.00472425000500030004x

Hewitt, A., Valcore, D., & Barry, T. (2001). Analyses of equipment, meteorology and other factors affecting drift from applications of sprays by ground rig sprayers. In A. Viets, R. Tann, & J. Mueninghoff (Eds.), *Pesticide formulations and application systems* (Vol. 20, pp. 44–56). ASTM International. https://doi.org/10.1520/STP10433S

Hewitt, A. J., Johnson, D. R., Fish, J. D., Hermansky, C. G., & Valcore, D. L. (2002). Development of the spray drift task force database for aerial applications. *Environmental Toxicology and Chemistry*, 21, 648–658.

Hilz, E., & Vermeer, A. W. P. (2013). Spray drift review: The extent to which a formulation can contribute to spray drift reduction. *Crop Protection*, 44, 75–83.

Holterman, H. J., van de Zande, J. C., Huijsmans, J. F. M., & Wenneker, M. (2017). An empirical model based on phenological growth stage for predicting pesticide spray drift in pome fruit orchards. *Biosystems Engineering*, 154, 46–61.

Holterman, H. J., van de Zande, J. C., Porskamp, H. A. J., & Huijsmans, J. F. M. (1997). Modelling spray drift from boom sprayers. *Computers and Electronics in Agriculture*, 19, 1–22. https://doi.org/10.1016/S0168-1699 (97)00018-5

Houbraken, M., Senaeve, D., Fevery, D., & Spanoghe, P. (2015). Influence of adjuvants on the dissipation of fenpropimorph, pyrimethanil, chlorpyrifos and lindane on the solid/gas interface. *Chemosphere*, 138, 357–363.

Houbraken, M., van den Berg, F., Ellis, C. M. B., Dekeyser, D., Nuyttens, D., de Schampheleire, M., & Spanoghea, P. (2015). Volatilisation of pesticides under field conditions: Inverse modelling and pesticide fate models. *Pest Management Science*, 72, 1309–1321.

Houot, S., Topp, E., Yassir, A., & Soulas, G. (2000). Dependence of accelerated degradation of atrazine on soil pH in French and Canadian soils. *Soil Biology and Biochemistry*, 32, 615–625.

Huang, W., Peng, P., Yua, Z., & Fu, J. (2003). Review: Effects of organic matter heterogeneity on sorption and desorption of organic contaminants by soils and sediments. *Applied Geochemistry*, 18, 955–972.

Hussain, S., Arshad, M., Springael, D., SøRensen, S. R., Bending, G. D., Devers-Lamrani, M., Maqbool, Z., & Martin-Laurent, F. (2015). Abiotic and biotic processes governing the fate of phenylurea herbicides in soils: A review. *Critical Reviews in Environmental Science and Technology*, 45, 1947–1998. https://dx.doi.org/10.1080/10643389.2014.1001141

Hwang, J.-I., Lee, S.-E., & Kim, J.-E. (2017). Comparison of theoretical and experimental values for plant uptake of pesticide from soil. *PLOS ONE*, 12(2), e0172254. https://dx.doi.org/10.1371/journal.pone.0172254

Jackson, S. H. (2014). Statistical means for proper determination of kinetic half-lives. *ACS Symposium Series*, 1174, 133–146. https://dx.doi.org/10.1021/bk-2014-1174.ch010

Janik, L. J., & Skjemstad, J. O. (1995). Characterization and analysis of soils using mid-infrared partial least-squares: 2. Correlations with some laboratory data. *Australian Journal of Soil Research*, 33, 637–650.

Jarvis, N. (2016). Extended sorption partitioning models for pesticide leaching risk assessments: Can we improve upon the k_{oc} concept? *Science of the Total Environment*, 539, 294–303. https://doi.org/10.1016/j.scitotenv.2015.09.002

Jarvis, N. J. & Dubus, I. G. (2006). *State-of-the-art review on preferential flow* (Report DL6 of the FP6 EU-funded FOOTPRINT project). www.eu-footprint.org

Jensen, P. K., & Olesen, M. H. (2014). Spray mass balance in pesticide application: A review. *Crop Protection*, 61, 23–31.

Joyce, B. A., Wallender, W. W., & Ginn, T. R. (2008). Modeling the transport of spray-applied pesticides from fields with vegetative cover. *Transactions of the ASABE*, 51, 1963–1976. https://dx.doi.org/10.13031/2013.25401

Joyce, B. A., Wallender, W. W., & Mailapalli, D. R. (2010). Application of pesticide transport model for simulating diazinon runoff in California's Central Valley. *Journal of Hydrology*, 395, 79–90. https://dx.doi.org/10.1016/j.jhydrol.20. https://doi.org/10.https://doi.org/10.017

Juraske, R., Antón, A., & Castells, F. (2008). Estimating half-lives of pesticides in/on vegetation for use in multimedia fate and exposure models. *Chemosphere*, 70, 1748–1755.

Jury, W. A., Farmer, W. J., & Spencer, W. F. (1984a). Behavior assessment model for trace organics in soil: II. Chemical classification and parameter sensitivity. *Journal of Environmental Quality*, 13, 567–572. https://doi.org/10.2134/jeq1984.00472425001300040012x

Jury, W. A., Spencer, W. F., & Farmer, W. J. (1983). Behavior assessment model for trace organics in soil: I. Model description. *Journal of Environmental Quality*, 12, 558–564. https://doi.org/10.2134/jeq1983.00472425001200040025x

Jury, W. A., Spencer, W. F., & Farmer, W. J. (1984b). Behavior assessment model for trace organics in soil. III. Application of screening model. *Journal of Environmental Quality*, 13, 573–579. https://doi.org/10.2134/jeq1984.00472425001300040013x

Jury, W. A., Spencer, W. F., & Farmer, W. J. (1984c). Behavior assessment model for trace organics in soil. IV. Review of experimental evidence. *Journal of Environmental Quality*, 13, 580–586. https://doi.org/10.2134/jeq1984.00472425001300040014x

Kah, M., & Brown, C. D. (2007a). Prediction of the adsorption of Ionizable pesticides in soils. *Journal of Agricultural and Food Chemistry*, 55, 2312–2322.

Kah, M., & Brown, C. D. (2007b). Changes in pesticide adsorption with time at high soil to solution ratios. *Chemosphere*, 68, 1335–1343.

Kan, A. T., Chen, W., & Tomson, M. B. (2000). Desorption kinetics of neutral hydrophobic organic compounds from field-contaminated sediment. *Environmental Pollution*, 108, 81–89.

Kästner, M., Nowak, K. M., Miltner, A., Trapp, S., & Schäffer, A. (2014). Classification and modelling of non-extractable residue (NER) formation of xenobiotics in soil: A synthesis. *Critical Reviews in Environmental Science and Technology*, 44, 2107–2171.

Katagi, T. (2013). Soil column leaching of pesticides. *Reviews of Environmental Contamination and Toxicology*, 221, 1–104.

Katagi, Y. (2004). Photodegradation of pesticides on plant and soil surfaces. *Reviews of Environmental Contamination and Toxicology*, 182, 1–195.

Katayama, A., Bhula, R., Burns, G. R., Carazo, E., Felsot, A., Hamilton, D., Harris, C., Kim, Y.-H., Kleter, G., Koedel, W., Linders, J., Peijnenburg, J. G. M. W., Sabljic, A., Stephenson, R. G., Racke, D. K., Rubin, B., Tanaka, K., Unsworth, J., & Wauchope, R. D. (2010). Bioavailability of xenobiotics in the soil environment. *Reviews of Environmental Contamination and Toxicology*, 203, 1–86.

Kitchen, N. R., Hughes, D. F., Donald, W. W., & Alberts, E. E. (1998). Agrichemical movement in the root-zone of claypan soils: Ridge- and mulch-tillage systems compared. *Soil and Tillage Research*, 48, 179–193.

Kladivko, E. J., Brown, L. V., & Baker, J. L. (2001). Pesticide transport to subsurface tile drains in humid regions of North America. *Critical Reviews in Environmental Science and Technology*, 31, 1–62. https://dx.doi.org/10.1080/20016491089163

Kleber, M., Sollins, P., & Sutton, R. (2007). A conceptual model of organo-mineral interactions in soils: Self-assembly of organic molecular fragments into zonal structures on mineral surfaces. *Biogeochemistry*, 85, 9–24.

Klein, M. (2011). *PELMO (Pesticide Leaching Model), Version 4.0: User manual*. Fraunhofer Institute for Molecular Biology and Applied Ecology.

Koch, H., Weisser, P., & Landfried, M. (2003). Effect of drift potential on drift exposure in terrestrial habitats. *Nachrichten Blatt des Deutschen Pflanzenschutzdienstes*, 55, 181–188.

Köhne, J. M., Köhne, S., & Šimůnek, J. (2009). A review of model applications for structured soils: B. Pesticide transport. *Journal of Contaminant Hydrology*, 104, 36–60.

Kookana, R., Holz, G., Barnes, C., Bubb, K., Fremlin, R., & Boardman, B. (2010). Impact of climatic and soil conditions on environmental fate of atrazine used under plantation forestry in Australia. *Journal of Environmental Management*, 91, 2649–2656. https://doi.org/10.1016/j.jenvman.20. https://doi.org/10.07.037

Kookana, R. S., Ahmad, R., & Farenhorst, A. (2014). Sorption of pesticides and its dependence on soil properties: Chemometrics approach for estimating sorption. *ACS Symposium Series*, 1174, 221–240. https://doi.org/10.1021/bk-2014-1174.ch012

Kookana, R. S., Aylmore, G., & Gerritse, R. G. (1992). Time-dependent sorption of pesticides during transport in soils. *Soil Science*, 154, 214–225.

Kozak, J. A., Ahuja, L. R., Green, T. R., & Ma, L. (2007). Modelling crop canopy and residue rainfall interception effects on soil hydrological components for semi-arid agriculture. *Hydrological Processes*, 21, 229–241.

Krutz, L. J., Shaner, D. L., Weaver, M. A., Webb, R. M. T., Zablotowicz, R. N., Reddy, K. N., Huang, Y., & Thomson, S. J. (2010a). Agronomic and environmental implications of enhanced s-triazine degradation. *Pest Management Science*, 66, 461–481.

Krutz, L. J., Shaner, D. L., & Zablotowicz, R. M. (2010b). Enhanced degradation and soil depth effects on the fate of atrazine and major metabolites in Colorado and Mississippi soils. *Journal of Environmental Quality*, 39, 1369–1377. https://doi.org/10.2134/jeq2009.0197

Kung, K.-J. S., Steenhuis, T. S., Kladivko, E. J., Gish, T. J., Bubenzer, G., & Helling, C. S. (2000). Impact of preferential flow on the transport of adsorbing and non-adsorbing tracers. *Soil Science Society of America Journal*, 64, 1290–1296. https://doi.org/10.2136/sssaj2000.6441290x

Laabs, V., Amelung, W., Pinto, A., & Zeeh, W. (2002). Fate of pesticides in tropical soils of Brazil under field conditions. *Journal of Environmental Quality*, 31, 256–268. https://doi.org/10.2134/jeq2002.2560

Lamshöft, M., Gao, Z., Resseler, H., Schriever, C., Sur, R., Sweeney, P., Webbd, S., Zillgense, B., Reitzf, M. U., & Reitz, M. (2018). Evaluation of a novel test design to determine uptake of chemicals by plant roots. *Science of the Total Environment*, 613–614, 10–19. https://doi.org/10.1016/j.scitotenv.2017.08.314

Lan, Y., Chen, S., & Fritz, B. (2017). Current status and future trends of precision agricultural aviation technologies. *International Journal of Agricultural and Biological Engineering*, 10(3), 1–17. https://doi.org/10.3965/j.ijabe.20171003.3088

Larson, S. J., Capel, P. D., & Majewski, M. S. (1999). *Pesticides in surface waters: Distribution, trends, and governing factors*. Ann Arbor Press.

Lavieille, D., Ter Halle, A., Bussiere, P. O., & Richard, C. (2009). Effect of a spreading adjuvant on mesotrione photolysis on wax films. *Journal of Agricultural and Food Chemistry, 57*, 9624–9628. https://doi.org/10.1021/jf901996d

Lavy, T. L., Mattice, J. D., Massey, J. H., Skulman, B. W., Senseman, S. A., Gbur, E. E., Jr., & Barrett, M. R. (1996). Long-term in situ leaching and degradation of six herbicides aged in subsoils. *Journal of Environmental Quality, 25*, 1268–1279. https://doi.org/10.2134/jeq1996.00472425002500060015x

Leboeuf, E. J., & Weber, W. J., Jr. (2000). Macromolecular characteristics of natural organic matter: 2. Sorption and desorption behavior. *Environmental Science & Technology, 34*, 3632–3640.

Leistra, M. (2005). *Estimating input data for computations on the volatilization of pesticides from plant canopies and competing processes* (Report 1256). Wageningen, the Netherlands: Alterra, Wageningen University and Research Centre. https://library.wur.nl/WebQuery/wurpubs/fulltext/25119

Leistra, M., & Matser, A. (2004). Adsorption, transformation, and bioavailability of the fungicides carbendazim and iprodione in soil, alone and in combination. *Journal of Environmental Science and Health, Part B, 39*, 1–17.

Leistra, M., & van den Berg, F. (2007). Volatilization of parathion and chlorothalonil from a potato crop simulated by the PEARL model. *Environmental Science & Technology, 41*, 2243–2248.

Leistra, M., van der Linden, A. M. A, Boesten, J. J. T. I., Tiktak, A., & van den Berg, F. (2001). *PEARL model for pesticide behaviour and emissions in soil-plant system: Description of the processes in FOCUS PEARL v 1.1.1* (RIVM Report 711401009/Alterra-rapport 013; 116). Bilthoven, the Netherlands: National Institute of Public Health and the Environment; Wageningen, the Netherlands: Alterra, Green World Research.

Leistra, M., Wolters, A., & van den Berg, F. (2008). Volatilisation and competing processes computed for a pesticide applied to plants in a wind tunnel system. *Pest Management Science, 64*, 669–675.

Leonard, R. A., Knisel, W. G., & Still, D. A. (1987). GLEAMS: Groundwater loading effects of agricultural management systems. *Transactions of the ASAE, 30*, 1403–1418.

Leonard, R. A., Langdale, G. W., & Fleming, W. G. (1979). Herbicide runoff from upland Piedmont watersheds: Data and implications for modeling pesticide transport. *Journal of Environmental Quality, 8*, 223–229. https://doi.org/10.2134/jeq1979.00472425000800020018x

Leterme, B., Vanclooster, M., van der Linden, T., Tiktak, A., & Rounsevell, M. D. A. (2007a). Including spatial variability in Monte Carlo simulations of pesticide leaching. *Environmental Science & Technology, 41*, 7444–7450.

Leterme, B., Vanclooster, M., van der Linden, T., Tiktak, A., & Rounsevell, M. D. A. (2007b). The consequences of interpolating or calculating first on the simulation of pesticide leaching at the regional scale. *Geoderma, 137*, 414–425.

Lewis, S. E., Silburn, D. M., Kookana, R. S., & Shaw, M. (2016). Pesticide behavior, fate, and effects in the tropics: An overview of the current state of knowledge. *Journal of Agricultural and Food Chemistry, 64*, 3917–3924. https://dx.doi.org/10.1021/acs.jafc.6b01320

Li, L.-L., He, X.-K., Song, J.-L., Liu, Y.-J., Zeng, A.-J., Liu, Y., Liu, C., & Liu, Z. (2018). Design and experiment of variable rate orchard sprayer based on laser scanning sensor. *International Journal of Agricultural and Biological Engineering, 11*, 101–108.

Lichiheb, N., Personne, E., Bedos, C., & Barriuso, E. (2014). Adaptation of a resistive model to pesticide volatilization from plants at the field scale: Comparison with a dataset. *Atmospheric Environment, 83*, 260–268.

Lichiheb, N., Personne, E., Bedos, C., Berg, F., & Barriuso, E. (2016). Implementation of the effects of physicochemical properties on the foliar penetration of pesticides and its potential for estimating pesticide volatilization from plants. *Science of the Total Environment, 550*, 1022–1031.

Lost Filho, F. H., Heldens, W. B., Kong, Z., & de Lange, E. S. (2019). Drones: Innovative technology for use in precision pest management. *Journal of Economic Entomology, 20*, 1–25. https://doi.org/10.1093/jee/toz268

Lowrance, R., Altier, L. S., Williams, R. G., Inamdar, S. P., Sheridan, J. M., Bosch, D. D., Hubbard, R. K., & Thomas, D. L. (2000). REMM: The Riparian Ecosystem Management Model. *Journal of Soil and Water Conservation, 55*, 27–34.

Ma, Q., Wauchope, R. D., Ma, L., Rojas, K. W., Malone, R. W., & Ahuja, L. R. (2004). Test of the Root Zone Water Quality Model (RZWQM) for predicting runoff of atrazine, alachlor and fenamiphos species from conventional-tillage corn mesoplots. *Pest Management Science, 60*, 267–276. https://dx.doi.org/10.1002/ps.846

Mackay, D., & van Wesenbeeck, I. (2014). Correlation of chemical evaporation rate with vapor pressure. *Environmental Science & Technology, 48*, 259–263. [erratum: *48*, 10269–10263]

Magri, A., Haith, D. A., Petrovic, A. M., Wu, L., & Green, R. L. (2009). Development and testing of a comprehensive model of pesticide losses from turf. *ACS Symposium Series, 1028*, 183–196. https://dx.doi.org/10.1021/bk-2009-1028.ch013

Maillard, E., & Imfeld, G. (2014). Pesticide mass budget in a stormwater wetland. *Environmental Science & Technology*, 48, 8603–8611.

Majewski, M. S. (1999). Micrometeorological methods for measuring the post-application volatilization of pesticides. *Water, Air, and Soil Pollution*, 115, 83–113.

Majewski, M. S., & Capel, P. D. (1995). *Pesticides in the atmosphere: Distribution, trends, and governing factors*. Ann Arbor Press.

Majewski, M. S., Foreman, W. T., Coupe, R. H., Goolsby, D. A., & Wiebe, F. W. (2007). *Pesticides in air and rainwater in the midcontinental united states, 1995: Methods and data* (Open-File Report 2007-1369). Reston, VA: USGS, National Water-Quality Assessment Program. http://pubs.water.usgs.gov/ofr/20071369

Majewski, M. S., Foreman, W. T., & Goolsby, D. A. (2000). Pesticides in the atmosphere of the Mississippi River valley: I. Rain. *Science of the Total Environment*, 248, 201–212.

Malone, R. W., Logsdon, S., Shipitalo, M. J., Weatherington-Rice, J., Ahuja, L., & Ma, L. (2003). Tillage effect on macroporosity and herbicide transport in percolate. *Geoderma*, 116, 191–215.

Malone, R. W., Shipitalo, M. J., & Meek, D. W. (2004a). Relationship between herbicide concentration in percolate, percolate breakthrough time, and number of active macropores. *Transactions of the ASAE*, 47, 1453–1456.

Malone, R. W., Weatherington-Rice, J., Shipitalo, M. J., Fausey, N., Ma, L., Ahuja, L. R., Wauchope, R. D., & Ma, Q. (2004b). Herbicide leaching as affected by macropore flow and within-storm rainfall intensity variation: A RZWQM simulation. *Pest Management Science*, 60, 277–285.

Mamy, L., Patureau, D., Barriuso, E., Bedos, C., Bessac, F., Louchart, X., Martin-Laurent, F., Miege, C., & Benoit, P. (2015). Prediction of the fate of organic compounds in the environment from their molecular properties: A review. *Critical Reviews in Environmental Science and Technology*, 45, 1277–1377.

Mangala, P., De Silva, C. S., Pathiratne, A., van Straalen, N. M., & van Gestel, C. A. M. (2010). Chlorpyrifos causes decreased organic matter decomposition by suppressing earthworm and termite communities in tropical soil. *Environmental Pollution*, 158, 3041–3047. https://doi.org/10.1016/j.envpol.20. https://doi.org/10.06.032

Mao, M., Cryer, S. A., Altieri, A., & Havens, P. (2018). Predicting pesticide volatility through coupled above- and belowground multiphysics modeling. *Environmental Modeling and Assessment*, 23, 569–582.

Massinon, M., De Cock, N., Forster, W. A., Nairn, J. J., McCue, S. W., Zabkiewicz, J. A., & Lebeau, F. (2017). Spray droplet impaction outcomes for different plant species and spray formulations. *Crop Protection*, 99, 65–75.

Matocha, M. A., Krutz, L. J., Reddy, K. N., Senseman, S. A., Locke, M. A., Steinriede, R. W., Jr., & Palmer, E. W. (2006). Foliar washoff potential and simulated surface runoff losses of trifloxysulfuron in cotton. *Journal of Agricultural and Food Chemistry*, 54, 5498–5502.

Melland, A. R., Silburn, D. M., McHugh, A. D., Fillols, E., Rojas-Ponce, S., Baillie, C., & Lewis, S. (2016). Spot spraying reduces herbicide concentrations in runoff. *Journal of Agricultural and Food Chemistry*, 64, 4009–4020.

Mills, M. S., Hill, I. R., Newcombe, A. C., Simmons, N. D., Vaughan, P. C., & Verity, A. A. (2001). Quantification of acetochlor degradation in the unsaturated zone using two novel in situ field techniques: Comparisons with laboratory generated data and implications for groundwater risk assessments. *Pest Management Science*, 57, 351–359.

Motoki, Y., Iwafune, T., Seike, N., Otani, T., & Asano, M. (2014). Effects of organic carbon quality on the sorption behavior of pesticides in Japanese soils. *Journal of Pest Science*, 39, 105–114. https://dx.doi.org/10.1584/jpestics.D13-067

Mueller, T. C., & Steckel, L. E. (2019). Spray mixture pH as affected by dicamba, glyphosate, and spray additives. *Weed Technology*, 33, 547–554.

Muir, D., Zhang, X., de Wit, C. A., Vorkamp, K., & Wilson, S. (2019). Identifying further chemicals of emerging arctic concern based on "in silico" screening of chemical inventories. *Emerging Contaminants*, 5, 201–210. https://doi.org/10.1016/j.emcon.2019.05.005

Muir, D. C. G., & Howard, P. H. (2006). Are there other persistent organic pollutants? A challenge for environmental chemists. *Environmental Science & Technology*, 40, 7157–7166.

Muller, K., Trolove, T., James, T. K., & Rahman, A. (2004). Herbicide loss in runoff: Effects of herbicide properties slope and rainfall intensity. *Australian Journal of Soil Research*, 42, 17–27.

Munira, S., Farenhorst, A., Sapkota, K., Nilsson, D., & Sheedy, C. (2018). Auxin herbicides and pesticide mixtures in groundwater of a Canadian Prairie Province. *Journal of Environmental Quality*, 47, 1462–1467. https://dx.doi.org/10.2134/jeq2018.05.0202

Muñoz-Carpena, R., Ritter, A., Fox, G. A., & Perez-Ovilla, O. (2015). Does mechanistic modeling of filter strip pesticide mass balance and degradation processes affect environmental exposure assessments? *Chemosphere*, 139, 410–421.

Negley, T., Allen, R., Tang, J., Dyer, D., & Geh, K. (2014). The significance of time-dependent sorption on leaching potential: A comparison of measured field results and modeled estimates. *ACS Symposium Series*, 1174, 337–356. https://dx.doi.org/10.1021/bk-2014-1174.ch015

Nelson, S. D., Ajwa, H. A., Trout, T., Stromberger, M., Yates, S. R., & Sharma, S. (2013). Water and methyl isothiocyanate distribution in soil after drip fumigation. *Journal of Environmental Quality*, 42, 1555–1564. https://doi.org/10.2134/jeq2013.03.0072

Nfon, E., Armitage, J. M., & Cousins, I. T. (2011). Development of a dynamic model for estimating the food web transfer of chemicals in small aquatic ecosystems. *Science of the Total Environment*, 409, 5416–5422.

NORMAN Network. (2020). *NORMAN: Network of reference laboratories, research centres and related organisations for monitoring of emerging environmental substances and pollutant lists.* https://www.norman-network.net/?q=Home

Northcott, G. L., & Jones, K. C. (2000). Experimental approaches and analytical techniques for determining organic compound bound residues in soil and sediment. *Environmental Pollution*, 108, 19–43.

Nowell, L. H., Capel, P. D., & Dileanis, P. D. (1999). *Pesticides in stream sediment and aquatic biota: Distribution, trends, and governing factors.* CRC Press.

Nsibande, S. A., Dabrowski, J. M., van der Walt, E., Venter, A., & Forbes, P. B. C. (2015). Validation of the AGDISP model for predicting airborne atrazine spray drift: A South African ground application case study. *Chemosphere*, 138, 454–461.

Paradelo, M., Norgaard, T., Møldrup, P., Ferré, T. P. A., Kumari, K. G. I. D., Arthur, E., & de Jonge, L. W. (2015). Prediction of the glyphosate sorption coefficient across two loamy agricultural fields. *Geoderma*, 259–260, 224–232. https://dx.doi.org/10.1016/j.geoderma.2015.06.011

Pedit, J. A., & Miller, C. T. (1994). Heterogeneous sorption processes in subsurface systems: 1. Model formulations and applications. *Environmental Science & Technology*, 28, 2094–2104.

Pedit, J. A., & Miller, C. T. (1995). Heterogeneous sorption processes in subsurface systems: 2. Diffusion modeling approaches. *Environmental Science & Technology*, 29, 1766–1772.

Phillips, K. L., Di Toro, D. M., & Sandler, S. I. (2011). Prediction of soil sorption coefficients using model molecular structures for organic matter and the quantum mechanical COSMO-SAC model. *Environmental Science & Technology*, 45, 1021–1027.

Potter, T. L., Lowrance, R. R., Bosch, D. D., & Williams, R. G. (2012). Estimating pesticide retention efficacy for edge-of-field buffers using the Riparian Ecosystem Management Model (REMM) in a southeastern plains landscape. *ACS Symposium Series*, 1075, 259–271. https://dx.doi.org/10.1021/bk-2011-1075.ch016

Potter, T. L., Truman, C. C., Bosch, D. D., & Bednarz, C. W. (2003). Cotton defoliant runoff as a function of active ingredient and tillage. *Journal of Environmental Quality*, 32, 2180–2188. https://doi.org/10.2134/jeq2003.2180

Potter, T. L., Truman, C. C., Strickland, T. C., Bosch, D. D., Webster, T. M., Franklin, D. H., & Bednarz, C. W. (2006). Combined effects of constant versus variable intensity simulated rainfall and reduced tillage management on cotton preemergence herbicide runoff. *Journal of Environmental Quality*, 35, 1894–1902. https://doi.org/10.2134/jeq2005.0444

Potter, T. L., Truman, C. C., Webster, T. M., Bosch, D. D., & Strickland, T. C. (2011). Tillage, cover-crop residue management, and irrigation incorporation impact on Fomesafen runoff. *Journal of Agricultural and Food Chemistry*, 59, 7910–7915. https://dx.doi.org/10.1021/jf201731u

Prueger, J. H., Alfieri, J., Gish, T. J., Kustas, W. P., Daughtry, C. S. T., Hatfield, J. L., & McKee, L. G. (2017). Multi-year measurements of field-scale metolachlor volatilization. *Water, Air, and Soil Pollution*, 228, 84–94.

Purdy, J. R., & Cheplick, M. (2014). Nonlinear soil dissipation kinetics: The use of a set of simple first-order processes to describe a biphasic degradation pattern. *ACS Symposium Series*, 1174, 133–146. https://dx.doi.org/10.1021/bk-2014-1174.ch009

Racke, K. D., Skidmore, M. W., Hamilton, D. J., Unsworth, J. B., Miyamoto, J., & Cohen, S. Z. (1997). Pesticide fate in tropical soils. *Pure and Applied Chemistry*, 69, 1349–1371.

Remucal, C. K. (2014). The role of indirect photochemical degradation in the environmental fate of pesticides: A review. *Environmental Science: Processes and Impacts*, 16, 628–653.

Rice, C. P., Nochetto, C. B., & Zaro, P. (2002). Volatilization of trifluralin, atrazine, metolachlor, chlorpyrifos α-endosulfan, and β-endosulfan from freshly tilled soil. *Journal of Agricultural and Food Chemistry*, 50, 4009–4017.

Roberts, T., Klein, W., Still, G. G., Kearney, P. C., Drescher, N., Desmoras, J., & Vonk, J. E. (1984). Non-extractable pesticide residues in soils and plants. *Pure and Applied Chemistry*, 56, 945–956.

Römbke, J., Waichman, A. V., & Garcia, M. V. B. (2008). Risk assessment of pesticides for soils of the central amazon, Brazil: Comparing outcomes with temperate and tropical data. *Integrated Environmental Assessment and Management*, 4, 94–104.

Rose, M. T., Sanchez-Bayo, F., Crossan, A. N., & Kennedy, I. (2006). Pesticide removal from cotton farm tail-water by a pilot-scale ponded wetland. *Chemosphere*, 63, 1849–1858.

Rosenbom, A. E., Binning, P. J., Aamand, J., Dechesne, A., Smets, B. F., & Johnsen, A. R. (2014). Does microbial centimeter-scale heterogeneity impact MCPA degradation in and leaching from a loamy agricultural soil? *Science of the Total Environment*, 472, 90–98.

Rüdel, H. (1997). Volatilization of pesticides from plant and soil surfaces. *Chemosphere*, 35, 143–152.

Sabljic, A., & Nakagawa, Y. (2014). Sorption and quantitative structure–activity relationship (QSAR). *ACS Symposium Series*, 1174, 85–118. https://dx.doi.org/10.1021/bk-2014-1174.ch004

Sanchez-Bayo, F., & Hyne, R. V. (2011). Comparison of environmental risks of pesticides between tropical and nontropical regions. *Integrated Environmental Assessment and Management*, 7, 577–586.

Sander, M., & Pignatello, J. J. (2005). An isotope exchange technique to assess mechanisms of sorption hysteresis applied to naphthalene in kerogenous organic matter. *Environmental Science & Technology*, 39, 7476–7484.

Sander, M., & Pignatello, J. J. (2009). Sorption irreversibility of 1, 4-dichlorobenzene in two natural organic matter-rich geosorbents. *Environmental Toxicology and Chemistry*, 28, 447–457.

Sandin-España, P., & Sevilla-Morán, B. (2012). Pesticide degradation in water. In K. S. Rathore & L. M. L. Nollet (Eds.), *Pesticides: Evaluation of environmental pollution* (pp. 79–13). CRC Press.

Sarmah, A. J., Müller, K., & Ahmad, R. (2004). Fate and behavior of pesticides in the agroecosystem: A review with a New Zealand perspective. *Australian Journal of Soil Research*, 42, 125–154.

Sassman, S. A., & Lee, L. S. (2005). Sorption of three tetracyclines by several soils: Assessing the role of pH and cation exchange. *Environmental Science & Technology*, 39, 7452–7459. https://doi.org/10.1021/es0480217

Schäffer, A. S., Kästner, M., & Trapp, S. (2018). A unified approach for including non-extractable residues (NER) of chemicals and pesticides in the assessment of persistence. *Environmental Sciences Europe*, 30, 51–65. https://dx.doi.org/10.1186/s12302-018-0181-x

Schymanski, E. L., Singer, H. P., Slobodnik, J., Ipolyi, I. M., Oswald, P., Krauss, M., Schulze, T., Haglund, P., Letzel, T., Grosse, S., Thomaidis, N. S., Bletsou, A., Zwiener, C., Ibáñez, M., Portolés, T., de Boer, R., Reid, M. J., Onghena, M., Kunkel, U., & Hollender, J. (2015). Non-target screening with high-resolution mass spectrometry: Critical review using a collaborative trial on water analysis. *Analytical and Bioanalytical Chemistry*, 407, 6237–6255. https://dx.doi.org/10.1007/s00216-015-8681-7

Senesi, N., & Loffredo, E. (1999). The chemistry of soil organic matter. In D. L. Sparks (Ed.), *Soil physical chemistry* (2nd ed., pp. 239–368). CRC Press.

Shaner, D., Brunk, G., Nissen, S., Westra, P., & Chen, W. (2012). Role of soil sorption and microbial degradation on dissipation of mesotrione in plant-available soil water. *Journal of Environmental Quality*, 41, 170–178. https://doi.org/10.2134/jeq2011.0187

Shi, S., & Bending, G. D. (2007). Changes to the structure of *Sphingomonas* spp. communities associated with biodegradation of the herbicide isoproturon in soil. *FEMS Microbiology Letters*, 269, 110–116. https://dx.doi.org/10.1111/j.1574-6968.2006.00621.x

Sidoli, P., Baran, N., & Angulo-Jaramillo, R. (2016). Glyphosate and AMPA adsorption in soils: Laboratory experiments and pedotransfer rules. *Environmental Science and Pollution Research*, 23, 5733–5742.

Silburn, D. M. (2003). *Characterising pesticide runoff from soil on cotton farms using a rainfall simulator* (Doctoral dissertation, University of Sydney). SeS Repository. https://hdl.handle.net/2123/24339

Silburn, D. M., & Kennedy, I. R. (2007). Rain simulation to estimate pesticide transport in runoff. *ACS Symposium Series*, 966. https://dx.doi.org/10.1021/bk-2007-0966

Šimůnek, J., Jarvis, N. J., van Genuchten, M. T., & Gärdenäs, A. (2003). Review and comparison of models for describing non-equilibrium and preferential flow and transport in the vadose zone. *Journal of Hydrology*, 272, 14–35.

Singh, B., Farenhorst, A., Gaultier, J., Pennock, D., Degenhardt, D., & McQueen, R. (2014). Soil characteristics and herbicide sorption coefficients in 140 soil profiles of two irregular undulating to hummocky terrains of western Canada. *Geoderma*, 232–234, 107–116.

Smith, C. N., Carsel, R. F., & Parrish, R. S. (1987). Estimating sample requirements for field evaluations of pesticide leaching. *Environmental Toxicology and Chemistry*, 6, 343–357.

Smith, M. C., Shaw, D. R., Massey, J. H., Boyette, M., & Kingery, W. (2003). Using nonequilibrium thin-disc and batch equilibrium techniques to evaluate herbicide sorption. *Journal of Environmental Quality*, 32, 1393–1404. https://doi.org/10.2134/jeq2003.1393

Soares, A., Paradelo, P. M., Møldrup, P., Delerue-Matos, C., & de Jonge, L. W. (2015). Predictivity strength of the spatial variability of phenanthrene sorption across two sandy loam fields. *Water, Air, and Soil Pollution*, 226, 36–48.

Spadotto, C. A., & Hornsby, A. G. (2003). Soil sorption of acidic pesticides: Modeling pH effects. *Journal of Environmental Quality*, 32, 949–956. https://doi.org/10.2134/jeq2003.9490

Spadotto, C. A., Locke, M. A., Bingner, R. L., & Mingoti, R. (2020). Estimating sorption of monovalent acidic herbicides at different pH levels using a single sorption coefficient. *Pest Management Science*, 76, 2693–2698. https://dx.doi.org/10.1002/ps.5815

Spencer, W. F., Cliath, M. M., Davis, K. R., Spear, R. C., & Popendorf, W. J. (1975). Persistence of parathion and its oxidation to paraoxon on the soil surface as related to worker reentry into treated crops. *Bulletin of Environmental Contamination and Toxicology*, 14, 265–272.

Spencer, W. F., Farmer, W. J., & Jury, W. A. (1982). Review: Behavior of organic chemicals at soil, air, water interfaces as related to predicting the transport and volatilization of organic pollutants. *Environmental Toxicology and Chemistry*, 1, 17–26.

Stellman, S. D., & Stellman, J. D. (2004). Exposure opportunity models for Agent Orange, dioxin, and other military herbicides used in Vietnam, 1961–1971. *Journal of Exposure Analysis and Environmental Epidemiology*, 14, 354–362.

Suddaby, L., Beulke, S., van Beinum, W., Oliver, R., Kuet, S., Celis, R., & Koskinen, W. (2014). Experiments and modeling to quantify irreversibility of pesticide sorption–desorption in soil. *ACS Symposium Series*, 1174, 199–219. https://dx.doi.org/10.1021/bk-2014-1174.ch011

Suddaby, L. A., (2012). *Investigation into irreversible sorption of pesticides to soil* (Doctoral dissertation, University of York). White Rose eTheses. https://etheses.whiterose.ac.uk/2725/

Suddaby, L. A., Beulke, S., van Beinum, W., Celis, R., Koskinen, W. C., & Brown, C. D. (2013). Reanalysis of experiments to quantify irreversibility of pesticide sorption–desorption in soil. *Journal of Agricultural and Food Chemistry*, 61, 2033–2038.

Sur, R. (2014). Terrestrial field degradation based on soil, climatic, and geographic factors. *ACS Symposium Series*, 1174, 39–56. https://dx.doi.org/10.1021/bk-2014-1174.ch003

Sur, R., Tang, J., Jones, R. L., Dyer, D. G., & Coody, P. N. (2014). Effect of refined environmental fate properties on groundwater concentrations calculated with PRZM. *ACS Symposium Series*, 1174, 299–336. https://dx.doi.org/10.1021/bk-2014-1174.ch016

Suszter, G. K., & Ambrus, A. (2017). Distribution of pesticide residues in soil and uncertainty of sampling. *Journal of Environmental Science and Health, Part B*, 52, 557–563.

Tang, J., Jones, R. L., Huang, M., Chen, W., Allen, R., Hayes, S., & Sur, R. (2014). Evaluations of regulatory kinetics analysis approaches. *ACS Symposium Series*, 1174, 119–132. https://dx.doi.org/10.1021/bk-2014-1174.ch006

Taylor, M., Lyons, S. M., Davie-Martin, C. L., Geohegan, T. S., & Hageman, K. J. (2020). Understanding trends in pesticide volatilization from agricultural fields using the pesticide loss via volatilization model. *Environmental Science & Technology*, 54, 2202–2209.

Taylor, P. (2011). The wetting of leaf surfaces. *Current Opinion in Colloid Interface Science*, 16, 326–334.

Ter Horst, M. M. S., Boesten, J. J. T. S., van Beinum, W., & Beulke, S. (2013). Acceptability of inversely-modelled parameters for non-equilibrium sorption of pesticides in soil. *Environmental Modelling and Software*, 46, 260–270.

Teske, M. E., Bird, S. L., Esterly, D. M., Curbishley, T. B., Ray, S. L., & Perry, S. G. (2002). AgDRIFT: A model for estimating near-field spray drift from aerial applications. *Environmental Toxicology and Chemistry*, 21, 659–671.

The Economist. (2020). Bit-Lit: A new AI language model generates poetry and prose. *The Economist*, 8 August, p. 63.

Thorbek, P., & Hyder, K. (2006). Relationship between physicochemical properties and maximum residue levels and tolerances of crop-protection products for crops set by the USA, European Union and Codex. *Food Additives and Contaminants*, 23, 764–776.

Thorburn, P. J., Wilkinson, S. N., & Silburn, D. M. (2013). Water quality in agricultural lands draining to the Great Barrier Reef: A review of causes, management and priorities. *Agriculture, Ecosystems & Environment*, 180, 4–20. https://doi.org/10.1016/j.agee.2013.07.006

Trapp, S. (2000). Modeling uptake into roots and subsequent translocation of neutral and ionisable organic compounds. *Pest Management Science*, 56, 767–778.

Trapp, S. (2007). Fruit tree model for uptake of organic compounds from soil and air. *SAR and QSAR in Environmental Research*, 18, 367–387. https://doi.org/10.1080/10629360701303693

Trapp, S. (2009). Bioaccumulation of polar and ionizable compounds in plants. In J. Devillers (Ed.), *Ecotoxicology modeling: Emerging topics in ecotoxicology (principles, approaches and perspectives)* (Vol. 2, pp. 299–353). Springer.

Tureli, F. C., Ok, S. S., & Goldberg, S. (2015). Specific surface area effect on adsorption of chlorpyrifos and TCP by soils and modelling. *Soil and Sediment Contamination*, 24, 64–75.

van Beinum, W., Beulke, S., & Brown, C. D. (2005). Pesticide sorption and diffusion in natural clay loam aggregates. *Journal of Agricultural and Food Chemistry*, 53, 9146–9154. https://doi.org/10.1021/jf050928g

van de Zande, J. C., Butler Ellis, M. C. B., Wenneker, M., Walklate, P. J., & Kennedy, M. (2014). Spray drift and bystander risk from fruit crop spraying. *Aspects of Applied Biology*, 122, 177–185.

van den Berg, F., Tiktak, A., Boesten, J. J. T. I., & van der Linden, A. M. A. (2016). PEARL model for pesticide behaviour and emissions in soil-plant systems: Description of processes (WOt-Technical Report 61). Wageningen, the Netherlands: Alterra.

van der Linden, A. M. A., Tiktak, A., Boesten, J. J. T. I., & Leijnse, A. (2009). Influence of pH-dependent sorption and transformation on simulated pesticide leaching. *Science of the Total Environment*, 407, 3415–3420.

van Wesenbeeck, I. J., Driver, J., & Ross, J. (2008). Relationship between the evaporation rate and vapor pressure of moderately and highly volatile chemicals. *Bulletin of Environmental Contamination and Toxicology*, 80, 315–318. https://dx.doi.org/10.1007/s00128-008-9380-2

Vereecken, H. (2005). Mobility and leaching of glyphosate: A review. *Pest Management Science*, 61, 1139–1151.

Vereecken, H., Vanderborght, J., Kasteel, R., Spiteller, M., Schäffer, A., & Close, M. (2011). Do lab-derived distribution coefficient values of pesticides match distribution coefficient values determined from column and field-scale experiments? A critical analysis of relevant literature. *Journal of Environmental Quality*, 40, 879–898. https://doi.org/10.2134/jeq20. https://doi.org/10.0404

Vitoratos, A., Fois, C. D., Danias, P., & Likudis, Z. (2016). Investigation of the soil sorption of neutral and basic pesticides. *Water, Air, and Soil Pollution*, 227, 397–413.

Vogel, J. R., Majewski, M. S., & Capel, P. D. (2008). Pesticides in rain in four agricultural watersheds in the United States. *Journal of Environmental Quality*, 37, 1101–1115. https://doi.org/10.2134/jeq2007.0079

Wang, C., He, X., Wang, X., Wang, Z., Wang, S., Li, L., Bonds, J., & Wang, J. (2018). Testing method and distribution characteristics of spatial pesticide spraying deposition quality balance for unmanned aerial vehicle. *International Journal of Agricultural and Biological Engineering*, 11, 18–26.

Wang, D., Yates, S. R., & Gao, S. (2011). Chloropicrin emissions after shank injection: Two-dimensional analytical and numerical model simulations of different source methods and field measurements. *Journal of Environmental Quality*, 40, 1443–1449. https://doi.org/10.2134/jeq20. https://doi.org/10.0233

Wang, G., Lan, Y., Yuan, H., Qi, H., Chen, P., Ouyang, F., & Han, Y. (2019). Comparison of spray deposition, control efficacy on wheat aphids and working efficiency in the wheat field of the unmanned aerial vehicle with boom sprayer and two conventional knapsack sprayers. *Applied Sciences*, 9, 218–234. https://dx.doi.org/10.3390/app9020218

Watanabe, H., & Grismer, M. E. (2003). Numerical modeling of diazinon transport through inter-row vegetative filter strips. *Journal of Environmental Management*, 69, 157–168.

Wauchope, R. D. (2006). Pest management practices: pesticide mitigation. In M. Schnepf & C. Cox (Eds.), *Environmental benefits of conservation on cropland: The status of our knowledge* (pp. 195–242). Soil and Water Conservation Society.

Wauchope, R. D., & Myers, R. S. (1985). Adsorption–desorption kinetics of atrazine and linuron in freshwater-sediment aqueous slurries. *Journal of Environmental Quality*, 14, 132–136. https://doi.org/10.2134/jeq1985.00472425001400010027x

Wauchope, R. D., Nash, R. G., Rojas, K. W., Ahuja, L. R., Willis, G. H., Ma, Q. L., LL, M. D., & Moorman, T. B. (2000). Pesticide processes. In L. R. Ahuja, K. W. Rojas, J. D. Hanson, M. J. Shaffer, & L. Ma (Eds.), *Root Zone Water Quality Model: Modeling management effects on water quality and crop production* (pp. 163–244). Water Resources Publications.

Wauchope, R. D., Rojas, K. W., Ahuja, L. R., Ma, Q., Malone, R. W., & Ma, L. (2004). Documenting the pesticide processes module of the ARS RZWQM agroecosystem model. *Pest Management Science*, 60, 222–239.

Wauchope, R. D., Yeh, S., Linders, J. B. H. J., Kloskowski, R., Tanaka, K., Rubin, B., Katayama, A., Kördel, W., Gerstl, Z., Lane, M., & Unsworth, J. B. (2002). Pesticide soil sorption parameters: Theory, measurement, uses, limitations and reliability. *Pest Management Science*, 58, 419–445.

Williams, A. J., Grulke, M., Edwards, J., McEachran, A. D., Mansouri, K., Baker, N. C., Patlewicz, G., Shah, I., Wambaugh, J. F., Judson, R. S., & Richard, A. M. (2020). The CompTox Chemistry Dashboard: A community data resource for environmental chemistry. *Journal of Cheminformatics*, 9, 61–88. https://dx.doi.org/10.1186/s13321-017-0247-6

Williams, C. O., Lowrance, R., Potter, T., Bosch, D. D., & Strickland, T. (2016). Atrazine transport within a coastal zone in southeastern Puerto Rico: A sensitivity analysis of an agricultural field model and riparian zone management model. *Environmental Modeling and Assessment*, 21, 751–761. https://dx.doi.org/10.1007/s10666-016-9508-4

Williams, J. R., Arnold, J. G., & Srinivasan, R. (2000). *The APEX model* (BREC Rep. No. 00–06). Temple, TX: Texas A&M University, Blackland Research and Extension Center.

Willis, G. H., Spencer, W. F., & McDowell, L. L. (1980). The interception of applied pesticides by foliage and their persistence and runoff potential. In W. G. Knisel (Ed.), *CREAMS: A field scale model for Chemicals, Runoff and Erosion from Agricultural Management Systems* (USDA–SEA Conservation Research Report 26, pp. 595–606).

Winchell, M. F., Pai, N., Brayden, B. H., Stone, C., Whatling, P., Hanzas, J. P., & Stryker, J. L. (2018). Evaluation of watershed-scale simulations of in-stream pesticide concentrations from off-target spray drift. *Journal of Environmental Quality*, 47, 79–87. https://doi.org/10.2134/jeq2017.06.0238

Wolters, A., Leistra, M., Linnemann, V., Klein, M., Schäffer, A., & Vereecken, H. (2004). Pesticide volatilization from plants: Improvement of the PEC model PELMO based on a boundary-layer concept. *Environmental Science & Technology*, 38, 2885–2893.

Woodrow, J. E., Seiber, J. N., & Baker, L. W. (1997). Correlation techniques for estimating pesticide volatilization flux and downwind concentrations. *Environmental Science & Technology*, 31, 523–529. https://dx.doi.org/10.1021/es960357w

Woodrow, J. E., Seiber, J. N., & Miller, G. C. (2011). Correlation to estimate emission rates for soil-applied fumigants. *Journal of Agricultural and Food Chemistry*, 59, 939–943. https://dx.doi.org/10.1021/jf103868k

Woodward, S. J. R., Connell, R. J., Zabkiewicz, J. A., Steele, K. D., & Praat, J. P. (2008). Evaluation of the AGDISP ground boom spray drift model. *New Zealand Plant Protection*, 61, 164–168.

Xin, F., Zhao, J., Zhou, Y., Wang, G., Han, X., Fu, W., & Lan, Y. (2018). Effects of dosage and spraying volume on cotton defoliants efficacy: A case study based on application of unmanned aerial vehicles. *Agronomy*, 8(6), 85. https://dx.doi.org/10.3390/agronomy8060085

Yates, S. R. (2006a). Measuring herbicide volatilization from bare soil. *Environmental Science & Technology*, 40, 3223–3228.

Yates, S. R. (2006b). Simulating herbicide volatilization from bare soil affected by atmospheric conditions and limited solubility in water. *Environmental Science & Technology*, 40, 6963–6968.

Yates, S. R., Wang, D., Papiernik, S. K., & Gan, J. (2002). Predicting pesticide volatilization from soils. *Environmetrics*, 13, 569–578.

Ye, J., Zhao, M., Liu, J., & Liu, W. (2010). Enantioselectivity in environmental risk assessment of modern chiral pesticides. *Environmental Pollution*, 158, 2371–2383.

Zabkiewicz, J. A. (2007). Spray formulation efficacy: Holistic and futuristic perspectives. *Crop Protection*, 26, 312–319.

8

Puyu Feng,
Bin Wang,
De Li Liu,
Qiang Yu, and
Kelin Hu

Coupling Machine Learning with APSIM Model Improves the Evaluation of Climate Extremes Impact on Wheat Yield in Australia

Abstract

Accurately assessing the impacts of extreme climate events (ECEs) on crop yield can help develop effective agronomic practices to deal with climate change. Process-based crop models are useful tools to evaluate the impacts of climate change on crop productivity, but they are usually limited in modeling the effects of ECEs due to over-simplification, vague descriptions of certain process, or uncertainties in parameterization. In this chapter, we describe a hybrid model that was developed by incorporating the Agricultural Production System sIMulator (APSIM) model outputs and growth stage-specific ECEs indicators (i.e., frost, drought and heat stress) into the Random Forest (RF) model, with the multiple linear regression (MLR) model as a benchmark. This APSIM + RF hybrid model could explain 81% of the observed yield variations in the New South Wales wheat belt of south-eastern Australia, which was a 33% improvement in modeling accuracy compared with the APSIM model alone and a 19% improvement compared with the APSIM + MLR hybrid model. Droughts during grain-filling, droughts during the vegetative stages, and heat events immediately prior to anthesis were identified as the three most serious ECEs causing yield loss. The APSIM + RF hybrid model was then compared with the APSIM model to estimate

Modeling Processes and Their Interactions in Cropping Systems: Challenges for the 21st Century, First Edition.
Edited by Lajpat R. Ahuja, Kurt C. Kersebaum, and Ole Wendroth.

the effects of future climate change on wheat yield. It was interesting to find that future yield projected from single APSIM model might be overestimated by 1–10% compared with the APSIM + RF hybrid model. The APSIM + RF hybrid model indicated that former models were underestimating the effects of climate change; future yields might be lower than predicted using the single APSIM informed model due to inadequate accounts of ECEs-induced yield loss. An increasing number of heat events around the anthesis and grain-filling periods were identified as major factors causing yield loss in the future. Therefore, the effects of ECEs on crop yield should be included in models to accurately assess the impacts of climate change. We expect this proposed hybrid-modeling approach can be applied to other regions and crops and offer new insights of the effects of ECEs on crop yield.

Introduction

As the global population and living standards increase, the demand for stable foods such as wheat is expected to increase by 60% by the middle of the 21st century (Godfray & Toulmin, 2010). Sustainable improvements in crop production are urgently needed to meet this demand. However, the ongoing impacts of climate change will increase the risk of meeting this demand through crop production (Howden et al., 2007). In particular, climate change-induced increases of extreme climatic events (ECEs) are recognized as the major threat to crop production (Trnka et al., 2014; Watson et al., 2017; Wheeler & Von Braun, 2013). In recent decades, ECEs have resulted in increased yield losses around the world (Lesk et al., 2016). For example, in south-eastern Australia, drought and co-occurring heat stress reduced the agricultural gross national product by approximately 30% in 1994, 2002, and 2006 (Kirono et al., 2011). Therefore, accurately estimating current and future ECEs-induced yield losses is urgently needed to assess the sustainability of agricultural production. This chapter is an expanded and reorganized version of work conducted at the University of Technology in Sydney, Australia (Feng et al., 2019a).

Climate Change and Extreme Climate Events

It is certain that climate is changing all over the world based on large-scale and long-term observations. According to the Intergovernmental Panel on Climate Change (IPCC), the global mean surface temperature has been increasing during the last two centuries, with an increasingly steeper rate (IPCC, 2012). Precipitation change is inconsistent around the world. In most mid-latitude regions, the average precipitation has been increasing in recent decades. Moreover, all factors of the climate system will continue changing if emissions of greenhouse gases are not well controlled.

ECEs are usually defined as atypical precipitation, temperature, and other weather factors compared with historical distributions (IPCC, 2012). It has been emphasized that climate change will inevitably alter climate extremes, as the shift of the distribution of a climatic factor toward one side will indeed enhance occurrences of extremes on that same side (Planton et al., 2008). The Coupled Model Intercomparison Project Phase 5 (CMIP5) multi-model ensemble indicates that, in all regions of the world, there are likely to be more heat extremes and less cold extremes (Kharin et al., 2013). In particular, the intensity of heat extremes is projected to reach unprecedented levels under

Representative Concentration Pathway (RCP) 8.5 scenario by the end of the 21st century, especially over tropical and subtropical areas.

A precipitation increase will also be more common around the world, according to CMIP5 projections. However, occurrences of dry spells or droughts also show an increasing tendency. This is mainly due to two reasons: an uneven distribution of precipitation and an increase in temperature. Rainfall events are likely to decrease, but heavy rainfall events are projected to increase, thus the number of consecutive dry days will increase (Li et al., 2012). Moreover, heavy rainfall amounts often surpass the bearing capacities of soils, and excessive water forms runoff which cannot be used by high rainfall areas (Yasufuku et al., 2015). On the other hand, increased temperature will lead to increased evaporation, which can aggravate drought occurrence (Venkataraman et al., 2016). Feng et al. (2019b) assessed impacts of climate change on drought features across the wheat belt of south-eastern Australia using 28 downscaled global climate models (GCMs). They revealed a significant increase in the intensity and frequency of future spring and winter drought events in the wheat belt. The combined effects of reduced precipitation and increased temperature during future spring and winter seasons were the main reasons causing these changes of drought.

Impacts of Climate Extremes on Crops

Heat stress has long been a big challenge for agricultural producers. Short-term heat stress can adversely reduce enzymatic activity and genetic expression, while long-term heat stress can cut down carbon sequestration and thus reduce growth velocity. Owing to these reasons, heat stress tends to accelerate the maturation processes. Shorter growth periods directly lead to reductions in carbon assimilation, which then lead to insufficient supplies of photo-assimilates in grain (Stone et al., 1994). Heat stress can also result in infertile grain and reduce grain number, thereby reducing yield (Liu et al., 2016). In addition, the actual effects of heat stress on crops are dependent on occurrence time because the endurance abilities of crops against high temperatures are different at different growth stages. For example, wheat (*Triticum aestivum* L.) plants are most vulnerable to heat stress at their reproductive stages, especially from anthesis to grain filling. Winter wheat is generally flowering in the late spring, and heat stress is most common during this period (Yang et al., 2017). Wollenweber et al. (2003) reported that wheat yield would decrease severely if wheat plants experienced a short-term heat stress (35 °C) during anthesis.

Drought is currently one of the main constraints to crop production in rainfed systems throughout the world. Drought adversely affects crop growth, mainly through water deficit. Crop growth can be restrained in all growing periods, but the extent of damage caused by drought mainly depends on its time of occurrence. For wheat, terminal droughts during the reproductive stages are the most harmful (Pradhan et al., 2012). Reproductive stages of wheat are key stages that determine eventual yields. Wheat plants during these stages require large amounts of water for carbon sequestration and assimilate translocation (Ji et al., 2012). Water deficit during these stages can not only reduce production of photosynthate through accelerating processes of leaf senescence and lessening photosynthesis (Chaves et al., 2002), but also decrease the distribution of assimilates to grains through reducing the rate of translocating and sink capacity (Liang et al., 2001).

As with drought and heat, frost can also restrain crop growth at all growing periods. For some types of wheat, during early stages, the seedlings are vulnerable to very cold conditions. Leaves might become withered, and seedlings might even die (Fuller et al., 2007). This will eventually lead to yield reduction. The magnitude of yield reduction caused by frost from pre-flowering to ripening is much larger than at any other growing periods. During the different stages of reproduction, frost can damage various parts of the plant, which all can result in serious effects on crop growth. Around booting, the stem is the most vulnerable to frost (Crimp et al., 2016). The most common injury to the stem occurs below the head which might lead to loss of the head. Stem injury is irreclaimable, and the translocation of photoassimilates in later growth periods will be affected greatly. When spikes have formed, but before flowering, a single frost event can lead to injury to the spike by causing sterile flowers, thereby decreasing grain number (Barlow et al., 2015). During the heading period, frost can also lead to sterile flowers by damaging anthers and embryos (Al-Issawi et al., 2013; Marcellos & Single, 1984). At the grain filling stage, frost can lead to shriveled and shrunken grains by killing partially filled grains (Perry et al., 2017). It should be noted that although climate warming has significantly reduced occurrences of cold extremes, wheat is still highly vulnerable to frost. This is mainly because a higher mean temperature can speed up the plants' growth rate and shorten whole growth duration, resulting in a shift of reproductive stages to earlier dates (He et al., 2015). Thus, frost-induced yield reduction should be focused on now as well as in the future.

Modeling Methods

Crop yield is affected by multiple factors, such as climate, soil, and farming management. To explore relationships between climate and crop yield, the complex impacts of climate and non-climate factors on crop growth should be systematically analyzed. Process-based crop models and statistical models are two principal methods to carry out this task.

Process-based crop models have been developed to account for the complex interactions among the local environment, the crop genotype, and management practices (Chenu et al., 2017). In recent years, crop models have been widely used to characterize the effects of historical and future ECEs on crop yield in multiple regions around the world (Cammarano & Tian, 2018; Harrison et al., 2014; Jin et al., 2017; Lobell et al., 2015). While process-based crop models can provide a comprehensive understanding of the timing, frequency, and intensity of ECEs on crop growth (Watson et al., 2017), they have limitations. Some of the limitations relate to oversimplification or vague description of certain process and uncertainties in parameterization, which can lead to inaccurate results (Eitzinger et al., 2013). These limitations are especially obvious in simulating ECEs (Barlow et al., 2015). For example, heat stress impacts are particularly poorly captured in crop models (Fischer, 2011). Most crop models simulate the effects of high temperature on leaf senescence and stem carbohydrate accumulation and distribution, rather than directly model damage to reproductive organs and processes. This raises uncertainty over the application of crop models to properly account for yield losses resulting from ECEs and the validity in assessing the long-term impacts of ECEs under climate change (Schauberger et al., 2017).

Statistical models use various regression methods to link historical yields to histor-ical climate data which are then used to make predictions about yields under altered climate conditions (Schlenker & Roberts, 2009). They are easy to handle and relatively easy to compute. With the increasing availability and improved quality of observed data, statistical models usually have a high level of accuracy (Folberth et al., 2019; Innes et al., 2015). Moreover, newly emerging machine learning algorithms may improve the ability of statistical models to explore climate–yield relationships (Chlingaryan et al., 2018). Machine learning algorithms are capable of disentangling the effects of co-linear climate variables and analyzing hierarchical and nonlinear relationships between the predictors and the response variable (Shalev-Shwartz & Ben-David, 2014), which usu-ally results in better model performance compared with conventional linear regression models (Feng et al., 2018; Jeong et al., 2016). However, a major limitation of statistical models is that they usually only provide empirical associations among variables, rather than provide a causal understanding of the physiological constraints required to inform adaptation strategies (Roberts et al., 2017).

Research Objectives

In recent years, the value of combining both process-based crop models and statistical models is gaining recognition. Pagani et al. (2017) incorporated outputs from the sugar-cane model Canegro (Inman-Bamber, 1991) and agro-climatic indicators into multiple linear regressions (MLRs) to reproduce recorded yield. Their results showed that the combined model increased prediction accuracy by approximately 20% compared with each individual model. Everingham et al. (2016) obtained similarly higher levels of accuracy by combining a process-based crop model, the Agricultural Production System sIMulator (APSIM) model (Holzworth et al., 2014), with the random forest (RF) algorithm (Breiman, 2001). Guzmán et al. (2018) combined the DSSAT model (Jones et al., 2003) and a support vector regression model for a comprehensive assess-ment of groundwater variability to demonstrate that the hybrid model performed better in characterizing groundwater variability. In evaluating the effects of ECEs on wheat yield, the existing state-of-art studies are still based on either crop models (Cammarano & Tian, 2018) or statistical models (García-León et al., 2019).

In this study, we combined the APSIM model and the RF algorithm to build a hybrid model to evaluate impacts of ECEs on wheat yields. The main objectives were to (a) develop a hybrid model to reproduce historical observed wheat yields in the New South Wales wheat belt, (b) quantify the relative importance of growth stage-specific drought, heat, and frost events in determining wheat yields, and (c) compare the yield differences projected by the APSIM alone and the hybrid model under future climate change.

Materials and Methods

Study Sites

The New South Wales wheat belt (Figure 8.1) is located in south-eastern Australia, with its western border bounded by the semi-arid interior. It accounts for nearly 30% of the total area planted to wheat in Australia (www.abares.gov.au, 2013–2014), making it important for both domestic and international food security (Ray et al., 2015.

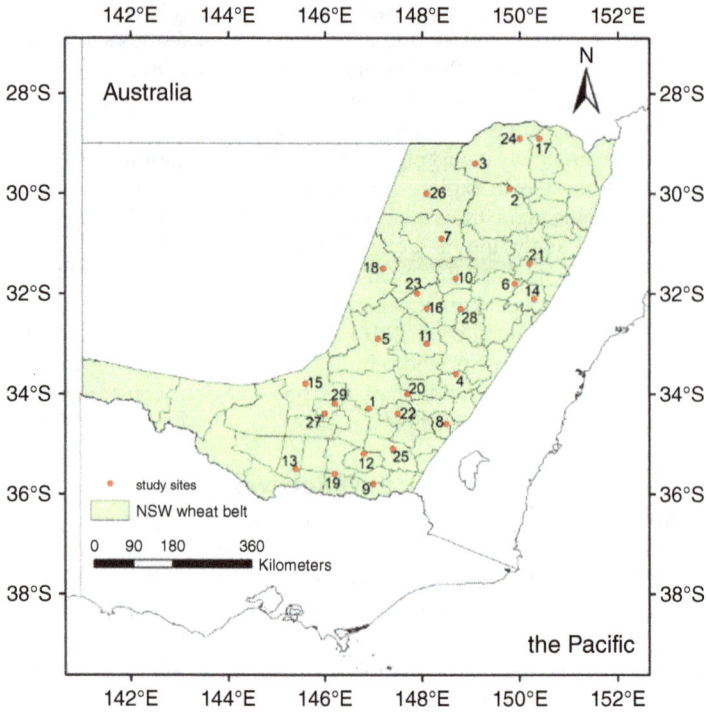

Figure 8.1 Locations of the 29 study sites in the New South Wales wheat belt in south-eastern Australia.

Generally, wheat in this region is grown under rainfed conditions, and the typical growing season is from May to November (Gomez-Macpherson & Richards, 1995).

The New South Wales wheat belt is characterized by variable topography and climatic conditions. There is an east–west gradient in both elevation and precipitation/temperature. The eastern part of the wheat belt consists of mountains with elevation up to 1,100 m, and the western areas are mainly plains. The average growing season temperature ranges from 8.3 °C in the south-east to 17.1 °C in the north-west, and the average growing season precipitation ranges from 171 mm in the south-west to 763 mm in the south-east; these data are from 1961–2000 (Wang et al., 2017a). In addition, the climate is characterized by large inter-annual variability mainly due to El Niño Southern Oscillation (Murphy & Timbal, 2008; Power et al., 1998). We used 29 sites that are listed in the Grains Research and Development Corporation National Variety Trials (GRDC-NVT, www.nvtonline.com.au) and located in the wheat belt. These sites are scattered throughout most of the wheat belt (Figure 8.1) to represent the range of agro-climatic zones across this area. We used NVT datasets because these trials were conducted in recent years and had detailed experimental records to calibrate and validate our model. A brief description of the 29 study sites, including climate and annual mean wheat yield, is shown in Table 8.1. In addition, there are hundreds of reference sites across Australia with soil information files available in the APSoil database (http://www.asris.csiro.au), which can be used as inputs to the APSIM model. We selected 27 soil files (Table 8.1) that were identified to be geographically closest to the 29 study sites.

Table 8.1 A Brief Description of the 29 Study Sites, Including Location, Soil Number (Details at http://www.asris.csiro.au), GSR (mm), GST (°C), HY, HDR, and AMWY (t ha⁻¹)

ID	Site	Soil no.	GSR	GST	HY	HDR	AMWY
			mm	°C			t ha⁻¹
1	Beckom	543	298	12.9	2009–2016	—	3.12
2	Bellata	83	307	15.9	2009–2011, 2013–2016	—	3.77
3	Bullarah	126	252	17.1	2009, 2010, 2013, 2016	—	3.48
4	Canowindra	703	342	12.6	2008, 2011, 2013, 2014, 2017	—	3.73
5	Condobolin	688	290	12.3	2008, 2010–2014, 2016	2016	2.66
6	Coolah	868	507	10.9	2008–2012, 2014–2016	2008–2010	4.82
7	Coonamble	247	267	15.4	2008–2014, 2016, 2017	—	3.44
8	Galong	545	392	10.9	2008, 2010, 2011, 2013–2016	2016	4.52
9	Gerogery	176	386	12.2	2008–2010, 2013, 2014, 2017	2017	5.15
10	Gilgandra	249	313	14.3	2008–2013, 2015–2017	—	3.35
11	Goonumbla	193	331	13.6	2008–2013, 2015, 2016	—	4.08
12	Lockhart	539	312	12.9	2008, 2009, 2011–2013, 2015	2016	3.53
13	Mayrung	538	251	13.1	2010, 2016	—	5.64
14	Merriwa	868	308	13.3	2008–2016	—	3.91
15	Merriwagga	696	233	14.2	2010, 2012, 2015, 2016	—	3.13
16	Narromine	686	289	14.5	2016	—	5.36
17	North Star	237	270	16.3	2010, 2011, 2016	—	4.16
18	Nyngan	246	253	15.5	2011–2017	—	2.80
19	Oaklands	186	306	12.8	2008–2017	—	3.60
20	Quandialla	693	317	13.1	2008, 2010–2016	—	3.91
21	Spring Ridge	127	314	14.4	2008, 2012, 2013, 2015, 2016	2009–2012	4.47
22	Temora	913	305	12.2	2010, 2011, 2013–2016	—	4.16
23	Trangie	683	273	14.7	2010–2013, 2015, 2017	2008	2.97
24	Tulloona	865	263	16.7	2009–2013, 2015, 2016	—	3.41
25	Wagga wagga	498	364	12.1	2010–2016	2010–2012	3.52
26	Walgett	1,016	239	16.3	2014–2016	—	3.17
27	Willbriggie	697	249	13.7	2010	—	5.66
28	Wongarbon	685	347	13.5	2008–2011, 2016	2008	3.54
29	Yenda	697	265	13.8	2015, 2016	—	4.22

Note: GSR, growing season rainfall; GST, growing season temperature; HY, harvest year; HDR, heading date record year; AMWY, and annual mean wheat yield.

APSIM Descriptions

APSIM version 7.7 (http://www.apsim.info) was used to simulate historical and future wheat phenology, final biomass, and yield at the 29 sites. The APSIM wheat module was developed in Australia and has been applied to numerous studies across the Australian wheat belt (Asseng et al., 2011; Chenu et al., 2013; Lobell et al., 2015). In particular, the wheat module has been shown to adequately simulate a number of processes at a daily time step—including phenological development, biomass accumulation, and yield formation—for multiple cultivars, soil moisture and nutrient status, weather conditions, and farming management practices (Jin et al., 2017). Phenological development is determined by temperature and cultivar, as mentioned above. Biomass accumulation is determined by both radiation interception and soil water limitation, while yield formation is calculated based on a simple assimilate partitioning rule (Tao et al., 2017).

In the APSIM wheat module, several simplified descriptions of certain processes have been defined to regulate the effects of ECEs on crop growth (Barlow et al., 2015). Frost and heat events are incorporated by stress functions which can lead to leaf senescence, while drought (water stress) events are defined as functions that can restrain leaf expansion and biomass accumulation. Detailed descriptions of these functions have been described by Zheng et al. (2014). It should be noted that these functions are mostly simple and linear. Moreover, many other forms of damage to crop growth caused by ECEs are not considered in APSIM, such as frost and heat-induced sterility around anthesis. In other words, APSIM simplifies several processes and ignores some limitations on yield, which may result in poorly modeling ECEs (Barlow et al., 2015).

Climate Data

Historical (2008–2017) daily climate data (rainfall, maximum and minimum air temperature, and solar radiation) for the 29 study sites were downloaded from Scientific Information for Land Owners patched point dataset (SILO-PPD; https://www.long-paddock.qld.gov.au/silo/) (Jeffrey et al., 2001).

Future (2020–2100) climate data were obtained for each of the 29 sites from 34 different GCMs (Table 8.2). The monthly climate data from these GCMs are provided by different climate modeling institutions all over the world. Detailed descriptions of these GCMs can be found at the CMIP5 website (https://pcmdi.llnl.gov/mips/cmip5/). Generally, raw GCMs are at coarse temporal (monthly) and spatial (100–300 km grid spacing) resolutions and therefore cannot be directly used to feed site-based crop models. Here we used a statistical downscaling method (NWAI-WG;

Table 8.2 List of 34 Global Climate Models (GCMs) Under RCP4.5 and RCP8.5 Future Climate Scenarios used in this Study for Statistical Downscaling Outputs of the 29 Sites over the New South Wales Wheat Belt in South-Eastern Australia

Model ID	Name of GCM	Abbr. of GCM	Institute ID	Country
1	ACCESS1-0	AC1	CSIRO and BoM	Australia
2	ACCESS1-3	AC2	CSIRO and BoM	Australia
3	BCC-CSM1-1	BC1	BCC	China
4	BCC-CSM1-1-m	BC2	BCC	China
5	BNU-ESM	BNU	GCESS	China
6	CanESM2	CaE	CCCMA	Canada
7	CCSM4	CCS	NCAR	USA
8	CESM1-BGC	CE1	NSF-DOE-NCAR	USA
9	CESM1-CAM5	CE2	NSF-DOE-NCAR	USA
10	CESM1-WACCM	CE5	NSF-DOE-NCAR	USA
11	CMCC-CM	CM2	CMCC	Europe
12	CMCC-CMS	CM3	CMCC	Europe
13	CNRM-CM5	CN1	CNRM-GAME	France
14	CSIRO-Mk3-6-0	CSI	CSIRO-QCCCE	Australia
15	EC-EARTH	ECE	EC-EARTH	Europe
16	FIO-ESM	FIO	FIO	China
17	GISS-E2-H	GE1	NASA GISS	USA
18	GISS-E2-H-CC	GE2	NASA GISS	USA
19	GISS-E2-R	GE3	NASA GISS	USA
20	GFDL-CM3	GF2	NOAA GFDL	USA
21	GFDL-ESM2G	GF3	NOAA GFDL	USA
22	GFDL-ESM2M	GF4	NOAA GFDL	USA
23	HadGEM2-AO	Ha5	MOHC	UK
24	INM-CM4	INC	INM	Russia
25	IPSL-CM5A-LR	IP1	IPSL	France
26	IPSL-CM5A-MR	IP2	IPSL	France
27	IPSL-CM5B-LR	IP3	IPSL	France
28	MIROC5	MI2	MIROC	Japan
29	MIROC-ESM	MI3	MIROC	Japan
30	MIROC-ESM-CHEM	MI4	MIROC	Japan
31	MPI-ESM-LR	MP1	MPI-M	Germany
32	MRI-CGCM3	MR3	MRI	Japan
33	NorESM1-M	NE1	NCC	Norway
34	NorESM1-ME	NE2	NCC	Norway

Note. Details of the 34 GCMs can be found at https://pcmdi.llnl.gov/mips/cmip5/availability.html.

Liu & Zuo, 2012), to downscale the monthly gridded data simulated by raw GCMs to daily climate data for each of the 29 sites. This approach has been frequently used in recent climate change research (Liu et al., 2017; Wang et al., 2015, 2017b). In addition, two representative concentration pathways (RCP4.5 and RCP8.5) are available in the GCMs dataset and were utilized in this study.

In-Situ Trial Data

The GRDC-NVT is a national program of comparative crop variety testing with standardized trial management, data generation, collection, and dissemination. Crop variety trial data from the GRDC-NVT have been frequently used by the scientific community in recent years (Dreccer et al., 2018; Shen et al., 2018; Zheng et al., 2012). In the present study, we used in-situ wheat trial data (2008–2017) for the 29 sites across the wheat belt. These data include sowing date, heading date (only available at several sites and years; Table 8.1), and yield for dozens of wheat varieties. In addition, soil nutrient status (including total nitrogen, phosphorous, organic carbon, pH, and conductivity) and fertilization practice (including date and fertilizer type) were also available. We chose four wheat varieties—Suntop, Sunvale, Ventura, and Wallup—which are widely cultivated across the wheat belt and have also been well parameterized on the vernalization and photoperiod response in the APSIM model. The parameters for vernalization sensitivity were 2.0, 2.8, 1.5, and 1.5 for each variety, respectively, and for photoperiod sensitivity, 3.5, 3.0, 3.0, and 3.5, respectively. In the APSIM-Wheat module, photoperiod and vernalization are two important factors that determine wheat phenology (Keating et al., 2003). As a result, the four selected varieties tended to have different dates of growing stages which would result in differences in responses to growth stage-specific ECEs. Thus, using the four varieties will enable a comprehensive evaluation of impacts of various ECEs on wheat yields. In addition, as each variety was only available in parts of years, we eventually collected 516 trials data. The yield data did not show an obvious skewed distribution and could be used for regression analysis. In addition, as these data were experimental data and very recent, no significant technological trend was detected after examination. Thus, no de-trending method was applied to remove factors (e.g., changes in management practices, pesticide application) not reproduced by the modeling solution (Pagani et al., 2017).

APSIM Simulations

The four varieties used in this study are available in the APSIM variety bank. In addition, there are more than 800 soil profiles in the APSoil database (http://www.asris.csiro.au) available for Australian agricultural areas. Most of these soils have already been parameterized for modeling wheat. We selected 28 soil profiles (Table 8.1) that are geographically closest to the study sites. The same soil was used for the Coolah and Merriwa sites. Detailed hydraulic properties of all these soils can be found at http://www.asris.csiro.au.

For the calibration dataset (2008–2017), we set up the APSIM simulations strictly according to the NVT trial data (variety, sowing date, soil nutrient status, and fertilization practice). The output yield and phenology data were then directly used for comparison with observed data. For the climate change impacts (1961–2100), APSIM simulations were driven by the 34 downscaled GCMs for each of the 29 sites. The four varieties were simulated for each site and for each GCM. For the management options, a sowing window starting on 1 May and ending on 30 June was used with the option "must sow." The fertilizer at sowing was 100 kg ha^{-1} of urea (equivalent to 46 kg ha^{-1} of N) under each future climate scenario. All other options were left as the defaults.

Effects of elevated CO_2 concentration were considered in simulations of future scenarios. In APSIM, CO_2 influences plant growth through regulating transpiration efficiency, radiation use efficiency, and critical leaf nitrogen concentration. However, APSIM has no function to generate time-varying values of CO_2 concentration. Thus, we added a function to APSIM so that it could calculate yearly atmospheric CO_2 concentrations through empirical relations of calendar year (Liu et al., 2014, 2017). For the RCP4.5 scenario, the atmospheric CO_2 concentration was calculated as:

$$[CO_2]_{year} = 650.18 + \frac{0.000075326 \times y - 0.16276}{0.00022299 - \frac{727.97}{y^2}} - 0.00018747 \times (y - 2045)^3 \tag{8.1}$$

For RCP8.5, it was fitted as:

$$[CO_2]_{year} = 1034.3 + \frac{267.78 - 1.6188 \times y}{4.0143 + \frac{53.342}{y^{3.2822}}} + 21.746 \times \left(\frac{y - 2010}{100}\right)^3 + 100.65 \times \left(\frac{y - 1911}{100}\right)^3 \tag{8.2}$$

where y is the calendar year from 1900 to 2100 (i.e., y = 1900, 1901, ..., 2100).

Climate Extremes Indicators

In APSIM, wheat cultivation is divided into 11 stages: sowing, germination, emergence, end of juvenile (EJ), floral initiation (FI), flowering (F), start of grain filling (SGF), end of grain filling, maturity, harvest rips, and end crop. In this study, we took EJ, FI, F, and SGF into consideration, as they represent the four main growing stages. In this study, we intended to evaluate the impacts of three kinds of ECEs (Table 8.3) on wheat yield at the four main growing stages. The indicators for heat and frost are simple counts of days with maximum/minimum temperatures above/below fixed thresholds (Tashiro & Wardlaw, 1989; Zheng et al., 2012). The impact of water deficit was assessed using the Agricultural Reference Index for Drought (ARID) (Woli et al., 2012). This drought index is a simple, general, soil–plant–atmosphere metric. It usually performs better than many other drought indices in agricultural drought evaluations (Woli et al., 2013).

$$ARID_i = 1 - \frac{T_i}{ET_{o,i}} \tag{8.3}$$

where i represents the ith day, T_i is the transpiration during the ith day (mm d^{-1}), and $ET_{o,i}$ is the reference evapotranspiration on the ith day (mm d^{-1}). When calculating ARID, $ET_{o,i}$ is assumed to be equal to potential evapotranspiration and can be

Table 8.3 List of Extreme Climate Events used in this Study

Extreme event	Description	Growth stage
Heat	Number of days with daily maximum temperature >27 °C	FI, F, SGF
Frost	Number of days with daily minimum temperature <0 °C	EJ, FI
Drought	Number of days with ARID >0.6	EJ, FI, F, SGF

Note. Heat events were calculated at FI, F, and SGF stages. Frost events were calculated at EJ and FI stages. Drought events were calculated at EJ, FI, F, and SGF stages. Thus, a total of nine weather extreme indicators were used in this study.

Note. EJ, end of juvenile; FI, floral initiation; F, flowering; SGF, start of grain filling; ARID, Agricultural Reference Index for Drought.

estimated using the Priestley and Taylor (1972) method. T_i is estimated through a macroscopic modeling approach which is based on the water content. Detailed descriptions and calculation processes can be found in Woli et al. (2012). ARID values fall between 0 and 1. Values higher than 0.6 are usually recognized as high water stress, thus we chose 0.6 as the threshold to evaluate daily drought condition.

To calculate these indicators, we first ran APSIM simulations and obtained wheat phenology information, including dates and duration of wheat growing stage. Then, according to the phenology information, we calculated or counted out three ECEs indicators for each of the four growth stages. Thus, we got a total 12 indicators as expected. However, we found that heat events rarely happened at the EJ stage and frost events rarely happened at the F and SGF stages during the historical period (2008–2017). Thus, we eventually selected 9 extreme events at the four growing stages (Table 8.3).

Statistical Models

Random Forest is a nonparametric and ensemble learning algorithm originated from classification and regression trees (Breiman, 2001). It is a nonparametric technique which builds multiple decision trees and combines them together to obtain a prediction. Thus, the RF usually presents good accuracy despite the possibility of missing values and outliers (Elavarasan et al., 2018). Moreover, the RF can approximate functions with both linear and non-linear relations and can also identify the relationship between the response and a variable, which is conditional upon other variables (Hoffman et al., 2018). Given that the effects of ECEs on crop yields are often nonlinear (Lobell et al., 2011; Schlenker & Roberts, 2009), RF is expected to perform well in assessing the nonlinear relationship. Our previous studies (Feng et al., 2018; 2019c; Wang et al., 2018) have demonstrated that the RF model usually performed better than many other machine learning techniques, in agricultural-based applications. In addition, RF is able to provide the relative importance of each predictor in determining the response variable. Therefore, in the present study, we intended to take advantage of RF to enhance the ability of APSIM in simulating the effects of ECEs on wheat yields.

We also used the MLR to build the hybrid model with the APSIM model. The MLR is a commonly used regression method to model the linear relationship between the independent variables and the dependent variable. It is considered to be the extension of ordinary least-squares that involves more than one explanatory variable. It is easy to understand and implement, but usually limited in disentangling the nonlinear relationships between the predictors and the response.

Hybrid-Modeling Approach

Figure 8.2 illustrates the processes of combining the APSIM and RF (or MLR) models in our study. First, the APSIM wheat module was run to simulate wheat phenology, biomass and yield based on NVT trial datasets. The outputs of phenology date were then used as references for the calculation of the nine indicators at the four growth stages. Lastly, APSIM simulated biomass and the nine indicators were applied as input predictors in RF (or MLR) for estimating wheat yield. In this study, we proposed the RF (or MLR) model as a post-hoc bias correction which was expected to help improve the performance of APSIM model in simulating the effects of growth stage-specific ECEs.

We performed the RF model using the R package "RandomForest" (Liaw & Wiener, 2002). Two parameters were needed to be determined before the implement of the model, that is, n_{tree} (the number of trees to grow in the forest) and m_{try} (the number of randomly selected predictor variables at each node). We set the n_{tree} as the default

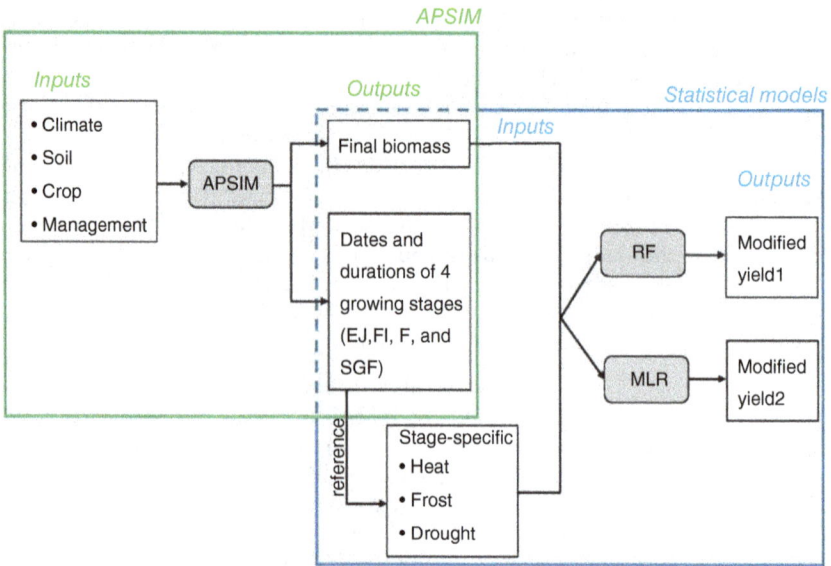

Figure 8.2 Diagram of the input and output per model for the APSIM + RF (or MLR) hybrid model applied in this study. EJ, end of juvenile; FI, floral initiation; F, flowering; SGF, start of grain filling; RF, random forest; MLR, multiple linear regression.

value of 500. As m_{try} could affect the model accuracy, we adopted a trial-and-error analysis to determine the value of m_{try}. Values of 1–10 were tried, and 5 was chosen as it led to a little higher model accuracy. The relative importance of variables was assessed through the "%IncMSE" metric in the RF model. In addition, the MLR model was performed using the R package "Rattle" (Williams, 2011).

Model Performance Assessment

The NVT trial data (516 trials, 2008–2017) were used to calibrate the models. The output yields from the APSIM model were directly used for comparison with the observed data. For the RF (or MLR) model, a 10-fold cross validation approach was applied to the 516 data points. The coefficient of determination (R^2) and root mean square error (RMSE) were used for model evaluation following:

$$R^2 = \left(\frac{\sum_{i=1}^{n} (O_i - \overline{O})(P_i - \overline{P})}{\sqrt{\sum_{i=1}^{n} (O_i - \overline{O})^2} \sqrt{\sum_{i=1}^{n} (P_i - \overline{P})^2}} \right)^2 \tag{8.4}$$

$$RMSE = \sqrt{\frac{\sum_{i=1}^{n} (O_i - P_i)^2}{n}} \tag{8.5}$$

where n is the number of samples, O_i and P_i denote observed and simulated values, respectively, and $\overline{O}$ represents the mean of observed values. Generally, the model with higher R^2 and lower RMSE is considered to be the more accurate model.

Results

Model Performance

The performance of the APSIM wheat module was evaluated by comparing the simulated wheat grain yields and flowering dates to the observed values (Figure 8.3). Agreement between simulations and observations was described by the RMSE, the coefficient of determination (R^2), and the slope of the regression lines. As shown in Figure 8.3a, the simulated flowering dates were consistent with observed dates, with an RMSE of 5.01 days ($R^2 = 0.82$, $y = 0.72x + 76.71$, $P < 0.01$), to suggest that the APSIM was able to provide a satisfactory estimation of wheat flowering dates. As flowering stage is generally viewed as the indicative stage of the entire wheat phenology, APSIM might also provide acceptable estimations of the other three growth stages. This was the foundation for our subsequent calculations of stage-specific ECEs indicators. For wheat yields, the model was able to explain 61% of the variation and the RMSE was 0.86 t ha^{-1}. This was a common and acceptable result for large-scale crop model simulations (Jin et al., 2017), even though the accuracy was not high. In addition, both slope coefficients (0.72 and 0.78) were lower than 1.0, indicating that the model tended to overestimate both flowering date and yield when the values were higher. In general, inaccurate simulations were due to the absent or rough assumptions around certain factors, including pest, diseases, and weather extremes as mentioned above. In the subsequent analyses, we managed to increase the accuracy through improving the ability of simulating ECEs.

We used the MLR model and the RF model for external modification on the APSIM model to create two hybrid models for predicting wheat yield. Compared with the APSIM model alone, both of these hybrid models showed higher accuracy in reproducing the observed yields (Figure 8.4). The APSIM + RF hybrid model explained 81% of the variation in observed yields, an increase of 33% compared with the APSIM model.

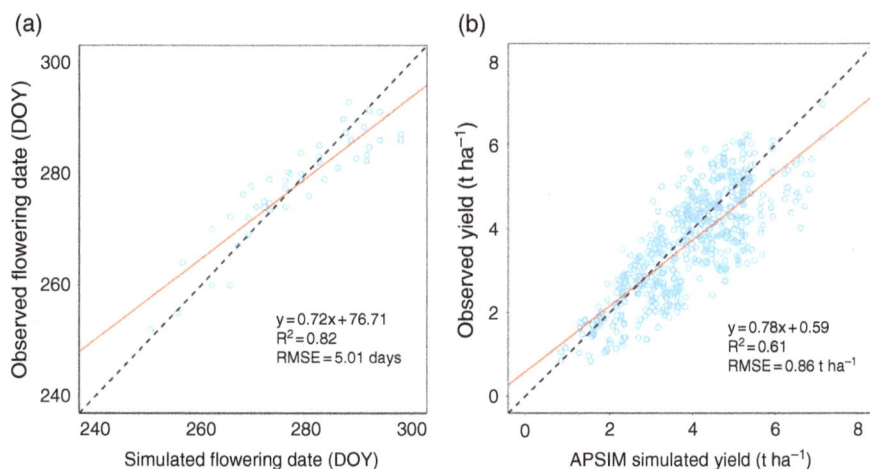

Figure 8.3 Comparison of observed and APSIM simulated values of grain yield and flowering date from 2008 to 2017 at the 29 sites across the New South Wales wheat belt. In total, 516 yield data points and 47 flowering date data points were collected from National Variety Trials of Australia (see text for more detail about this dataset). Dashed lines are the 1 to 1 ratio line. Red lines are the linear regression fit.

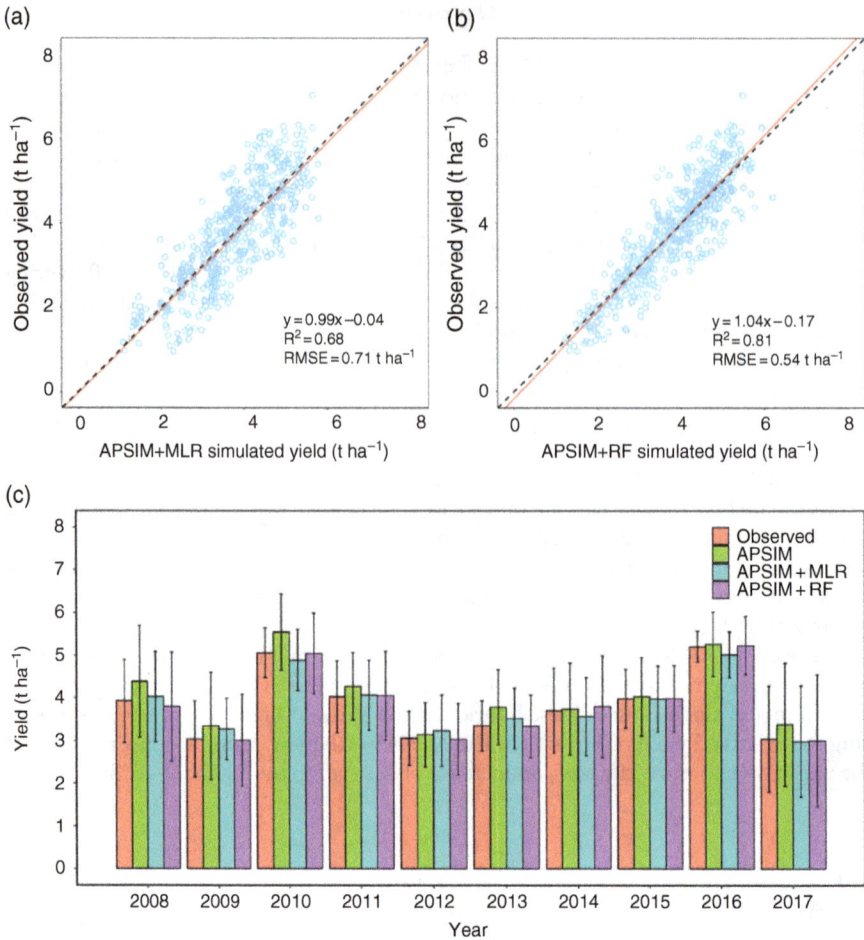

Figure 8.4 Comparison of observed, APSIM simulated, APSIM + MLR simulated, and APSIM + RF simulated wheat yields from 2008 to 2017 at the 29 sites across the New South Wales wheat belt. (a) Observed versus APSIM + MLR hybrid model simulated. (b) Observed versus APSIM + RF hybrid model simulated. (c) Time series of the four kinds of yields. In (c), error bars indicate the standard deviation from yields at the 29 sites. APSIM + RF (or MLR) simulated wheat yields were derived from the 10-fold cross validation approach.

It also reduced the RMSE by $0.32 \, \text{t ha}^{-1}$. Moreover, the slope of the regression function was close to 1.0, meaning that the APSIM + RF hybrid model was unbiased in its simulation of wheat yield for our study area. The accuracy of the APSIM + MLR hybrid model was only increased slightly compared with the APSIM model alone and was far below the APSIM + RF hybrid model. Thus, the external modification using the RF model with ECEs indicators could greatly improve the performance of the APSIM model.

Figure 8.4c shows the time series of the three kinds of wheat yields (i.e., observed, APSIM simulated, and the two hybrid models simulated) from 2008 to 2017.

The observed yields ranged from 3.0 to 5.2 t ha^{-1}, with the greatest variability on an interannual basis. In general, all of the model simulations successfully captured the temporal pattern of observed wheat yields. However, the APSIM model tended to slightly overestimate the yields in almost every year. This was particularly evident in 2008, 2010, and 2013, when an overestimation of up to 0.5 t ha^{-1} occurred. These overestimations could be attributed to an underestimation of the effects of ECEs on wheat yields. Through incorporating ECEs indicators, the APSIM + RF hybrid model succeeded in simulating yields more consistent with the observed yields.

Effects and Projected Changes of ECEs

The historical occurrence of ECEs is shown in Figure 8.5. Drought was the most common ECE during the historical period. In general, the four growth stages (i.e., EJ, FI, F, and SGF) usually last for 72–88, 37–48, 6–9, and 25–34 days, respectively, for the four cultivars. Thus, nearly one quarter, half, and half of the EJ, F, and SGF stages, respectively, experienced drought conditions. Heat commonly occurred around and post anthesis. The SGF stage was most vulnerable to heat stress, and up to two-thirds of this stage might be under heat threat. Frost events mainly occurred during the EJ stage, and a few frost events also occurred at the FI stage. Figures 8.4b and 8.5 show that there was an obvious and direct relationship between wheat yield and the occurrence of ECEs. Wheat yields were much lower during years with a higher occurrence of ECEs, such as in 2012 and 2017.

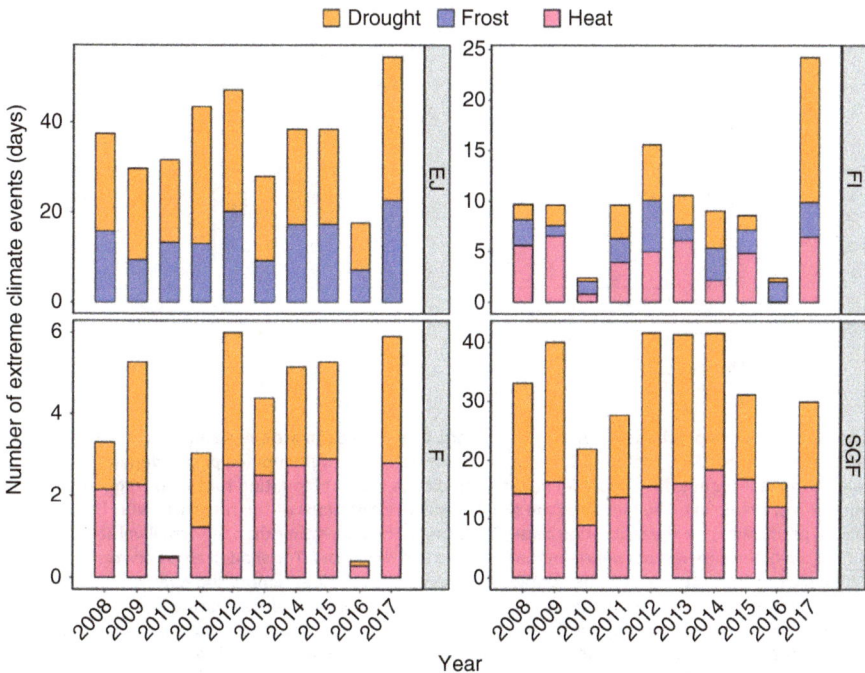

Figure 8.5 Number of extremes climate events occurred from 2008 to 2017. Values for each year were averaged values of the 29 study sites.

According to the above results, the RF model could potentially improve the performance of the APSIM model in simulating the effects of ECEs. We obtained the relative importance (percentage values in Figure 8.6) and the marginal effect (lines in Figure 8.6) of each predictor from the RF model. The trend of the line, rather than the actual values, describes the nature of the dependence between the response and the predictor variables. All ECEs, except frost during the EJ stage, had negative effects

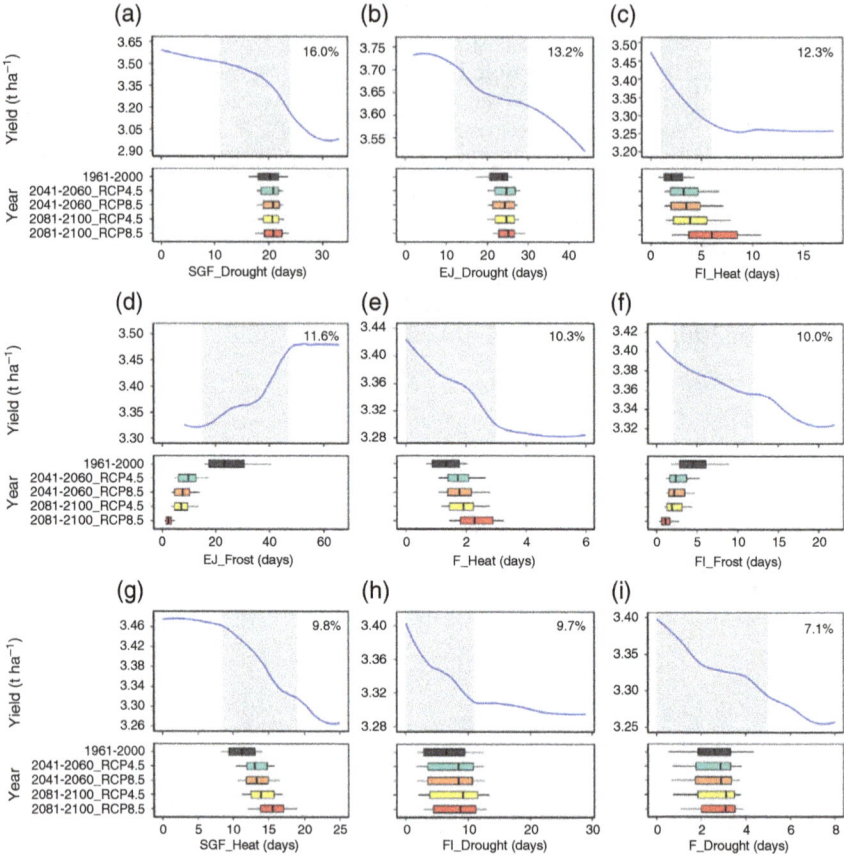

Figure 8.6 Partial dependence of wheat yield change on extreme climate events and projected changes in each event under RCP4.5 and RCP8.5. The random forest model could provide partial dependence of the change in the response (blue lines) for selected predictors, when accounting for the average effect of all other driver predictors. The blue lines are smoothed representations of the response, with fitted values (model predictions) for the calibration data. The trend of the line, rather than the actual values, describes the nature of the dependence between the response and predictors. The shaded area denotes calibration data between the 10th and 90th percentiles. The percentages values denote the relative importance of each predictor generated from the random forest model. The box plots indicate the occurrences of extreme climate events during the baseline (1961–2000) period and two future periods (2041–2060 and 2081–2100) based on the 34 downscaled GCMs. Box boundaries indicate the 25th and 75th percentiles across GCMs, whiskers below and above the box indicate the 10th and 90th percentiles. The black lines within each box indicate the multi-model median. Growth stages EJ, FI, F, and SGF indicate end of juvenile, floral initiation, flowering, and start of grain filling, respectively.

on wheat yield. Drought events which occurred at the SGF and EJ stages had high importance values, meaning that they were more harmful to wheat growth. The third impact was heat events at the FI stage. Thus, even though heat events were more common during the SGF stage (Figure 8.5), heat tended to cause more yield loss if it occurred at the FI stage. Frost events at the EJ stage had positive effects on wheat yield. This may be because winter wheat plant is capable of withstanding extreme cold before the initiation of flowering (Fowler & Carles, 1979). Moreover, the plants also require some exposure to cool temperatures for jarovization, which affects subsequent growth and development (Robertson et al., 1996). The responses of wheat yield to other ECEs at a same stage were different. For example, the F stage was more vulnerable to heat than drought, whereas the SGF stage was more vulnerable to drought than heat.

The shaded area of Figure 8.6 denotes the 10th to 90th percentiles of each variable in the calibration dataset. All lines have a sharp change in the shaded area, meaning that a small change of each ECE may have a large impact on wheat yield. As shown in the boxplots in Figure 8.6, drought events were projected to have small increase in occurrence at the four growth stages, but heat events were likely to increase significantly. In particular, under RCP8.5, heat events during the end of the 21st century may be doubled compared with the baseline period. Frost events were likely to decrease, such as by >10 days at the EJ stage and by 1–4 days at the FI stage. However, a decrease of frost events at the EJ stage might also cause yield reductions. Thus, in general, more yield losses were indicated as a result of changes in future ECEs.

APSIM Projected Versus Hybrid Model Projected Future Wheat Yields

The RF model performed better in improving the performance of the APSIM model compared with MLR. We used the APSIM model and the APSIM + RF model to evaluate the impacts of future climate change on wheat yield in the study area. Projected changes in simulated wheat yield from the APSIM model and the hybrid model for two of the study sites are shown in Figure 8.7. We calculated changes of simulated wheat yields for each site and found that trends from APSIM-simulated wheat yield differed among sites. However, the APIM + RF hybrid model-simulated yield was projected to decrease in all study sites. For example, Figure 8.7 shows the APSIM-simulated yields were projected to decrease by 0.5–3% at Bellata but increase by 1.5–3% at Mayrung. This might be due to different soil conditions and climate projections at the two sites. However, the APSIM + RF hybrid model-simulated wheat yields were projected to decrease at both sites. According to the multi-model ensemble mean values (2041–2060), the APSIM + RF hybrid model-simulated wheat yields were 4 and 3% lower than the baseline levels at Bellata and Mayrung, respectively. Moreover, the yield reductions magnified over time. The differences between the two models were mainly caused by different responses to ECEs. Climate change might result in various trends of the APSIM projections at different sites, but the ECEs changes, especially the increase of heat and drought events (Figure 8.6), were most likely to reduce the yields to lower levels.

Discussion

The comparison of results obtained from the APSIM model and the hybrid model showed that the APSIM + RF hybrid model is better at reproducing historical wheat yields. Using the RF algorithm as an external modification on the APSIM model outputs appeared to improve the performance of the individual APSIM model (Figures 8.3

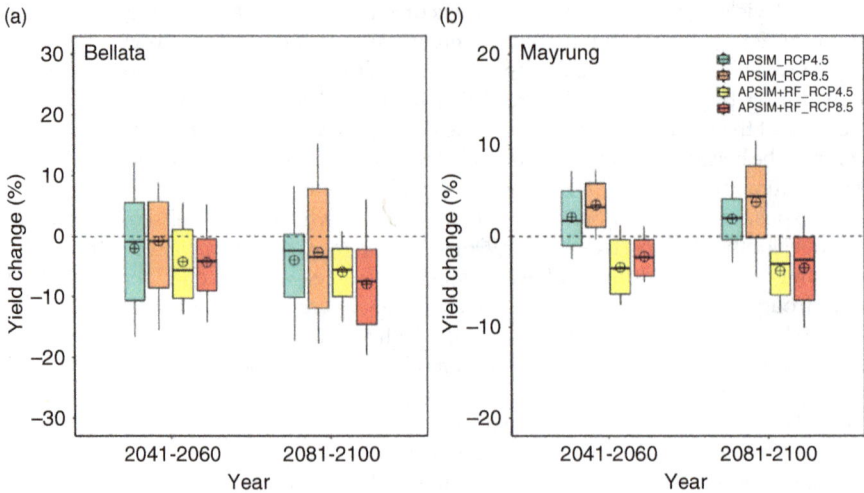

Figure 8.7 Projected changes in simulated wheat yield from the APSIM model and the APSIM + RF hybrid model for two of the study sites. Changes were estimated between two future periods (2041–2060 and 2081–2100) and the baseline period (1961–2000) under RCP4.5 and RCP8.5 based on the 34 downscaled GCMs. Box boundaries indicate the 25th and 75th percentiles across GCMs, whiskers below and above the box indicated the 10th and 90th percentiles. The black lines and crosshairs within each box indicated the multi-model median and mean respectively.

and 8.4). Moreover, the RF model also outperformed the MLR model. Everingham et al. (2016) conducted a similar study by incorporating the biomass simulated by the APSIM model and several climate indices into the RF algorithm to simulate sugarcane yield and also obtained a high R^2 of ~0.8. The most likely explanation is that the external statistical model may help improve the performance of the crop model by simulating the effects of these selected climate indices. Using the crop model outputs as predictors may also improve the ability of the statistical model to consider more agro-climatic processes. Keating and Thorburn (2018) proposed the possible research trend of blending mechanistic and empirical statistical models. The approach outlined in this paper can be viewed a feasible method which may be easily extended to other wheat growing regions to obtain new insights to guide agricultural practices.

In recent years, researchers have been concerned with ECEs because of their remarkable damage to crops (Lesk et al., 2016; Rezaei et al., 2015). It was reported that nearly one-quarter of all damage and losses in the agriculture sector are caused by ECEs in many countries (FAO, 2015). As ECEs may occur during different stages of a crop cycle and cause varying degrees of yield losses, it is necessary to identify the most harmful ECEs in a particular region. In our study, drought events during grain-filling and vegetative stages had the greatest adverse effects on wheat yields across a wide range of agro-climatic zones. Drought events around anthesis may have relatively small effects on wheat yields. One probable reason is that yield formation mainly occurs at the grain filling stage (Royo et al., 2006). Moreover, the number of drought events occurring around anthesis is relatively low. A short-term drought event is not likely to severely reduce final yield (Clarke et al., 1992).

In contrast, the impacts of heat and frost events on wheat yield are usually direct and rapid (Barlow et al., 2015). We found heat events around anthesis (FI_Heat and F_Heat) resulted in a large negative impact on wheat yields, even though they might only occur over a few days in the growing season; this was consistent with the studies of Balla et al. (2009) and Hays et al. (2007). The sensitivity of wheat to various ECEs varies with different growth stages (Hlaváčová et al., 2018). During anthesis, the awns or spikes start to emerge from the flag leaf as the grain begins to form. However, even a single frost or heat event may cause sterility and aborted grains, thereby reducing the number of grains in the inflorescence. The adverse impacts from heat events will be reduced during the grain-filling period, as grains have already formed (Barlow et al., 2015).

The ECEs that occurred during the historical period have already resulted in large yield losses, so their future impact through climate change needs to be assessed. In our study, we found that drought frequency is expected to have a slight increase, while heat events around anthesis and grain-filling periods will likely be more common. Semenov and Shewry (2011) obtained similar results in a study in Europe and demonstrated that wheat plants are expected to suffer more heat stress than drought in the future. Given the severe and rapid damage caused by heat stress around anthesis, a small increase in heat events is likely to cause great wheat yield losses. On the other hand, while fewer frost days are expected because of global warming, we found that frost events during vegetative stages had a positive effect on wheat yield across our study region. Thus, in general, ECE changes will result in future increased risk for wheat production in the study area.

Given the more unfavorable weather conditions in the future, the possible trend of future wheat yields is frequently discussed among researchers. The most commonly used method to assess climate impacts is a combination of process-based crop models and GCMs. For example, Qian et al. (2016) reported that an average increase in wheat yields of 26–37% is to be expected during 2041–2070 compared with a baseline period (1971–2000) in the Canadian Prairies using the CERES-Wheat model and two GCMs. Wang, Liu, O'leary, et al. (2017b) conducted a study in Australia using the APSIM model and 11 GCMs to demonstrate that there would be a decrease in the yield in the south-eastern Australian wheat belt throughout the 21st century. The majority of these and similar studies used one single crop model. However, model comparison studies, especially for climate change studies, have emphasized the limited ability of crop models to account for ECEs. Sánchez et al. (2014) observed that a number of crop models together adequately predicted mean yields, but were less able to predict extreme low yields, due to their inability to handle ECEs. Hochman et al. (2013) reported that the APSIM model poorly accounted for ECEs such as severe frost and was also overly optimistic about water limited yield impacts in some seasons and locations. Therefore, it is questionable whether crop model-based projections accurately reflect the direction and magnitude of the effects of climate change on yield.

In our study, we used the hybrid model which systematically incorporated the APSIM model output and ECEs indicators through RF algorithm to predict future wheat yields; we found that a single APSIM model may have 1–10% overestimation for future yield projections compared with the APSIM + RF hybrid model. This overestimation was mainly caused by underestimating the ECEs-induced yield losses. This is likely to be a common phenomenon in crop model projections, as the most popular crop models poorly account for impacts of ECEs (Eitzinger et al., 2013). However, as future ECEs are projected to increase in most parts of the world (IPCC, 2012), previous projections based on crop models might overestimate the most likely yield level in the future. Thus, regardless of whether an increase or a decrease of wheat yield is projected in a particular region using crop models, the most likely achieved yields may be lower due to the underestimation of ECEs-induced yield losses.

Appropriate adaptation strategies can be developed in order to maintain and improve wheat yields in the face of current and future increasing ECEs in the New South Wales wheat belt. The two kinds of strategies frequently discussed among researchers are to either minimize or escape the adverse effects of ECEs. In terms of minimizing the adverse effects of ECEs, the main approach is to increase the resistance of crops through breeding. Our study suggests that heat tolerance traits will be important. Stratonovitch and Semenov (2015) also reported heat tolerance around flowering in wheat as a key trait for increased yield potential in Europe under climate change. On the other hand, escaping the adverse effects of climate extremes is also a potential approach, which mainly aims to stagger the reproductive stages to avoid suffering ECEs. Adjusting sowing date is currently the most effective farming management practice to avoid negative impacts of ECEs. Optimizing sowing dates can lead to suitable duration of the pre-anthesis period for biomass accumulation and can lead to suitable flowering and grain-filling windows to avoid frost, heat, and terminal drought (Bell et al., 2014). Shortening the whole growth period may also help avoid suffering terminal heat and drought events, but yield might subsequently decline. Therefore, more studies that account for both breeding and selection approaches as well as farming practices are required to maintain and improve wheat yields in face of increasing ECEs.

Conclusions and Limitations

Impacts of heat, frost, and drought events on current and future wheat yields were analyzed in this study based on a hybrid model which incorporated the APSIM model output and climate extremes indicators into the RF algorithm. The likely changes of the climate extremes were also discussed. The major conclusions and limitations are as follows:

1. The APSIM + RF hybrid model could explain 81% of the observed yield variations in the New South Wales wheat belt, which had a 33% improvement in modeling accuracy compared with the APSIM model alone. We believe this proposed model is worthy of consideration for future research and would provide some useful information for local farmers and policy makers with respect to development of adaptation strategies in face of increased climate extremes under climate change.

2. Drought and pre-anthesis heat events are the major climate extremes causing current wheat yield loss in the New South Wales wheat belt of south-eastern Australia. The future climate in the New South Wales wheat belt is expected to be increasingly less favorable. Drought events are projected to remain at historical levels, while heat events are projected to increase by 100% at the end of the twenty-first century. Frost events during the vegetative and pre-anthesis stages will decrease by >10 days and by 1–4 days.

3. Future yield projections from conventional process-based crop models might have a 1–10% overestimation because of underestimated climate extreme-induced yield loss.

4. The hybrid yield monitoring model satisfactorily estimates yield in the New South Wales wheat belt. However, this model requires a large amount of data from different sources, including soil, climate, crop, and farming management information. Thus, it may not be suitable for data-poor areas. Nevertheless, with societal and

economic developments, more and more areas will gain sufficient data to implement this yield monitoring method. In addition, ENSO-related indices are also common indicators for yield monitoring but are not considered in this project. Future studies using a similar modeling approach may use additional information (such as ENSO-related indices) and potentially achieve even greater monitoring accuracy.

5. Multiple GCMs are used in this study to provide projections of wheat yield to the end of the twenty-first century. Projected yield under future climate scenarios could be significantly different due to the differences in GCMs, RCPs, and simulation models, indicating large uncertainties in future yield projections. Future research should quantify the uncertainties originating from various sources in order to evaluate the reliability of yield projections.

6. In this study, machine learning models show better performance in wheat yield monitoring compared with traditional regression models. However, it should be noted that machine learning as applied in this study is a post-hoc bias correction, which does not directly improve the performance of the APSIM model itself. As a possibility, the information obtained from machine learning application can be used as some empirical formulas within the process-based model.

References

Al-Issawi, M., Rihan, H. Z., El-Sarkassy, N., & Fuller, M. P. (2013). Frost hardiness expression and characterisation in wheat at ear emergence. *Journal of Agronomy and Crop Science*, 199(1), 66–74.

Asseng, S., Foster, I., & Turner, N. C. (2011). The impact of temperature variability on wheat yields. *Global Change Biology*, 17(2), 997–1012.

Balla, K., Bencze, S., Janda, T., & Veisz, O. (2009). Analysis of heat stress tolerance in winter wheat. *Acta Agronomica Hungarica*, 57(4), 437–444.

Barlow, K. M., Christy, B. P., O'Leary, G. J., Riffkin, P. A., & Nuttall, J. G. (2015). Simulating the impact of extreme heat and frost events on wheat crop production: A review. *Field Crops Research*, 171, 109–119.

Bell, L., Lilley, J., Hunt, J., & Kirkegaard, J. (2014). Optimising grain yield and grazing potential of crops across Australia's high rainfall zone: A simulation analysis. 1. Wheat. Crop and Pasture. *Science*, 66(4), 332–348.

Breiman, L. (2001). Random forest. *Machine Learning*, 45, 5–32.

Cammarano, D., & Tian, D. (2018). The effects of projected climate and climate extremes on a winter and summer crop in the Southeast USA. *Agricultural and Forest Meteorology*, 248, 109–118.

Chaves, M. M., Pereira, J. S., Maroco, J., Rodrigues, M. L., Ricardo, C. P. P., Osório, M. L., Carvalho, I., Faria, T., & Pinheiro, C. (2002). How plants cope with water stress in the field? *Photosynthesis and growth Annals of Botany*, 89(7), 907–916.

Chenu, K., Deihimfard, R., & Chapman, S. C. (2013). Large-scale characterization of drought pattern: A continent-wide modelling approach applied to the Australian wheatbelt—Spatial and temporal trends. *New Phytologist*, 198(3), 801–820.

Chenu, K., Porter, J. R., Martre, P., Basso, B., Chapman, S. C., Ewert, F., Bindi, M., & Asseng, S. (2017). Contribution of crop models to adaptation in wheat. *Trends in Plant Science*, 22(6), 472–490.

Chlingaryan, A., Sukkarieh, S., & Whelan, B. (2018). Machine learning approaches for crop yield prediction and nitrogen status estimation in precision agriculture: A review. *Computers and Electronics in Agriculture*, 151, 61–69.

Clarke, J. M., DePauw, R. M., & Townley-Smith, T. F. (1992). Evaluation of methods for quantification of drought tolerance in wheat. *Crop Science*, 32(3), 723–728.

Crimp, S. J., Zheng, B., Khimashia, N., Gobbett, D. L., Chapman, S., Howden, M., & Nicholls, N. (2016). Recent changes in southern Australian frost occurrence: Implications for wheat production risk. *Crop and Pasture Science*, 67(8), 801–811.

Dreccer, M. F., Fainges, J., Whish, J., Ogbonnaya, F. C., & Sadras, V. O. (2018). Comparison of sensitive stages of wheat, barley, canola, chickpea and field pea to temperature and water stress across Australia. *Agricultural and Forest Meteorology*, 248, 275–294.

Eitzinger, J., Thaler, S., Schmid, E., Strauss, F., Ferrise, R., Moriondo, M., Bindi, M., Palosuo, T., Rötter, R., Kersebaum, K. C., & Olesen, J. E. (2013). Sensitivities of crop models to extreme weather conditions during flowering period demonstrated for maize and winter wheat in Austria. *The Journal of Agricultural Science,* 151(6), 813–835.

Elavarasan, D., Vincent, D. R., Sharma, V., Zomaya, A. Y., & Srinivasan, K. (2018). Forecasting yield by integrating agrarian factors and machine learning models: A survey. *Computers and Electronics in Agriculture,* 155, 257–282.

Everingham, Y., Sexton, J., Skocaj, D., & Inman-Bamber, G. (2016). Accurate prediction of sugarcane yield using a random forest algorithm. *Agronomy for Sustainable Development,* 36(2), 27.

FAO (2015). *The impact of natural hazards and disasters on agriculture and food and nutrition security – A call for action to build resilient livelihoods.* Food and Agriculture Organization of the United Nations.

Feng, P., Liu, D. L., Wang, B., Waters, C., Zhang, M., & Yu, Q. (2019a). Projected changes in drought across the wheat belt of southeastern Australia using a downscaled climate ensemble. *International Journal of Climatology,* 39(2), 1041–1053.

Feng, P., Wang, B., Liu, D. L., Waters, C., & Yu, Q. (2019b). Incorporating machine learning with biophysical model can improve the evaluation of climate extremes impacts on wheat yield in south-eastern Australia. *Agricultural and Forest Meteorology,* 275, 100–113.

Feng, P., Wang, B., Liu, D. L., Xing, H., Ji, F., Macadam, I., Ruan, H., & Yu, Q. (2018). Impacts of rainfall extremes on wheat yield in semi-arid cropping systems in eastern Australia. *Climatic Change,* 147(3–4), 555–569.

Feng, P., Wang, B., Liu, D. L., & Yu, Q. (2019c). Machine learning-based integration of remotely-sensed drought factors can improve the estimation of agricultural drought in south-eastern Australia. *Agricultural Systems,* 173, 303–316.

Fischer, R. (2011). Wheat physiology: A review of recent developments. *Crop and Pasture Science,* 62(2), 95–114.

Folberth, C., Baklanov, A., Balkovič, J., Skalský, R., Khabarov, N., & Obersteiner, M. (2019). Spatio-temporal downscaling of gridded crop model yield estimates based on machine learning. *Agricultural and Forest Meteorology,* 264, 1–15.

Fowler, D. B., & Carles, R. J. (1979). Growth, development, and cold tolerence of fall-acclimated cereal grains 1. *Crop Science,* 19(6), 915–922.

Fuller, M. P., Fuller, A. M., Kaniouras, S., Christophers, J., & Fredericks, T. (2007). The freezing characteristics of wheat at ear emergence. *European Journal of Agronomy,* 26(4), 435–441.

García-León, D., Contreras, S., & Hunink, J. (2019). Comparison of meteorological and satellite-based drought indices as yield predictors of Spanish cereals. *Agricultural Water Management,* 213, 388–396.

Godfray, H. C. J., & Toulmin, C. (2010). Food security: The challenge of feeding 9 billion people. *Science,* 327 (5967), 812–818.

Gomez-Macpherson, H., & Richards, R. (1995). Effect of sowing time on yield and agronomic characteristics of wheat in south-eastern Australia. *Australian Journal of Agricultural Research,* 46(7), 1381–1399.

Guzmán, S. M., Paz, J. O., Tagert, M. L. M., Mercer, A. E., & Pote, J. W. (2018). An integrated SVR and crop model to estimate the impacts of irrigation on daily groundwater levels. *Agricultural Systems,* 159, 248–259.

Harrison, M. T., Tardieu, F., Dong, Z., Messina, C. D., & Hammer, G. L. (2014). Characterizing drought stress and trait influence on maize yield under current and future conditions. *Global Change Biology,* 20(3), 867–878.

Hays, D. B., Do, J. H., Mason, R. E., Morgan, G., & Finlayson, S. A. (2007). Heat stress induced ethylene production in developing wheat grains induces kernel abortion and increased maturation in a susceptible cultivar. *Plant Science,* 172(6), 1113–1123.

He, L., Asseng, S., Zhao, G., Wu, D., Yang, X., Zhuang, W., … Yu, Q. (2015). Impacts of recent climate warming, cultivar changes, and crop management on winter wheat phenology across the loess plateau of China. *Agricultural and Forest Meteorology,* 200, 135–143.

Hlaváčová, M., Klem, K., Rapantová, B., Novotná, K., Urban, O., Hlavinka, P., Smutná, P., Horáková, V., Škarpa, P., Pohanková, E., & Wimmerová, M. (2018). Interactive effects of high temperature and drought stress during stem elongation, anthesis and early grain filling on the yield formation and photosynthesis of winter wheat. *Field Crops Research,* 221, 182–195.

Hochman, Z., Gobbett, D., Holzworth, D., McClelland, T., van Rees, H., Marinoni, O., Garcia, J. N., & Horan, H. (2013). Reprint of "Quantifying yield gaps in rainfed cropping systems: A case study of wheat in Australia". *Field Crops Research,* 143, 65–75.

Hoffman, A. L., Kemanian, A. R., & Forest, C. E. (2018). Analysis of climate signals in the crop yield record of sub-Saharan Africa. *Global Change Biology,* 24(1), 143–157.

Holzworth, D. P., Huth, N. I., Zurcher, E. J., Herrmann, N. I., McLean, G., Chenu, K., van Oosterom, E. J., Snow, V., Murphy, C., Moore, A. D., & Brown, H. (2014). APSIM–evolution towards a new generation of agricultural systems simulation. *Environmental Modelling and Software,* 62, 327–350.

Howden, S. M., Soussana, J. F., Tubiello, F. N., Chhetri, N., Dunlop, M., & Meinke, H. (2007). Adapting agriculture to climate change. *Proceedings of the National Academy of Sciences*, 104(50), 19691–19696.

Inman-Bamber, N. (1991). A growth model for sugar-cane based on a simple carbon balance and the CERES-maize water balance. *South African Journal of Plant and Soil*, 8(2), 93–99.

Innes, P. J., Tan, D. K. Y., Ogtrop, F. V., & Amthor, J. S. (2015). Effects of high-temperature episodes on wheat yields in New South Wales, Australia. *Agricultural and Forest Meteorology*, 208, 95–107.

IPCC (2012). Managing the risks of extreme events and disasters to advance climate change adaptation. Special Report of the Intergovernmental Panel on Climate Change. *Journal of Clinical Endocrinology and Metabolism*, 18(6), 586–599.

Jeffrey, S. J., Carter, J. O., Moodie, K. B., & Beswick, A. R. (2001). Using spatial interpolation to construct a comprehensive archive of Australian climate data. *Environmental Modelling and Software*, 16(4), 309–330.

Jeong, J. H., Resop, J. P., Mueller, N. D., Fleisher, D. H., Yun, K., Butler, E. E., Timlin, D. J., Shim, K. M., Gerber, J. S., Reddy, V. R., & Kim, S. H. (2016). Random forests for global and regional crop yield predictions. *PLoS One*, 11(6), e0156571.

Ji, K., Wang, Y., Sun, W., Lou, Q., Mei, H., Shen, S., & Chen, H. (2012). Drought-responsive mechanisms in rice genotypes with contrasting drought tolerance during reproductive stage. *Journal of Plant Physiology*, 169(4), 336–344.

Jin, Z., Zhuang, Q., Wang, J., Archontoulis, S. V., Zobel, Z., & Kotamarthi, V. R. (2017). The combined and separate impacts of climate extremes on the current and future US rainfed maize and soybean production under elevated CO_2. *Global Change Biology*, 23(7), 2687–2704.

Jones, J. W., Hoogenboom, G., Porter, C. H., Boote, K. J., Batchelor, W. D., Hunt, L. A., Wilkens, P. W., Singh, U., Gijsman, A. J., & Ritchie, J. T. (2003). The DSSAT cropping system model. *European Journal of Agronomy*, 18 (3–4), 235–265.

Keating, B. A., Carberry, P. S., Hammer, G. L., Probert, M. E., Robertson, M. J., Holzworth, D., Huth, N. I., Hargreaves, J. N., Meinke, H., Hochman, Z., & McLean, G. (2003). An overview of APSIM, a model designed for farming systems simulation. *European Journal of Agronomy*, 18(3–4), 267–288.

Keating, B. A., & Thorburn, P. J. (2018). Modelling crops and cropping systems—Evolving purpose, practice and prospects. *European Journal of Agronomy*, 100, 163–176.

Kharin, V. V., Zwiers, F. W., Zhang, X., & Hegerl, G. C. (2013). Changes in temperature and precipitation extremes in the IPCC ensemble of global coupled model simulations. *Climatic Change*, 119(2), 345–357.

Kirono, D. G. C., Kent, D. M., Hennessy, K. J., & Mpelasoka, F. (2011). Characteristics of Australian droughts under enhanced greenhouse conditions: Results from 14 global climate models. *Journal of Arid Environments*, 75(6), 566–575.

Lesk, C., Rowhani, P., & Ramankutty, N. (2016). Influence of extreme weather disasters on global crop production. *Nature*, 529(7584), 84–87.

Li, Z., Zheng, F. L., Liu, W. Z., & Jiang, D. J. (2012). Spatially downscaling GCMs outputs to project changes in extreme precipitation and temperature events on the loess plateau of China during the 21st century. *Global and Planetary Change*, 82–83, 65–73.

Liang, J., Zhang, J., & Cao, X. (2001). Grain sink strength may be related to the poor grain filling of indica-japonica rice (*Oryza sativa*) hybrids. *Physiologia Plantarum*, 112(4), 470–477.

Liaw, A., & Wiener, M. (2002). Classification and regression by random forest. *R News*, 2(3), 18–22.

Liu, B., Asseng, S., Liu, L., Tang, L., Cao, W., & Zhu, Y. (2016). Testing the responses of four wheat crop models to heat stress at anthesis and grain filling. *Global Change Biology*, 22(5), 1890–1903.

Liu, D. L., Anwar, M. R., O'Leary, G., & Conyers, M. K. (2014). Managing wheat stubble as an effective approach to sequester soil carbon in a semi-arid environment: Spatial modelling. *Geoderma*, 214, 50–61.

Liu, D. L., Zeleke, K. T., Wang, B., Macadam, I., Scott, F., & Martin, R. J. (2017). Crop residue incorporation can mitigate negative climate change impacts on crop yield and improve water use efficiency in a semiarid environment. *European Journal of Agronomy*, 85, 51–68.

Liu, D. L., & Zuo, H. (2012). Statistical downscaling of daily climate variables for climate change impact assessment over New South Wales, Australia, *Climatic Change*, 115(3–4), 629–666.

Lobell, D. B., Bänziger, M., Magorokosho, C., & Vivek, B. (2011). Nonlinear heat effects on African maize as evidenced by historical yield trials. *Nature Climate Change*, 1(1), 42.

Lobell, D. B., Hammer, G. L., Chenu, K., Zheng, B., McLean, G., & Chapman, S. C. (2015). The shifting influence of drought and heat stress for crops in northeast Australia. *Global Change Biology*, 21(11), 4115–4127.

Marcellos, H., & Single, W. V. (1984). Frost injury in wheat ears after ear emergence. *Functional Plant Biology*, 11 (2), 7–15.

Murphy, B. F., & Timbal, B. (2008). A review of recent climate variability and climate change in southeastern Australia. *International Journal of Climatology*, 28(7), 859–879.

Pagani, V., Stella, T., Guarneri, T., Finotto, G., Van den Berg, M., Marin, F. R., Acutis, M., & Confalonieri, R. (2017). Forecasting sugarcane yields using agro-climatic indicators and Canegro model: A case study in the main production region in Brazil. *Agricultural Systems*, 154, 45–52.

Perry, E. M., Nuttall, J. G., Wallace, A. J., & Fitzgerald, G. J. (2017). In-field methods for rapid detection of frost damage in Australian dryland wheat during the reproductive and grain-filling phase Crop & Pasture. *Science*, 68, 516–526.

Planton, S., Déqué, M., Chauvin, F., & Terray, L. (2008). Expected impacts of climate change on extreme climate events. *Comptes Rendus Geosciences*, 340(9), 564–574.

Power, S., Tseitkin, F., Torok, S., Lavery, B., Dahni, R., & McAvaney, B. (1998). Australian temperature, Australian rainfall and the southern oscillation, 1910-1992: Coherent variability and recent changes. *Australian Meteorological Magazine*, 47(2), 85–101.

Pradhan, G. P., Prasad, P. V. V., Fritz, A. K., Kirkham, M. B., & Gill, B. S. (2012). Effects of drought and high temperature stress on synthetic hexaploid wheat. *Functional Plant Biology*, 39(3), 190–198.

Priestley, C. H. B., & Taylor, R. J. (1972). On the assessment of surface heat flux and evaporation using large-scale parameters. *Monthly Weather Review*, 100(2), 81–92.

Qian, B., De Jong, R., Huffman, T., Wang, H., & Yang, J. (2016). Projecting yield changes of spring wheat under future climate scenarios on the Canadian prairies. *Theoretical and Applied Climatology*, 123(3–4), 651–669.

Ray, D. K., Gerber, J. S., Macdonald, G. K., & West, P. C. (2015). Climate variation explains a third of global crop yield variability. Nature. *Communications*, 65989.

Rezaei, E. E., Webber, H., Gaiser, T., Naab, J., & Ewert, F. (2015). Heat stress in cereals: Mechanisms and modelling. *European Journal of Agronomy*, 64, 98–113.

Roberts, M. J., Braun, N. O., Sinclair, T. R., Lobell, D. B., & Schlenker, W. (2017). Comparing and combining process-based crop models and statistical models with some implications for climate change. *Environmental Research Letters*, 12(9), 095010.

Robertson, M. J., Brooking, I. R., & Ritchie, J. T. (1996). Temperature response of vernalization in wheat: Modelling the effect on the final number of mainstem leaves. *Annals of Botany*, 78(3), 371–381.

Royo, C., Villegas, D., Rharrabti, Y., Blanco, R., Martos, V., & García del Moral, L. (2006). Grain growth and yield formation of durum wheat grown at contrasting latitudes and water regimes in a Mediterranean environment. *Cereal Research Communications*, 34(2–3), 1021–1028.

Sánchez, B., Rasmussen, A., & Porter, J. R. (2014). Temperatures and the growth and development of maize and rice: A review. *Global Change Biology*, 20(2), 408–417.

Schauberger, B., Archontoulis, S., Arneth, A., Balkovic, J., Ciais, P., Deryng, D., Elliott, J., Folberth, C., Khabarov, N., Müller, C., & Pugh, T. A. (2017). Consistent negative response of US crops to high temperatures in observations and crop models. *Nature Communications*, 8, 13931.

Schlenker, W., & Roberts, M. J. (2009). Nonlinear temperature effects indicate severe damages to US crop yields under climate change. *Proceedings of the National Academy of Sciences of the United States of America*, 106(37), 15594–15598.

Semenov, M. A., & Shewry, P. R. (2011). Modelling predicts that heat stress, not drought, will increase vulnerability of wheat in Europe. *Scientific Reports*, 1, 66.

Shalev-Shwartz, S., & Ben-David, S. (2014). *Understanding machine learning: From theory to algorithms*. University of Cambridge: Cambridge University Press.

Shen, J., Huete, A., Tran, N. N., Devadas, R., Ma, X., Eamus, D., & Yu, Q. (2018). Diverse sensitivity of winter crops over the growing season to climate and land surface temperature across the rainfed cropland-belt of eastern Australia. *Agriculture, Ecosystems and Environment*, 254, 99–110.

Stone, P. J., Nicolas, M. E., Stone, P. J., & Nicolas, M. E. (1994). Wheat cultivars vary widely in their responses of grain yield and quality to short periods of post-anthesis heat stress. *Functional Plant Biology*, 21(6), 887–900.

Stratonovitch, P., & Semenov, M. A. (2015). Heat tolerance around flowering in wheat identified as a key trait for increased yield potential in Europe under climate change. *Journal of Experimental Botany*, 66(12), 3599–3609.

Tao, F., Rötter, R. P., Palosuo, T., Díaz-Ambrona, C. G. H., Mínguez, M. I., Semenov, M. A., Kersebaum, K. C., Nendel, C., Cammarano, D., Hoffmann, H., & Ewert, F. (2017). Designing future barley ideotypes using a crop model ensemble. *European Journal of Agronomy*, 82, 144–162.

Tashiro, T., & Wardlaw, I. F. (1989). A comparison of the effect of high temperature on grain development in wheat and rice. *Annals of Botany-London*, 64(1), 59–65.

Trnka, M., Rötter, R. P., Ruiz-Ramos, M., Kersebaum, K. C., Olesen, J. E., Žalud, Z., & Semenov, M. A. (2014). Adverse weather conditions for European wheat production will become more frequent with climate change. *Nature Climate Change*, 4(7), 637.

Venkataraman, K., Tummuri, S., Medina, A., & Perry, J. (2016). 21st century drought outlook for major climate divisions of Texas based on CMIP5 multimodel ensemble: Implications for water resource management. *Journal of Hydrology*, 534, 300–316.

Wang, B., Li Liu, D., Asseng, S., Macadam, I., & Yu, Q. (2015). Impact of climate change on wheat flowering time in eastern Australia. *Agricultural and Forest Meteorology*, 209, 11–21.

Wang, B., Liu, D. L., Asseng, S., Macadam, I., Yang, X., & Yu, Q. (2017a). Spatiotemporal changes in wheat phenology, yield and water use efficiency under the CMIP5 multimodel ensemble projections in eastern Australia. *Climate Research*, 72(2), 83–99.

Wang, B., Liu, D. L., O'leary, G. J., Asseng, S., Macadam, I., Lines-Kelly, R., Yang, X., Clark, A., Crean, J., Sides, T., & Xing, H. (2017b). Australian wheat production expected to decrease by the late 21st century. *Global Change Biology*, 24(6), 2403–2415.

Wang, B., Waters, C., Orgill, S., Cowie, A., Clark, A., Liu, D. L., Simpson, M., McGowen, I., & Sides, T. (2018). Estimating soil organic carbon stocks using different modelling techniques in the semi-arid rangelands of eastern Australia. *Ecological Indicators*, 88, 425–438.

Watson, J., Zheng, B., Chapman, S., & Chenu, K. (2017). Projected impact of future climate on water-stress patterns across the Australian wheatbelt. *Journal of Experimental Botany*, 68(21–22), 5907–5921.

Wheeler, T., & Von Braun, J. (2013). Climate change impacts on global food security. *Science*, 341(6145), 508–513.

Williams, G. (2011). *Data mining with rattle and R: The art of excavating data for knowledge discovery*. Springer Science and Business Media.

Woli, P., Jones, J. W., & Ingram, K. T. (2013). Assessing the agricultural reference index for drought (ARID) using uncertainty and sensitivity analyses. *Agronomy Journal*, 105(1), 150–160.

Woli, P., Jones, J. W., Ingram, K. T., & Fraisse, C. W. (2012). Agricultural reference index for drought (ARID). *Agronomy Journal*, 104(2), 287–300.

Wollenweber, B., Porter, J. R., & Schellberg, J. (2003). Lack of interaction between extreme high-temperature events at vegetative and reproductive growth stages in wheat. *Journal of Agronomy and Crop Science*, 189 (3), 142–150.

Yang, X., Tian, Z., Sun, L., Chen, B., Tubiello, F. N., & Xu, Y. (2017). The impacts of increased heat stress events on wheat yield under climate change in China. *Climatic Change*, 140(3–4), 605–620.

Yasufuku, N., Araki, K., & Omine, K. (2015). Evaluation of inhibitory effect by adaptation measures for red soil runoff from farmland due to heavy rainfall (special issue on adaptation measures for disasters due to climate change). *Journal of Disaster Research*, 10, 457–466.

Zheng, B., Chenu, K., Doherty, A., & Chapman, S., 2014. The APSIM-Wheat Module (7.5 R3008). Agricultural Production Systems Simulator (APSIM) Initiative, 615.

Zheng, B., Chenu, K., Fernanda, D. M., & Chapman, S. C. (2012). Breeding for the future: What are the potential impacts of future frost and heat events on sowing and flowering time requirements for Australian bread wheat (*Triticum aestivium*) varieties? *Global Change Biology*, 18(9), 2899–2914.

9

Jon I. Lizaso,
Vakhtang Shelia, and
Gerrit Hoogenboom

Simulating Maize Crop Processes in DSSAT: CSM-CERES-Maize vs. CSM-IXIM-Maize

Abstract

Evolving from the original Version 2.1 with four crop models to the current Version 4.8 with models for 37 crops, the Decision Support System for Agrotechnology Transfer (DSSAT) has been improving until it has become the most used crop modeling platform worldwide. DSSAT also has alternative crop models for some crops, providing users with the option to explore the range of uncertainty for simulated variables of interest and making more robust projections toward future events. This is the case for maize, which has two models: CERES-Maize and CSM-IXIM. In this chapter, we identify and analyze the various strategies followed by each model to simulate major processes: crop development; crop growth, focusing on the simulation of leaf area, light capture, and growth rate; growth partitioning; crop yield; and crop N demand. Required model inputs are also discussed. We conclude by emphasizing the advantage of having more than one crop model, in the same platform of simulation, for cropping systems analysis.

Introduction

Funded by the US Agency for International Development (USAID), the International Benchmark Sites Network for Agrotechnology Transfer (IBSNAT) Project created a research and application environment of international collaboration (IBSNAT,

Modeling Processes and Their Interactions in Cropping Systems: Challenges for the 21st Century, First Edition.
Edited by Lajpat R. Ahuja, Kurt C. Kersebaum, and Ole Wendroth.
© 2022 American Society of Agronomy, Inc. / Crop Science Society of America, Inc. / Soil Science Society of America, Inc.
Published 2022 by John Wiley & Sons, Inc.

1993; Tsuji et al., 1998). Resulting from these stimulating conditions, an innovative tool, the Decision Support System for Agrotechnology Transfer (DSSAT) was developed. The initial DSSAT v2.1 (IBSNAT, 1989) was comprised of four crop models including CERES-Maize (Jones & Kiniry, 1986; Ritchie, 1986), CERES-Wheat (Ritchie & Otter, 1985), SOYGRO (Hoogenboom et al., 1992; Wilkerson et al., 1983), and PNUTGRO (Boote et al., 1986; Hoogenboom et al., 1992). With time, DSSAT has evolved and has incorporated and added new crop models and supporting tools for creating input files and analyzing simulated outputs. The crop models have also been combined into one comprehensive model called the Cropping System Model (CSM) to address a cropping system rather than a single crop (Jones et al., 2003). The CSM includes the same approach for simulating the dynamics for soil water, N, P, and organic C, while the dynamics for the simulation of crop growth and development use different modules. Currently, DSSAT v4.8 (Hoogenboom et al., 2019, 2021; Jones et al., 2003) incorporates models for 37 crops. In addition, for several crops, the system provides duplicate models to conduct alternative simulations. This is the case for cassava, sugarcane, wheat, barley, and maize. For maize, CSM has two models: CERES-Maize (Ritchie et al., 1998) and CSM-IXIM (Lizaso et al., 2011). Sweet corn, which is grown and managed as a vegetable, is based on CERES-Maize but uses a different crop module (Vegetables) within DSSAT (Lizaso et al., 2007a).

The CERES-Maize model was originally released in 1986 and has become the most widely used maize model in the world (Basso et al., 2016). CSM-IXIM, released in 2011, resulted from consolidating a large number of modifications to CERES-Maize. The goal of this chapter is to provide a brief overview and analysis of the differences in the simulation approaches of CSM-CERES-Maize and CSM-IXIM-Maize.

Simulated Processes

Crop Development

Crop development, the ontogenetic progression of a crop through its life cycle, responds to the daily accumulation of heat, computed as thermal time rather than chronological time. Some crops are also responsive to vernalization, requiring exposure to low temperatures to reach a specific developmental stage, such as seed germination or flowering. Photoperiod, or the duration of daylight, is another environmental cue that a crop can detect through the phytochrome system of the leaves. This system reacts to exposure to red and far-red light and therefore accumulates the uninterrupted duration of the dark period (Connor et al., 2011).

The simulation of thermal time is an important component of crop models, and maize models use several procedures to estimate heat accumulation (Kumudini et al., 2014). In terms of photoperiod response, tropical hybrids in many cases retain the typical short-day photoperiod sensitivity during the phase of flower induction, when the apical meristem switches from the vegetative to the reproductive phase. In contrast, hybrids grown in temperate regions exhibit little or no response to photoperiod (Tollenaar et al., 2018).

In DSSAT, the crop-related coefficients are organized in three different files (Jones et al., 1994): species (SPE), ecotype (ECO), and cultivar (CUL). The coefficients that are defined in these three input files are associated with a unique crop model, such as

Table 9.1 Development Phases and Parameters Controlling the Duration of each Phase in CERES-Maize Version 4.8

Phase name	Controlled by	Read from file	Description
Juvenile (emergence to end juvenile)	P1	CUL	Thermal time duration of the phase, °C d
Flower induction (end juvenile to tassel initiation)	DJTI P2	ECO CUL	Duration of the phase under short photoperiods, d Photoperiod sensitivity, d
Tassel initiation to silking	Ph	CUL	Phyllochron determines number of leaves differentiated and ultimately silking date, °C d
Lag phase (anthesis to linear grain growth)	DSGFT	ECO	Thermal time duration of the phase, °C d
Linear grain filling to physiological maturity	P5	CUL	Thermal time duration of the phase, °C d

Note. CUL, cultivar coefficients file; ECO, ecotype coefficients file; °C d, degree days.

CERES-Maize or CSM-IXIM, and are required inputs (discussed in further detail below). The parameters are associated with development phases.

The CERES-Maize module distinguishes five developmental phases from emergence to physiological maturity (Table 9.1). CERES-Maize calculates the daily mean air temperature as the average of the maximum (T_{max}) and minimum (T_{min}) temperatures and then estimates the daily thermal time (DTT) between a base temperature (T_b) of 8 °C and an optimum temperature (T_o) of 34 °C:

$$DTT = \frac{T_{max} + T_{min}}{2} - T_b \tag{9.1}$$

Both T_b and T_o are read from the ECO file. When the minimum temperature falls below the base temperature, or the maximum exceeds the optimum, the model generates a sinusoidal distribution of hourly temperatures between minimum and maximum temperatures. Then values under the base are replaced by the base temperature, and values above the optimum are replaced by the optimum temperature. The new daily DTT is calculated as the average of these constrained hourly values.

The CSM-IXIM model also has five development phases from emergence to maturity, like the CERES-Maize model but with some additional components (Lizaso et al.2007b, 2017). The model simulates both flowering events in maize anthesis, that is, 50% of plants shedding pollen, and silking, that is, 50% of plants exhibiting visible silks on the apical ear, separated by a number of days under non-stress conditions (ASNS), which can either be positive (anthesis first) or negative (silking first) (Table 9.2).

Simulated crop development is driven by a three-parameter simplified β function (Yan & Hunt, 1999) to estimate DTT. The function operates with one set of parameters for the vegetative phase and another set of parameters for the reproductive phase (Lizaso et al., 2017):

$$DTT = TT_x \left(\frac{T_x - T}{T_x - T_o}\right) \left(\frac{T}{T_o}\right)^{T_o/(T_x - T_o)} \tag{9.2}$$

where T_o and T_x are the function parameters for optimum and maximum temperatures (not to be confounded with T_{max}, the daily maximum temperature in Equation 9.1), and TT_x is the effective DTT accumulated at T_o. In Equation 9.2, the mean daily temperature (T), is computed from hourly generated values (Parton & Logan, 1981) constrained

Table 9.2 Development Phases and Parameters Controlling the Duration of each Phase and Main Events in CSM-IXIM Version 4.8

Phase name	Controlled by	Read from file	Description
Juvenile (emergence to end juvenile)	P1	CUL	Thermal time duration of the phase, °C d
Flower induction (end juvenile to tassel initiation)	DJTI	ECO	Duration of the phase under short photoperiods, d
	P2	CUL	Photoperiod sensitivity, d
Tassel initiation to anthesis (50% of plants shedding pollen)	Ph	CUL	Phyllochron determines number of leaves differentiated and ultimately anthesis date, °C d
Lag phase (anthesis to linear grain growth)	DSGFT	ECO	Thermal time phase duration
	ASNS	CUL	Days between anthesis and silking (ASI) under no stress
	ASEN	CUL	Days ASI is increased under strong stress
Linear grain filling to physiological maturity	P5	CUL	Thermal time duration of the phase

Note. CUL, cultivar coefficients file; ECO, ecotype coefficients file; °C d, degree days.

between zero and T_x. The function produces a bell-shaped curve, accumulating DTT within T values of zero and T_x, reaching the maximum at T_o, and then falling back to zero at T_x (Lizaso et al., 2017, Figure 9.3).

The effect of photoperiod on crop development is simulated identically in both models. During the flower induction phase, crop development is entirely controlled by photoperiod. Both models assume a minimum duration of that phase in days, not in thermal time, which is defined in the ECO file with a default value of four days. The duration can be extended in response to the combined action of two factors: a daily photoperiod longer than the critical photoperiod and a cultivar exhibiting sensitivity to long photoperiods. The critical daily photoperiod length is defined in the ECO file, with a default value of 12.5 hours. The photoperiod sensitivity of each cultivar (P2) is defined as the number of days that development is delayed for each hour that the photoperiod increases above the critical photoperiod (12.5 hours). It is a user-calibrated parameter defined in the CUL file.

Crop Growth

Both models simulate crop growth on a single-plant basis. Here we mainly focus on three major simulation components of crop growth: (a) leaf area expansion, (b) light capture and growth rate, and (c) growth partitioning.

Leaf area

CERES-Maize assumes that only one leaf expands until reaching its final size, and then the growth of the next leaf begins. The model simulates the expansion of plant leaf area using five discrete functions of the currently expanding leaf number (LN):

$$\text{PLAG} = \begin{cases} 4.0 \times \text{LN} \times \text{TI}, \text{when } 1 \leq \text{LN} < 4 \\ 3.0 \times \text{LN}^2 \times \text{TI}, \text{when } 4 \leq \text{LN} - \text{End Juvenile} \\ 3.5 \times \text{LN}^2 \times \text{TI}, \text{End Juvenile} - \text{LN} < 12 \\ 3.5 \times 170 \times \text{TI}, 12 \leq \text{LN} < \text{Last three leaves} \\ \dfrac{3.5 \times 170 \times \text{TI}}{(\text{LN} + 3 - \text{TL})^{0.5}}, \text{Last three leaves} \end{cases} \quad (9.3)$$

where PLAG is the daily rate of leaf area expansion (cm^2 $plant^{-1}$), TL is the total number of leaves, and TI is the fraction of LN increase on the current day:

$$TI = \frac{DTT}{Ph} \tag{9.4}$$

where Ph is the phyllochron or thermal time between the appearance of two consecutive leaves (°C d). Every day LN is updated by adding TI.

CERES-Maize simulates two types of senescence: (a) time-driven senescence (SenT), a function of cumulative thermal time, and (b) stress-driven senescence (SenS), a cumulative function of the most limiting daily stress factor associated with water, light competition, low temperature, N, and P. The model then updates on daily bases the total plant senescence (TotSen, cm^2 $plant^{-1}$) by choosing the largest value between SenT and SenS. The value of TotSen is never allowed to be larger than the current green leaf area.

CSM-IXIM simulates leaf area on a per-leaf basis within a thermal time framework (Lizaso et al., 2003a). For each leaf, the model computes leaf expansion, longevity, and senescence. Leaf expansion for each ith leaf (LA_i) is described with a three-parameter logistic function:

$$LA_i = \frac{A_i}{1 + \exp\left[-k_i(t - t_{50i})\right]} \tag{9.5}$$

where A_i is the final surface area of the ith leaf (cm^2), t_{50i} is the thermal time when the leaf reaches 50% of A_i, and k_i is a slope parameter. The leaf longevity of each leaf (LL_i) is the thermal time separating 50% of leaf expansion from 50% of leaf senescence. Leaf senescence (SA_i) follows a similar logistic function parallel to leaf expansion, with the same A_i and k_i parameters, but occurring LL_i thermal units later:

$$SA_i = \frac{A_i}{1 + \exp\left\{-k_i[t - (t_{50i} + LL_i)]\right\}} \tag{9.6}$$

Equations 9.5 and 9.6 require sets of the three parameters for each leaf (A_i, k_i, and t_{50i}) and values of LL_i, which are calculated from the model-estimated final number of leaves (Lizaso et al., 2003a).

The most limiting stress factors due to temperature, water, N, P, and O_2 is used by CSM-IXIM to reduce daily leaf expansion. The same factors except temperature will accelerate leaf senescence by shortening the longevity of individual leaves.

After calculating leaf area expansion, both models check the source of C available for leaf growth and adjust the expansion of leaves as required. We examine these components further below.

Light capture and growth rate

CERES-Maize simulates the daily amount of light captured (Ipar, MJ $plant^{-1}$) by the canopy as

$$Ipar = \frac{PAR}{Pop} \times [1 - \exp(-k \times LAI)] \tag{9.7}$$

where PAR (MJ m^{-2}) is the photosynthetically active radiation calculated as 50% of the total incoming solar radiation, Pop is the plant population density (plants m^{-2}), k is a unitless light extinction coefficient, and LAI is the leaf area index (m^2 m^{-2}). Equation 9.7

is consistent with the adaptation of Beer's Law by Monsi and Saeki (Hirose, 2005). Next, the model calculates the potential growth rate (PG, g plant^{-1}) and the actual growth rate (G, g plant^{-1}):

$$PG = \text{Ipar} \times \text{RUE} \times F_{CO2} \qquad (9.8)$$

where RUE is the radiation use efficiency (g dry mass MJ^{-1} PAR intercepted by the canopy), and F_{CO2} is a relative factor correcting for the increased growth associated with the elevated CO_2 concentration in the atmosphere. In each model, the SPE file has a set of x:y values relating atmospheric CO_2 concentration (ppmv, 0–9,999) and growth relative to that corresponding to 330 ppmv. These responses differ for C_3 and C_4 crops. The models also have available a standard file with monthly Mauna Loa CO_2 observations (Keeling et al., 2001). With these components, the models estimate the relative growth (F_{CO2}) linked to the CO_2 concentration for the simulated dates. The actual growth rate, G, is estimated by multiplying PG by the most limiting stress factor of the day due to temperature, water, N, P, and K. In addition to these stresses, the CERES-Maize model allows for the simulation of unspecified growth-limiting factors mostly associated with soil conditions (e.g., acidity, compaction, or toxicity). This soil fertility (SF) factor, ranging from 0 to 1, summarizes the soil potential fertility and is defined in the soil file. The PG is multiplied by the SF factor to obtain G.

CSM-IXIM takes advantage of the per-leaf simulation of leaf area to implement a detailed light absorption, per-leaf gross assimilation and a canopy respiration model (Lizaso et al., 2005). The per-leaf light absorption procedures were based on the approach of Boote and Pickering (1994). Individual leaves assimilate CO_2 according to a distinct light response curve determined by leaf age and air temperature (Lizaso et al., 2005). The respiration components include maintenance respiration (Wilkerson et al., 1983) and growth respiration (Penning de Vries et al., 1989; Penning de Vries & van Laar, 1982). The model also incorporates an algorithm to calculate the fraction of PAR in the daily solar radiation (Lizaso et al., 2003b) and the effect of leaf rolling on light capture due to severe drought stress (Lizaso et al., 2011). The estimated PG from the photosynthesis and respiration modules is also corrected by the effect of elevated CO_2, F_{CO2}. The actual growth rate (G) is calculated like CERES-Maize, by multiplying PG for the most limiting stress factor by the SF factor.

Growth partitioning

During the vegetative phase before flower induction is completed, CERES-Maize allocates 75% of the daily growth to the leaves and the remaining 25% to the roots. During this phase, the biomass of stems, initialized at emergence at a default value of 0.2 g dry mass plant^{-1}, is not modified. Once the flower induction is completed and the final number of leaves is determined, dry mass partitioning is adjusted.

Dry mass partitioning to the leaves (LFgro) becomes a function of both the leaf area growth rate (PLAG) and the current leaf area (PLA):

$$\text{LFgro} = 0.00116 \times \text{PLAG} \times \text{PLA}^{0.25} \qquad (9.9)$$

Dry mass partitioning to the stem (STgro) becomes a fraction of the biomass allocated to the leaves, but modulated by the remaining number of leaves that must expand:

$$\text{STgro} = 0.0182 \times \text{LFgro} \times (\text{LT} - \text{LN})^2 \qquad (9.10)$$

where LT is the total number of leaves that are already differentiated in the apical meristem, calculated as a function of the phyllochron, and LN is the leaf number currently expanding. During the expansion of the last three leaves, stem partitioning is controlled by the daily fraction of the increase in LN (TI, Equation 9.4):

$$STgro = 3 \times 3.1 \times TI \tag{9.11}$$

During this period, the growth of the stem is reduced under stress conditions by multiplying STgro by the most limiting stress factor linked to water, O_2, and mineral nutrition (N, P, and K).

After tassel initiation, and specifically within a thermal time window bracketing silking, CERES-Maize simulates ear growth. The default value for this thermal time window is 250 growing degree days (GDD) before silking and 100 GDD after silking (Lizaso et al., 2001; Otegui & Bonhomme, 1998). However, since the default duration of the lag phase is 170 GDD after silking, for consistency with the model-defined development phases, the model keeps growing ears during the entire duration of the lag phase:

$$EARgro = \frac{0.81}{1 + \exp\left[-0.02\left(TTeg - 210\right)\right]} \tag{9.12}$$

where EARgro is the fraction of daily growth partitioned to the ear, and TTeg is the cumulative thermal time elapsed since the beginning of ear growth. Equation 9.12 allocates increasing fractions of daily crop growth to the ear, up to 80% at the end of the lag phase. In the presence of stresses, EARgro may be reduced by the most limiting stress factor for that day. During this post-silking lag phase, the model initially attempts to allocate up to 8% of the daily growth to roots and the remaining to stems. However, when this is not possible, ears maintain their growth share as calculated with Equation 9.12, and roots (RTgro) and stems (STgro) each obtain half of the remaining daily growth:

$$RTgro = \left(G - EARgro\right) \times 0.5 \tag{9.13}$$

$$STgro = \left(G - EARgro\right) \times 0.5 \tag{9.14}$$

CSM-IXIM incorporates a novel function to change partitioning in response to stresses:

$$Fst = 0.46 + 0.72 \times \exp\left[-2.83(1 - St)\right] \tag{9.15}$$

where Fst is the factor changing partitioning in the presence of the most limiting water or N stress factor, St. Equation 9.15 is constrained at a maximum value of 1. The function changes the non-stress partitioning factor by up to 50% in response to drought or N stress (St), fluctuating within the range of 0.9 (mild stress) to 0.5 (maximum stress). During the juvenile phase (Table 9.2), the model assumes that one-third of the daily growth is allocated to roots, while the remaining two-thirds are shared between the growing leaves and stems. The leaf area routine estimates the leaf surface growth for each individual leaf and also defines the biomass demand for each leaf based on new growth and the specific leaf area (SLA, $cm^2\ g^{-1}$). The value of SLA for each leaf accounts for the leaf position within the canopy (Tanaka & Yamaguchi, 1972; Thiagarajah & Hunt, 1982), PAR intensity (Warrington & Norton, 1991), and the mean air temperature (Thiagarajah & Hunt, 1982). When the daily C assimilation limits crop growth, leaf dry mass and leaf expansion are reduced accordingly. Stems are assigned 18.4% of daily growth prior to tassel initiation. Under drought or N stress, Equation 9.15 is used to increase partitioning to the roots at the expense of partitioning to the leaves and stem.

Once tassel initiation is complete (Phase 3, Table 9.2), the partitioning to the roots decreases linearly until the beginning of the linear grain filling (Phase 5, Table 9.2) when root growth terminates and the grain becomes the major sink. During Phase 3, ears begin to grow starting at a thermal time defined prior to anthesis (default 250 GDD). The rate of growth increases daily until the end of the lag phase, when ears accumulate their maximum fraction (EARp; default 0.15) of the daily net C assimilation, which is defined in the ECO file:

$$EARgro = \frac{EARp}{1 + \exp\left[-0.02(TTeg - 225)\right]} \times Fst \tag{9.16}$$

A major difference between the assumptions of the two models can be seen when comparing Equations 9.12 and 9.16. CERES-Maize assumes that the reserve of assimilates to supplement grain growth is mostly stored in the ear, thus partitioning up to 80% of the daily growth at the end of the lag phase into ears (Table 9.1). On the other hand, CSM-IXIM assumes that the storage is located in the stems, while partitioning only 15% of the daily growth to the ears at the end of the lag phase (Table 9.2). That fraction, however, can be modified by calibrating the final fraction allocated to ears (EARp) and concurrently to stems (STMp) defined in the ECO file.

Crop Yield

The simulation of crop grain yield has two major components: sink size and sink filling rate. CERES-Maize estimates the size of the sink, that is, the number of grains per plant, with a linear function of the average rate of photosynthesis (PS, mg carbohydrate plant^{-1}) during the post-silking lag phase (Table 9.1):

$$PS = \frac{Gsum \times 1,000}{Gdur} \times 0.68 \tag{9.17}$$

where Gsum and Gdur are the cumulative growth (g plant^{-1}) and duration (d) of the lag phase, respectively. The equation assumes a factor of 0.68 to convert biomass into carbohydrate (Lizaso et al., 2001). Next, the model estimates grain number per plant (GPP):

$$GPP = G2 \times \frac{PS}{7,200} + 50 \tag{9.18}$$

where G2 is the potential grain number per plant, a cultivar-specific coefficient read from the CUL file, and 7,200 is the estimated cumulative assimilation (mg carbohydrate plant^{-1}) during the lag phase, the integral of PS, making GPP equal to G2 + 50. Finally, the value of GPP is limited at a maximum of G2 and a minimum of 50 grains per plant. Equation 9.18 is based on the work of Keating and Wafula conducted in Kenya (B. Keating, personal communication, 2001).

The model simulates the growth of the grain (GRgro, g d^{-1}) based on a parameter that is defined in the CUL file, G3, representing the potential growth rate of an individual grain (mg grain^{-1} d^{-1}):

$$GRgro = \frac{T_f \times GPP \times G3}{1,000}$$ (9.19)

where T_f is a temperature function that is defined in the SPE file. The default parameter values assume an optimum temperature for grain filling between 16 and 27 °C, with minimum and maximum temperatures of 5.5 and 35 °C, respectively. The model also incorporates the impact of drought stress through a reduction of the grain growth rate.

CSM-IXIM estimates the sink size with a function of the average shoot growth rate (G_{avg}, g plant^{-1} d^{-1}) during a thermal time window around anthesis (tasseling). This window is delimited by two parameters that are stored in the SPE file, with default values of 250 GDD before anthesis and 100 GDD after anthesis. The approach, modified from Tollenaar et al. (1992), assumes that at low population densities, the maize crop will set kernels in a second ear and therefore plants accumulate the number of grains set on both ears:

$$GPP1 = G2 \times \left[1 - \exp\left(-0.347 \times G_{avg}\right)\right]$$ (9.20)

$$GPP2 = 1.8 \times G2\{1 - \exp\left[-0.895\left(G_{avg} - 3\right)\right]\}$$

where G2 is the potential kernel number in the apical ear, thus having a different meaning than the G2 used in CERES-Maize (Equation 9.18). The model also assumes that the second ear will set a maximum of 80% of the number of grains that are set in the apical ear (G2). The largest value between GPP1 and GPP2 is selected as the final size of the sink, GPP.

Finally, the CSM-IXIM model simulates grain growth following an identical procedure as in CERES-Maize (Equation 9.19).

Crop Nitrogen Demand

CERES-Maize simulates crop N demand as a function of crop development based on the approach of Jones (1983), while CSM-IXIM uses a function of crop growth modified from Plénet and Lemaire (2000). A relative-phenology scale (RP, 0–10; silking = 4.5) is used by CERES-Maize to estimate shoot N demand:

$$N_{crit} = \exp\left(1.52 - 0.16 \times RP\right)$$ (9.21)

where N_{crit} is the target non-stress N concentration (%) in aboveground tissues during the crop cycle. The corresponding target N concentration (fraction) for root tissue is a constant value that is defined in the SPE file (default 0.01 g N g^{-1} root dry mass). Based on these target concentrations for shoots and roots, the current simulated concentrations, and accounting for plant population, the model calculates the daily N demand (kg N ha^{-1}).

CSM-IXIM simulates the N demand of shoots and roots based on biomass (Plénet & Lemaire, 2000). Nitrogen that is taken up by the roots is allocated according to the target N concentration (Yakoub et al., 2017) of leaves, stems, and ears as fractions (g N g^{-1} of dry mass). These target N concentrations for leaves (N_{lf}) and stems (N_{st}) were derived from Lindquist and Mortensen (1999) and Lindquist et al. (2007). The N concentration for ears (N_{ear}) was modified from Plénet and Lemaire (2000):

$$N_{lf} = 8.35 \times b^{-0.23}$$

$$N_{st} = 118.56 \times b^{-0.54} \tag{9.22}$$

$$N_{ear} = 4.2 \times b_e^{-0.25}$$

where b is the shoot biomass (g plant^{-1}) and b_e is the ear biomass (g plant^{-1}). The value of N_{crit} is calculated by adding N_{lf} and N_{st}.

The maximum N concentration for leaves, stems, and ears is set at 5, 4, and 3.3%, respectively (Lindquist et al., 2007; Lindquist & Mortensen, 1999; Plénet & Lemaire, 2000).

For both models, whenever the current N concentration (N_{act}) in aboveground tissues is below N_{crit}, the model calculates three N stress factors, fine-tuning the computed N factor (NF) to the sensitivity of several processes to the N constraint:

$$NF = 1.0 - \frac{N_{crit} - N_{act}}{N_{crit} - N_{min}} \tag{9.23}$$

The value of NF is based on the extent of the N deficit ($N_{crit} - N_{act}$) weighted by the total range of N concentrations ($N_{crit} - N_{min}$) allowed by the models.

Model Inputs

The crop modules of CSM obtain the parameters related to genotype and crop physiology from three types of files: CUL, SPE, and ECO. The CUL file incorporates genotype-specific parameters that enable simulation of the performance of hybrids and cultivars under field conditions. In the file header for ECO and CUL, there are definitions and units for each coefficient. The ECO file in CSM-IXIM has two additional parameters, compared with CERES-Maize, to calibrate the growth partitioning to stems and ears (Equation 9.16). The SPE file also has additional coefficients involved in the further details incorporated in CSM-IXIM. These details involve the simulation of heat stress, leaf area, photosynthesis and respiration, and kernel numbers.

Tables 9.3 and 9.4 summarize the crop parameters included in the CUL files used by CERES-Maize and CSM-IXIM, respectively. These genotype-specific parameters should be calibrated by users when detailed field information on phenology, growth, and yield is available.

Additional input files include weather (WTH), soil (SOL), and management (MZX). The Mauna Loa monthly measurements of atmospheric CO_2 concentrations, used to calculate the daily growth rate, is in file CO2047.WDA.

Table 9.3 Cultivar Coefficients Required by CERES-Maize Version 4.8

Parameter	Definition
P1	Thermal time from emergence to the end of the juvenile phase, °C d
P2	Development delay for each additional hour of daylight above the critical photoperiod (default 12.5 hours), d
P5	Thermal time from silking to physiological maturity, °C d
G2	Potential number of grains per plant, no.
G3	Potential grain filling rate during the linear grain filling phase, mg grain^{-1} d^{-1}
PHINT	Thermal time between successive leaf tip appearance, °C d

Note. °C d, degree days.

Table 9.4 Cultivar Coefficients Required by CSM-IXIM Version 4.8

Parameter	Definition
P1	Thermal time from emergence to the end of the juvenile phase, °C d
P2	Development delay for each additional hour of daylight above the critical photoperiod (default 12.5 hours), d
P5	Thermal time from tasseling to physiological maturity, °C d
G2	Potential number of grains in the apical ear, no.
G3	Potential grain filling rate during the linear grain filling phase, mg grain^{-1} d^{-1}
PHINT	Thermal time between successive leaf tip appearance, °C d
AX	Leaf surface area of the largest leaf, cm^2 leaf^{-1}
ALL	Longevity of the most longevous leaf, usually the ear leaf (see Equation 9.6), °C d
ASNS	Days between 50% anthesis and 50% silking (ASI) in the absence of stresses, d
ASEN	Number of days ASI is increased under a strong stress during the critical period for kernel set, producing an average shoot growth rate of 1 g plant^{-1} d^{-1}, d

Note. °C d, degree days.

Model Performance

Figure 9.1 shows an example of the main simulated components of the two models discussed here. Two treatments were simulated including optimal growing conditions, where water and N fertilizer were supplied as required, and stress conditions, where both irrigation and fertilizer were limited, thus severely constraining crop performance throughout the season. The figure also shows that the intensity of the water stress was stronger than that of the N stress, especially during the period of 65–86 days after sowing. During this time, crop phenology was at the end of Phase 3 (tassel initiation to end of leaf growth) and Phase 4 (end of leaf growth to beginning of linear grain filling). The stress at this time must have affected pollen viability and the entire pollination process. This was evidenced by a 54% reduction in measured kernel numbers per plant for the stressed compared with the non-stressed treatments. Corresponding simulated reductions were on the order of 80% for CERES-Maize and 66% for CSM-IXIM. During grain filling, additional events of water and N stress increased these reductions, as shown by a 75% drop in the measured grain yield when stress and non-stress treatments were compared. Simulated reductions in grain yield were 82% for CERES-Maize and 77% for CSM-IXIM.

Future Research Needs

As suggested by Figure 9.1, both models seem to handle better crop responses under drought than under N limitations. The gap should be filled with additional research into root and shoot growth under various levels of N stress and incorporation and evaluation of improved or additional components to the current simulation approaches that we have detailed here. Another issue of special uncertainty is the performance of the models under heat conditions. This is an issue of major concern today and it will become more important as global warming progresses.

Conclusion

DSSAT v4.8 offers two alternative maize models, CERES-Maize and CSM-IXIM. We have described in detail the approaches followed by the two models to simulate the main processes involved in the growth, development, yield, and N demand of maize crops. Our purpose was to document the particular components of the models and to support the current validity of both approaches, providing users with two appropriate

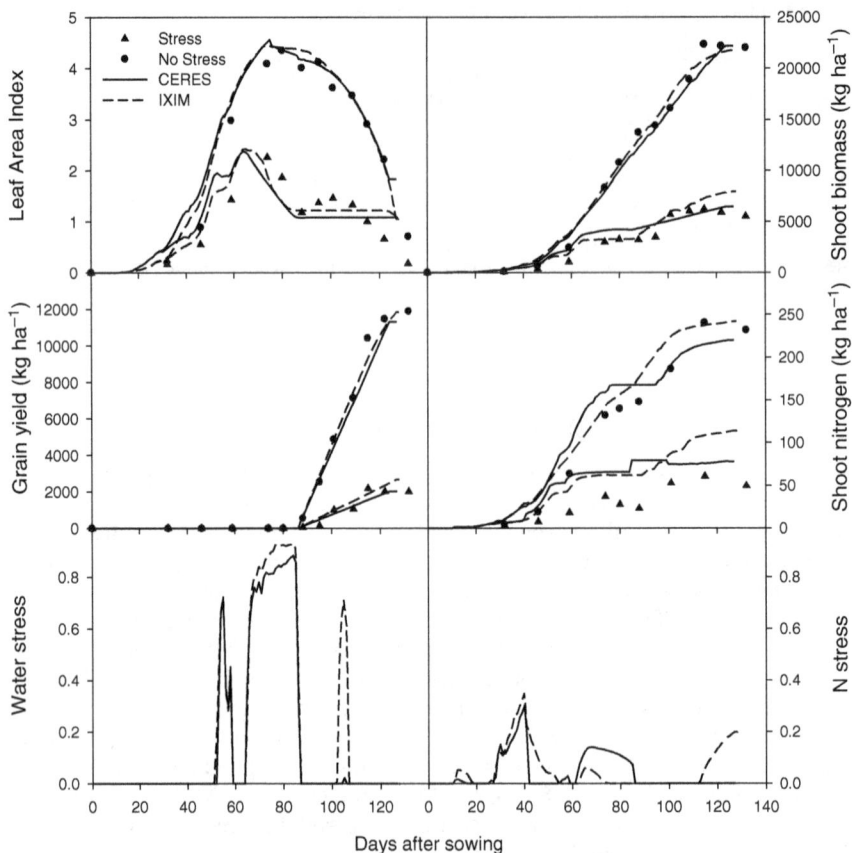

Figure 9.1 CERES-Maize and CSM-IXIM simulated leaf area index, shoot biomass, grain yield, and shoot N content under stress and non-stress conditions (top and middle), and model-simulated water and N stresses (bottom). Water and N limitations were experimentally achieved by reducing irrigation and fertilization. Symbols are measured values.

alternatives. Depending on the field data available and the purpose of the work, users may utilize this work to illustrate which model may better fit their specific needs. Whenever possible, we strongly suggest using both models to unveil possible ranges of outputs as well as levels of uncertainty in variables of main interest.

References

Basso, B., Liu, L., & Ritchie, J. T. (2016). A comprehensive review of the CERES-Wheat, -Maize and -Rice models' performances. *Advances in Agronomy*, 136, 27–132. https://doi.org/10.1016/bs.agron.2015.11.004

Boote, K. J., Jones, J. W., Mishoe, J. W., & Wilkerson, G. G. (1986). Modeling growth and yield of groundnut. In M. V. K. Sivakumar, S. M. Virmani, & S. R. Beckerman (Eds.), *Agrometeorology of groundnut: Proceedings of an International Symposium, Niamey, Niger* (pp. 243–254). ICRISAT.

Boote, K. J., & Pickering, N. B. (1994). Modeling photosynthesis of row crop canopies. *HortScience*, 29, 1423–1434.

Connor, D. J., Loomis, R. S., & Cassman, K. G. (2011). *Crop ecology: Productivity and management in agricultural systems* (2nd ed.). Cambridge University Press.

Hirose, T. (2005). Development of the Monsi–Saeki theory on canopy structure and function. *Annals of Botany*, 95, 483–494.

Hoogenboom, G., Jones, J. W., & Boote, K. J. (1992). Modeling growth, development and yield of grain legumes using SOYGRO, PNUTGRO, and BEANGRO: A review. *Transactions of the ASAE*, 35, 2043–2056.

Hoogenboom, G., Porter, C. H., Boote, K. J., Shelia, V., Wilkens, P. W., Singh, U., White, J. W., Lizaso, J. I., Moreno, L. P., Pavan, W., Ogoshi, R., Hunt, L. A., Tsuji, G. Y., & Jones, J. W. (2019). The DSSAT crop modeling ecosystem. In K. J. Boote (Ed.), *Advances in crop modeling for a sustainable agriculture* (pp. 173–216). Burleigh Dodds Science. http://dx.doi.org/10.19103/AS.2019.0061.10

Hoogenboom, G., Porter, C. H., Shelia, V., Boote, K. J., Singh, U., White, J. W., Hunt, R., Ogoshi, J. I., Lizaso, J., Koo, S., Asseng, A., Singels, L., Moreno, P., Jones, J. W., & Jones, J. W. (2021). *Decision Support System for Agrotechnology Transfer (DSSAT) Version 4.8*. DSSAT Foundation. https://DSSAT.net

IBSNAT. (1989). *Decision Support System for Agrotechnology Transfer, Version 2.1: User guide*. Department of Agronomy and Soil Science, College of Tropical Agriculture and Human Resources, University of Hawaii.

IBSNAT. (1993). *The IBSNAT decade. Honolulu, HI: Department of Agronomy and Soil Science, College of Tropical Agriculture and Human Resources*. University of Hawaii.

Jones, C. A. (1983). A survey of the variability in tissue nitrogen and phosphorus concentrations in maize and grain sorghum. *Field Crops Research*, 6, 133–147.

Jones, C. A., & Kiniry, J. R. (1986). *CERES-Maize: A simulation model of maize growth and development*. Texas A&M University Press.

Jones, J. W., Hoogenboom, G., Porter, C. H., Boote, K. J., Batchelor, W. D., Hunt, L. A., Wilkens, P. W., Singh, U., Gijsman, A. J., & Ritchie, J. T. (2003). DSSAT cropping system model. *European Journal of Agronomy*, 18, 235–265.

Jones, J. W., Hunt, L. A., Hoogenboom, G., Godwin, D. C., Singh, U., Tsuji, G. Y., Pickering, N. P., Thornton, P. K., Bowen, W. T., Boote, K. J., & Ritchie, J. T. (1994). Input and output files. In G. Y. Tsuji, J. W. Jones, & S. Balas (Eds.), *DSSAT v3* (pp. 1–94). University of Hawaii.

Keeling, C. D., Piper, S. C., Bacastow, R. B., Wahlen, M., Whorf, T. P., Heimann, M., & Meijer, H. A. (2001). *Exchanges of atmospheric CO_2 and $^{13}CO_2$ with the terrestrial biosphere and oceans from 1978 to 2000: I. Global aspects (SIO Reference Series, 01–06)*. Scripps Institution of Oceanography.

Kumudini, S., Andrade, F. H., Boote, K. J., Brown, G. A., Dzotsi, K. A., Edmeades, G. O., Gocken, T., Goodwin, M., Halter, A. L., Hammer, G. L., Hatfield, J. L., Jones, J. W., Kemanian, A. R., Kim, S.-H., Kiniry, J., Lizaso, J. I., Nendel, C., Nielsen, R. L., Parent, B., … Tollenaar, M. (2014). Predicting maize phenology: Intercomparison of functions for developmental response to temperature. *Agronomy Journal*, 106, 2087–2097. https://doi.org/10.2134/agronj14.0200

Lindquist, J. L., Barker, D. C., Knezevic, S. Z., Martin, A. R., & Walters, D. T. (2007). Comparative nitrogen uptake and distribution in corn and velvetleaf (*Abutilon theophrasti*). *Weed Science*, 55, 102–110.

Lindquist, J. L., & Mortensen, D. A. (1999). Ecophysiological characteristics of four maize hybrids and *Abutilon theophrasti*. *Weed Research*, 39, 271–285.

Lizaso, J. I., Batchelor, W. D., & Adams, S. S. (2001). Alternate approach to improve kernel number calculation in CERES-Maize. *Transactions of the ASAE*, 44, 1011–1018.

Lizaso, J. I., Batchelor, W. D., Boote, K. J., & Westgate, M. E. (2005). Development of a leaf-level canopy assimilation model for CERES-Maize. *Agronomy Journal*, 97, 722–733. https://doi.org/10.2134/agronj2004.0171

Lizaso, J. I., Batchelor, W. D., Westgate, M. E., & Echarte, L. (2003b). Enhancing the ability of CERES-Maize to compute light capture. *Agricultural Systems*, 76, 293–311.

Lizaso, J. I., Batchelor, W. D., & Westgate, M. E. (2003a). A leaf area model to simulate cultivar specific expansion and senescence of maize leaves. *Field Crops Research*, 80, 1–17. https://doi.org/10.1016/S0378-4290(02)00151-X

Lizaso, J. I., Boote, K. J., Cherr, C. M., Scholberg, J. M. S., Casanova, J. J., Judge, J., Jones, J. W., & Hoogenboom, G. (2007a). Developing a sweet corn simulation model to predict fresh market yield and quality of ears. *Journal of the American Society for Horticultural Science*, 132, 415–422.

Lizaso, J. I., Boote, K. J., Jones, J. W., Porter, C. H., Echarte, L., Westgate, M. E., & Sonohat, G. (2011). CSM-IXIM: A new maize simulation model for DSSAT version 4.5. *Agronomy Journal*, 103, 766–779. https://doi.org/10.2134/agronj2010.0423

Lizaso, J. I., Fonseca, A. E., & Westgate, M. E. (2007b). Simulating source-limited and sink-limited kernel set with CERES-Maize. *Crop Science*, 47, 2078–2088. https://doi.org/10.2135/cropsci2006.08.0533

Lizaso, J. I., Ruiz-Ramos, M., Rodríguez, L., Gabaldon-Leal, C., Oliveira, J. A., Lorite, I. J., Rodríguez, A., Maddonni, G. A., & Otegui, M. E. (2017). Modeling the response of maize phenology, kernel set, and yield components to heat stress and heat shock with CSM-IXIM. *Field Crops Research*, 214, 239–252.

Otegui, M. E., & Bonhomme, R. (1998). Grain yield components in maize: I. Ear growth and kernel set. *Field Crops Research*, 56, 247–256.

Parton, W. J., & Logan, J. A. (1981). A model for diurnal variation in soil and air temperature. *Agricultural Meteorology*, 23, 205–216.

Penning de Vries, F. W. T., Jansen, D. M., ten Berge, H. F. M., & Bakema, A. (1989). *Simulation of ecophysiological processes of growth in several annual crops* (Simulation Monograph 29). PUDOC.

Penning de Vries, F. W. T., & van Laar, H. H. (1982). Simulation of growth processes and the model BACROS. In F. W. T. Penning de Vries & H. H. van Laar (Eds.), *Simulation of plant growth and crop production* (pp. 114–135). PUDOC.

Plénet, D., & Lemaire, G. (2000). Relationships between dynamics of nitrogen uptake and dry matter accumulation in maize crops: Determination of critical N concentration. *Plant and Soil*, 216, 65–82. https://doi.org/10.1023/A:1004783431055

Ritchie, J. T. (1986). The CERES-Maize model. In C. A. Jones & J. R. Kiniry (Eds.), *CERES-Maize: A simulation model of maize growth and development* (pp. 1–6). Texas A&M University Press.

Ritchie, J. T., & Otter, S. (1985). Description and performance of CERES-Wheat: A user-oriented wheat yield model. In *ARS wheat yield project* (ARS-38) (pp. 159–175). National Technical Information Service.

Ritchie, J. T., Singh, U., Godwin, D. C., & Bowen, W. T. (1998). Cereal growth, development, and yield. In G. Y. Tsuji, G. Hoogenboom, & P. K. Thornton (Eds.), *Understanding options for agricultural production* (pp. 79–98). Kluwer Academic Press.

Tanaka, A., & Yamaguchi, J. (1972). Dry matter production, yield components and grain yield of the maize plant. *Journal of the Faculty of Agriculture, Hokkaido University*, 57, 71–132.

Thiagarajah, M. R., & Hunt, L. A. (1982). Effects of temperature on leaf growth in corn (*Zea mays*). *Canadian Journal of Botany*, 60, 1647–1652.

Tollenaar, M., Dwyer, L. M., & Stewart, D. W. (1992). Ear and kernel formation in maize hybrids representing three decades of grain yield improvement in Ontario. *Crop Science*, 32, 432–438. https://doi.org/10.2135/cropsci1992.0011183X003200020030x

Tollenaar, M., Dzotsi, K., Kumudini, S., Boote, K. J., Chen, K., Hatfield, J., Kumudini, S., Boote, K., Chen, K., Hatfield, J., Jones, J. W., Lizaso, J. I., Nielsen, R. L., Thomison, P., Timlin, D. J., Valentinuz, O., Vyn, T. J., & Yang, H. (2018). Modeling the effects of genotypic and environmental variation on maize phenology: The phenology subroutine of the AgMaize crop model. In J. L. Hatfield, M. V. K. Sivakumar, & J. H. Prueger (Eds.), *Agroclimatology: Linking agriculture to climate* (Agronomy Monograph 60 (pp. 173–200). ASA, CSSA, and SSSA.

Tsuji, G. Y., Hoogenboom, G., & Thornton, P. K. (Eds.) (1998). *Understanding options for agricultural production* (Systems approaches for sustainable agricultural development no. 7). Kluwer Academic Publishers.

Warrington, I. J., & Norton, R. A. (1991). An evaluation of plant growth and development under various daily quantum integrals. *Journal of the American Society for Horticultural Science*, 116, 544–551.

Wilkerson, G. G., Jones, J. W., Boote, K. J., Ingram, K. T., & Mishoe, J. W. (1983). Modeling soybean growth for crop management. *Transactions of the ASAE*, 26, 63–73.

Yakoub, A., Lloveras, J., Biau, A., Lindquist, J. L., & Lizaso, J. I. (2017). Testing and improving the maize models in DSSAT: Development, growth, yield, and N uptake. *Field Crops Research*, 212, 95–106.

Yan, W., & Hunt, L. A. (1999). An equation for modelling the temperature response of plants using only the cardinal temperatures. *Annals of Botany*, 84, 607–614.

10

Kurt C. Kersebaum and
Claudio O. Stöckle

Simulation of Climate Change Effects on Evapotranspiration, Photosynthesis, Crop Growth, and Yield Processes in Crop Models

Abstract

Climate change is a challenge for global food security. Crop models are indispensable tools to assess the impact of changing climatic boundary conditions on food production. However, comparisons among international models have revealed significant uncertainty; there is a wide range of model results, likely due to large variation in the description and parametrization of important processes in each model. In this chapter, we provide an overview of process approaches used within crop models that help with impact assessments, and which are relevant to specific responses to climate change. This effort focuses on evapotranspiration, crop growth and development, and crop responses to water availability and elevated CO_2. For process comparisons, we use the model approaches and parametrization for wheat, which is one of the best investigated and most frequently simulated crops. We also look at processes that are currently not well represented in crop models due to the lack of suitable data.

Introduction

Global agriculture is facing the challenge of feeding nine billion people by 2050 (Godfray et al., 2010), while at the same time mitigating further negative impact of intensive agricultural production on the environment and climate. Moreover,

Modeling Processes and Their Interactions in Cropping Systems: Challenges for the 21st Century, First Edition.
Edited by Lajpat R. Ahuja, Kurt C. Kersebaum, and Ole Wendroth.
© 2022 American Society of Agronomy, Inc. / Crop Science Society of America, Inc. / Soil Science Society of America, Inc.
Published 2022 by John Wiley & Sons, Inc.

agriculture is among the sectors most vulnerable to climate change (e.g., Godfray et al., 2010; Teixeira et al., 2013). Agriculture is directly exposed to variable weather conditions, has strong linkages to water resource management, and is tied to the mitigation of greenhouse gas (GHG) emissions (IPCC, 2014) since it contributes significantly to the emission of the greenhouse gases carbon dioxide (CO_2), nitrous oxide (N_2O), and methane (CH_4). How climate change and extreme weather events translate into agricultural risk depends on the magnitude and frequency of the impacts, as well as the system's vulnerability, meaning its ability to cope with its consequences (e.g., Parry, 2007). To cope with global change, the adaptation of agricultural production systems is required to minimize the risk of crop failures and yield losses, and to exploit new opportunities derived from changing climatic conditions. Risks due to inter-annual weather variability are usually managed through changing cultivation practices, such as by shifting sowing dates, the choice of genotypes, the introduction of new crops or changing the contribution and frequency of crop species in a rotation, or changing tillage practice and crop residue handling (Olesen et al., 2011; Porter et al., 2014). Therefore, it is essential to develop and evaluate methods under changing climatic boundary conditions since present best management practices may no longer be optimal under future conditions. Climate change impact assessments have to consider genotype × environment × management (GxExM) interactions (Beres et al., 2020). These assessments may also require integrated assessment and modeling (IAM) approaches linking biophysical and economic models to assess the direction and magnitude of projected climate change impacts (Ewert et al., 2015).

The wide variability of sites and agricultural systems suggests it is difficult to assume that results from a specific location are transferable (Powlson et al., 2016). Therefore, process-oriented modeling that considers the various interactions is suggested to provide site-specific evaluation of crop performance and recommendations of management options. Agro-ecosystem models considering the complex interactions in the atmosphere–plant–soil system are essential tools for an integrated impact assessment (Ewert et al., 2015) to reckon the effects of projected climate change on various ecosystem services, including crop production (e.g., Lesk et al., 2016; Lobell & Gourdji, 2012; Wheeler & von Braun, 2013), environmental effects such as water use and pollution (Jiménez Cisneros et al., 2014), GHG emissions (Lynch et al., 2021), resource use efficiency (Schimel et al., 2008), and long-term effects on soil properties, for example, soil carbon stocks (Yigini & Panagos, 2016). These models have the main advantage of conducting many interactive simulations under various site and climatic conditions beyond their experimental basis of development (in silico experiments) within a relatively short time to evaluate site-specific impacts and adaptation options.

Changing the climatic boundary conditions has a comprehensive impact on energy and matter fluxes in the atmosphere–crop–soil–water system. Figure 10.1 illustrates climatic effects on and relationships among the main cycles of water, carbon, and nutrients. In this chapter, we describe the approaches of selected process-oriented crop models that are often applied in climate impact assessment studies. Specifically, we focus on the processes of evapotranspiration (ET), crop carbon acquisition, growth, development, yield, and their responses to changing climatic conditions. For process comparisons we used mainly model approaches and parametrization for wheat, which

climatic drivers	affected basic processes	balance effects	effects on nutrient dynamics

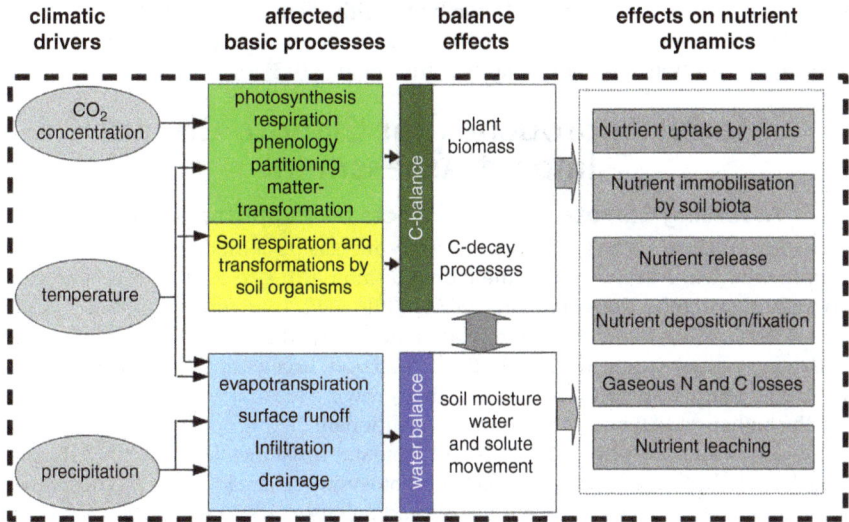

Figure 10.1 General relations between main climate change variables, water and carbon cycle a nutrient dynamics (Niklas et al. 2007, modified).

is one of the best investigated and most frequently simulated crops. However, we try to show deviating approaches for other relevant crops as well.

General Trends in Projected Climate Change

Observed climate trends show an increase in land temperature (2006–2015) by ~1.5 °C compared with pre-industrial levels (EEA, 2017; IPCC, 2013) and an increase of the frequency, intensity and duration of heat waves (Alexander et al., 2006; Christidis et al., 2015; Gourdji et al., 2013; Lorenz et al., 2019). Precipitation shows a more spatially differentiated picture with increased probability of dry spells (Alexander et al., 2006; Dai, 2013; Dai & Zhao, 2017; Lobell et al., 2011; Spinoni et al., 2014; Trnka et al., 2019) in the midwestern United States and southern European regions, and an increase of precipitation during winter in the northern latitudes (Davenport et al., 2021; Lobell & Gourdji, 2012; Trnka et al., 2011). Warming has prolonged the thermal growing season and frost-free period for crops across all of Europe since the 1980s (EEA, 2017) and has significantly reduced the number of extreme cold days (Lorenz et al., 2019). Extended growing seasons facilitate the introduction of new crops or the extension of these crops to higher latitudes and altitudes, which were previously limited by low temperature or shorter growing seasons; this facilitates adaptation via crop avoidance of extreme heat through earlier planting. The probability of multiple adverse climate events during a growing season including heavy rains (Degaetano, 2009; Lehmann et al., 2015) and storms has increased (Trnka et al., 2014). Cropping systems have already been affected by climate change (Olesen et al., 2011), which influenced the tendency toward stagnation of cereal grain yields and increased yield variability (Brisson et al., 2010; Grassini et al., 2013). Projected climate trends are likely to enhance the risk posed by extreme weather events under climate change scenarios (Chen & Ford, 2021; Forzieri et al., 2016; Grillakis, 2019; Trnka et al., 2014; Trnka et al., 2019). The projected increase in the variability of dry and wet events combined with event uncertainty (particularly in the United States) poses a challenge for selecting the most appropriate adaptation strategies and for setting adequate policies (Trnka et al., 2013).

Model Approaches for Climate Change Impact Assessment

Assessing Climate Change Effects on Evapotranspiration

Global warming will affect the water cycle by increasing temperature and changing precipitation, while at the same time global dimming due to an increase of aerosols may reduce radiation (Ramanathan, 2007). While a warmer atmosphere can take up more water leading to a decrease in relative humidity, the change in radiation alters the driving energy input for ET. On the other hand, acceleration of crop phenology by elevated temperature may lead to a shorter growing season, which may counterbalance the higher daily water consumption. Additionally, rising concentration of atmospheric CO_2 influences photosynthesis and stomatal resistance (Ainsworth & Rogers, 2007). Elevated CO_2 has a positive effect on photosynthesis, mainly for C3 crops, resulting in higher biomass production, which could enhance crop transpiration through a higher crop surface. On the other hand, increased stomatal resistance will lead to lower water consumption per unit leaf area per unit mass (Manderscheid et al., 2015; Manderscheid & Weigel, 2007), while CO_2 flux remains unchanged due to the higher gradient. Therefore, the net effect on water consumption will depend on the dominant response, whether that be larger crop surface or increased stomatal resistance (Kimball, 2016).

Calculating Crop Potential Evapotranspiration

Although many formulas exist for the calculation of potential ET, most crop models applied for climate impact assessment focus on just a few algorithms following the global ranking of existing formula provided by Smith et al. (1998) based on Jensen et al. (1990). Nevertheless, a comparison of 29 maize crop models showed a surprisingly high range of actual ET estimates, which varied by a factor of more than two even after intensive calibration (Kimball et al., 2019). The authors identified that the estimation of potential ET was the main reason for the high variance between models although 45% of the modelers used a Penman–Monteith approach, and 31% the formula from Priestley and Taylor (1972). Other methods were based on Penman (1948), Shuttleworth and Wallace (1985) or Hargreaves (1975). The reasons as to why different ET formulas were specifically implemented in models might be related to the availability of standard weather variables or a better performance of the selected approach under the climatic conditions of their main application region. Smith et al. (1998) showed that ranking of ET formula changed if applied in arid or humid regions. In some cases, slightly modified versions of the FAO-56 (Allen et al., 1998) or Priestley–Taylor methods were applied. Moreover, modelers used different methods to estimate potential crop ET (unstressed crop transpiration plus evaporation from wet soil) from a reference crop ET (either short grass or alfalfa references), mostly following the standard methods described in the FAO-56 guideline (Allen et al., 1998) based on a single or dual crop coefficient (ratio of crop to reference ET). However, some models used more dynamic approaches, which considered other factors such as leaf area index (LAI), dynamic canopy conductance depending on photosynthesis, canopy temperature, or CO_2 concentration. Details can be found in Kimball et al. (2019). Since some models linked the crop coefficient phases to the phenological development simulated by the model, the individual time courses of crop coefficients created a temporal variability in the ET simulation between models even when they use the same approach. The Penman–Monteith equation, derived by combining the energy balance and vapor and heat transfer equations, is the most complete biophysical equation which often performs the best in comparisons with measured ET (e.g., Allen, 1986; Benli et al., 2010; Jensen et al., 1990; Smith et al., 1998). Most other equations are simplifications of Penman–Monteith, intended to relax input requirements, but requiring local calibration and decreasing accuracy. Implementation of Penman–Monteith ET has been standardized, incorporating the use of a reference crop with specified parameters that relates to potential crop ET via the use of crop specific parameters, but other biophysically complete approaches of varying degrees of complexity are also available. Given the large number of approaches to calculate ET, it is essential for modelers and model users to be aware of their assumptions, limitations, and the need for proper calibration.

Role of Atmospheric CO_2 Concentration on Crop Transpiration

The effect of increasing atmospheric CO_2 concentrations has been studied in several experiments either using closed or open chambers, or Free-Air-Carbon-Enrichment (FACE) experiments (Ainsworth & Long, 2005), which were thought to provide more realistic production conditions (O'Leary et al., 2015; Tubiello & Ewert, 2002) than closed-chamber or open-chamber experiments. However, there is a contentious debate about the apparent discrepancies between FACE and enclosure studies. While some authors state that crop responses to CO_2 as observed in enclosure systems overdraw realistic field responses (e.g., Long et al., 2006), others are convinced that responses in FACE experiments are underestimating real responses due to uncertainties in maintaining CO_2 concentration treatments (e.g., Bunce, 2012). Nonetheless, Ziska and Bunce (2007) demonstrated that relative yield stimulations in response to future CO_2 concentrations obtained with a number of enclosure methodologies were quantitatively consistent with FACE results for rice (*Oryza sativa* L.), soybean (*Glycine max* L.) and wheat (*Triticum aestivum* L.), when properly scaled. Since the crop models applied for climate change impact assessment are usually designed to operate at the field or regional scale with variable meteorological inputs, modelers may favor to calibrate and test their models against FACE experimental data, since adjusting the model to a completely controlled environment would require additional efforts.

The observation that stomata respond to CO_2 concentration has been made many times since the earliest work by Linsbauer (1916) and Freudenberger (1940) (cited in Morison, 1985). A decline in stomatal conductance ranging 16–41% has been reported for crops and grasses for an increase in CO_2 from 366–567 $\mu mol\ mol^{-1}$ (often expressed as "ppm") on average in FACE environments (Ainsworth & Rogers, 2007). However, Boote et al. (2013) emphasized, that reduction of stomata conductance by 40% does not correspond to a similar reduction in transpiration due to changes in the foliar energy balance leading to an increase of leaf temperature and consequently to a relative increase in canopy transpiration. Therefore, only 10–12% reduction for C3 were observed in canopy transpiration with a 40% reduction in leaf conductance for doubled CO_2 concentration (Bernacchi et al., 2007; Boote et al., 1997; Hatfield et al., 2008, 2011). Reviewing FACE experimental data, Kimball (2016) found reductions of ET of zero (cotton) to 13% (sorghum) under sufficient N and water supply. Manderscheid and Weigel (2007) reported a reduction of wheat water use of about 20 mm per growing season in a humid environment at 550 $\mu mol\ mol^{-1}$ CO_2 compared with 350 $\mu mol\ mol^{-1}$.

It is well documented that the sensitivity of stomata to CO_2 varies with the conditions during growth and the conditions during measurement such as the leaf age, light intensity, humidity, and temperature (Morison, 1985). The early findings that stomata of C3 species are less sensitive to CO_2 than those of C4 species (e.g., Akita & Moss, 1973) were confirmed by a larger reduction in canopy transpiration for C4 crops with doubled CO_2. Kim et al. (2006) and Chun et al. (2011) found a 22% reduction in transpiration for maize; reductions of 18 and 20% were reported for sorghum and maize, respectively, by Allen et al. (2011), all performed in sunlit soil–plant–atmosphere research (SPAR) chambers. Triggs et al. (2004) found a 13% reduction in ET of sorghum under FACE in Arizona by increasing CO_2 from 368 to 561 $\mu mol\ mol^{-1}$. Results of a FACE-experiment on maize in Germany reported little yield sensitivity of maize on elevated CO_2 under good water supply, but a distinct yield increase of nearly 20% under dry conditions compared with the ambient concentration (Manderscheid et al., 2014; Manderscheid et al., 2015). Although a direct effect of increasing CO_2 concentration on C4 crop photosynthesis is not expected, elevated CO_2 consistently ameliorates reductions in net CO_2 assimilation caused by drought stress in C4 species (Ghannoum et al., 2000; Leakey, 2009; Samarakoon & Gifford, 1996; Wall et al., 2001), but how stomatal and non-stomatal factors contribute to the response is still uncertain (Markelz et al., 2011).

Tubiello and Ewert (2002) reviewed models and their algorithms used for describing CO_2 impact on crops and identified differences between algorithms and their parameterization as an additional source of uncertainty. Consequently, they emphasized the need for intensive evaluation of CO_2 concentration algorithms implemented in biophysical crop models with experimentally obtained crop responses. However, Vanuytrecht and Thorburn (2017) had to conclude, fifteen years later, that transpiration and stomatal responses of model approaches varied widely and showed a decrease of stomatal conductance between 35 and 90% among models when doubling CO_2 to 700 $\mu mol\ mol^{-1}$. Moreover, model algorithms to reflect the effect on stomatal resistance are mainly based on findings for C3 crops (e.g., Durand et al., 2018; Kersebaum et al., 2009; Tubiello & Ewert, 2002). While several controlled environment experiments exist for C4 crops as mentioned above, there is still a lack of field experimental studies addressing the response of C4 crops to changes in atmospheric CO_2 concentration and water availability to test model approaches under nearly natural environmental conditions (Leakey, 2009). Therefore, uncertainty in modeling the effect of elevated CO_2 on crop water use of C4 crops is still high and positive yield response under dry conditions at elevated CO_2 was clearly underestimated by all models in an ensemble simulation of the above-mentioned German maize FACE experiment (Durand et al., 2018).

Several approaches ranging from simplistic empirical relations to more process based semi-complex approaches are implemented in current versions of crop models (Vanuytrecht & Thorburn, 2017). Only few models employ leaf-level energy balance approaches and up-scaling to crop canopy simulating instantaneous (hourly) energy balance of equilibrium canopy and soil temperature and leaf conductance, and thus do not require computing potential ET (Fleisher et al., 2015; Grant & Baldocchi, 1992; Kim et al., 2012; Pickering et al., 1995; Twine et al., 2013; Yang et al., 2009; Yin & van Laar, 2005). Some model platforms can select between different response types depending on the crop to be simulated. Some selected approaches are presented below.

The APSIM model (Holzworth et al., 2014) system has implemented several approaches. The most simplistic one is a linear relationship of transpiration efficiency (TE, g mm^{-1}) to atmospheric CO_2 concentration (Vanuytrecht & Thorburn, 2017), which is implemented in the algorithm for daily transpiration Ta (mm) of C3 plants:

$$Ta = \min\left(\frac{\Delta B}{TE}, Ws\right) = \min\left(\frac{\Delta B \cdot VPD}{f_{TE} \cdot ft_{CO2}}, Ws\right) \tag{10.1}$$

where Ws is the soil water supply for that day depending on soil water content, soil and root properties (mm); ΔB is the daily biomass production (g m^{-2}); VPD is the vapor pressure deficit (kPa); f_{TE} is the dependency of the transpiration efficiency TE (g mm^{-1}) on VPD (g kPa mm^{-1}); and ft_{CO2} is the TE correction factor for atmospheric CO_2, where TE increases linearly from 1 to 1.37 with an increase of CO_2 from 350 to 700 µmol mol^{-1}.

A more complex approach in APSIM is the formula provided to simulate C4 grass ET, where stomatal conductance is dynamically adapted by a modifier (f_{CO2}), different than the one introduced in Eq. 10.1:

$$f_{CO2} = g_{c,min} + \left(g_{c,max} - g_{c,min}\right) \cdot \frac{\left(1 - g_{c,min}\right) \cdot [CO2]_a^b}{\left(g_{c,max} - 1\right) \cdot [CO2]_e^b + \left(1 - g_{c,min}\right) \cdot [CO2]_a^b} \tag{10.2}$$

where $(CO_2)_e$ is the atmospheric CO_2 concentration (µmol mol^{-1}); $(CO_2)_a$ is the reference CO_2 concentration (380 µmol mol^{-1} by default); $g_{c,min}$ and $g_{c,max}$ are minimum and maximum canopy conductance, respectively (0.2 and 1.25 cm s^{-1} by default); and b is a curvature coefficient (2.5 by default).

In the EPIC model (Stöckle et al., 1992), the effect of elevated CO_2 (C_a) on crop transpiration is simulated by modifying the stomatal resistance as a reciprocal value of leaf conductance g_d. Since leaf conductance depends on the VPD, the VPD depending leaf conductance during a daily time step is first calculated on an hourly basis (g_{1h}) during the daytime neglecting transpiration during the night.

$$g_{1h}(VPD, CO_2) = g_{1h}(VPD) \cdot \left(1.4 - 0.4 \cdot \frac{C_a}{330}\right) \tag{10.3}$$

The daily leaf resistance r_l is calculated by the reciprocal value of daily stomata conductance, which results from the sum of hourly $g_{1h}(VPD, CO_2)$ by weighting them by the share of hourly reference ET to the daily reference ET. This is used to calculate the canopy resistance within the Penman–Monteith equation (Monteith, 1965). Canopy resistance r_c is calculated according to Allen et al. (1998) as

$$r_c = \frac{r_l}{0.5 \cdot LAI} \tag{10.4}$$

The STICS model (Brisson et al., 2003) used the approach from Stöckle et al. (1992) as well to calculate stomatal resistance, but applies an ET formula adapted from Shuttleworth and Wallace (1985). The latter formula is also used in RZWQM2 (Ma et al., 2012), which has implemented the DSSAT approach to consider CO_2 effects.

In CropSyst crop transpiration is calculated based on a modification of the FAO-56 ET approach (Allen et al., 1998), using a standardized short-grass reference crop with specified parameters and a crop coefficient that assumes the hypothetical condition of a crop canopy intercepting all incoming solar radiation, which is then scaled down using the actual canopy fraction of solar radiation interception to estimate potential crop transpiration. The response to CO_2 consists of two steps: (a) change of stomatal conductance as a function of CO_2, and (b) modification of crop transpiration. Stomatal conductance reduction with increased CO_2 has been found linear in the range 330–660 μmol mol^{-1}, with the relative responses being similar for C3 and C4 species (Morison, 1985). Nevertheless, the apparent linearity breaks down at larger CO_2 and the stomatal conductance response is better represented by a non-linear relationship as the quadratic equation given by Allen (1990), which was adopted in CropSyst. The potential crop transpiration determined for baseline conditions is multiplied by a factor F to account for the increased canopy resistance (reduced conductance) in response to elevated CO_2.

$$F = \frac{\Delta + \gamma \cdot \left(1 + \frac{r_{co}}{r_a}\right)}{\Delta + \gamma \cdot \left(1 + \frac{r_{ce}}{r_a}\right)} \tag{10.5}$$

where Δ and γ are the slope of the saturation vapor pressure function of temperature (°C kPa^{-1}) and the psychrometric constant (°C kPa^{-1}), respectively. R_a is the aerodynamic resistance, and r_{co} and r_{ce} are the canopy resistance for baseline and a given elevated CO_2, respectively.

The WOFOST model (Boogaard et al., 1998) uses a simple linear reduction factor to adjust the calculated potential ET to elevated CO_2 assuming no reduction (f = 1) at 365 μmol mol^{-1} and a maximum reduction of 10% (f = 0.9) at 840 μmol mol^{-1} (Rötter et al., 2011).

The HERMES (Kersebaum & Nendel, 2014) and the MONICA (Nendel et al., 2011) models both modify the stomatal resistance in the FAO-56 Penman–Monteith equation (Allen et al., 1998) using an equation suggested by Yu et al. (2001) as

$$r_s = [CO_2] \cdot \frac{\left(1 + \frac{VPD}{VPD_0}\right)}{a \cdot A_g} \tag{10.6}$$

where a is a constant; A_g denotes the gross photosynthesis rate; VPD_0 is a parameter describing the response of r_s to air water vapor deficit VPD; and CO_2 is the CO_2 concentration at leaf level, which was set equal to the elevated atmospheric CO_2 concentration (μmol mol^{-1}). VPD_0 and a were used for parameter calibration using data at two CO_2 concentration levels of the German FACE experiment (Weigel & Dämmgen, 2000). Since the formula is applied to the calculation of the grass reference ET, the relation is used for all crops generically.

Uncertainty from Model Ensembles

Many widely used crop models have been tested against the same limited set of CO_2 response data from either open-top chambers or FACE experiments. However, their responses to changes of temperature, precipitation and atmospheric CO_2 vary substantially. A recent study of 29 maize models (Kimball et al., 2019) showed large differences in simulated potential ET significantly affecting differences in actual model ET, with partitioning among soil evaporation and transpiration affecting transpiration variability. The ensemble included a few models calculating the energy balance at canopy level (Agro-IBIS [Twine et al., 2013], MAIZSIM [Kim et al., 2012], Jules-crop [Best et al.,

2011], SIMPLACE-Lintul5_Tcanopy [Webber et al., 2016], Expert-N-Gecros [Yin & van Laar, 2005]), which performed mostly well in estimating crop ET. Ahmed et al. (2019) used a small ensemble of five models to evaluate the uncertainty in CO_2 response. However, they documented just the differences in water use efficiency (WUE), which did not distinguish between the CO_2 effect on transpiration and photosynthesis. Kersebaum et al. (2016) estimated the uncertainty caused by different models to estimate the water footprint, which is the inverse value of WUE. Using data of two seasons of winter wheat and two nitrogen levels from a FACE experiment in Germany at 374 and 550 $\mu mol \, mol^{-1}$, they found that the simulated response to elevated CO_2 of six models was different. While AQUACROP, HERMES and APSIM showed a decrease in transpiration for the elevated CO_2 concentration of 34, 40, and 18 mm (9, 12, and 10% less than their individual baseline transpiration), respectively, the two DSSAT (Versions 4.5 and 4.6, which differed mainly in their way of crop parameter calibration) simulations, and DAISY (Hansen et al., 2012) showed nearly no response and CROP-SYST and SWAP/WOFOST (Kroes et al., 2009) showed an increase of 19 and 15 mm (10 and 5% more than their individual baseline transpiration), respectively. Inter-model variability was higher than the inter-annual variability and CV% was 18% for ET under both CO_2 concentrations and 29 and 25% for crop transpiration T_r at 374 and 550 $\mu mol \, mol^{-1}$ CO_2, respectively. Cammarano et al. (2016) used 16 wheat simulation models to quantify sources of model uncertainty and to estimate the relative changes and variability between models for simulated water use, water use efficiency (WUE was defined as water use per unit of grain dry mass), transpiration efficiency (Teff defined as transpiration per kilogram of grain yield dry mass), grain yield, crop transpiration and soil evaporation at increased temperatures and elevated atmospheric CO_2. The greatest uncertainty in simulating water use, potential ET, crop transpiration and soil evaporation was caused by differences in modeling, crop transpiration that accounted for 50% of the total variability among models. Although differences between models are partly related to the ET formula they used, the ensemble study revealed that partitioning of ET between soil evaporation and crop transpiration contributed more to the variability in simulated water use. While the response to increasing temperature showed high variance among models, the responses to elevated CO_2 were rather consistent. This may not come as a surprise since typically models include many processes that respond to temperature (e.g., phenology, growth, ET) while the response to CO_2 is often simplified at a higher level of integration (Kersebaum & Nendel, 2014). In contrast, the ensemble employed by Vanuytrecht and Thorburn (2017) showed that production responses revealed some consistency among models up to moderately high CO_2 (around 700 μ $mol \, mol^{-1}$), whereas simulated transpiration and stomatal responses varied more widely (e.g., a decrease in stomatal conductance varying between 35 and 90% for CO_2 doubling to 700 $\mu mol \, mol^{-1}$).

Simulation of Climate Change Effects on Crop Growth and Yield

Climate change is expected to affect crop growth mainly by increasing temperatures, shifting distribution of precipitation, changing amount and intensity of precipitation, and rising atmospheric carbon dioxide concentration. Properly describing the interactions of crop growth, soil processes, and weather variables in simulation models is essential to interpret downscaled global climate model (GCM) outputs for climate impact assessment in agriculture. However, several crop model ensemble studies during the last decade reveal distinct differences among models regarding their responses to changing climate variables (Asseng et al., 2013, 2015; Bassu et al., 2014; Palosuo et al.,

2011; Rötter et al., 2012). The variation in responses is related to the diversity of response functions to main weather variables in crop models, but also to differences in the description of soil–crop interactions, since vulnerability of crops to climate change is also related to the buffering capacity of soils against adverse conditions, such as plant available water in the root zone bridging dry periods (Abdallah et al., 2021; He & Wang, 2019; Kersebaum & Nendel, 2014; Williams et al., 2016).

Biomass Production

Model approaches for biomass production can be mainly divided into models which follow the radiation-use efficiency approach (RUE), models that use transpiration-use efficiency (TUE or WP), and models using gross photosynthesis minus maintenance and growth respiration approach.

The RUE-based approach is the most frequently used in crop models (e.g., APSIM, DSSAT, EPIC, STICS), with RUE describing the conversion from incoming photosynthetic active radiation (PAR) intercepted by the canopy to biomass formation (g dry mass MJ^{-1} of PAR intercepted by the canopy), normally using one seasonal RUE parameter or two specified for vegetative and reproductive phases, which can fluctuate as a function of temperature and CO_2.

The gross photosynthesis minus respiration approach was described in early crop model developments from the Wageningen group, for example, de Wit and Goudriaan (1978), which was later adopted by several models such as SUCROS, WOFOST, DAISY, HERMES, MONICA, and others. The approach uses a light response curve, which is mainly characterized by the initial light use efficiency at the compensation point and the maximum assimilation rate at light saturation Amax. Light interception within the crop canopy is integrated to calculate the gross assimilation rate per time step from incoming radiation (Goudriaan & van Laar, 1978). Assimilates are reduced by maintenance and growth respiration. Maintenance respiration depends on the composition of biomass; enzymes, for instance, need more maintenance than cellulose or lignin (Amthor et al., 2019), which means that crop organs with a larger contribution of structural biomass have lower respiration rates than organs with higher protein content. Although composition varies during crop development, typical respiration rates for specific crop organs are often used. Typical rates at 20 °C for leaves are 0.03 d^{-1} for leaves, 0.015 d^{-1} for stems, and 0.01 d^{-1} for roots across different crops (Van Heemst, 1988). For storage organs 0.01 d^{-1} was estimated for the majority of crops. However, due to differences in their composition maintenance rates for storage organs can vary between 0.003 (cassava) and 0.023 d^{-1} (sunflower). While gross photosynthesis shows an optimum function on temperature, respiration rates defined for specific temperatures are following an exponential increase with temperature with a doubling of the respiration rate for every 10 °C temperature increase (Penning de Vries et al., 1979). Therefore, resulting net biomass production shows a well-defined temperature optimum.

Only few models (e.g., CropSyst and AQUACROP) use the conversion of actual crop transpiration to biomass production based on concepts formalized by Tanner and Sinclair (1983) and extended by Steduto et al. (2007). This approach requires the models to provide reasonable estimations of daily crop transpiration. The CropSyst model calculates daily TUE based on a standardized parameter (TUE*) representing TUE at a daylight-hours average VPD (D) of 1 kPa, which is modified by daily D to the power of a β coefficient (~0.5) (Eq. 10.7). D is approximated as 66% of the maximum

VPD of the day. The value of TUE* is modified yearly as a function of CO_2. The daily biomass gain (B) is calculated as shown in Eq. 10.8.

$$TUE = \frac{TUE^*}{D^\beta} \tag{10.7}$$

$$B = Tr\,TUE \tag{10.8}$$

The AQUACROP model uses a normalized biomass water productivity WP* parameter (Steduto et al., 2007) accumulating B to the end of the season from daily transpiration Tr and the corresponding daily grass reference evapotranspiration ET_o.

$$B = K_{sb} \cdot WP^* \sum \left(\frac{Tr}{ET_0}\right) \tag{10.9}$$

where K_{sb} is a cold temperature stress coefficient for biomass, which is used to reduce biomass production (Vanuytrecht et al., 2014). The normalization of WP involves two environmental factors: evaporative demand of the atmosphere as represented by ET_o estimated by the FAO-56 Penman–Monteith formula (Allen et al., 1998) and CO_2 for the reference year 2000. Normalization makes WP* applicable to diverse locations and seasons, accounting for ETo variations, and over a time span of years, accounting for rising CO_2. Crop specific values are provided by Raes et al. (2018) and range from $15\,\text{g m}^{-2}$ (wheat, barley, cotton, soybean), $17–20\,\text{g m}^{-2}$ (sugar beet, sunflower, rice, potato) to $33.7\,\text{g m}^{-2}$ (maize, sorghum). However, the level of calibration varies since proper calibration requires well designed field experiments providing conditions for crop growth close to optimum.

Temperature Effects

We introduce now the main approaches implemented in some models accounting for responses to environmental conditions, using wheat as a main example. A good description on temperature functions used by various models for wheat can be found in Alderman et al. (2013).

Temperature affects crop via two main effects. One is the advancement of crop phenology: Elevated temperatures may shorten the duration of the growing season, and change the relative duration of vegetative and reproductive phases of the crop cycle. On the other hand, net biomass production is affected by the temperature dependency of photosynthesis and respiration. While photosynthesis shows an optimum dependency within the range of usual air temperatures, respiration increases exponentially with temperature (Taiz & Zeiger, 1991). Additionally, extreme temperatures beyond the optimum reduce grain number and grain weight, and accelerate leaf senescence (Asseng et al., 2011; Porter & Gawith, 1999; Rezaei et al., 2015). Globally, the total area of wheat (~77 Mha) and rice (~57 Mha) cultivated land affected by heat stress during 1971–2000 was estimated by Teixeira et al. (2013) using a heat intensity index above crop specific temperature thresholds calculated throughout the thermal-sensitive period in the reproductive phase. Global yield loss caused by heat for wheat under climate change was estimated at 6% per 1 °C increase of global warming by Asseng et al. (2015).

Grain number is largely affected by sterility and abortion of grains, which lead to a non-reversible reduction in the potential yield of the crop (Barnabas et al., 2008; Spiertz, 1974). Grain size is affected by accelerated phenological development shortening the time of grain filling, while extreme temperatures may lead to cellular damage during periods with maximum sensitivity to heat stress, for example, for wheat, around booting (Ugarte et al., 2007). Critical high temperature thresholds for yield reduction (Larcher, 1995; Wang et al., 2017) differ by crop cultivar and vary with development stage, which was documented for rice (e.g., Horie et al., 1995), wheat (Wollenweber et al., 2003), and maize (Sanchez et al., 2014). The highest impact of extreme heat events on cereal crop yield is apparently manifested in reductions in number of grains when heat stress occurs at flowering or reductions in grain weight when high temperatures occur during grain filling (e.g., Talukder et al., 2014).

Many biological processes such as assimilation of carbon by photosynthesis are enzymatic processes and such processes are generally temperature dependent. Therefore, temperature functions, for example, for photosynthesis or RUE are based on early basic leaf-level experiments under controlled conditions determining CO_2 exchange at different levels of light intensions and temperatures, which were simplified and upscaled to the canopy to be used under real field conditions. The overall temperature response of CSM-CROPSIM Wheat and CERES Wheat integrated in DSSAT is determined by the integration of several individual process responses. The general shape of the relative temperature response $r(T)$ is trapezoidal (Figure 10.2c), normalized between 0 and 1, and most processes are driven by daily mean air temperature. The actual response curve for each process is defined by four cardinal temperatures: a minimum temperature at which the process is activated; a lower and upper threshold for the optimum rate of the process, and a maximum temperature above which the activity stops. Maximum specific process rates are multiplied with the relative process rates, which are determined by linear interpolation between the cardinal temperatures:

$$Ra_p = Rmax_p \cdot r(T)_p \tag{10.10}$$

where Ra_p is the actual process rate of process p at temperature T, $Rmax_p$ is the maximal process rate of process p at optimum temperature, for example, RUE, and $r(T)_p$ is the relative process rate (ranging from 1 at optimum and zero, when the process is stopped) of process p at temperature T. Cardinal temperatures can be specified for each process and development stage. A similar approach with the same shape of temperature function is used in APSIM Wheat (V7.4), APSIM-Nwheat, DSSAT-CERES, CROPSYST, Salus and Lintul-4 for adjusting RUE to the daily mean temperature (Figure 10.2c). Details for the temperature response of maize for the DSSAT modules CERES and IXIM can be found in Chapter 8 (Lizaso et al., 2021). However, computational result of daily mean temperature varies between models and in some cases models use a simulated canopy temperature for scaling RUE or gross photosynthesis.

A second type of response is used by models such as FASSET, EPIC-Wheat or AQUACROP, which use a linear or sigmoidal (AQUACROP) increase between a minimum and an optimum temperature, reaching a plateau without reduction at higher temperatures (Figure 10.2b).

Within the HERMES model a linearized temperature function from the SUCROS model (Van Keulen et al., 1982) is applied for gross photosynthesis using the relative process rate $r(T)_p$ to modify the maximum gross assimilation rate at light saturation $Amax$, which is defined for an optimal temperature at 25 °C for C3 crops. Net assimilation, which can be compared with RUE, results from gross photosynthesis minus carbon losses by growth and maintenance respiration (Figure 10.2f). While growth respiration provides the carbon skeletons, ATP, and reducing power to drive growth, including the synthesis of cellular components and ion pumping, and hence relates

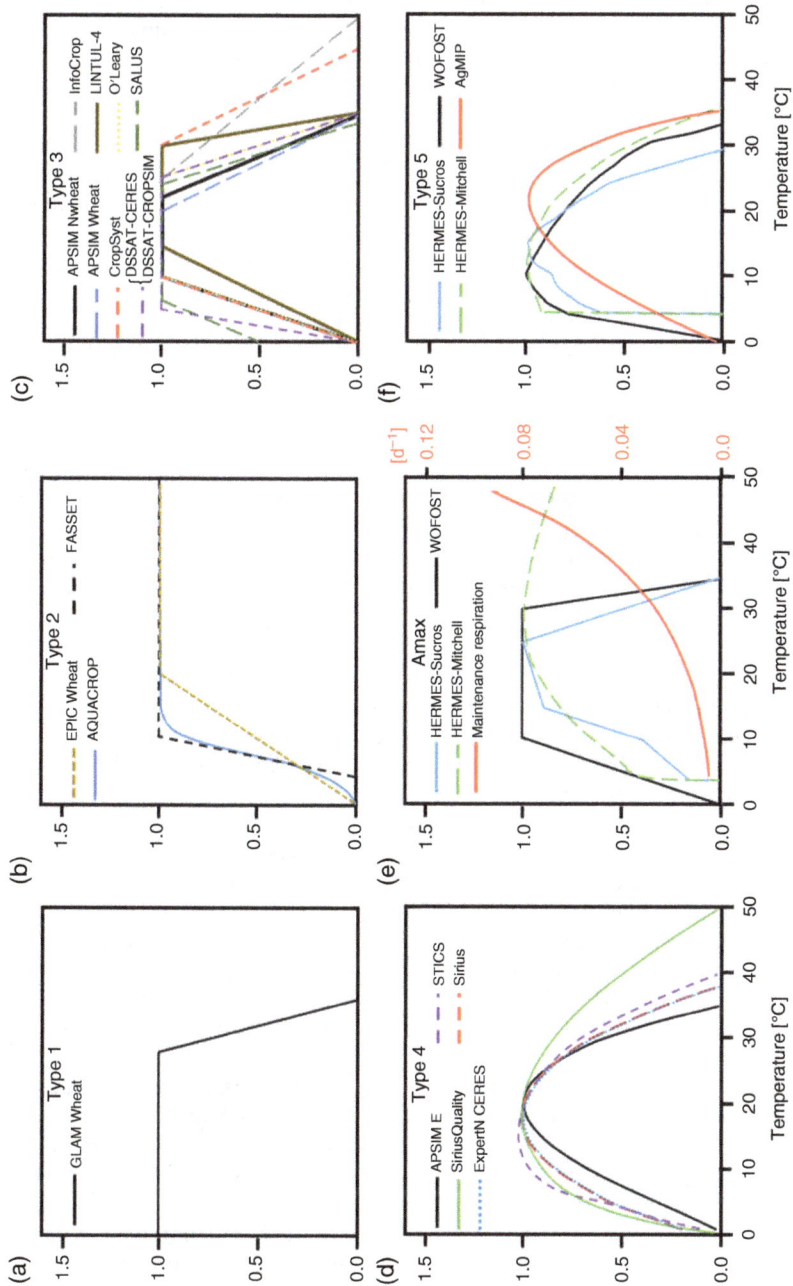

Figure 10.2 Temperature response of different crop models for biomass production resp. RUE (a)–(d) and maximum CO_2 assimilation rate Amax (scaling factor without unit) and maintenance respiration (d^{-1}) (e) and resulting net biomass production (f) (Wang et al., 2017 / with permission of Springer Nature.).

to process stoichiometry, maintenance respiration fuels ongoing activities, such as macromolecule turnover and countering ion leakage, and hence relates to cyclic enzymatic processes that are not directly related to new biomass accumulation, but are needed to counter accumulation of entropy for already existing biomass (Amthor et al., 2019). The maintenance respiration coefficient (d^{-1}) is therefore related to the existing biomass and follows an exponential function for temperature response (Figure 10.2e). Consequently, growth respiration is the dominant component in fast-growing young plants, while maintenance respiration is more important in older and bigger plants. HERMES includes different options to consider CO_2 response of photosynthesis (see below). While two of the options are using the default temperature function, the implemented routine from Mitchell et al. (1995) employs its own optimum temperature dependency (Figure 10.2e). A simpler trapezoidal form for the temperature dependence of *Amax* is applied by WOFOST, while the model uses the same response curve for respiration (Figure 10.2e). Different temperature responses are used for C3 and C4 crops with a higher temperature optimum for C4 crops. The resulting relative net production from gross assimilation minus respiration for a C3 crop is shown in Figure 10.2f.

Continuous curves for temperature responses are used by APSIM-E, SIRIUS, SIRIUSQuality, STICS, MONICA, and the Expert-N-CERES model (see Wang et al., 2017) (Figure 10.2d), and more recently in CropSyst as well. Wang et al. (2017) optimized a generalized temperature function (AgMIP curve in Figure 10.2f) for wheat based on a comprehensive analysis how different temperature response functions affected simulations of wheat growth compared with observations. Experimental data of the Hot Serial Cereal (HSC) experiment at Maricopa, AZ, USA (Wall et al., 2011; White et al., 2011) in which well-fertilized and irrigated wheat grew under contrasting sowing dates and temperature environments, and an additional global dataset from the International Heat Stress Genotpye Experiment (IHSGE) carried out by the International Maize and Wheat Improvement Center (CIMMYT) (Martre et al., 2017) were used to derive the functions and to test their implementation in two RUE based crop models (APSIM and SiriusQuality).

Phenological development of crops strongly depends on temperature. However, development can be additionally influenced by photoperiodic responses and by vernalization requirements, usually during the vegetative development phases. Temperature effects on crop phenology are considered in models differently using linear (Figure 10.3a), plateau (Figure 10.3b), triangular (Figure 10.3d), or trapezoidal functions (Figure 10.3c), but also non-linear curves (Figure 10.3e). The generalized functions developed by Wang et al. (2017) differentiated two slightly different temperature functions for the crop development in the vegetative and the reproductive phase of wheat (Figure 10.3f).

The effect of heat stress on crop yield is implemented in several models, although in a model inter-comparison of wheat models using the Maricopa heat data set (Kimball et al., 2018) those without explicit heat stress algorithms were among the best performing models (Asseng et al., 2015). An overview of mechanisms and model approaches for cereals is provided by Rezaei et al. (2015).

The simplest approach to account for negative impacts of high temperature on crop yield is realized by reducing the final yield based on the daily accumulated thermal time of the mean or maximum temperature above a temperature threshold during the sensitive period of the crop, for example, around flowering and grain filling. This approach was used in global scale modeling (Teixeira et al., 2013) and has been

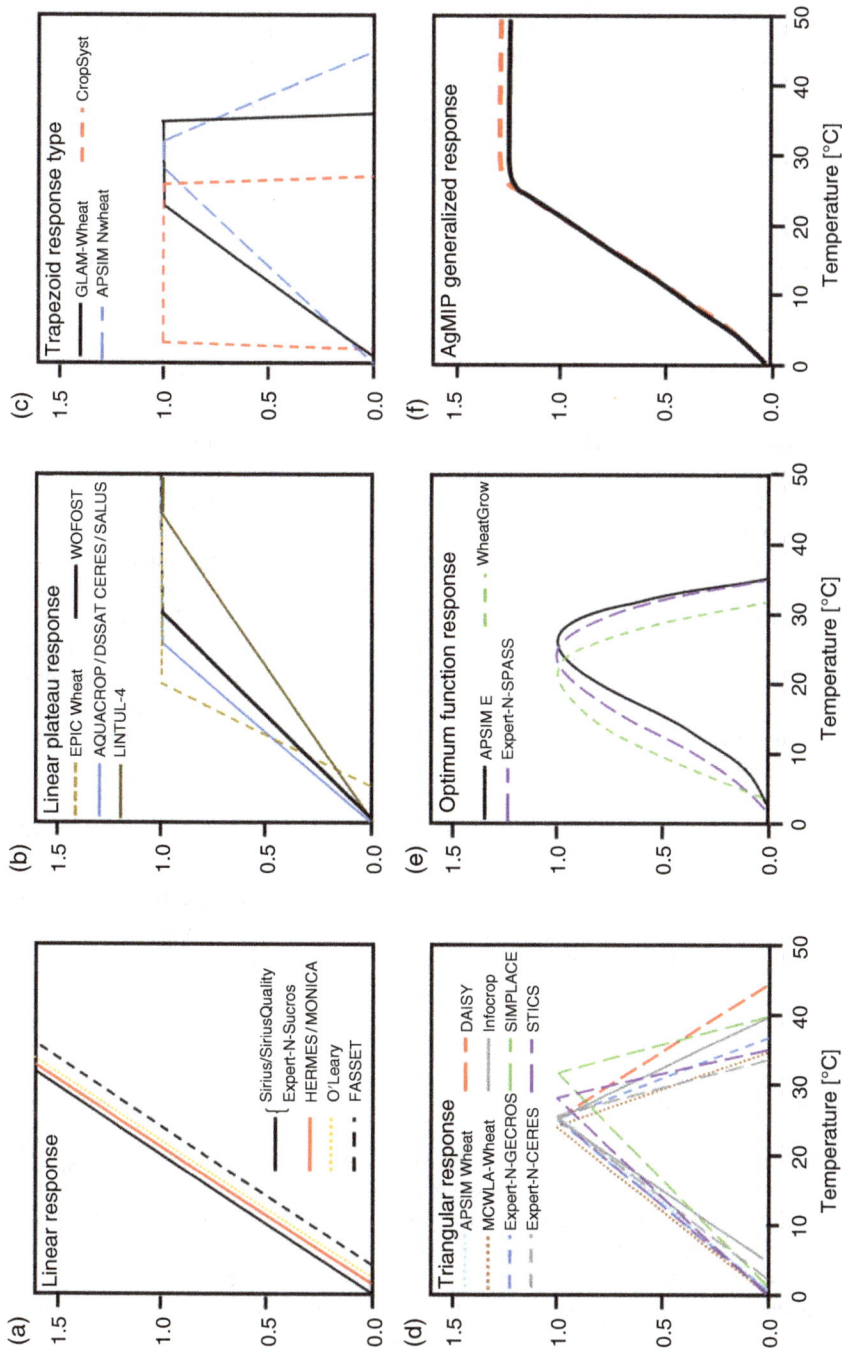

Figure 10.3 Temperature response (unitless scaling factor) of different crop models for crop phenological development (a)–(e). Generalized functions from Wang et al., 2017 are shown in (f) (Wang et al., 2017 / with permission of Springer Nature.).

implemented in the SIMPLACE model by Rezaei et al. (2013). However, the approach cannot distinguish between any processes (e.g., enhanced flag leaf senescence, failure of flowering, grain sterility). An empirical approach reducing the daily increment of harvest index was used by the AQUACROP model, which reduces the daily increment in harvest index by a fraction weighted according to the fraction of flowers that are opening on days when heat stress occurs (Raes et al., 2009). However, grain yield is reduced by crop exposure to high temperatures before and after flowering, not only at the time of flower opening or pollination. Limiting heat impacts to the time of flower opening is likely to underestimate impacts of extreme temperatures (Rezaei et al., 2015).

Moriondo et al. (2011) modified the CropSyst model by implementing a similar approach using a constant value of daily increment in harvest index during the grain filling period (t_{gf})

$$HI = \frac{dHI}{dt} \cdot t_{gf} \tag{10.11}$$

which is also reduced by high temperature periods during flowering and lead to a reduction on final yield. In the determination of the most limiting heat stress event that reduces the daily harvest index increment, CropSyst weights the reduction factor by the fraction of flowers opening on that day. Heat stress phases were identified by comparing the mean temperature between 8 a.m. and 2 p.m. (T_{dm}) using a sinusoidal function in the range of T_{min} and T_{max} to a critical value (T_{cr}) above which grain-set starts to decline up to the minimum level nominated as severe heat shock (T_{lim}), where the grain-set is completely stopped. Any high temperature episodes ($T_{am} > T_{cr}$) occurring during the sensitive period from −5 to +12 days from the onset of anthesis were identified and evaluated according to duration and time of occurrence. The impact of high temperature on the grain set was calculated for a given day by a stress factor P in the range of 1 (=no stress) to 0 (high stress)

$$P = 1 - \frac{T_{dm} - T_{cr}}{T_{lim} - T_{cr}} \tag{10.12}$$

For each event, the reduction in the total grain-set was calculated as the sum of the daily impact of that episode during the flowering developmental stage (N_f). Hence, for each episode the fractional grain-set was given by the product

$$P_{tot} = \sum_{i=1}^{N_f} P(i) \cdot F_d(i) \tag{10.13}$$

where the flowering distribution F_d indicates the fraction of total flowers opening on day i.

A similar approach is implemented in the MONICA model, which does not explicitly simulate grain number but uses the weighted reduction factor based on the accumulated heat stress above a defined threshold based on the daily average temperature to reduce assimilates partitioned to the ears. The approach was also implemented in a HERMES version within a European study on climate change effects on wheat and maize yields (Webber et al. 2018a), whereas the model can switch to use internally simulated canopy temperature instead of daily mean air temperature.

Models that explicitly simulate the number of grains can use empirical functions to reduce the number of grains due to high temperature. Under optimal conditions, the number of grains set in maize is directly related to photosynthetic rate (Lizaso et al., 2007), represented by either curvilinear (Andrade et al., 1993) or linear (Otegui, 1997) models in response to intercepted PAR. An empirical reduction factor is used under heat stress conditions to reduce grain number by accounting for the failure of flowering or abortion. APSIM v7.4 for cereals (Keating et al., 2003) simulates the direct impact of high temperatures near flowering on grain number reduction. In CERES-Maize and related approaches in APSIM maize (Carberry et al., 1989) and millet, grain number is reduced at a set rate per accumulated degree days of maximum daily temperatures beyond a species-based threshold between flag leaf and the last day of flowering.

Although increase of temperature is the main characteristic of climate projections, the risk of crop damages by frost might be enhanced in some regions due to an earlier crop cultivation and faster development, causing coincidence of late frost with susceptible development phases. Damage to wheat from frost has been observed in all stages of growth from seedlings throughout maturity (Fuller et al., 2007; Porter & Gawith, 1999; Shroyer et al., 1995). During vegetative growth frost affects seedling survival (Fuller et al., 2007) and leaf cell structures might be damaged (Shroyer et al., 1995). Additionally, frost without snow cover can create soil cracks leading to mechanic rupture of crop roots. However, yield impact resulting from frost damage for wheat at the reproductive stage of growth is much more relevant than at any other stage (Frederiks et al., 2012). Seedling death, sterility and the irreversible abortion of formed grains has by far the highest impact on cereal yield (Barlow et al., 2015). Post-head-emergence frost damage to winter-habit cereals has been periodically reported for regions in Australia, Canada, and United States (Frederiks et al., 2015), whereas losses of 10% of wheat yield are common in parts of Australia (Fuller et al., 2007). Crop models currently account for frost damage with varying complexity, applying a range of mechanisms including crop death, as well as reducing seedling density, crop biomass or leaf area. The direct impacts of frost around anthesis on grain number are not widely captured in contemporary process-based crop models (Barlow et al., 2015).

EPIC employs a frost function which reduces total biomass by a percentage for each day below a certain threshold temperature. The model includes a frost kill function which is linked to a snow cover factor. Frost damage is considered greatest in seedlings and vulnerability approaches zero at maturity (Williams, 2002). However, the model does not consider a higher vulnerability of wheat to frost damage around anthesis. CropSyst (Stöckle et al., 2003) considers some freezing parameters whose effect is mainly on soil freezing and hydrology. More recently, a frost damage linear function from no damage to frost kill was added, which affects existing green leaves, but no effect on harvest index has been considered. CERES-Wheat (Jones et al., 2003) and CAT (DPI, 2009) have implemented a winter-kill function similar to EPIC, which is applied at any growth stage prior to anthesis. APSIM incorporates a frost stress function which induces leaf senescence, which is the maximum of five factors (age, water stress, light intensity, frost, and heat). While frost is included as a factor in APSIM the default value is zero, which means there is no frost stress (Zheng et al., 2014). The function for leaf area senescence caused by frost (ΔLAI_{frost}) is:

$$\Delta LAI_{frost} = LAI \cdot k_{frost} \qquad (10.14)$$

where k_{frost} is a function of daily minimum temperature and is defined by the linear interpolation between two parameters x_{temp} and $y_{senescence}$, which are linearly interpolated.

The STICS model has implemented separate relations for frost impact on seedling density, leaf senescence, and grain number calculating a frost stress index ranging from 0 to 1 (Brisson et al., 2003). The indices are defined as a function of four minimum crop temperatures. The cardinal temperatures used as threshold values need to be adjusted as a function of genetic tolerance (Brisson et al., 2008). For seedling density the index is applied in a multiplicative way which reduces plant density.

Typically models such as CERES-Wheat, EPIC, and CropSyst were designed to account for very cold temperatures such as in continental Europe and Northern America, however they are not necessarily suited to more temperate climates, where spring radiant frosts are the primary risk (Barlow et al., 2015).

Water Deficit Effects on Crop Yields

Lack of water supply to crops during drought reduces transpiration rate through the reduction of stomatal conductance when crop turgor is reduced (e.g., Brodribb et al., 2003). Consequently, CO_2 flux into the intracellular space is reduced as well and limits photosynthesis resulting in lower green area for assimilation. Moreover, the cooling effect of transpiration is reduced, which will lead to an increase of the canopy temperature (Choudhury, 1998) causing higher respiration of assimilates (Brooks & Farquhar, 1985), an acceleration of crop development and senescence (Roberts & Summerfield, 1987). Shortening of the growing period especially during the establishment of storage organs may cause yield reduction (Porter & Gawith, 1999). Furthermore, crop nutrient uptake is reduced since mass flow as well as diffusive transport of nutrients to the roots is increasingly hampered with soil moisture depletion (Baldwin et al., 1973). Both timing and duration of a dry spell period of low soil moisture determine the degree of yield penalties or crop failure (Araus et al., 2002; Cattivelli et al., 2008). For crops like wheat, maize or barley, the period around flowering has been identified to be most sensitive to drought (Rezaei et al., 2015). However, crop yields are not only dependent on the stress sensitivity of the reproductive and grain filling stages, but on the overall life cycle plant growth and development. Water stress can reduce canopy development (leaf area index), and therefore the interception of solar radiation, thus affecting canopy photosynthesis and biomass production. Efficient photosynthesis and stem reserve accumulation during the vegetative phase have an impact on the formation of reproductive organs, which consequently may affect final yield (Barnabas et al., 2008).

Model approaches mainly use the relation of actual to potential ET or transpiration to consider water limitation by drought. However, models differ in the way at which threshold of soil moisture water uptake is reduced and at which rate crop growth is reduced. The response functions may differ depending on the climatic conditions where a model was originally developed. Wallor et al. (2018) showed in a model inter-comparison, that models developed primarily for semi-arid to arid conditions might over-react for moderate dry conditions in a temperate climate zone. Water balance models commonly include plant water stress functions (e.g., Feddes et al., 1978), where critical limits of soil water supply can be adjusted according to the vegetation characteristics. Although it is widely accepted that root water uptake is determined by water potential gradients and hydraulic resistances in the soil–plant system (Steudle & Peterson, 1998), this principle is rarely included in crop models (Cai et al., 2017). Stöckle and Jara (1998) compared two water uptake/transpiration models based on water potential gradients, hydraulic resistances, root fraction and soil water potential by layers, and atmospheric evaporative demand, with daily (as included in CropSyst) and 20-minute time steps. The two model implementations were tested in maize, with and without water deficit, against measurements performed with a Bowen Ratio system (ET), sap flow devices (equated to transpiration by neglecting plant water storage), and soil water content (neutron probe), with both implementations performing similarly in the simulation of transpiration (normalized root mean square errors of 7.2 and 10.5%) and soil water content, concluding that the use of daily time steps in crop models was sufficient from this perspective. Jara and Stöckle (1999) compared daily models of water uptake with different levels of process detail, with the more process-oriented algorithm in CropSyst showing an increasing advantage as water stress severity increased. In many models the root water uptake is described as a sink term in the soil compartments without solving mechanistically water flow toward roots. Vertical distribution of water uptake is typically assumed to be proportional to root length densities and local water availability. The parameters and concepts used in these models are to a large extent empirical. Feddes et al. (2001) showed that the onset of water stress in crops varies depending on the evaporative demand of the atmosphere resulting in an onset at higher water contents under high evaporative demand compared with moderate VPD. However, few models exist that use leaf water and soil matric potentials as analogous to water stress for multiple crops (CropSyst; Jara & Stöckle, 1999), maize (MAIZSIM; Yang et al., 2009), potato (SPUDSIM; Fleisher et al., 2015). Empirical relations implemented in models may include inherently the conditions under which the models were primarily developed.

Moderate water stress, for example, during early vegetative stages may enhance drought resilience of rainfed crops inducing a preference for root biomass formation and deeper penetration of roots, which may be beneficial to cope with drought periods in later periods (Liu et al., 2004). However, the response varied among varieties (Manschadi et al., 2006). If a stronger root system may be beneficial during later stages, for example, grain filling, depends on the temporal pattern of water availability during the season (Palta et al., 2011). While the prevailing assumption that a large root system would prevent negative impacts of drought by exploring a higher soil volume for soil water extraction, Passioura (1983) suggested in the 1980s that the early extraction of too much water might be counterproductive. Palta et al. (2011) and Lynch (2018) recently expanded on this idea by suggesting that the optimal size of a crop root system depends on the typical pattern of soil moisture and input into the cropping system, with a parsimonious root system likely to be more efficient, especially under high input conditions (Lynch, 2018).

Temperature and Drought Relationships

The debate continues as to whether heat or drought episodes will have a dominant impact on crop yields. Several studies have assessed the impact of heat or drought stress on crop yield using crop models and/or data synthesis (e.g., Asseng et al., 2015; Daryanto et al., 2016; Liu et al., 2016; Rezaei et al., 2018). However, the occurrence of these two climatic extremes is highly interlaced since heat stress is frequently associated with reduced availability of water (Lesk et al., 2016; Mazdiyasni & AghaKouchak, 2015; Sharma & Mujumdar, 2017; Wang et al., 2018). Concurrent drought and heat stress events are projected to increase in the future (IPCC, 2013). Very few studies have attempted to quantify the impacts of concurrent heat and drought events (Zampieri et al., 2017). Attempts to disentangle the future impact of both extremes on wheat or maize production were made by Semenov and Shewry (2011) and Webber et al. (2018a).

There are several interactions between heat and drought that can amplify yield reduction. As mentioned in the previous chapter, reduced transpiration increases canopy temperature and VPD will rise (Idso et al., 1981a, 1981b; Lobell et al., 2013; Van Oort et al., 2014). Siebert et al. (2014) found that stress thermal time during crop sensitive phases using canopy temperature was a better indicator of heat stress than air temperature. Furthermore, an increase in temperature also increases unproductive water loss through evaporation, thereby aggravating water deficits (Lobell et al., 2013; Passioura, 1994; Siebert et al., 2017). However, most crop models ignore complex interactions among air temperature, crop and soil water status, CO_2 concentration, and atmospheric conditions that influence crop canopy temperature (Webber et al. 2018a). Most models use air temperature (Tair) in their heat stress responses although studies showed evidence that crop canopy temperature (Tc) may better explain grain yield losses (e.g., Siebert et al., 2014). Canopy temperature (Tc) can deviate significantly from Tair based on climatic factors and the water status of the crop. Ayeneh et al. (2002) found for wheat that spike temperature was generally higher than leaf temperature, but lower than ambient air temperature. However, they concluded that the use of canopy temperature did not mask confounding interactions between organ temperatures and thus could be used reliably to measure temperature effects during grain filling under heat stress conditions.

In a first model inter-comparison Webber et al. (2017) used an ensemble of nine crop models to simulate wheat canopy temperature for the HSC experiment (Kimball et al., 2015; Kimball et al., 2018; Wall et al., 2011) in Maricopa, AZ, USA. Models showed a wide range of approaches to simulate canopy temperature. Generally, models used either (a) empirical relations to convert air temperature into canopy temperature (EMP), (b) process-based energy balance approaches assuming neutral atmospheric stability (EBN) or (c) energy balance approaches correcting for the atmospheric stability conditions (EBSC). The participating models varied widely in their ability to reproduce observed Tc. Surprisingly, the empirical approaches performed similarly well as the EBSC approach, while the commonly used EBN models failed in many cases. Use of canopy temperature to account for heat stress effects only slightly improved simulations compared to using Tair. This might be due to uncertainties related to soil information and rooting depth, since soil water availability strongly affects transpiration and consequently canopy temperature.

To assess the stability of different approaches Webber et al. (2018b) organized another model inter-comparison with a slightly different model ensemble of nine crop models at six locations across environmental and production conditions with various water, nitrogen and CO_2 treatments. Generally, the models using EBSC approaches performed best, although one EMP model calibrated on the data showed a good overall performance as well. Here we provide a brief description of main approaches used in various models.

Empirical approaches were used in the APSIM NWheat (Asseng et al., 2011), FASSET (Olesen et al., 2002), HUMEWheat (Ratjen & Kage, 2015), and STICS (Brisson et al., 2003; De Sanctis et al., 2013) and CROPSYST (Stöckle, 2013). In APSIM NWheat Tc varies between 6 °C higher than Tair when the crop is fully water stressed and 6 °C cooler than Tair on average when the crop has ample water. Between these limits, the basis of the expression for Tc is the relationship of ΔT (Tc – Tair) and the ratio of actual and potential ET and VPD (Idso et al., 1981a; Jackson et al., 1981). The FASSET and the STICS model employed an empirical relationship between midday crop temperature, ET and net radiation (Seguin & Itier, 1983). Maximum and minimum Tc are calculated on a daily time step. While Tc_{min} is set equal to Tmin, the maximum canopy temperature is estimated by:

$$Tc_{max} = \frac{\left(\frac{Rn}{2.46} - ET - 1.27\right)}{\frac{1.68}{\ln\frac{1}{2}}} \qquad (10.15)$$

where

$$Z = MIN(0.13 \cdot H_c, 0.001) \qquad (10.16)$$

with the daily net radiation Rn (MJ m^{-2}), the daily evapotranspiration ET (mm) and the canopy height H_c (m). Tc_{max} cannot be lower than Tmax.

In CROPSYST the effect of water stress on crop development is considered by correcting the daily Tmax values to a maximum canopy temperature Tc_{max} using a relation from Idso (1982) of measured canopy temperature to crop water stress. In this case Tc_{max} is defined as

$$Tc_{max} = Tmax \cdot [1 + (1.5 \cdot VPD \cdot SI_w)] \cdot PSWS \qquad (10.17)$$

where Tmax (°C) is the daily maximum air temperature, SI_w is the daily water stress index ranging from 0 (no stress) to 1 (maximum stress), VPD is the input of the daily maximum VPD (kPa). PSWS represents a parameter for the phenologic sensitivity to water stress.

The HUMEWHeat model was calibrated within the study of Webber et al. (2018b) and was therefore used as reference model. A statistical approach was used to determine regression models for daily minimum, mean and maximum Tc as described in Neukam et al. (2016). Tc observations performed around midday in this study were considered to be the daily maximum. During pre- and post-heading phase different parameter sets for the covariates (p1, p2, p3, p4, and p5) are used in the multiple linear regression model to predict daily maximum Tc (Tc_{max}):

$$Tc_{max} = p1 + p2 \cdot R_{inc} + p3 \cdot Tmax + p4 \cdot \log(LAI) + p5 \cdot \left(VPD \cdot \frac{Tr_a}{Tr_{pot}}\right) \qquad (10.18)$$

where R_{inc} is the amount of daily incident solar radiation, Tmax is the daily maximum air temperature, LAI is the leaf area index, VPD is the mean daily VPD, and Tr_a and Tr_{pot} are the actual and potential transpiration.

The group of crop models using energy balance approaches with neutral atmospheric stability comprises Sirius (Jamieson et al., 1998) and SiriusQuality (Martre et al., 2006), HERMES (Kersebaum, 2011), SSM-Wheat (Soltani et al., 2013), and a modified STICS version. In the STICS model the empirical function was replaced by two instantaneous energy balances, which were calculated by an iterative convergence procedure to estimate Tc_{max} and Tc_{min}, assumed to occur at midday and at the end of the night, respectively. Minimum and maximum values of net radiation, soil heat, ET and the aerodynamic resistance r_a are taken into account. This method requires wind speed and humidity data.

The Sirius and SiriusQuality models, SSM-Wheat are all using an approach for the energy balance based on the calculation of the sensible heat flux H according to Choudhury (1983) and Choudhury and Idso (1985) as:

$$H = \frac{\rho \cdot cp \cdot (Tc - Ta)}{r_a} \tag{10.19}$$

where ρ is air density, cp is the specific heat of air, Ta is air temperature, and r_a is the aerodynamic resistance, which was estimated according to Monteith (1973) as

$$r_a = \frac{k^2 \cdot u}{[\ln(z - d)/z_0]^2} \tag{10.20}$$

where k is the von Karman constant equal to 0.40, u is the wind speed at reference height z, d is the zero-displacement height equal to 0.13× h, z_0 is the roughness length for momentum and heat transfers each equal to 0.63 × h, and h is the crop height.

HERMES uses a similar approach that calculates Tc from the hourly energy balance by considering net solar radiation and crop transpiration and solving Tc from Ta:

$$Tc = Ta + [Rn \cdot (1 - \tau) - \lambda \cdot Tr_a] \cdot \frac{r_a}{\rho \cdot cp} \tag{10.21}$$

where Tr_a is the actual transpiration, τ is the radiation transmission coefficient of the canopy to account for the radiation transmitted through the canopy to the soil surface, and λ is the heat of vaporization. Hourly temperature and radiation values were determined from daily values following Hoogenboom and Huck (1986). R_a was calculated according to Thom and Oliver (1977) as

$$r_a = \left(\frac{4.72 \cdot [\ln((z - d + z_0)/z_0)]^2}{(1 + 0.54 \cdot u)} \right) \tag{10.22}$$

where u is wind speed at reference height z, d is the zero-displacement height equal to $1.04 \times h^{0.88}$, z_0 is the roughness length for momentum and heat transfer each equal to $z_0 = 0.062 \times h^{1.08}$, and h is the crop height. For the hourly energy balance Tr_a is based on ET_0 from the Penman–Monteith equation with a single crop coefficient (Allen et al., 1998), where the stomatal resistance term is adjusted to consider different CO_2 levels (Kersebaum & Nendel, 2014).

The above-mentioned energy balance approaches are based the assumptions of neutral atmospheric stability conditions, which likely render them inappropriate when canopy temperature deviates much from air temperature (Liu et al., 2007). This would be expected with combinations of highwater stress (Allen et al., 1998), low wind speeds and high radiative heating. The models ECOSYS (Grant et al., 2011) and SIMPLACE_LINTUL (Gaiser et al., 2013; Webber et al., 2016) use energy balance terms that consider atmospheric stability conditions. In ECOSYS, Tc is the temperature that closes the canopy energy balance (net radiation [Rn] + sensible heat flux [H] + latent heat flux [LEc] + change in heat storage [Gc = 0]). These calculations get more complex with sparse canopies that affect resistances and with soils becoming source and sink of heat. Aerodynamic resistance (ra) is affected by Tc – Ta through the Richardson number. Canopy resistance (rc) is affected by the canopy water status through canopy water and turgor potentials solved from soil-root-canopy-atmosphere hydraulic scheme. In SIMPLACE Tc is calculated from an hourly energy balance by summing incident solar radiation, soil, latent and sensible (H) heat fluxes and solving Tc from H as described above. Atmospheric stability is considered by using Monin-Obukhov

Similarity Theory (MOST) (Monin & Obukhov, 1954) and empirical stability correction factors to solve for ra. However, using MOST to account for atmospheric stability requires iteration. Following MOST, when either forced or mixed convection dominates, the general expression for ra is:

$$ra = \frac{1}{uk^2} \left[ln \left(\frac{z_m - d}{z_{0m}} \right) - \Psi_m \right] \left[ln \left(\frac{z_h - d}{z_{0h}} \right) - \Psi_h \right] \tag{10.23}$$

where u is the wind speed over the crop surface, k is the von Karmin constant, z_m and z_h are the heights at which wind speed and air temperature are measured, respectively, d is the zero-displacement height, z_{0m} and z_{0h} are the roughness lengths for momentum and heat transfer, respectively, and Ψ_m and Ψ_h are the respective stability correction factors. The expressions used for d, z_{0m} and z_{0h} are functions of crop height and LAI, which are calculated daily in this SIMPLACE version. Details to estimate Ψ_m and Ψ_h are provided by Webber et al. (2016).

Under very unstable conditions free convection dominates, and the corresponding equation for ra for a flat plate with characteristic dimension D (=3.0 m, assumed to be equivalent to the horizontal fetch of the canopy at uniform Tc) and parameters from Monteith and Unsworth (1990) for the Nusselt, Prandtl and Grashof numbers for under laminar conditions (Grashof number less than 105), is used:

$$ra = \frac{\rho C_\rho}{n \, |T_c - T_a|^{0.25}} \tag{10.24}$$

with n = 100.

Tc is calculated for two bounding extremes: upper (Tc_U, no transpiration) and lower (T_{cL}, full transpiration) limits of Tc, avoiding the need to specify canopy resistance terms at intermediate transpiration rates. With these two extreme potential values of Tc, actual Tc is estimated as:

$$Tc = Tc_L + (1 - SI_w) \cdot (Tc_U - Tc_L) \tag{10.25}$$

where SI_w is the soil water stress index. A full description is given in Webber et al. (2016).

Figure 10.4 provides selected results from the study of Webber et al. (2018b) showing the effect of elevated CO_2 on canopy temperature for the Maricopa FACE

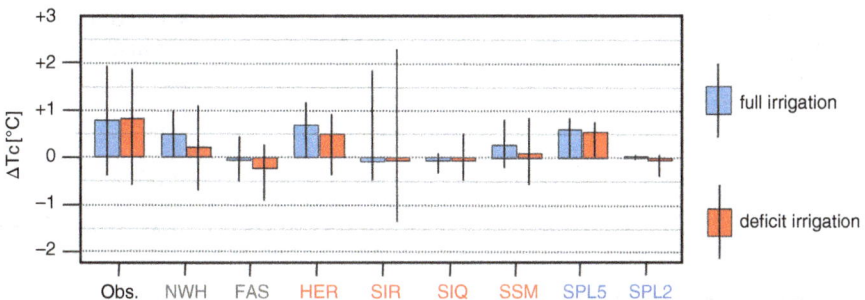

Figure 10.4 Differences of canopy temperature Tc between ambient and elevated CO_2 concentration in a FACE experiment (Obs.) and modeled by different models (NWH = NWheat, FAS = FASSET, HER = HERMES, SIR = Sirius, SIQ = SiriusQuality, SSM = SSM Wheat, SPL5 = SIMPLACE Lintul-5, SPL2 = SIMPLACE Lintul-2). Gray color of model name indicate empirical approaches, red energy balance approaches, and blue energy balance approaches correcting for atmospheric stability.

experiment. Reduction of transpiration through elevated CO_2 led to an average increase of the canopy temperature by $0.8\,°C$ across two growing periods of wheat. The effect was only partly reflected by the models employed in this simulation study, where the results of an ENB model (HERMES) and the model SIMPLACE Lintul-5 from the group of EBSC model were closer to the observed increase than the other models.

Growth Response to Elevated CO_2

Elevated CO_2 directly affects crop growth via changes in photosynthetic rates and stomatal conductance, and indirectly via changes in transpiration. It may also affect crop development via changes in canopy temperature.

In the APSIM wheat model (v7.4) the dynamic RUE response with varying CO_2 is a non-linear relationship using a multiplier (f_{CO2}) increasing RUE with elevated CO_2:

$$f_{CO2} = \frac{(C_a - \Gamma) \cdot (C_{ref} + 2 \cdot \Gamma)}{(C_a + 2 \cdot \Gamma) \cdot (C_{ref} - \Gamma)} \tag{10.26}$$

where C_a is the elevated CO_2 concentration ($\mu mol\ mol^{-1}$), C_{ref} refers to the reference CO_2 concentration ($350\ \mu mol\ mol^{-1}$), and Γ is the CO_2 compensation point with

$$\Gamma = \frac{163 - T_{av}}{5 - 0.1 \cdot T_{av}} \tag{10.27}$$

where T_{av} is the daily average temperature ($°C$) (Bykov et al., 1981).

In CROPSYST, the consideration of elevated (CO_2) for biomass production is based on experiments with controlled CO_2 (Stöckle et al., 2010). A ratio between growth under elevated CO_2 (C_{eo}) to growth under baseline CO_2 (C_{bo}) obtained under experimental conditions is used. Relative growth of biomass (G_r) is calculated by a Michaelis–Menten type equation

$$G_r = \frac{C_a \cdot G_x}{C_a + s} \tag{10.28}$$

where Gr will be less than 1.0 if elevated CO_2 (Ca) $< C_{bo}$ and vice versa. G_x represents the maximum growth increase relative to baseline conditions and s (Ca for half maximum growth increase) can be obtained after considering $Gr = 1.0$, when $C_a = C_{bo}$ and equal to G_{ro}, when $C_a = C_{eo}$:

$$G_x = \frac{C_{bo} + s}{C_{bo}} \tag{10.29}$$

$$s = \frac{C_{eo} \cdot C_{bo} \cdot (G_{ro} - 1)}{C_{eo} - G_{ro} \cdot C_{bo}} \tag{10.30}$$

RUE at any given CO_2 can then be calculated as

$$RUE_{CO2} = RUE_{base} \cdot G_r \tag{10.31}$$

The value of TUE* must account for the effect of elevated CO_2 on Gr and on increased stomatal resistance:

$$TUE^*_{CO2} = \frac{TUE^* \, G_r}{F} \tag{10.32}$$

with TUE* and F previously defined in the biomass production section for Eq. 10.7 and 10.5, respectively.

In Aquacrop water consumption and biomass production are closely interlinked since canopy resistance for water vapor flux r_w and CO_2 flux r_c at leaf level have a quite robust constant relationship of $r_c/r_w \approx 0.625$. Therefore, the CO_2 effect on crop growth is considered by the multiplier (f_{CO2}), which depends on the ratio and difference between CO_2 in the arbitrary reference year (2000) and the ambient CO_2 in the environment, where and when the crop is grown.

The CERES Wheat model in DSSAT (v4.7) uses an asymptotic look-up multiplier on RUE for the relative growth response to elevated CO_2. Values for the multiplier (f_{RUE}) are provided as part of the software WHCER045.spe package. Ahmed et al. (2019) fitted a regression line to describe the CO_2 ($\mu mol \, mol^{-1}$) response of RUE in CERES Wheat as

$$f_{RUE} = 0.4574 \, ln \, ([CO_2]) - 1.713 \tag{10.33}$$

The EPIC model uses a logistic function to simulate the effect of elevated CO_2 on radiation use efficiency RUE:

$$RUE = \frac{[CO_2] - 100}{[CO_2] + b_1 \cdot \exp\left(-b_2 \cdot [CO_2]\right)} \tag{10.34}$$

Solving the equation for two known pairs of RUE and CO_2 taken from experimental data or literature can be used to estimate the parameters b_1 and b_2 (Stöckle et al., 1992). Once b1 and b2 are estimated from two points, Eq. 10.34 can be used to estimate RUE at any other CO_2. The EPIC approach has been adapted also for the STICS model (Brisson et al., 2003).

The WOFOST model considers rising CO_2 by a simple relationship increasing the maximum assimilation rate at light saturation A_{max} by a factor, which increases linearly from 1.0 at 355 to 1.57 at 840 $\mu mol \, mol^{-1}$ (Rötter et al., 2011). Additionally, the specific leaf area (SLA) is affected by elevated CO_2, which decreases SLA by 10% leading to thicker and smaller leaves.

In the HERMES model three algorithms were implemented to describe the impact of elevated CO_2 on carbon assimilation, which were tested against data of a FACE crop rotation in Germany (Manderscheid & Weigel, 2007; Weigel & Dämmgen, 2000). The mechanistic and partly empirical character of the HERMES model determined the range of complexity the response algorithms have to match (Kersebaum et al., 2009). The following approaches are only applied for C3 crops, while for C4 crops, the effect is simply neglected in HERMES.

I.   The Mitchell approach (Mitchell et al., 1995) used a set of algorithms based on concepts by Farquhar and von Caemmerer (1982) and Long (1991), calculating the maximum photosynthesis rate

$$A_{max} = \frac{(C_i - \Gamma) \cdot Vc_{max}}{C_i + K_c \cdot \left(1 + \frac{O_i}{K_o}\right)} \tag{10.35}$$

where C_i and O_i are the intercellular CO_2 and O_2 concentrations, respectively; Γ is the CO_2 compensation point of photosynthesis in absence of dark respiration; Vc_{max} is the maximum Rubisco saturated rate of carboxylation; and K_c and K_o are Michaelis–Menten constants for CO_2 and O_2. The calculation of the latter four parameters is carried out according to Long (1991). Some modifications were applied to simplify the algorithms for suboptimal light conditions and light use efficiency. For the transition between photosynthetic light use efficiency ε and light saturated photosynthesis Mitchell et al. (1995) suggested:

$$\varepsilon = \frac{0.37 \cdot (C_i - \Gamma)}{4.5 \cdot C_i + 10.5 \cdot \Gamma} \tag{10.36}$$

II.   The Nonhebel approach is a much simpler approach extracted from the SUCROS87 model (Nonhebel, 1996). Here, the initial light use efficiency EFF as the slope of the light saturation curve at the compensation point is directly affected by CO_2 as:

$$EFF = \left(\frac{C_a - \Gamma}{C_a + 2\Gamma}\right) \cdot E_0 \tag{10.37}$$

where C_a denotes the actual atmospheric CO_2 concentration, and E_0 is the quantum use efficiency. Additionally, the maximum gross photosynthesis rate is influenced by CO_2 using

$$A_{max}(CO_2) = \frac{C_a - \Gamma}{350 - \Gamma} \cdot A_{max}(350) \tag{10.38}$$

III. The Hoffmann approach (Hoffmann, 1995) was similar to Nonhebel (1996) based on his own work with sugar beet and tree species, and on data previously obtained by Gaastra (1959). He adjusted A_{max} by the factor

$$A_{max} = \frac{\left(\frac{(C_a - \Gamma)}{(k_1 + C_a - \Gamma)}\right)}{\left(\frac{(C_{a0} - \Gamma)}{(k_1 + C_{a0} - \Gamma)}\right)} \tag{10.39}$$

where C_{a0} denotes the ambient reference CO_2 concentration, which is set to $350\ \mu mol\ mol^{-1}$, and C_a is the elevated CO_2. Furthermore, $k_1 = 220 + 0.158 \cdot Ig$ and $\Gamma = 80 - 0.0036 \cdot Ig$, with Ig being the daily global radiation (W m^{-2}). The MONICA model uses solely the approach of Mitchell et al. (1995).

The response of selected models to increasing CO_2 is shown for RUE for wheat corresponding to biomass production (Figure 10.5a) and for the maximum assimilation rate at light saturation Amax (Figure 10.5b) for those models, which use the photosynthesis minus respiration method. All curves are normalized to $350\ \mu mol\ mol^{-1}\ CO_2$ although some approaches are using different reference CO_2. Increase for Amax is generally higher than for RUE since respiration still has to be deducted from gross assimilation to transfer into biomass production. Therefore, resulting curves for the HERMES model are implemented in Figure 10.5a assuming a temperature of 20 °C

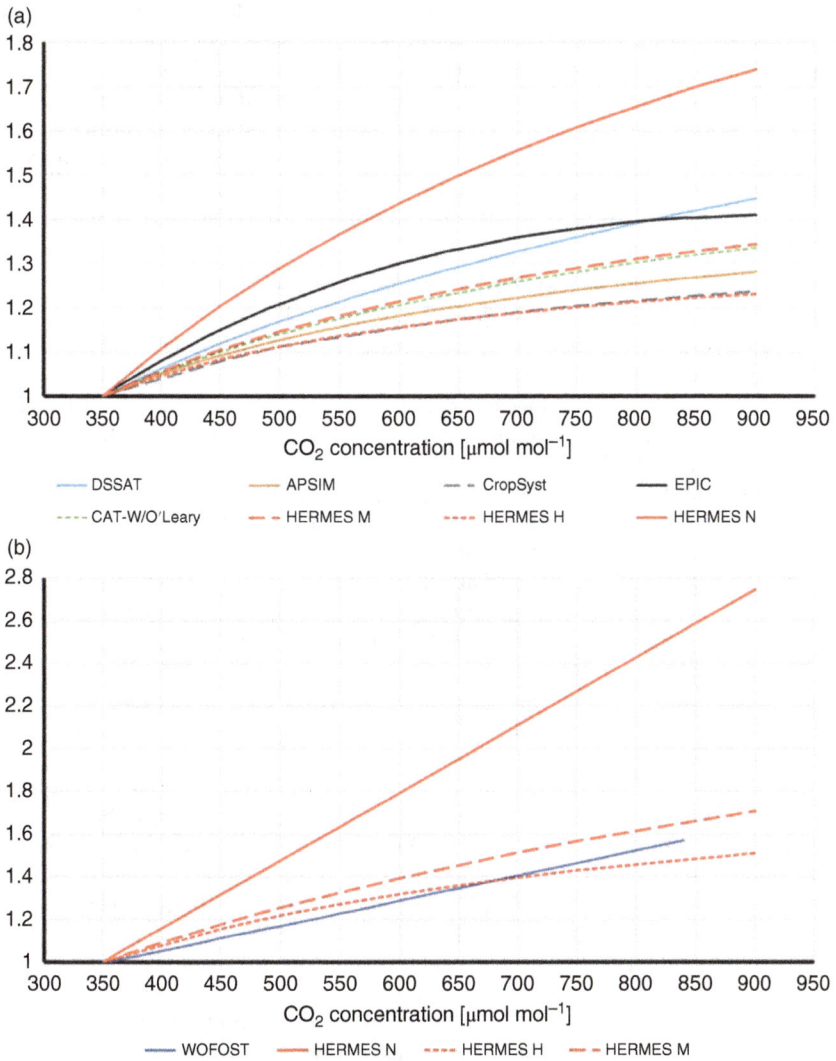

Figure 10.5 CO_2 response functions for (a) RUE resp. biomass production and (b) maximum CO_2 assimilation rate Amax. Results are related to $350\,\mu mol\,mol^{-1}$ CO_2 concentration (=1).

and a biomass for maintenance respiration of 5,000 kg dry matter per hectare. For growth respiration generally 30% of gross assimilation is deduced.

Model uncertainties can partly be explained by the different formalisms used in the models to represent the CO_2 fertilization effect, often based on empirical functions (Yin, 2013) or developed and/or calibrated using data from experiments that failed to represent open-field responses (Ainsworth & Long, 2005; Long et al., 2006). Kersebaum and Nendel (2014) quantified the uncertainty by using different approaches within the model HERMES on regional estimates of yield change due to climate change

(350–550 μmol mol^{-1}) for wheat in Germany in the magnitude of 0.4 t ha^{-1}, which was mainly caused by the relative strong response of the Nonhebel approach, while the other two algorithms were close together. Results also reveal that site properties had a strong influence especially on the contribution of reduced transpiration to the beneficial effect of elevated CO_2. While the transpiration effect had a strong positive effect on sites with moderate water stress, the effect was of minor importance for yield formation on sites with ample water supply. On sites with severe water stress the transpiration effect might alleviate water stress just for a few days and was less important than in the case of moderate stress.

Underrepresented Processes in Climate Change Impact Simulations

As mentioned earlier, climate change is more than an increase in average temperature, but also increases the inter-annual variability of weather variables and their temporal patterns within years. It is projected that crop production will likely be exposed to increase risks from adverse weather (Chen & Ford, 2021; Forzieri et al., 2016; Grillakis, 2019; Trnka et al., 2014; Trnka et al., 2019). Adverse agro-climatic extremes considered most detrimental to crop production include drought and heat events, heavy rains/ hail and storms, flooding and frost, and combinations of them, which are so far not well represented in crop models (Rötter et al., 2018).

Heavy rain, hail and strong wind affect crop yield in different ways depending on the specific susceptibility during crop development. Lodging, which is defined as "the permanent displacement of plant stems from their vertical position" (Pinthus, 1973), is a problem for many crop species across the world at sites where plants are exposed to heavy rain and strong wind especially during their final yield formation phase. An overview on yield reductions due to lodging was provided by Sha et al. (2017), who reported yield losses of 31–80% in wheat, 4–65% in barley, 37–40% in oats, 5–20% in maize, and 5–84% in rice. Crops are prone to lodging at two points: at the stem and at the root. When wind pushes the stem horizontally, the stem may fail due to bending or buckling of the lower stem internodes (Neenan & Spencer-Smith, 1975). Roots may be more affected during heavy rain and may lose their contact and anchorage in soil when soil strength is reduced and increases the load which the plant must bear. Wind then acts as the force which pushes the plant over (Sterling et al., 2003). Several mechanisms were discussed to understand yield loss from lodging and stand-alone model solutions were elaborated to test hypotheses (e.g., Baker et al., 1998; 2014; Berry & Spink, 2012; Sterling et al., 2003). Main mechanisms reported are reduced translocation of mineral nutrients and carbon for grain filling, reduced carbon assimilation within the canopy due to self-shading, increased respiration, rapid senescence and greater susceptibility to pests and diseases, such as for rice (Hitaka, 1968; Setter et al., 1997). Yield loss for wheat can be mainly explained by inefficient radiation use by the canopy due to a reduction of leaf area that is sunlit (Berry & Spink, 2012). Additionally, lodging of cereals creates a favorable environment for leaf diseases and causes harvesting losses (Baker et al., 1998; Malik et al., 2002; Tripathi et al., 2003). Grain quality is also negatively affected by lodging as it might for example inhibit grain drying due to reduced air circulation and increased humidity (Gardiner et al., 2016). Other physical damages reducing crop growth and yield can be caused by hail especially in the juvenile phase, which can hardly be assessed in model simulations.

Excessive rainfall exceeding soil infiltration or drainage leads to water logging or even flooding. Water excess may affect wheat production in several countries, especially in tropical regions and in some regions of the mid to high latitudes, and explains yield anomalies in main wheat producing countries like China and India (Zampieri et al., 2017). Crops suffer from oxygen deficiency when soils are close to water saturation (Bartholomeus et al., 2008) resulting in yield loss (Dasberg & Bakker, 1970). Oxygen stress at high soil moisture contents impede the metabolic activity of plants by decreased root respiration (De Willigen & Van Noordwijk, 1984). There is evidence that oxygen stress under the concurrence of low aeration and high temperature is more pronounced showing a higher reduction in root dry weight (Thompson & Fick, 1981) and the root growth rate (Tsukahara & Kozlowski, 1986), a situation which is projected to occur more frequently in a future climate (Solomon et al., 2007). Moreover, root water uptake in water-logged soils is reduced (Feddes et al., 1978), which is considered depending on the extent and duration of water logging in the WOFOST model (Supit et al., 1994) and also adapted in HERMES. Further side effects of excess water are nutrient deficiency due to leaching and denitrification. Mechanical damage is likely to occur on sloping land as a consequence of sheet or gully erosion, and in extreme cases may wash away plants or parts of the crop. Finally, an aspect rarely considered is the detrimental effect on agronomic management, for example, trafficability at sowing or harvest (Trnka et al., 2014).

Final Remarks

Despite much recent research and advancements made in modeling agricultural impacts of and adaptations to climate change, the success in true crop model improvements is still limited. One step in the direction of model improvement was the establishment of international model activities like AgMIP or MACSUR, which aim to compare models and analyze their behavior to find out strengths and weaknesses of their approaches for different crops. Testing models over a wide range of conditions is essential to stress their approaches and evaluate their performance. One example is the development of new general temperature functions for wheat growth and development by Wang et al. (2018). However, advances in crop modeling are strongly depending on good data suitable to analyze processes and their interactions. This requires a close interaction between experimentalists and modelers. Although models often use similar shapes of responses, their results in climate impact studies often differ substantially. Main source of uncertainty might not be the general form of response functions, but their parameterization or differences in the interpretation of inputs, or lack of balance in the representation of soil, plant, atmosphere processes. For instance, if soil water dynamics are not simulated properly, even a sophisticated model for canopy temperature might not give acceptable results. Deficiencies in model improvement are still related to the simulation of extreme event impacts. This might be related to the fact that field experiments are often canceled if affected by extremes without the estimation of damages and crop yield losses. Likewise, there has been little change in terms of the focal crops, farming systems and regions that agricultural research is invested in (Rötter et al., 2018), requiring massive re-direction of research toward crops/farming systems most vulnerable and likely to be exposed to more severe agro-climatic extremes such as in many tropical regions.

References

Abdallah, A. M., Jat, H. S., Choudhary, M., Abdelaty, E. F., Sharma, P. C., & Jat, M. L. (2021). Conservation agriculture effects on soil water holding capacity and water-saving varied with management practices and agroecological conditions: A review. *Agronomy*, 11, 1681.

Ahmed, M., Stöckle, C. O., Nelson, R., Higgins, S., Ahmad, S., & Raza, M. A. (2019). Novel multimodel ensemble approach to evaluate the sole effect of elevated CO_2 on winter wheat productivity. *Scientific Reports*, 9, 7813.

Ainsworth, E. A., & Long, S. P. (2005). What have we learned from 15 years of free-air CO_2 enrichment (FACE)? A meta-analytic review of the responses of photosynthesis, canopy properties and plant production to rising CO_2. *The New Phytologist*, 165, 351–372.

Ainsworth, E. A., & Rogers, A. (2007). The response of photosynthesis and stomatal conductance to rising $[CO_2]$: Mechanisms and environmental interactions. *Plant, Cell & Environment*, 30, 258–270.

Akita, S., & Moss, D. N. (1973). Photosynthetic responses to CO_2 and light by maize and wheat leaves adjusted for constant stomatal apertures. *Crop Science*, 13, 234–237.

Alderman, P. D., Quilligan, E., Asseng, S., Ewert, F., & Reynolds, M.P. 2013. *Proceedings of the Workshop on Modeling Wheat Response to High Temperature*. CIMMYT Workshop, El Batán, Mexico. 19–21 June 2013. El Batán, Mexico: CIMMYT.

Alexander, L. V., Zhang, X., Peterson, T. C., Caesar, J., Gleason, B., Klein Tank, A. M. G., Haylock, M., Collins, D., Trewin, B., Rahimzadeh, F., Tagipour, A., Rupa Kumar, K., Revadekar, J., Griffiths, G., Vincent, L., Stephenson, D. B., Burn, J., Aguilar, J., Brunet, E., … Vazquez-Aguirre, J. L. (2006). Global observed changes in daily climate extremes of temperature and precipitation. *Journal of Geophysical Research*, 111, D05109.

Allen, L. H. (1990). Plant responses to rising carbon dioxide and potential interactions with air pollutants. *Environmental Quality*, 19, 15–34.

Allen, L. H., Kakani, V. G., Vu, J. C., & Boote, K. J. (2011). Elevated CO_2 increases water use efficiency by sustaining photosynthesis of water-limited maize and sorghum. *Journal of Plant Physiology*, 168, 1909–1918.

Allen, R. G. (1986). A Penman for all seasons. *Journal of Irrigation and Drainage Engineering*, 112, 348–368.

Allen, R. G., Pereira, L. S., Raes, D., & Smith, M. 1998. *Crop evapotranspiration—Guidelines for computing crop water requirements*. FAO Irrigation and Drainage Paper 56. Rome, Italy: Food and Agriculture Organization of the United Nations.

Amthor, J. S., Bar-Even, A., Hanson, A. D., Millar, A. H., Stitt, M., Sweetlove, L. J., & Tyerman, S. D. (2019). Engineering strategies to boost crop productivity by cutting respiratory carbon loss. *Plant Cell*, 31(2), 297–314.

Andrade, F. H., Uhart, S. A., & Frugone, M. I. (1993). Intercepted radiation at flowering and kernel number in maize: Shade versus plant density effects. *Crop Science*, 33, 482–485.

Araus, J. L., Slafer, G. A., Reynolds, M. P., & Royo, C. (2002). Plant breeding and drought in C3 cereals: What should we breed for? *Annals of Botany*, 89, 925–940.

Asseng, S., Ewert, F., Martre, P., Rötter, R. P., Lobell, D. B., Cammarano, D., Kimball, B. A., Ottman, M. J., Wall, G. W., White, J. W., Reynolds, M. P., Alderman, P. D., Prasad, P. V. V., Aggarwal, P. K., Anothai, J., Basso, B., Biernath, C., Challinor, A. J., De Sanctis, G., … Zhu, Y. (2015). Rising temperatures reduce global wheat production. *Nature Climate Change*, 5(2), 143–147.

Asseng, S., Ewert, F., Rosenzweig, C., Jones, J. W., Hatfield, J. L., Ruane, A., Boote, K. J., Thorburn, P., Rötter, R. P., Cammarano, D., Brisson, N., Basso, B., Martre, P., Aggarwal, P. K., Angulo, C., Bertuzzi, P., Biernath, C., Challinor, A., Doltra, J., … Wolf, J. (2013). Quantifying uncertainties in simulating wheat yields under climate change. *Nature Climate Change*, 3, 827–832.

Asseng, S., Foster, I., & Turner, N. C. (2011). The impact of temperature variability on wheat yields. *Global Change Biology*, 17, 997–1012.

Ayeneh, A., van Ginkel, M., Reynolds, M. P., & Ammar, K. (2002). Comparison of leaf, spike, peduncle and canopy temperature depression in wheat under heat stress. *Field Crops Research*, 79, 173–184.

Baker, C. J., Berry, P. M., Spink, J. H., Sylvester-Bradley, R., Griffin, J. M., Scott, R. K., & Clare, R. W. (1998). A method for the assessment of the risk of wheat lodging. *Journal of Theoretical Biology*, 194, 587–603.

Baker, C. J., Sterling, M., & Berry, P. (2014). A generalized model of crop lodging. *Journal of Theoretical Biology*, 363, 1–12.

Baldwin, J. P., Nye, P. H., & Tinker, P. B. (1973). Uptake of solutes by multiple root systems from soil. III. A model for calculating the solute uptake by a randomly dispersed root system developing in a finite volume of soil. *Plant and Soil*, 38, 621–635.

Barlow, K. M., Christy, B. P., O'Leary, G. J., Riffkin, P. A., & Nuttall, J. G. (2015). Simulating the impact of extreme heat and frost events on wheat crop production. A review. *Field Crops Research*, 71, 109–119.

Barnabas, B., Jäger, K., & Feher, A. (2008). The effect of drought and heat stress on reproductive processes in cereals. *Plant, Cell & Environment*, 31, 11–38.

Bartholomeus, R. P., Witte, J. P. M., van Bodegom, P. M., van Dam, J. C., & Aerts, R. (2008). Critical soil conditions for oxygen stress to plant roots: Substituting the Feddes function by a process-based model. *Journal of Hydrology*, 360, 147–165.

Bassu, S., Brisson, N., Durand, J.-L., Boote, K. J., Lizaso, J., Jones, J. W., Rosenzweig, C., Ruane, A. C., Adam, M., Baron, C., Basso, B., Biernath, C., Boogaard, H., Conijn, S., Corbeels, M., Deryng, D., De Sanctis, G., Gayler, S., Grassini, P., ... Waha, K. (2014). How do various maize crop models vary in their responses to climate change factors? *Global Change Biology*, 20(7), 2301–2320.

Benli, B., Bruggeman, A., Oweis, T., & Ustun, H. (2010). Performance of Penman-Monteith FAO56 in a semiarid Highland environment. *Journal of Irrigation and Drainage Engineering*, 136, 757–765.

Beres, B. L., Hatfield, J. L., Kirkegaard, J. A., Eigenbrode, S. D., Pan, W. L., Lollato, R. P., Hunt, J. R., Strydhorst, S., Porker, K., Lyon, D., Ransom, J., & Wiersma, J. (2020). Toward a better understanding of genotype × environment × management interactions—A global wheat initiative agronomic research strategy. *Frontiers in Plant Science*, 11, 828.

Bernacchi, C. J., Kimball, B. A., Quarles, D. R., Long, S. P., & Ort, D. R. (2007). Decreases in stomatal conductance of soybean under open-air elevation of CO_2 are closely coupled with decreases in ecosystem evapotranspiration. *Plant Physiology*, 143, 134–144.

Berry, P. M., & Spink, J. (2012). Predicting yield losses caused by lodging in wheat. *Field Crops Research Field Crops Res.*, 137, 19–26.

Best, M. J., Pryor, M., Clark, D. B., Rooney, G. G., Essery, R., Ménard, C. B., Edwards, J. M., Hendry, M. A., Porson, A., Gedney, A. N., Mercado, L. M., Sitch, S., Blyth, E., Boucher, O., Cox, P. M., & Grimmond, C. S. B. (2011). The Joint UK Land Environment Simulator (JULES), model description part 1: Energy and water fluxes. *Geoscientific Model Development*, 4(3), 677–699.

Boogaard, H. L., van Diepen, C. A., Rötter, R. P., Cabrera, J. M. C. A., & van Laar, H. H. (1998). *WOFOST 7.1. User's guide for the WOFOST 7.1 crop growth simulation model and WOFOST control center 1.5.52.* DLO Winand Staring Centre.

Boote, K. J., Jones, J. W., White, J. W., Asseng, S., & Lizaso, J. I. (2013). Putting mechanisms into crop production models. *Plant, Cell & Environment*, 36, 1658–1672.

Boote, K. J., Pickering, N. B., & Allen, L. H., Jr. (1997). Plant modeling: advances and gaps in our capability to predict future crop growth and yield in response to global climate change. In L. H. Allen, Jr., M. B. Kirkham, D. M. Olszyk, & C. E. Whitman (Eds.), *Advances in carbon dioxide effects research*. ASA Special Publication No. 61 (pp. 179–228). ASA, CSSA, and SSSA.

Brisson, N., Gary, C., Justes, E., Roche, R., Mary, B., Ripoche, D., Zimmer, D., Sierra, J., Bertuzzi, P., Burger, P., Bussiere, F., Cabidoche, Y. M., Cellier, P., Debaeke, P., Gaudillere, J. P., Henault, C., Maraux, F., Seguin, B., & Sinoquet, H. (2003). An overview of the crop model STICS. *European Journal of Agronomy*, 18, 309–332.

Brisson, N., Gate, P., Gouache, D., Charmet, G., Oury, F.-X., & Huard, F. (2010). Why are wheat yields stagnating in Europe? A comprehensive data analysis for France. *Field Crops Research*, 119, 201–212.

Brisson, N., Launay, M., Mary, B., & Beaudoin, N. (2008). *Conceptual basis, formalizations and parameterization of the STICS crop model*. Éditions Quæ, INRA.

Brodribb, T. J., Holbrook, N. M., Edwards, E. J., & Gutiérrez, M. V. (2003). Relations between stomatal closure, leaf turgor and xylem vulnerability in eight tropical dry forest trees. *Plant, Cell & Environment*, 26, 443–450.

Brooks, A., & Farquhar, G. (1985). Effect of temperature on the CO_2/O_2 specificity of ribulose-1, 5-bisphosphate carboxylase/oxygenase and the rate of respiration in the light. *Planta*, 165, 397–406.

Bunce, J. A. (2012). Responses of cotton and wheat photosynthesis and growth to cyclic variation in carbon dioxide concentration. *Photosynthetica*, 50, 395–400.

Bykov, O. D., Koshkin, V. A., & Catsky, J. (1981). Carbon dioxide compensation concentration of C3 and C4 plants: Dependence on temperature. *Photosynthetica*, 15, 114–121.

Cai, G., Vanderborght, J., Couvreur, V., Mboh, C. M., & Vereecken, H. (2017). Parameterization of root water uptake models considering dynamic root distributions and water uptake compensation. *Vadose Zone Journal*, 17, 160125.

Cammarano, D., Rötter, R. P., Asseng, S., Ewert, F., Wallach, D., Martre, P., Hatfield, J. L., Jones, J. W., Rosenzweig, C., Ruane, A. C., Boote, K. J., Thorburn, P. J., Kersebaum, K. C., Aggarwal, P. K., Angulo, C., Basso, B., Bertuzzi, P., Biernath, C., Brisson, N., ... Wolf, J. (2016). Water use of wheat: Simulated patterns and sensitivity to temperature and CO_2. *Field Crops Research*, 198, 80–92.

Carberry, P., Muchow, R., & McCown, R. (1989). Testing the CERES–maize simulation model in a semi-arid tropical environment. *Field Crops Research*, 20, 297–315.

Cattivelli, L., Rizza, F., Badeck, F. W., Mazzucotelli, E., Mastrangelo, A. M., Francia, E., Mare, C., Tondelli, A., & Stanca, A. M. (2008). Drought tolerance improvement in crop plants: An integrated view from breeding to genomics. *Field Crops Research*, 115, 1–14.

Chen, L., & Ford, T. W. (2021). Effects of 0.5 °C less global warming on climate extremes in the contiguous United States. *Climate Dynamics*, 57, 303–319.

Choudhury, B. (1983). Simulating the effects of weather variables and soil water potential on a corn canopy temperature. *Agricultural Meteorology*, 29 (3), 169–182.

Choudhury, B., & Idso, S. B. (1985). An empirical model for stomatal resistance of field- grown wheat. *Agricultural and Forest Meteorology*, 36, 65–82.

Christidis, N., Jones, G. S., & Stott, P. A. (2015). Dramatically increasing chance of extremely hot summers since the 2003 European heatwave. *Nature Climate Change*, 5, 56.

Chun, J. A., Wang, Q., Timlin, D., Fleisher, D., & Reddy, V. R. (2011). Effect of elevated carbon dioxide and water stress on gas exchange and water use efficiency in corn. *Agricultural and Forest Meteorology*, 151, 378–384.

Dai, A. (2013). Increasing drought under global warming in observations and models. *Nature Climate Change*, 3, 52–58.

Dai, A., & Zhao, T. (2017). Uncertainties in historical changes and future projections of drought. Part I: Estimates of historical drought changes. *Climatic Change*, 144, 519–533.

Daryanto, S., Wang, L., & Jacinthe, P.-A. (2016). Global synthesis of drought effects on maize and wheat production. *PLoS One*, 11, e0156362.

Dasberg, S., & Bakker, J. W. (1970). Characterizing soil aeration under changing soil moisture conditions for bean growth. *Agronomy Journal*, 62, 689–692.

Davenport, F. V., Burke, M., & Diffenbaugh, N. S. (2021). Contribution of historical precipitation change to US flood damages. *PNAS*, 118(4), e2017524118.

De Sanctis, G., de Cortazar, G., Atauri, I., Launay, M., Ruget, F., Ripoche, D., & Bertuzzi, P. (2013). Temperature effects in the STICS model: Theoretical basis and essential routines for annual crops. In P. Alderman, E. Quilligan, S. Asseng, F. Ewert, & M. P. Reynolds (Eds.), *Modeling wheat response to high temperature* (pp. 120–122). CIMMYT.

De Willigen, P., & Van Noordwijk, W. (1984). Mathematical models on diffusion of oxygen to and within plant roots, with special emphasis on effects of soil-root contact I. Derivation of the models. *Plant Soil*, 77, 215–231.

de Wit, C. T., & Goudriaan, J. (1978). *Simulation of ecological processes*. Simulation Monographs. Pudoc.

Degaetano, A. T. (2009). Time-dependent changes in extreme-precipitation return-period amounts in the continental United States. *Journal of Applied Meteorology and Climatology*, 48, 2086–2099.

DPI. (2009). *Models of the Catchment Analysis Tool (CAT1D version 32). Future farming systems research*. Author.

Durand, J.-L., Delusca, K., Boote, K., Lizaso, J., Manderscheid, R., Weigel, H.-J., Ruane, A. C., Rosenzweig, C., Jones, J. W., Ahuja, L., Anapalli, S., Basso, B., Baron, C., Bertuzzi, P., Biernath, C., Deryng, D., Ewert, F., Gaiser, T., … Zhao, Z. (2018). How accurately do maize crop models simulate the interactions of atmospheric CO_2 concentration levels with limited water supply on water use and yield? *European Journal of Agronomy*, 100, 67–75.

EEA. (2017). *Climate change, impacts, and vulnerability in Europe 2016*. EEA Report No. 1/2017. Author.

Ewert, F., Rötter, R. P., Bindi, M., Webber, H., Trnka, M., Kersebaum, K.-C., Olesen, J. E., van Ittersum, M. K., Janssen, S., Rivington, M., Semenov, M., Wallach, D., Porter, J. R., Stewart, D., Verhagen, J., Gaiser, T., Palosuo, T., Tao, F., … Asseng, S. (2015). Crop modelling for integrated assessment of risk to food production from climate change. *Environmental Modelling and Software*, 72, 287–303.

Farquhar, G. D., & von Caemmerer, S. (1982). Modelling of photosynthetic response to environmental conditions. In O. L. Lange, P. S. Nobel, C. B. Osmond, & H. Ziegler (Eds.), *Encyclopedia of plant physiology. New series. Physiological plant ecology. II. Water relations and carbon assimilation* (Vol. 12B, pp. 549–587). Springer.

Feddes, R. A., Hoff, H., Bruen, M., Dawson, T., de Rosnay, P., Dirmeyer, P., Jackson, R. B., Kabat, P., Kleidon, A., Lilly, A., & Pitman, A. J. (2001). Modeling root water uptake in hydrological and climate models. *Bulletin of the American Meteorological Society*, 82, 2797–2809.

Feddes, R. A., Kowalik, P. J., & Zaradny, H. (1978). *Simulation of field water use and crop yield*. Simulation Monographs. Pudoc.

Fleisher, D. H., Dathe, A., Timlin, D. J., & Reddy, V. R. (2015). Improving potato drought simulations: Assessing water stress factors using a coupled model. *Agricultural and Forest Meteorology*, 200, 144–155.

Forzieri, G., Feyen, L., Russo, S., Vousdoukas, M., Alfieri, L., Outten, S., Migliavacca, M., Bianchi, A., Rojas, R., & Cid, A. (2016). Multi-hazard assessment in Europe under climate change. *Climatic Change*, 137, 105–119.

Frederiks, T. M., Christopher, J. T., Harvey, G. L., Sutherland, M. W., & Borrell, A. K. (2012). Current and emerging screening methods to identify post-head-emergence frost adaptation in wheat and barley. *Journal of Experimental Botany*, 63, 5405–5416.

Frederiks, T. M., Christopher, J. T., Sutherland, M. W., & Borrell, A. K. (2015). Post-head emergence frost in wheat and barley: Defining the problem, assessing the damage, and identifying resistance. *Journal of Experimental Botany*, 66, 3487–3498.

Freudenberger, H. (1940). Die Reaktion der Schliesszellen auf Kohlensaure and Sauerstoffentzug. *Protoplasma*, 35, 15–54.

Fuller, M. P., Fuller, A. M., Kaniouras, S., Christophers, J., & Fredericks, T. (2007). The freezing characteristics of wheat at ear emergence. *European Journal of Agronomy*, 26, 435–441.

Gaastra, P. (1959). Photosynthesis of crop plants as influenced by light, carbon dioxide, temperature, and stomatal diffusion resistance. *Mededel Landbouwhogeschool Wageningen*, 59(13), 1–68.

Gaiser, T., Perkons, U., Küpper, P. M., Kautz, T., Uteau-Puschmann, D., Ewert, F., Enders, A., & Krauss, G. (2013). Modeling biopore effects on root growth and biomass production on soils with pronounced sub-soil clay accumulation. *Ecological Modelling*, 256, 6–15.

Gardiner, B., Berry, P., & Moulia, B. (2016). Review: Wind impacts on plant growth, mechanics and damage. *Plant Science*, 245, 94–118.

Ghannoum, O., Caemmerer, S. V., Ziska, L. H., & Conroy, J. P. (2000). The growth response of C4 plants to rising atmospheric CO_2 partial pressure: A reassessment. *Plant, Cell & Environment*, 23, 931–942.

Godfray, H. C. J., Beddington, J. R., Crute, I. R., Haddad, L., Lawrence, D., Muir, J. F., Pretty, J., Robinson, S., Thomas, S. M., & Toulmin, C. (2010). Food security: The challenge of feeding 9 billion people. *Science*, 327, 812–818.

Goudriaan, J., & van Laar, H. H. (1978). Calculation of daily totals of the gross CO_2 assimilation of leaf canopies. *Netherlands Journal of Agricultural Science*, 26, 416–425.

Gourdji, S. M., Sibley, A. M., & Lobell, D. B. (2013). Global crop exposure to critical high temperatures in the reproductive period: Historical trends and future projections. *Environmental Research Letters*, 8, 024041.

Grant, R. F., & Baldocchi, D. D. (1992). Energy-transfer over crop canopies—Simulation and experimental-verification. *Agricultural and Forest Meteorology*, 61, 129–149.

Grant, R. F., Kimball, B. A., Conley, M. M., White, J. W., Wall, G. W., & Ottman, M. J. (2011). Controlled warming effects on wheat growth and yield: Field measurements and modelling. *Agronomy Journal*, 103, 1742–1754.

Grassini, P., Eskridge, K. M., & Cassman, K. G. (2013). Distinguishing between yield advances and yield plateaus in historical crop production trends. *Nature Communications*, 4, 2918.

Grillakis, M. G. (2019). Increase in severe and extreme soil moisture droughts for Europe under climate change. *Sciences of the Total Environment*, 660, 245–1255.

Hansen, S., Abrahamsen, P., Petersen, C. T., & Styczen, M. (2012). Daisy: Model use, calibration and validation. *Transactions of the ASABE*, 55, 1315–1333.

Hargreaves, G. H. (1975). Moisture availability and crop production. *Transactions of ASAE*, 18, 980–984.

Hatfield, J., Boote, K., Fay, P., Hahn, L., Izaurralde, C., Kimball, B. A., Mader, T., Morgan, J., Ort, D., Polley, W., Thomson, A., & Wolfe, D. (2008). *The effects of climate change on agriculture, land resources, water resources, and biodiversity*. US Climate Change Science Program and the Subcommittee on Global Change Research.

Hatfield, J. L., Boote, K. J., Kimball, B. A., Ziska, L. H., Izaurralde, R. C., Ort, D., Thomson, A. M., & Wolfe, D. (2011). Climate impacts on agriculture: Implications for crop production. *Agronomy Journal*, 103, 351–370.

He, D., & Wang, E. (2019). On the relation between soil water holding capacity and dryland crop productivity. *Geoderma*, 353, 11–24.

Hitaka, H. (1968). Experimental studies on the mechanisms of lodging and its effect on the yield of rice plants. *Bulletein of National Institute of Agricultural Sciences Tokyo*, 15, 1–175.

Hoffmann, F. (1995). Fagus, a model for growth and development of beech. *Ecological Modelling*, 83, 327–348.

Holzworth, D. P., Huth, N. I., de Voil, P. G., Zurcher, E. J., Herrmann, N. I., McLean, G., Chenu, K., van Oosterom, E. J., Snow, V., Murphy, C., Moore, A. D., Brown, H., Whish, J. P. M., Verrall, S., Fainges, J., Bell, L. W., Peake, A. S., Poulton, P. L., Hochman, Z., … Keating, B. A. (2014). APSIM—Evolution towards a new generation of agricultural systems simulation. *Environmental Modelling and Software*, 62, 327–350.

Hoogenboom, G. & Huck, M.G. 1986. *Rootsimu v4. 0: A dynamic simulation of root growth, water uptake, and biomass partitioning in a soil-plant-atmosphere continuum: Update and documentation*. Agronomy and Soils Departmental Series. Auburn University, Alabama Agricultural Experiment Station.

Horie, T., Nakagawa, H., Centeno, H. G. S., & Kropff, M. J. (1995). The rice crop simulation model SIMRIW and its testing. In R. B. Matthews, M. J. Kropff, D. Bachelet, & H. H. van Laar (Eds.), *Modelling the impact of climate change* (pp. 51–66). International Rice Research Institute.

Idso, S., Jackson, R., Pinter, P., Reginato, R., & Hatfield, J. (1981a). Normalizing the stress-degree-day parameter for environmental variability. *Agricultural Meteorology*, 24, 45–55.

Idso, S. B., Reginato, R., Reicosky, D., & Hatfield, J. (1981b). Determining soil induced plant water potential depression in alfalfa by means of infrared thermometer. *Agronomy Journal*, 73, 826–830.

Idso, S. B. (1982). Non-water-stressed baselines: A key to measuring and interpreting plant water stress. *Agricultural Meteorology*, 27, 59–70.

IPCC. (2013). Climate change 2013: The physical science basis. In T. F. Stocker, D. Qin, G.-K. Plattner, M. Tignor, S. K. Allen, J. Boschung, A. Nauels, Y. Xia, V. Bex, & P. M. Midgley (Eds.), *Contribution of Working Group I to the Fifth Assessment Report of the Intergovernmental Panel on Climate Change*. Cambridge University Press.

IPCC. (2014). Climate change 2014: Impacts, adaptation, and vulnerability. Part A: Global and sectoral aspects. In C. B. Field, V. R. Barros, D. J. Dokken, K. J. Mach, M. D. Mastrandrea, T. E. Bilir, M. Chatterjee, K. L. Ebi, Y. O. Estrada, R. C. Genova, B. Girma, E. S. Kissel, A. N. Levy, S. MacCracken, P. R. Mastrandrea, & L. L. White (Eds.), *Contribution of Working Group II to the Fifth Assessment Report of the Intergovernmental Panel on Climate Change*. Cambridge University Press.

Jackson, R. D., Idso, S., Reginato, R., & Pinter, P., Jr. (1981). Canopy temperature as a crop water stress indicator. *Water Resources Research*, 17, 1133–1138.

Jamieson, P. D., Porter, J. R., Goudriaan, J., Ritchie, J. T., van Keulen, H., & Stol, W. (1998). A comparison of the models AFRCWHEAT2 CERES-wheat, sirius, SUCROS2 and SWHEAT with measurements from wheat grown under drought. *Field Crops Research*, 55, 23–44.

Jara, J., & Stöckle, C. O. (1999). Simulation of water uptake in maize using different levels of process detail. *Agronomy Journal*, 91, 256–265.

Jensen, M. E., Burman, R. D., & Allen, R. G. 1990. *Evapotranspiration and irrigation water requirements*. ASCE Manuals and Reports on Engineering Practice No. 70, American Society of Civil Engineering.

Jiménez Cisneros, B. E., Oki, T., Arnell, N. W., Benito, G., Cogley, J. G., Döll, P., Jiang, T., & Mwakalila, S. S. (2014). Freshwater resources. In C. B. Field, V. R. Barros, D. J. Dokken, K. J. Mach, M. D. Mastrandrea, T. E. Bilir, M. Chatterjee, K. L. Ebi, Y. O. Estrada, R. C. Genova, B. Girma, E. S. Kissel, A. N. Levy, S. MacCracken, P. R. Mastrandrea, & L. L. White (Eds.), *Climate change 2014: Impacts, adaptation, and vulnerability. Part A: Global and sectoral aspects. Contribution of Working Group II to the Fifth Assessment Report of the Intergovernmental Panel on Climate Change* (pp. 229–269). Cambridge University Press.

Jones, J. W., Hoogenboom, G., Porter, C. H., Boote, K. J., Batchelor, W. D., Hunt, L. A., Wilkens, P. W., Singh, U., Gijsman, A. J., & Ritchie, J. T. (2003). The DSSAT cropping system model. *European Journal of Agronomy*, 18, 235–265.

Keating, B. A., Carberry, P., Hammer, G., Probert, M. E., Robertson, M., Holzworth, D., Huth, N., Hargreaves, J., Meinke, H., & Hochman, Z. (2003). An overview of APSIM, a model designed for farming systems simulation. *European Journal of Agronomy*, 18, 267–288.

Kersebaum, K. C. (2011). Special features of the HERMES model and additional procedures for parameterization, calibration, validation, and applications. In L. R. Ahuja & L. Ma (Eds.), *Methods introducing system models into agricultural research* (pp. 65–94). ASA, CSSA, and SSSA.

Kersebaum, K. C., Kroes, J., Gobin, A., Takáč, J., Hlavinka, P., Trnka, M., Ventrella, D., Giglio, L., Ferrise, R., Moriondo, M., Dalla Marta, A., Luo, Q., Eitzinger, J., Mirschel, W., Weigel, H.-J., Manderscheid, R., Hoffmann, M., Nejedlik, P., & Hösch, J. (2016). Assessing the uncertainty of model based water footprint estimation using an ensemble of crop growth models on winter wheat. *Water*, 8, 571.

Kersebaum, K. C., & Nendel, C. (2014). Site-specific impacts of climate change on wheat production across regions of Germany using different CO_2 response functions. *European Journal of Agronomy*, 52, 22–32.

Kersebaum, K. C., Nendel, C., Mirschel, W., Manderscheid, R., Weigel, H.-J., & Wenkel, K.-O. (2009). Testing different CO_2 response algorithms against a face crop rotation experiment and application for climate change impact assessment at different sites in Germany. *Időjárás*, 113, 79–88.

Kim, S.-H., Sicher, R. C., Bae, H., Gitz, D. C., Baker, J. T., Timlin, D. J., & Reddy, V. R. (2006). Canopy photosynthesis, evapotranspiration, leaf nitrogen, and transcription profiles of maize in response to CO_2 enrichment. *Global Change Biology*, 12, 588–600.

Kim, S.-H., Yang, Y., Timlin, D. J., Fleisher, D., Dathe, A., & Reddy, V. R. (2012). Modeling nonlinear temperature responses of leaf growth, development, and biomass in MAIZSIM. *Agronomy Journal*, 104, 1523–1537.

Kimball, B., White, J., Ottman, M., Wall, G., Bernacchi, C., Morgan, J., & Smith, D. (2015). Predicting canopy temperatures and infrared heater energy requirements for warming field plots. *Agronomy Journal*, 107, 129–141.

Kimball, B. A. (2016). Crop responses to elevated CO_2 and interactions with H_2O, N, and temperature. *Current Opinion in Plant Biology*, 31, 36–43.

Kimball, B. A., Boote, K. J., Hatfield, J. L., Ahuja, L. R., Stöckle, C., Archontoulis, S., Baron, C., Basso, B., Bertuzzi, P., Constantin, J., Deryng, D., Dumont, B., Durand, J.-L., Ewert, F., Gaiser, T., Gayler, S., Hoffmann, M. P., Jiang, Q., Kim, S.-H., ... Williams, K. (2019). Simulation of maize evapotranspiration: An inter-comparison among 29 maize models. *Agricultural and Forest Meteorology*, 271, 264–284.

Kimball, B. A., White, J. W., Wall, G. W., Ottman, M. J., & Martre, P. (2018). Wheat response to a wide range of temperatures, as determined from the Hot Serial Cereal (HSC) experiment. *Open Data Journal for Agricultural Research*, 4, 16–21.

Kroes, J. G., van Dam, J. C., Groenendijk, P., Hendriks, R. F. A., Jacobs, C. M. J., 2009. *SWAP Version 3.2(26).* Theory description and user manual. Alterra-Report 1649. Wageningen, The Netherlands: Alterra Research Institute.

Larcher, W. (1995). *Physiological plant ecology* (3rd ed., pp. 506). Springer.

Leakey, A. D. B. (2009). Rising atmospheric carbon dioxide concentration and the future of C4 crops for food and fuel. *Proceedings of Royal Society of London B: Biological Sciences*, 276, 2333–2343.

Lehmann, J., Coumou, D., & Frieler, K. (2015). Increased record-breaking precipitation events under global warming. *Climatic Change*, 132, 501–515.

Lesk, C., Rowhani, P., & Ramankutty, N. (2016). Influence of extreme weather disasters on global crop production. *Nature*, 529, 84–87.

Linsbauer, K. (1916). Beitrage zur Kenntnis der Spaltöffnungsbewegung. *Flora*, 9, 100–143.

Liu, B., Asseng, S., Müller, C., Ewert, F., Elliott, J., Lobell, D., Martre, P., Ruane, A., Wallach, D., Jones, J. W., Rosenzweig, C., Aggarwal, P., Alderman, P., Anothai, J., Basso, B., Biernath, C., Cammarano, D., Challinor, A., Deryng, D., ... Yan, Z. (2016). Similar estimates of temperature impacts on global wheat yield by three independent methods. *Nature Climate Change*, 6, 1130–1136.

Liu, H. S., Li, F. M., & Xu, H. (2004). Deficiency of water can enhance root respiration rate of drought-sensitive but not drought-tolerant spring wheat. *Agricultural Water Management*, 64, 41–48.

Liu, S., Lu, L., Mao, D., & Jia, L. (2007). Evaluating parameterizations of aerodynamic resistance to heat transfer using field measurements. *Hydrology and Earth Systems Sciences*, 11, 769–783.

Lizaso, J., Fonseca, A., & Westgate, M. (2007). Simulating source-limited and sink limited kernel set with CERES-maize. *Crop Science*, 47, 2078–2088.

Lizaso, J. I., Shelia, V., Hoogenboom, G., 2021. *Simulating maize crop processes in DSSAT v4.7: CERES-Maize vs. IXIM-Maize.* In l. Ahuja (Ed.), *Modeling Processes and Their Interactions in Cropping Systems: Challenges for the 21st Century* (pp. 277–290). ASA, CSSA, SSSA. DOI 10.1002/9780891183907.

Lobell, D. B., & Gourdji, S. M. (2012). The influence of climate change on global crop productivity. *Plant Physiology*, 160, 1686–1697.

Lobell, D. B., Hammer, G. L., McLean, G., Messina, C., Roberts, M. J., & Schlenker, W. (2013). The critical role of extreme heat for maize production in the United States. *Nature Climate Change*, 3, 497–501.

Lobell, D. B., Schlenker, W., & Costa-Roberts, J. (2011). Climate trends and global crop production since 1980. *Science*, 333, 616–620.

Long, S. P. (1991). Modification of the response of photosynthetic productivity to rising temperature by atmospheric CO_2 concentrations—Has its importance been underestimated? *Plant, Cell & Environment*, 14, 729–739.

Long, S. P., Ainsworth, E. A., Leakey, A. D. B., Nosberger, J., & Ort, D. R. (2006). Food for thought: Lower than expected crop yield stimulation with rising CO_2 concentrations. *Science*, 312, 1918–1921.

Lorenz, R., Stalhandske, Z., & Fischer, E. M. (2019). Detection of a climate change signal in extreme heat, heat stress, and cold in Europe from observations. *Geophysical Research Letters*, 46, 8363–8374.

Lynch, J., Cain, M., Frame, D., & Pierrehumbert, R. (2021). Agriculture's contribution to climate change and role in mitigation is distinct from predominantly fossil CO_2-emitting sectors. *Frontiers in Sustainable Food Systems*, 4, 518039.

Lynch, J. P. (2018). Rightsizing root phenotypes for drought resistance. *Journal of Experimental Botany*, 69, 3279–3292.

Ma, L., Ahuja, L. R., Nolan, B. T., Malone, R. W., Trout, T. J., & Qi, Z. (2012). Root Zone Water Quality Model (RZWQM2): Model use, calibration, and validation. *Transactions of the ASABE*, 55(4), 1425–1446.

Malik, A. I., Colmer, T. D., Lambers, H., Setter, T. L., & Schortemeyer, M. (2002). Short-term waterlogging has long-term effects on the growth and physiology of wheat. *The New Phytologist*, 153, 225–236.

Manderscheid, R., Erbs, M., Burkart, S., Wittich, K.-P., Löpmeier, F.-J., & Weigel, H.-J. (2015). Effects of free air carbon dioxide enrichment on sap flow and canopy microclimate of maize grown under different water supply. *Journal of Agronomy and Crop Science*, 202, 255–268.

Manderscheid, R., Erbs, M., & Weigel, H.-J. (2014). Interactive effects of free-air CO_2 enrichment and drought stress on maize growth. *European Journal of Agronomy*, 52, 11–21.

Manderscheid, R., & Weigel, H. J. (2007). Drought stress effects on wheat are mitigated by atmospheric CO_2 enrichment. *Agronomy for Sustainable Development*, 27(2), 79–87.

Manschadi, A. M., Christopher, J., Devoil, P., & Hammer, G. L. (2006). Dynamics of root architectural traits in adaptation of wheat to water-limited environments. *Functional Plant Biology*, 33, 823–837.

Markelz, R. C., Strellner, R. S., & Leakey, A. D. (2011). Impairment of C4 photosynthesis by drought is exacerbated by limiting nitrogen and ameliorated by elevated $[CO_2]$ in maize. *Journal of Experimental Botany*, 65, 3235–3246.

Martre, P., Jamieson, P. D., Semenov, M. A., Zyskowski, R. F., Porter, J. R., & Triboi, E. (2006). Modelling protein content and composition in relation to crop nitrogen dynamics for wheat. *European Journal of Agronomy*, 25, 138–154.

Martre, P., Reynolds, M. P., Asseng, S., Ewert, F., Alderman, P. D., Cammarano, D., Maiorano, A., Ruane, A. C., Aggarwal, P. K., Anothai, J., Basso, B., Biernath, C., Challinor, A. J., De Sanctis, G., Doltra, J., Dumont, B., Fereres, E., Garcia-Vila, M., Gayler, S., … Zhu, Y. (2017). The international heat stress genotype experiment for modeling wheat response to heat: Field experiments and AgMIP-wheat multi-model simulations. *Open Data Journal for Agricultural Research*, 3, 23–28.

Mazdiyasni, O., & AghaKouchak, A. (2015). Substantial increase in concurrent droughts and heatwaves in the United States. *PNAS*, 112, 11484–11489.

Mitchell, R. A. C., Lawlor, D. W., Mitchell, V. J., Gibbard, C. L., White, E. M., & Porter, J. R. (1995). Effects of elevated CO_2 concentration and increased temperature on winter-wheat—Test of ARCWHEAT1 simulation model. *Plant, Cell & Environment*, 18, 736–748.

Monin, A., & Obukhov, A. (1954). Basic laws of turbulent mixing in the surface layer of the atmosphere. *Contribution of Geophysical Institute of Academy of the Sciences USSR*, 151, 163–187.

Monteith, J., & Unsworth, M. (1990). *Principles of environmental physics* (2nd ed., pp. 304). Edward Arnold.

Monteith, J. L. 1965. *Evaporation and the environment*. In: The state and movement of water in living organisms. XIXth Symposium. Soc. for Exp. Biol., Swansea. Cambridge University Press. pp. 205–234.

Monteith, J. L. (1973). *Principles of environmental physics* (pp. 241). Edward Arnold.

Moriondo, M., Giannakopoulos, C., & Bindi, M. (2011). Climate change impact assessment: The role of climate extremes in crop yield simulation. *Climatic Change*, 104, 679–701.

Morison, J. I. L. (1985). Sensitivity of stomata and water use efficiency to high CO_2. *Plant, Cell & Environment*, 8, 467–474.

Neenan, M., & Spencer-Smith, J. L. (1975). An analysis of the problem of lodging with particular reference to wheat and barley. *Journal of Agriculture Science (Cambridge)*, 85, 494–507.

Nendel, C., Berg, M., Kersebaum, K. C., Mirschel, W., Specka, X., Wegehenkel, M., Wenkel, K. O., & Wieland, R. (2011). The MONICA model: Testing predictability for crop growth, soil moisture and nitrogen dynamics. *Ecological Modelling*, 222, 1614–1625.

Neukam, D., Ahrends, H., Luig, A., Manderscheid, R., & Kage, H. (2016). Integrating wheat canopy temperatures in crop system models. *Agronomy*, 6, 7.

Nonhebel, S. (1996). Effects of temperature rise and increase in CO_2 concentration on simulated wheat yields in Europe. *Climatic Change*, 34, 73–90.

O'Leary, G. J., Christy, B., Nuttall, J., Huth, N., Cammarano, D., Stöckle, C., Basso, B., Shcherbak, I., Fitzgerald, G., Luo, Q., Farre-Codina, I., Palta, J., & Asseng, S. (2015). Response of wheat growth: Grain yield and water use to elevated CO_2 under a Free-Air CO_2 Enrichment (FACE) experiment and modelling in a semi-arid environment. *Global Change Biol.*, 21, 2670–2686.

Olesen, J. E., Petersen, B. M., Berntsen, J., Hansen, S., Jamieson, P., & Thomsen, A. (2002). Comparison of methods for simulating effects of nitrogen on green area index and dry matter growth in winter wheat. *Field Crops Research*, 74, 131–149.

Olesen, J. E., Trnka, M., Kersebaum, K. C., Skjelvag, A. O., Seguin, B., Peltonen-Sainio, P., Rossi, F., Kozyra, J., & Micale, F. (2011). Impacts and adaptation of European crop production systems to climate change. *European Journal of Agronomy*, 43(2), 96–112.

Otegui, M. E. (1997). Kernel set and flower synchrony within the ear of maize: II. Plant population effects. *Crop Science*, 37, 448–455.

Palosuo, T., Kersebaum, K. C., Angulo, C., Hlavinka, P., Moriondo, M., Olesen, J., Patil, R., Ruget, F., Rumbaur, C., Takac, J., Trnka, M., Bindi, M., Caldag, B., Ewert, F., Ferrise, R., Mirschel, W., Saylan, L., Siska, B., & Rötter, R. (2011). Simulation of winter wheat yield and its variability in different climates of Europe: A comparison of eight crop growth models. *European Journal of Agronomy*, 35(3), 103–114.

Palta, J. A., Chen, X., Milroy, S. P., Rebetzke, G. J., Dreccer, M. F., & Watt, M. (2011). Large root systems: Are they useful in adapting wheat to dry environments? *Functional Plant Biology*, 38, 347–354.

Parry, M. L. (2007). *Climate change 2007: Impacts, adaptation and vulnerability. Contribution of Working Group II to the Fourth Assessment Report of the Intergovernmental Panel on Climate Change*. Cambridge University Press.

Passioura, J. B. (1983). Roots and drought resistance. *Agricultural Water Management*, 7, 265–280.

Passioura, J. B. (1994). The yield of crops in relation to drought. In K. J. Boote, J. M. Bennett, T. R. Sinclair, & G. M. Paulsen (Eds.), *Physiology and determination of crop yield* (pp. 343–359). ASA, CSSA, and SSSA.

Penman, H. C. (1948). Natural evaporation from open water, bare soil and grass. *Proceedings of the Royal Society of London Series A*, 193, 120–145.

Penning de Vries, F. W. T., Witlage, J. M., & Kremer, D. (1979). Rates of respiration and of increase in structural dry matter in young wheat, ryegrass and maize plants in relation to temperature, to water stress and their sugar content. *Annals of Botany*, 44, 595–609.

Pickering, N. B., Jones, J. W., & Boote, K. J. (1995). Adapting SOYGRO V5.42 for prediction under climate change conditions. In C. Rosenzweig, J. W. Jones, & L. H. Allen, Jr. (Eds.), *Climate change and agriculture: Analysis of potential international impacts*. ASA Spec. Pub. No. 59 (pp. 77–98). ASA, CSSA, and SSSA.

Pinthus, M. J. (1973). Lodging in wheat, barley and oats: The phenomenon: Its causes and preventative measures. *Advances in Agronomy*, 25, 209–263.

Porter, J., & Gawith, M. (1999). Temperatures and the growth and development of wheat: A review. *European Journal of Agronomy*, 10, 23–36.

Porter, J. R., Xie, L., Challinor, A. J., Cochrane, K., Howden, S. M., Iqbal, M. M., Lobell, D. B., & Travasso, M. I. (2014). Food security and food production systems. In C. B. Field, V. R. Barros, D. J. Dokken, K. J. Mach, M. D. Mastrandrea, T. E. Bilir, M. Chatterjee, K. L. Ebi, Y. O. Estrada, R. C. Genova, B. Girma, E. S. Kissel, A. N. Levy, S. MacCracken, P. R. Mastrandrea, & L. L. White (Eds.), *Climate Change 2014: Impacts, adaptation, and vulnerability. Part A: Global and sectoral aspects. Contribution of Working Group II to the Fifth Assessment Report of the Intergovernmental Panel on Climate Change* (pp. 485–533). Cambridge University Press.

Powlson, D. S., Stirling, C. M., Thierfelder, C., White, R. P., & Jat, M. L. (2016). Does conservation agriculture deliver climate change mitigation through soil carbon sequestration in tropical agro-ecosystems? *Agriculture, Ecosystems & Environment*, 220, 164–174.

Priestley, C. H. B., & Taylor, R. J. (1972). On the assessment of surface heat flux and evaporation using large-scale parameters. *Monthly Weather Review*, 100, 81–92.

Raes, D., Steduto, P., Hsiao, T. C., & Fereres, E. (2009). AquaCrop the FAO crop model to simulate yield response to water: II. Main algorithms and software description. *Agronomy Journal*, 101, 438–447.

Raes, D., Steduto, P., Hsiao, T. C., & Fereres, E. 2018. *AquaCrop version 6.0–6.1 reference manual*. Rome, Italy: Food and Agriculture Organization of the United Nations. https://www.fao.org/3/BR244E/br244e.pdf

Ramanathan, V. (2007). Global dimming by air pollution and global warming by greenhouse gases: Global and regional perspectives. In C. D. O'Dowd & P. E. Wagner (Eds.), *Nucleation and atmospheric aerosols* (pp. 473–483). Springer.

Ratjen, A., & Kage, H. (2015). Forecasting yield via reference-and scenario calculations. *Computers and Electronics in Agriculture*, 114, 212–124.

Rezaei, E. E., Siebert, S., & Ewert, F. (2013). Temperature routines in SIMPLACE <LINTUL2-CC-HEAT>. In P. Alderman, E. Quilligan, S. Asseng, F. Ewert, & M. P. Reynolds (Eds.), *Modeling wheat response to high temperature* (pp. 111–113). CIMMYT.

Rezaei, E. E., Siebert, S., Manderscheid, R., Müller, J., Mahrookashani, A., Ehrenpfordt, B., Haensch, J., Weigel, J.-J., & Ewert, F. (2018). Quantifying the response of wheat yields to heat stress: The role of the experimental setup. *Field Crops Research*, 217, 93–103.

Rezaei, E. E., Webber, H., Gaiser, T., Naab, J., & Ewert, F. (2015). Heat stress in cereals: Mechanisms and modeling. *European Journal of Agronomy*, 64, 98–113.

Roberts, E., & Summerfield, R. (1987). Measurement and prediction of flowering in annual crops. In J. G. Atherton (Ed.), *Manipulation of flowering* (pp. 17–50). Butterworths.

Rötter, R. P., Appiah, M., Fichtler, E., Kersebaum, K. C., Trnka, M., & Hoffmann, M. P. (2018). Linking modelling and experimentation to better capture crop impacts of agroclimatic extremes—A review. *Field Crops Research*, 221, 152–156.

Rötter, R. P., Palosuo, T., Kersebaum, K. C., Angulo, C., Bindi, M., Ewert, F., Ferrise, R., Hlavinka, P., Moriondo, M., Nendel, C., Olesen, J. E., Patil, R., Ruget, F., Takác, J., & Trnka, M. (2012). Simulation of spring barley yield in different climatic zones of Northern and Central Europe: A comparison of nine crop models. *Field Crops Research*, 133, 23–36.

Rötter, R. P., Palosuo, T., Pirttioja, N. K., Dubrovsky, M., Salo, T., Fronzek, S., Aikasalo, R., Trnka, M., Ristolainen, A., & Carter, T. R. (2011). What would happen to barley production in Finland if global warming exceeded 4 °C? A model-based assessment. *European Journal of Agronomy*, 35, 205–214.

Samarakoon, A. B., & Gifford, R. M. (1996). Elevated CO_2 effects on eater use and growth of maize in wet and drying soil. *Functional Plant Biology*, 23, 53–62.

Sanchez, B., Rasmussen, A., & Porter, J. (2014). Temperature and the development of maize and rice: A review. *Global Change Biology*, 20, 408–417.

Schimel, D., Janetos, A., Backlund, P., Hatfield, J., Ryan, M., Archer, S., & Lettenmaier, D., 2008. *Synthesis*. In: The effects of climate change on agriculture, land resources, water resources, and biodiversity. A Report by the U.S. Climate Change Science Program and the subcommittee on Global change Research. Washington, DC.

Seguin, B., & Itier, B. (1983). Using midday surface temperature to estimate daily evaporation from satellite thermal IR data. *International Journal of Remote Sensing*, 4, 371–383.

Semenov, M. A., & Shewry, P. R. (2011). Modelling predicts that heat stress, not drought, will increase vulnerability of wheat in Europe. *Scientific Reports*, 1, 66.

Setter, T. L., Laureles, E. V., & Mazaredo, A. M. (1997). Lodging reduces the yield of rice by self-shading and reductions in canopy photosynthesis. *Field Crops Research*, 49, 95–106.

Sha, A. N., Tanveer, M., Rehman, A. U., Anjum, S. A., Iqbal, J., & Ahmad, R. (2017). Lodging stress in cereal—Effects and management: An overview. *Environemental Science and Pollution Research*, 24, 5222–5237.

Sharma, S., & Mujumdar, P. P. (2017). Increasing frequency and spatial extent of concurrent meteorological droughts and heatwaves in India. *Scientific Reports*, 7, 15582.

Shroyer, J., Mikesell, M., & Paulsen, G. (1995). *Spring freeze injury to Kansas wheat*. Kansas State University.

Shuttleworth, W. J., & Wallace, J. S. (1985). Evaporation from sparse crops-an energy combination theory. *Quarterly Journal of the Royal Meteorological Society*, 111, 839–855.

Siebert, S., Ewert, F., Rezaei, E. E., Kage, H., & Graß, R. (2014). Impact of heat stress on crop yield—On the importance of considering canopy temperature. *Environmental Research Letters*, 9, 044012.

Siebert, S., Webber, H., & Rezaei, E. E. (2017). Weather impacts on crop yields—Searching for simple answers to a complex problem. *Environmental Research Letters*, 12, 081001.

Smith, M., Allen, R., & Pereira, L. (1998). *Revised FAO methodology for crop-water requirements (IAEA-TECDOC-1026)*. International Atomic Energy Agency.

Solomon, S., Qin, D., Manning, M., Alley, R.B., Berntsen, T., Bindoff, N.L., Chen, Z., Chidthaisong, A., Gregory, J.M., Hegerl, G.C., Heimann, M., Hewitson, B., Hoskins, B.J., Joos, F., Jouzel, J., Kattsov, V., Lohmann, U., Matsuno, T., Molina, M., …, Wratt, D., 2007. Technical summary. In: Solomon, S., Qin, D., Manning, M., Chen, Z., Marquis, M., Averyt, K.B., Tignor, M., Miller, H.L. (Eds.), *Climate change 2007: The physical science basis. Contribution of Working Group I to the Fourth Assessment Report of the Intergovernmental Panel on Climate Change*. Cambridge University Press.

Soltani, A., Maddah, V., & Sinclair, T. (2013). SSM-Wheat: A simulation model for wheat development, growth and yield. *International Journal of Plant Production*, 7, 711–740.

Spiertz, J. (1974). Grain growth and distribution of dry matter in the wheat plant as influenced by temperature: Light energy and ear size. *Netherlands Journal of Agricultural Science*, 22, 207–220.

Spinoni, J., Naumann, G., Carrao, H., Barbosa, P., & Vogt, J. (2014). World drought frequency, duration, and severity for 1951–2010. *International Journal of Climatology*, 34(8), 2792–2804.

Steduto, P., Hsiao, T. C., & Fereres, E. (2007). On the conservative behavior of biomass water productivity. *Irrigation Science*, 25, 189–207.

Sterling, M., Baker, C. J., Berry, P. M., & Wade, A. (2003). An experimental investigation of the lodging of wheat. *Agricultural and Forest Meteorology*, 119, 149–165.

Steudle, E., & Peterson, C. A. (1998). How does water get through roots? *Journal of Experimental Botany*, 49, 775–788.

Stöckle, C. O. (2013). Temperature routines in CropSyst. In P. Alderman, E. Quilligan, S. Asseng, F. Ewert, & M. P. Reynolds (Eds.), *Modeling wheat response to high temperature* (pp. 47–49). CIMMYT.

Stöckle, C. O., Donatelli, M., & Nelson, R. (2003). CropSyst, a cropping systems simulation model. *European Journal of Agronomy*, 18, 289–307.

Stöckle, C. O., & Jara, J. (1998). Modeling transpiration and soil water content from a corn field: 20 min vs. daytime integration step. *Agricultural and Forest Meteorology*, 92, 119–130.

Stöckle, C. O., Nelson, R. L., Higgins, S., Brunner, J., Grove, G., Boydston, R., Whiting, M., & Kruger, C. (2010). Assessment of climate change impact on Eastern Washington agriculture. *Climatic Change*, 102, 77–102.

Stöckle, C. O., Williams, J. R., Rosenberg, N. J., & Jones, C. A. (1992). A method for estimating the direct and climatic effects of rising atmospheric carbon dioxide on growth and yield of crops: Part I. modification of the EPIC model for climate change analysis. *Agricultural Systems*, 38, 225–238.

Supit, I., Hooijer, A. A., van Diepen, C. A. 1994. *System description of the WOFOST 6.0 crop simulation model implemented in CGMS*. CGMS Publication 15956. EUR 15956 EN of the Office for Official Publications of the E.U. Luxembourg.

Taiz, L., & Zeiger, E. (1991). *Plant physiology*. The Benjamin Cummings Publishing Company.

Talukder, A. S. M. H. M., McDonald, G. K., & Gill, G. S. (2014). Effect of short-term heat stress prior to flowering and early grain set on the grain yield of wheat. *Field Crops Research*, 160, 54–63.

Tanner, C. B., & Sinclair, T. R. (1983). Efficient water use in crop production: Research or re-search. In H. M. Taylor, W. R. Jordan, & T. R. Sinclair (Eds.), *Limitations to efficient water use in crop production*. ASA, CSSA, and SSSA.

Teixeira, E. I., Fischer, G., van Velthuizen, H., Walter, C., & Ewert, F. (2013). Global hot-spots on heat stress on agricultural crops due to climate change. *Agricultural and Forest Meteorology*, 170, 206–215.

Thom, A., & Oliver, H. (1977). On Penman's equation for estimating regional evaporation. *Quarterly Journal of the Royal Meteorological Society*, 103, 345–357.

Thompson, T. E., & Fick, G. W. (1981). Growth response of alfalfa to duration of soil flooding and to temperature. *Agronomy Journal*, 73, 329–332.

Triggs, J. M., Kimball, B. A., Pinter, P. J., Wall, G. W., Conley, M. M., Brooks, T. J., LaMort, R. L., Adam, N. R., Ottman, M. J., Matthias, A. D., Leavitt, S. W., & Cerveny (2004). Free-air CO$_2$ enrichment effects on the energy balance and evapotranspiration of sorghum. *Agricultural and Forest Meteorology*, 124, 63–79.

Tripathi, S. C., Sayre, K. D., Kaul, J. N., & Narang, R. S. (2003). Growth and morphology of spring wheat (*Triticum aestivum* L.) culms and their association with lodging: Effects of genotypes, N levels and ethephon. *Field Crops Research*, 84, 271–290.

Trnka, M., Feng, S., Semenov, M. A., Olesen, J. E., Kersebaum, K. C., Rötter, R. P., Semerádová, D., Klem, K., Huang, W., Ruiz-Ramos, M., Hlavinka, P., Meitner, J., Balek, J., Havlik, P., & Büntgen, U. (2019). Mitigation efforts will not fully alleviate the increase in the water scarcity occurrence probability in wheat-producing areas. *Science Advances*, 5, eaau2406.

Trnka, M., Kersebaum, K. C., Eitzinger, J., Hayes, M., Hlavinka, P., Svoboda, M., Dubrovský, M., Semerádová, D., Wardlow, B., Pokorný, E., Možný, M., Wilhite, D., & Žalud, Z. (2013). Consequences of climate change for the soil climate in central Europe and the central plains of the United States. *Climate Change*, 120, 405–418 (+14 suppl.).

Trnka, M., Olesen, J. E., Kersebaum, K. C., Skjelvåg, A. O., Eitzinger, J., Seguin, B., Peltonen-Sainio, P., Rötter, R. P., Iglesias, A., Orlandini, S., Dubrovský, M., Hlavinka, P., Balek, J., Eckersten, H., Cloppet, E., Calanca, P., Gobin, A., Vucetic, V., Nejedlik, P., … Zalud, Z. (2011). Agroclimatic conditions in Europe under climate change. *Global Change Biology*, 17, 2415–2427.

Trnka, M., Rötter, R. P., Ruiz-Ramos, M., Kersebaum, K.-C., Olesen, J. E., & Semenov, M. A. (2014). Adverse weather conditions for wheat production in Europe will become more frequent with climate change. *Nature Climate Change*, 4, 637–643.

Tsukahara, H., & Kozlowski, T. (1986). Effect of flooding and temperature regime on growth and stomatal resistance of *Betula platyphylla* var. japonica seedlings. *Plant and Soil*, 92, 103–112.

Tubiello, F. N., & Ewert, F. (2002). Simulating the effects of elevated CO$_2$ on crops: Approaches and applications for climate change. *European Journal of Agronomy*, 18, 57–74.

Twine, T. E., Bryant, J. J., Richter, K. T., Bernacchi, C. J., Mcconnaughay, K. D., Morris, S. J., & Leakey, A. D. B. (2013). Impacts of elevated CO$_2$ concentration on the productivity and surface energy budget of the soybean and maize agroecosystem in the Midwest USA. *Global Change Biology*, 19, 2838–2852.

Ugarte, C., Calderini, D. F., & Slafer, G. A. (2007). Grain weight and grain number responsiveness to pre-anthesis temperature in wheat barley and triticale. *Field Crops Research*, 100, 240–248.

Van Heemst, H. D. J. 1988. *Plant data values required for simple crop growth simulation models: Review and bibliography*. Simulation Report CABO-TT no. 17. Wageningen, The Netherlands: CABO.

Van Keulen, H., Penning de Vries, F. W. T., & Drees, E. M. (1982). A summary model for crop growth. In F. W. T. Penning de Vries & H. H. van Laar (Eds.), *Simulation of plant growth and crop production* (pp. 87–94). Pudoc.

Van Oort, P. A. J., Saito, K., Zwart, S. J., & Shrestha, S. (2014). A simple model for simulating heat induced sterility in rice as a function of flowering time and transpirational cooling. *Field Crops Research*, 156, 303–312.

Vanuytrecht, E., Raes, D., Steduto, P., Hsiao, T. C., Fereres, E., Heng, L. K., Garcia Vila, M., & Mejias Moreno, P. (2014). AquaCrop: FAO's crop water productivity and yield response model. *Environmental Modelling and Software*, 62, 351e360.

Vanuytrecht, E., & Thorburn, P. J. (2017). Responses to atmospheric CO_2 concentrations in crop simulation models: A review of current simple and semi complex representations and options for model development. *Global Change Biology*, 23, 1806–1820.

Wall, G. W., Brooks, T. J., Adam, N. R., Cousins, A. B., Kimball, B. A., Pinter, P. J., LaMorte, R. L., Triggs, J., Ottman, M. J., Leavitt, S. W., Matthias, A. D., Williams, D. G., & Webber, A. N. (2001). Elevated atmospheric CO_2 improved sorghum plant water status by ameliorating the adverse effects of drought. *The New Phytologist*, 152, 231–248.

Wall, G. W., Kimball, B. A., White, J. W., & Ottman, M. J. (2011). Gas exchange and water relations of spring wheat under full-season infrared warming. *Global Change Biology*, 17, 2113–2133.

Wallor, E., Kersebaum, K. C., Ventrella, D., Bindi, M., Cammarano, D., Coucheney, E., Gaiser, T., Garofalo, P., Giglio, L., Giola, P., Hoffmann, M. P., Iocola, I., Lana, M., Lewan, E., Maharjan, G. R., Moriondo, M., Mula, L., Nendel, C., Pohankova, E., ... Trombi, G. (2018). The response of process-based agro-ecosystem models to within-field variability in site conditions. *Field Crops Research*, 228, 1–19.

Wang, E., Martre, P., Zhao, Z., Ewert, F., Maiorano, A., Rötter, R. P., Kimball, B. A., Ottmann, M. J., Wall, G. W., White, J. W., Reynolds, M. P., Alderman, P. D., Aggarwal, P., Anothais, J., Basso, B., Biernath, C., Cammarano, D., Challinor, A. J., De Sanctis, G., ... Asseng, S. (2017). The uncertainty of crop yield projections is reduced by improved temperature response functions. *Nature Plants*, 3, 1–11.

Wang, L., Liao, S., Huang, S., Ming, B., Meng, Q., & Wang, P. (2018). Increasing concurrent drought and heat during the summer maize season in Huang–Huai–Hai plain, China. *International Journal of Climatology*, 38 (7), 3177–3190.

Webber, H., Ewert, F., Kimball, B. A., Siebert, S., White, J. W., Wall, G. W., Ottman, M. J., Trawally, D. N. A., & Gaiser, T. (2016). Simulating canopy temperature for modelling heat stress in cereals. *Environmental Modelling and Software*, 77, 143–155.

Webber, H., Ewert, F., Olesen, J. E., Müller, C., Fronzek, S., Ruane, A., Bourgault, M., Martre, P., Ababaei, B., Bindi, M., Ferrise, R., Finger, R., Fodor, N., Gabaldón-Leal, C., Gaiser, T., Jabloun, M., Kersebaum, K. C., Lizaso, J. I., Lorite, I., ... Wallach, D. (2018a). Diverging importance of drought stress for maize and winter wheat in Europe. *Nature Communications*, 9, 4249.

Webber, H., White, J. W., Kimball, B. A., Ewert, F., Asseng, S., Rezaei, E. E., Pinter, P. J., Jr., Hatfield, J. L., Reynolds, M. P., Ababaei, B., Bindi, M., Doltra, J., Ferrise, R., Kage, H., Kassie, B. T., Kersebaum, K. C., Luig, A., Olesen, J. E., Semenov, M. A., ... Martre, P. (2018b). Physical robustness of canopy temperature models for crop heat stress simulation across environments and production conditions. *Field Crops Research*, 216, 75–88.

Webber, H., Martre, P., Asseng, S., Kimball, B., White, J., Ottman, M., Wall, G. W., De Sanctis, G., Doltra, J., Grant, R., Kassie, B., Maiorano, A., Olesen, J. E., Ripoche, D., Rezaei, E. E., Semenov, M. A., Stratonovitch, P., & Ewert, F. (2017). Canopy temperature for simulation of heat stress in irrigated wheat in a semi-arid environment: A multimodel comparison. *Field Crops Research*, 202, 21–35.

Weigel, H. J., & Dämmgen, U. (2000). The Braunschweig carbon project: Atmospheric flux monitoring and free air carbon dioxide enrichment (FACE). *Journal of Applied Botany*, 74, 55–60.

Wheeler, T., & von Braun, J. (2013). Climate change impacts on global food security. *Science*, 341, 508–513.

White, J. W., Kimball, B. A., Wall, G. W., Ottman, M. J., & Hunt, L. A. (2011). Responses of time of anthesis and maturity to sowing dates and infrared warming in spring wheat. *Field Crops Research*, 124, 213–222.

Williams, A., Hunter, M. C., Kammerer, M., Kane, D. A., Jordan, N. R., Mortensen, D. A., Smith, R. G., Snapp, S., & Davis, A. S. (2016). Soil water holding capacity mitigates downside risk and volatility in US rainfed maize: Time to invest in soil organic matter? *PLoS One*, 11, e0160974.

Williams, J. (2002). The EPIC model. In V. Singh (Ed.), *Computer models of watershed hydrology* (pp. 909–1000). Water Resources Publications.

Wollenweber, B., Porter, J. R., & Schellberg, J. (2003). Lack of interaction between extreme high-temperature events at vegetative and reproductive growth stages in wheat. *Journal of Agronomy and Crop Science*, 189, 142–150.

Yang, Y., Kim, S.-H., Timlin, D. J., Fleisher, D. H., Quebedeaux, B., & Reddy, V. R. (2009). Simulating canopy evapotranspiration and photosynthesis of corn plants under different water status using a coupled Maize-Sim+2DSOIL model. *Transactions of the ASAEB*, 52(3), 1011–1024.

Yigini, Y., & Panagos, P. (2016). Assessment of soil organic carbon stocks under future climate and land cover changes in Europe. *Science of the Total Environment*, 557–558, 838–850.

Yin, X. (2013). Improving ecophysiological simulation models to predict the impact of elevated atmospheric CO_2 concentration on crop productivity. *Annals of Botany*, 112, 465–475.

Yin, X., & van Laar, H. H. (2005). *Crop systems dynamics: An ecophysiological simulation model for genotype-by-environment interactions.* Wageningen Academic Publishers.

Yu, Q., Goudriaan, J., & Wang, T. D. (2001). Modelling diurnal courses of photosynthesis and transpiration of leaves on the basis of stomatal and non-stomatal responses, including photoinhibition. *Photosynthetica*, 39 (1), 43–51.

Zampieri, M., Ceglar, A., Dentener, F., & Toreti, A. (2017). Wheat yield loss attributable to heat waves, drought and water excess at the global, national and subnational scales. *Environmental Research Letters*, 12, 064008.

Zheng, B., Chenu, K., Doherty, A., & Chapman, S., 2014. *The APSIM-Wheat Module (7.5 R3008)*, http://www.apsim.info/Portals/0/Documentation/Crops/WheatDocumentation.pdf

Ziska, L. H., & Bunce, J. A. (2007). Predicting the impact of changing CO_2 on crop yields: Some thoughts on food. *The New Phytologist*, 175, 607–617.

11

Lajpat R. Ahuja and
Timothy R. Green

Quantifying and Modeling Management Effects on Soil Properties: Tillage, Reconsolidation, Crop Residues, and Crop Management

Abstract

Numerous field studies have shown evidence of significant management effects on soil, water, nutrient, and plant properties and/or processes. However, the results vary widely across locations, climates, soils, and experiments. Very little effort has been made to synthesize the site-specific effects into hypotheses and theories, design well-controlled experiments to evaluate these effects, and quantify the effects for practical applications. Yet, it is only through such synthesis and quantification that we can develop suitable management practices to solve complex problems. Here, we summarize the state-of-the-science of quantifications of tillage, crop residue, crop management, and related effects on important soil properties and identify knowledge gaps for future predictive research.

Introduction

In the last 40 years, soil scientists have taken on the challenge of addressing real-world problems, moving from laboratory-scale studies to field and landscape scales. These studies have shown substantial evidence of significant spatial and temporal variability of soil physical properties and processes. The temporal variability in soil properties

Modeling Processes and Their Interactions in Cropping Systems: Challenges for the 21st Century, First Edition.
Edited by Lajpat R. Ahuja, Kurt C. Kersebaum, and Ole Wendroth.
© 2022 American Society of Agronomy, Inc. / Crop Science Society of America, Inc. / Soil Science Society of America, Inc.
Published 2022 by John Wiley & Sons, Inc.

and processes may be greater than the spatial variability (van Es et al., 1999). Major sources of temporal variability in soil properties and/or processes are agricultural management practices such as tillage (and reconsolidation after tillage), cultivation, planting, other field operations with heavy machinery, surface residues, plants and crop rotations, irrigation, manuring and fertilization, and grazing. The changes in soil properties and processes, in turn, impact soil water, mass transport, plant growth dynamics, soil microbial processes, and the environment. Weather-related factors, such as freezing–thawing and wetting–drying, may modify the management effects. Numerous field studies have shown evidence of significant management effects on soil, water, nutrient, and plant properties and/or processes (Blanco-Canqui & Ruis, 2018; Maharjan et al., 2018; Strudley et al., 2008). However, the results vary widely across locations, soils, and experiments. Mankin et al. (1996) presented a synthesis of expert knowledge on the effects of management on several soil properties, where the opinions of the 32 experts demonstrated considerable variability across the United States. The results may often be inconsistent, probably due to variability of the controlling factors. Very little effort has been made to synthesize the site-specific effects into hypotheses and theories, design well-controlled experiments to evaluate the controlling factors and treatment effects, and quantify the effects for practical applications. Yet, it is only through such synthesis and quantification that we can develop suitable management practices to solve complex problems.

Primary tillage is most commonly done to prepare the seedbed before planting, to break crust and suppress weeds and pests, reduce surface residues, and manage soil moisture, whereas secondary tillage is applied before planting and during inter-cultivation and mixing of crop residues and surface-applied manures, fertilizers, and pesticides. Primary tillage is generally deeper than secondary tillage. These tillage operations commonly result in two tilled horizons at the top of the soil profile (Figure 11.1), and mixing with each tillage operation removes any soil textural and organic matter differences within the tilled depth. Below, we summarize the state-of-the-science of quantifications of tillage, residue, crop, and related effects on important soil properties and identify where knowledge gaps exist for future predictive research.

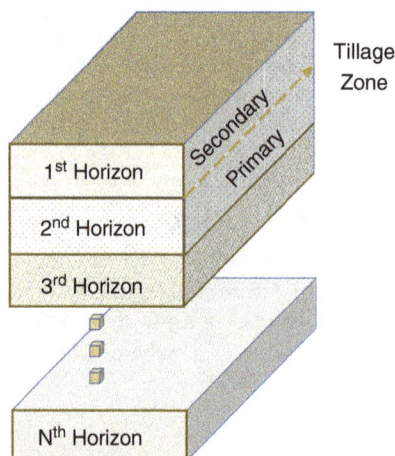

Figure 11.1 Relationship among soil horizons and the primary and secondary tillage zones. Modified from Rojas and Ahuja (2000).

The presentation is built on our earlier work (Ahuja et al., 2000a, 2000b, 2006; Green et al., 2003; Rojas & Ahuja, 2000).

The detailed focus of this chapter is on the effects of tillage and natural reconsolidation, soil compaction and plow pan formation with tillage, crop residues under no-till and tillage, and crop management on soil physical and hydraulic properties, which then affect a variety of soil–plant–atmosphere processes. The mixing of surface crop residues and surface-applied manures, fertilizers, and pesticides with the soil in the tilled horizon is estimated using simple algorithms, as described by Rojas and Ahuja (2000) and Maharjan et al. (2018).

Predicting Effects of Tillage and Reconsolidation

Tillage and subsequent reconsolidation due to wetting and drying can change the soil bulk density and porosity, soil hydraulic properties, surface roughness, and depression storage (DS) of rain.

Soil bulk density and porosity

Tillage initially decreases the soil bulk density (the term *soil bulk density* or simply *bulk density* is used throughout this chapter for dry soil bulk density) and increases the porosity of the tilled zone, which later gradually reverts back to the original state due to reconsolidation during cycles of wetting and drying (Cassel, 1983; Mapa et al., 1986; Onstad et al., 1984; Rousseva et al., 1988). Importantly, soil reconsolidation processes are driven by both drying and wetting, but hysteresis in soil-water retention and unsaturated hydraulic conductivity are generally ignored. None of the methods developed for characterizing dynamic soil hydraulic properties consider hysteresis explicitly and mechanistically. Zhang et al. (2017) concluded, "Further field studies are required to investigate the soil water retention characteristics (SWRCs) soil water retention characteristics of various soil textures in the complete range of saturation under both wetting and drying processes." Thus the studies and methods summarized below are all limited by this knowledge gap and current state-of-the-science.

The depth of the tilled zone and decrease in bulk density due to tillage depend on the tillage intensity, soil water content at the time of tillage, soil type, and cropping and management history. Allmaras et al. (1966) and Allmaras et al. (1967) presented data on such changes in three different soil associations. Williams et al. (1984) used an approximate equation in their EPIC model to estimate soil bulk density after tillage:

$$\rho(I) = \rho_o - \left(\rho_o - \frac{2}{3} \rho_c \right) I \tag{11.1}$$

where $\rho(I)$ is the bulk density (g cm^{-3}) after tillage, I is the mixing efficiency or the fraction of soil mixed, ρ_o is the bulk density before tillage, and ρ_c is the bulk density of the soil when it is completely settled. The ρ_c value may be related to soil clay and organic matter contents (Chen et al., 1998). The tillage intensity, I, is a factor ranging from 0 to 1 that depends on the implement used and surface soil conditions. Alberts et al. (1995) provided values of the tillage intensity for 78 different tillage implements or operations. For a tillage operation, such as a moldboard plow–disk–harrow, that completely mixes the soil ($I = 1$), the ρ after tillage will equal 2/3 ρ_c. Williams et al. (1984) did not give the basis or a test of this equation. Our calculation using the data of Allmaras et al.

(1966) indicated that it underpredicted the bulk density for a plow–disk–harrow operation. Similarly, Mankin et al. (1996) estimated bulk density reductions of 18% in the top 0–10 cm and 16% in the 10–20-cm layer for fine soils after moldboard plow—less than the 33% reduction from Equation 11.1. These studies illustrate the need for improved model skill, which has not been addressed adequately to date to our knowledge. Nonetheless, Equation 11.1 is being used in several agricultural system models (Ahuja et al., 2000b; Flanagan & Nearing, 1995; Green et al., 2019). The change in total porosity is calculated from soil bulk density by the commonly accepted relationship. Consequently, models may need to allow for dynamic geometry of soil units. For example, does tillage significantly change the thickness of the tilled horizon and how important is this?

After tillage, the soil bulk density increases with time due to reconsolidation by successive wetting and drying. Onstad et al. (1984) assumed that the amount of rainfall after tillage is a major factor governing this change and described these changes by the following empirical equation:

$$\rho(t) = \rho_i + a\frac{P(t)}{1 + P(t)} \tag{11.2}$$

where $\rho(t)$ is the bulk density at time t after tillage (g cm^{-3}), ρ_i is the bulk density right after tillage, $P(t)$ is the cumulative rainfall (mm), and a is a constant (g cm^{-3}). The results from simulated rainfall experiments using Equation 11.2 showed that the bulk density reached a near-maximum value at about 100 mm of rainfall. However, Rousseva et al. (1988) showed that the bulk density continued to change with rainfall beyond 100 mm. Equation 11.2 may be modified to allow this behavior by introducing another parameter. On the other hand, Linden et al. (1987) assumed that both rainfall amount and rainfall energy were the major factors governing reconsolidation and gave the following empirical equation in terms of soil porosity (ϕ, cm^3 cm^{-3}) changes with time:

$$\phi = \phi_i - (\phi_i - \phi_c)[1 - \exp(-0.15P - 1.5E)] \tag{11.3}$$

where ϕ_i is the initial porosity, ϕ_c is the final stable porosity, E is the cumulative rainfall energy (J cm^{-2}), and P is the cumulative rainfall (cm). This equation is used in the ARS Root Zone Water Quality Model (RZWQM) (Ahuja et al., 2000b). With slight modification of the coefficients for the E and P terms, this equation has given good results in the application of this model. Basso et al. (2006) used a similar equation with only E, taking into account the reduction in rainfall kinetic energy by surface residues and soil depth. However, further research is needed for testing the physical basis and improving these equations and developing new equations.

Soil Water Retention Characteristics and Hydraulic Conductivity

Gupta and Larson (1979), Rawls et al. (1982), Rawls et al. (2007), and Saxton and Rawls (2006) were the leaders in developing regression equations to estimate the soil water retention curve, $\theta(h)$, from soil texture, bulk density, and organic matter content. Here, θ is the volumetric soil water content (cm^3 cm^{-3}) and h is the soil water pressure head (cm). Saturated soil water content (θ_s) is estimated as the total porosity from the bulk density and the commonly used value for particle density of 2.65. The effect of changes in the stable soil organic matter (humus) on porosity can be obtained by using the volume-averaged particle density for mineral and humus components. A commonly used method to obtain the change in soil bulk density with the addition of stable organic matter was given by Maharjan et al. (2018). For field conditions, the porosity could be adjusted for an average entrapped air content between 5 and 10%. In the initial work of Gupta and Larson (1979) and Rawls et al. (1982), the regression equations predicted the unsaturated soil water content, θ, at fixed matric potentials, h. Subsequent investigators (Saxton et al., 1986; Wosten & van Genuchten, 1988) extended this approach to estimate parameters of the commonly used functional forms of $\theta(h)$, for example, the Brooks–Corey and van Genuchten equations. Ahuja and Williams (1991) obtained improved estimates of the water retention curve from soil porosity and soil water content at −33 kPa matric potential. Evaluation of these regression equations to estimate changes in $\theta(h)$ curves due to changes in soil bulk density with tillage and reconsolidation, and their adoption for dynamics of soil structure and biochannels, requires further research.

Ahuja et al. (1998) reviewed the experimental evidence for tillage effects on the pore-size distribution and $\theta(h)$. In general, the field data indicated that the changes occurred mostly in larger pores, at the wet end of the soil water retention curve (Hamblin & Tennant, 1981; Lindstrom & Onstad, 1984; Mapa et al., 1986). Some data also indicated that the air-entry or bubbling pressure value was not significantly affected by tillage (Powers et al., 1992). Ahuja et al. (1998) first showed that the application of the extended similar-media approach (e.g., Ahuja et al., 1985) did not correctly represent these field observations. They then proposed two semi-empirical methods for determining changes caused by tillage in the parameters of the Brooks and Corey (1964) form of the $\theta(h)$ curve:

$$\theta(h) = \theta_r \quad \forall |h| \leq |h_b| \tag{11.4a}$$

$$\theta(h) = (\theta_s - \theta_r)\left(\frac{h}{h_b}\right)^{-\lambda} + \theta_r \quad \forall |h| > |h_b| \tag{11.4b}$$

where θ_r is the so-called residual soil water content (cm^3 cm^{-3}), h_b is the air-entry or bubbling pressure head (negative values, −cm), λ is the pore size distribution (PSD) index, and $\forall$ denotes "for all" conditions. In the first method, Ahuja et al. (1998) assumed that: (a) the changes in soil bulk density and hence soil porosity, ϕ or θ_s, due to tillage were known (Equation 11.1); (b) the soil properties θ_r and h_b were not influenced by tillage; (c) the pore-size distribution index, λ, increased with tillage in the wet range, between $h = h_b$ and $h = 10h_b$, and in this range of h the tilled soil's λ value, λ_{till}, was computed from the tilled soil's saturated water content, $\theta_{s,till}$:

$$\lambda_{till} = \frac{\log(\theta_{s,till} - \theta_r) - \log[\theta(10h_b) - \theta_r]}{\log|h_b| - \log|10h_b|} \tag{11.5}$$

and (d) below this range, that is, for $h < 10h_b$, the λ value remained unchanged.

The second method was based on similar assumptions, except that between h_b and $10h_b$ the value of θ was assumed to change inversely with the h value, an intuitively appealing assumption. Tests of these methods on four datasets gave good results. Method 2 (Figure 11.2) was slightly better than Method 1. Ahuja et al. (1998) proposed that the above methods can also be used to estimate changes in $\theta(h)$ during reconsolidation. Zhang et al. (2006) showed that compaction of two loess silty loam soils decreased their water retention only in the wet range, above $h = -100$ kPa (−1 bar). An analysis similar to that done above for tillage effects on the water retention curve could also be done for compaction effects on a natural soil. Assouline (2006a) expressed

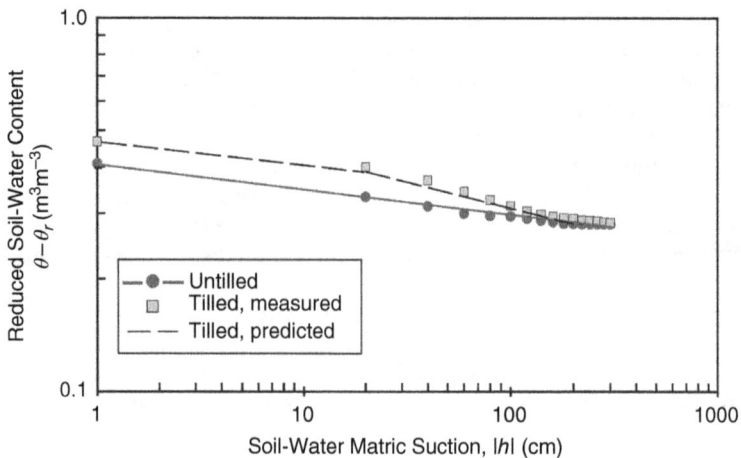

Figure 11.2 Measured and predicted (Method 2) soil water retention curves for Molakai silty clay under tilled and untilled conditions (Ahuja et al., 1998) / with permissions of John Wiley & Sons.

$\theta(h)$ of non-compacted and compacted soils using empirical functions and obtained their parameters by calibration with the experimental data using knowledge of soil bulk densities. The calibrated equations gave good results for the effect of an increase in soil bulk density on the water retention curve of a compacted soil. Hopefully, these methods will spur the development of more physically based methods.

To estimate the effects of tillage and reconsolidation on the saturated hydraulic conductivity, K_s (cm h^{-1}), we could consider the modified Kozeny–Carman equation (Ahuja et al., 1984, 1989):

$$K_s = B\phi_e^n \tag{11.6}$$

where ϕ_e is the effective porosity of larger pores, calculated as the saturated water content (θ_s) minus the water content at -33 kPa matric potential, and B and n are empirical coefficients. Equation 11.6 exhibited a degree of universality in that it was applicable to a wide range of soils from the southern region of the United States, Hawaii, and Arizona to several soils from Korea (Ahuja et al., 1989) and a variety of soils from Indiana (Franzmeier, 1991). Messing's (1989) results for some Norwegian soils showed that Equation 11.6 fit data for individual soils well, although the coefficients varied slightly with soil type. However, some of these soils had high clay contents and probably exhibited shrink–swell behavior, which could possibly affect the values of the fitted coefficients. Rawls et al. (1998) found that $n = 3 - \lambda$ for the textural class mean K_s values. Timlin et al. (1999) presented a slightly improved version of Equation 11.6 by incorporating the additional effect of λ on K_s. Ahuja et al. (2000a) and Green et al. (2003) found Equation 11.6 to be applicable to K_s data from both wheel-tracked and non-wheel-tracked portions of fields (Figure 11.3). The log–log relationship explained 67% of the variance in log(K_s), but deviations of measured K_s can exceed an order of magnitude. Extreme positive deviations of measured $K_s \gg$ estimated $K_s(\phi_e)$ may be affected by soil structure and continuous macropores. Equation 11.6 will estimate changes in K_s due to tillage from changes in ϕ_e resulting from changes in θ_s and θ_{33kPa}. The complete unsaturated hydraulic conductivity curve, $K(h)$, of the tilled soil can be determined from K_s and the

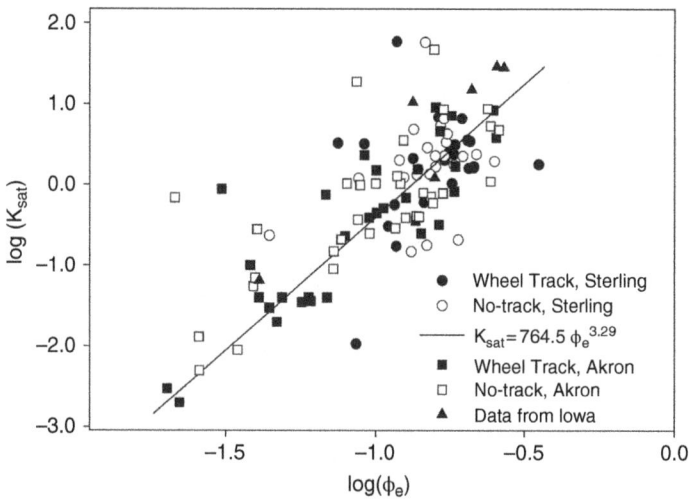

Figure 11.3 Log–log linear relationship between saturated hydraulic conductivity, K_{sat}, and effective porosity, ϕ_e, for Sterling and Akron, CO, and data from Iowa. The regression line was derived by Ahuja et al. (1989) for nine other soils. Figure from Green et al. (2003).

parameters of the new $\theta(h)$ curve of the tilled soil based on the work of Campbell (1974) and Schaap and Leij (2000). The method should also apply during natural reconsolidation after tillage. Subsequently, Assouline (2006b) showed somewhat varying results for relative compacted/non-compacted (K_{cs}/K_s) values of K_s for a number of soils at different bulk densities with several different versions of the original Kozeny–Carmen equation (Figure 11.4). Good results for both saturated and unsaturated K were obtained using the Assouline (2006b) model for the hydraulic conductivity function incorporating the bulk density effects.

Recently, Vanderlinden et al. (2021) measured spatial distributions of soil water retention curves and field soil water states in a topsoil (0–0.05 m) under long-term direct drill (DD, no tillage) and conventional tillage (CT). Water retention was significantly larger under DD than CT for absolute values of the soil water pressure head >63 cm. Field soil water states were controlled by the specific water retention curve and pore-size distribution ranges for both DD and CT. Silva et al. (2019) found that the short-term pore class distribution after tillage was closely related to soil CO_2 emission. Wilson et al. (2020) measured natural reconsolidation effects on soil bulk density, saturated hydraulic conductivity, water content, soil penetration strength, and surface shear strength following a series of wetting and drying cycles under simulated rainfall. The largest increases in bulk density, penetration resistance, and shear strength, and the largest decrease in saturated hydraulic conductivity, occurred after the first wetting and drying cycle. Bauer et al. (2015) reported that soil consolidation after tillage was different for long-term managements of conventional versus reduced tillage.

Tillage along with macropores created by soil organisms or soil cracks can result in a bimodal PSD. Kreiselmeier et al. (2019) fitted the bimodal soil water retention model of Kosugi (1996) to measured data for three different tillage levels to obtain their bimodal parameters. The bimodal water retention curve and measured K_s can then also be used to obtain bimodal unsaturated hydraulic conductivity characteristics.

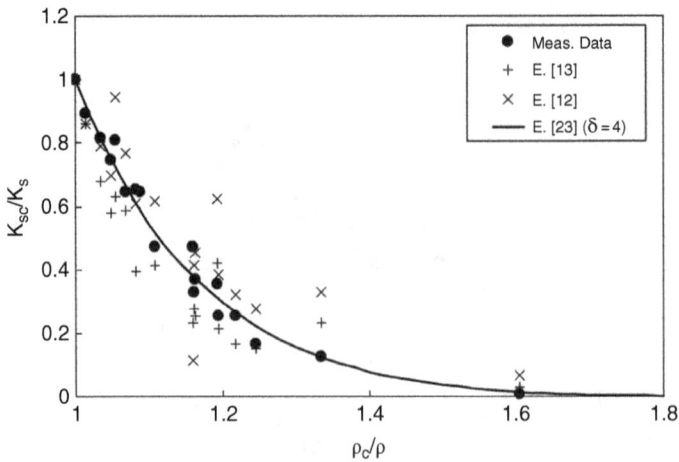

Figure 11.4 Ratio of predicted compacted to non-compacted initial saturated hydraulic conductivity (K_{sc}/K_s) versus the ratio of compacted to non-compacted initial soil bulk density (ρ_c/ρ) based on Assouline (2006b, Equation 23) compared with measured data. Figure from Assouline (2006b).

Or et al. (2000) presented a stochastic model that represented changes in PSDs with the Fokker–Planck (FP) advection–diffusion type equation. The model gave changes in total porosity, mean pore radius, and variance of the pore-size distribution. The wetting and drying were shown to affect the soil water retention and hydraulic conductivity. Leij et al. (2002a, 2002b) derived analytical solutions to the governing FP equation using known temporal functions for the displacement of the mean pore size. Pelak and Porporato (2019) proposed a modeling framework for the time evolution of the PSD (hence, soil water retention) in response to tillage, reconsolidation, and changes in soil organic matter and a time-varying power-law PSD for certain conditions. Such mechanistic approaches may or may not help to further improve the prediction of $\theta(h)$ and $K(h)$ changes due to tillage in field soils.

Considerable soil physics research has been conducted on the development of pedotransfer functions for estimating soil hydraulic properties from easily obtained data for soil physical properties (see Pachepsky & Rawls, 2004). Some research has also dealt with characterizing hydraulic properties of dual- and triple-porosity soils (e.g., Mohanty et al., 1997; Wilson et al., 1992). These developments also need to address the effects of tillage and related management on soil hydraulic properties, a focus of this chapter.

Surface Roughness and Depression Storage

Surface roughness retards overland flow and increases the transient depth and duration of this flow. These effects can increase cumulative infiltration in the field (Darboux & Huang, 2005). The surface roughness also affects soil erosion, soil thermal properties, and energy balance (Allmaras et al., 1972, 1977; Cogo, 1981; Cruse et al., 1980). Tillage increases surface roughness, the magnitude of which depends on the implements used and the soil condition (e.g., soil water content and degree of compaction) at the time of tillage. Based on some limited data, Alberts et al. (1995) assigned potential random roughness, RR, values to 78 different tillage implements. The RR changes due to tillage were described by

$$RR_{till} = RR_i T_i + RR_o (1 - T_i) \qquad (11.7)$$

where RR_i is the potential RR (cm) for the ith implement, T_i is the fraction of area tilled, and RR_o is the RR (cm) before tillage. Onstad et al. (1984) modeled the degradation of RR with rainfall after tillage using an equation similar to Equation 11.2 but with a different constant. Zobeck and Onstad (1987) proposed an exponential decline with rainfall amount, similar to Equation 11.3. Earlier investigators (Zobeck & Onstad, 1987) tried an exponential decline with cumulative rainfall energy, but the results were not conclusive. Bauer et al. (2015) reported that soil surface roughness depended on precipitation and soil tillage history. Further experimental investigations are needed.

Surface depression storage is a function of surface roughness and topographic slope. It may be derived mathematically by assuming an appropriate representation of the geometry of the depressions. Onstad (1984) improved on the conceptualizations of Mitchell and Jones (1976) and Moore and Larson (1979) for calculating DS from surface roughness (RR) and gave simple regression equations for calculating the maximum DS (cm):

$$DS = 0.0112RR + 0.031RR^2 - 0.012RR \times Slope \qquad (11.8)$$

where Slope is the land surface slope (rise over run). Onstad et al. (1984) also developed a regression equation to calculate the precipitation excess required to satisfy this DS based on the concept, supported by experimental data, that all DS is not filled before runoff begins due to connectivities. Hansen (2000) related DS to a mean surface depression depth calculated from elevation data taken from a microrelief meter. Huang and Bradford (1990) used Markov–Gaussian random fields to represent microtopography. In all cases, the depression storage decreases as the soil slope increases; thus, an increasing slope significantly decreases infiltration. This is, therefore, an important area for further research to help enhance water infiltration. For example, DS may also be influenced by soil surface curvature and topographic depressions at scales greater than surface RR.

Effects of Wheel-Track Compaction

Heavy vehicles used for tillage and other operations can compact the soil, resulting in an increase in bulk density, a decrease in porosity, and a change in the PSD (Ahuja et al., 2000a; Warkentin, 1971). The amount of soil compaction depends on the applied load, soil type, soil water status, and landscape position but may also vary from year to year (Alakukku, 1996; Liebig et al., 1993; Lindstrom & Voorhees, 1995). These changes in soil properties change the soil hydraulic properties and the amount of infiltration and available soil water. A great deal of knowledge exists on the stress–strain processes involved in soil deformation by compaction and shearing under load (Ghezzehei & Or, 2001; Horn, 2003; Horn et al., 2000; Or & Ghezzehei, 2002) and on modeling these processes and the soil compaction (Baumgartl & Kock, 2004; Braudeau et al., 2004a; Défossez et al., 2003; Défossez & Richard, 2002; Dorner et al., 2011; Ghezzehei, 2019; Gupta & Raper, 1994; Koolen et al., 1992; O'Sullivan et al., 1999; O'Sullivan & Simota, 1995; Smith, 1985; van den Akker & van Wijk, 1987; Young & Fattah, 1976). The modeling may involve different approaches depending on the objectives. To illustrate with a simple example, we look at the COMPISOIL model used by Défossez et al. (2003) to calculate the dry soil bulk density profile under the centerline of a wheel track and the resulting rut depth using an analytical method.

Soil compactness is defined as the specific volume v (volume per unit weight) as

$$v = \frac{\rho_s}{\rho_b} \qquad (11.9)$$

where ρ_s is the density of soil solids and ρ_b is the soil bulk density. For a given soil, this degree of compaction may also be taken as the ratio of the soil bulk density divided by a reference soil density. The value of v is given by an empirical virgin compression line (Défossez et al., 2003):

$$v = N - CI \ln (p) \qquad (11.10)$$

where CI is the compression index, p is the mean normal stress (mean of vertical, perpendicular, and transverse stresses), and N is the specific volume at $p = 1$ kPa. This equation gives a logarithmic decrease in specific volume with an increase in stress applied. The model allows for some rebounding when unloaded. The parameters for each soil depth are estimated from empirical relationships with the known initial soil bulk density and water content profiles. The final output is the dry bulk density profile under the centerline of the wheel track and rut depth.

For practical management purposes, long-term effects of wheel track compaction on the changes in properties of the soil horizons are more important, and these changes in the physical properties of the soil profile can be obtained by occasional manual sampling rather than yearly modeling. The long-term changes depend on the history of tillage intensity, crop and residue management, and weather conditions. However, the knowledge of long-term compaction effects on changes in soil profile properties under certain given conditions will help guide management decisions.

There are several reports of the effects of wheel track compaction on soil properties under varying conditions (Benjamin, 1993; Croney & Coleman, 1954; Culley et al., 1987; Gupta et al., 1989; Hill & Meza-Montalvo, 1990; Hill & Sumner, 1967; Lindstrom & Voorhees, 1995; Sillon et al., 2003; Wu et al., 1992). Lipiec and Hatano (2003) reviewed the compaction effects on various soil processes. The effects varied widely depending on the prevailing conditions, such as soil texture, initial bulk density, surface roughness, and soil water content. Hill and Sumner (1967) found that the changes in soil water retention for different soils artificially compacted to different bulk densities varied by soil textural class. Logsdon et al. (1992) found wheel compaction decreased K_s and ponded-water infiltration in a clay loam soil, particularly under higher values of the axle loads tested (4.5, 9, and 18 Mg) for both wet and dry soil conditions. Benjamin (1993) measured detailed water retention curves and K_s in wheel-track and no-track areas under dryland field conditions for three different Iowa soil types. The wheel tracks caused significant changes in the $\theta(h)$ curves (Figure 11.5). The λ value of the water retention curves roughly increased linearly with increasing bulk density (Ahuja et al., 2000a). Ahuja et al. (2000a) also showed that under semi-arid conditions with sandy loam to silt loam soils of the central Great Plains of the United States, wheel tracks did not cause a significant effect on the average soil hydraulic property curves. Sillon et al. (2003) found higher unsaturated hydraulic conductivity in a compacted calcareous soil but no difference in a loess soil. Destain et al. (2016) reported the highest

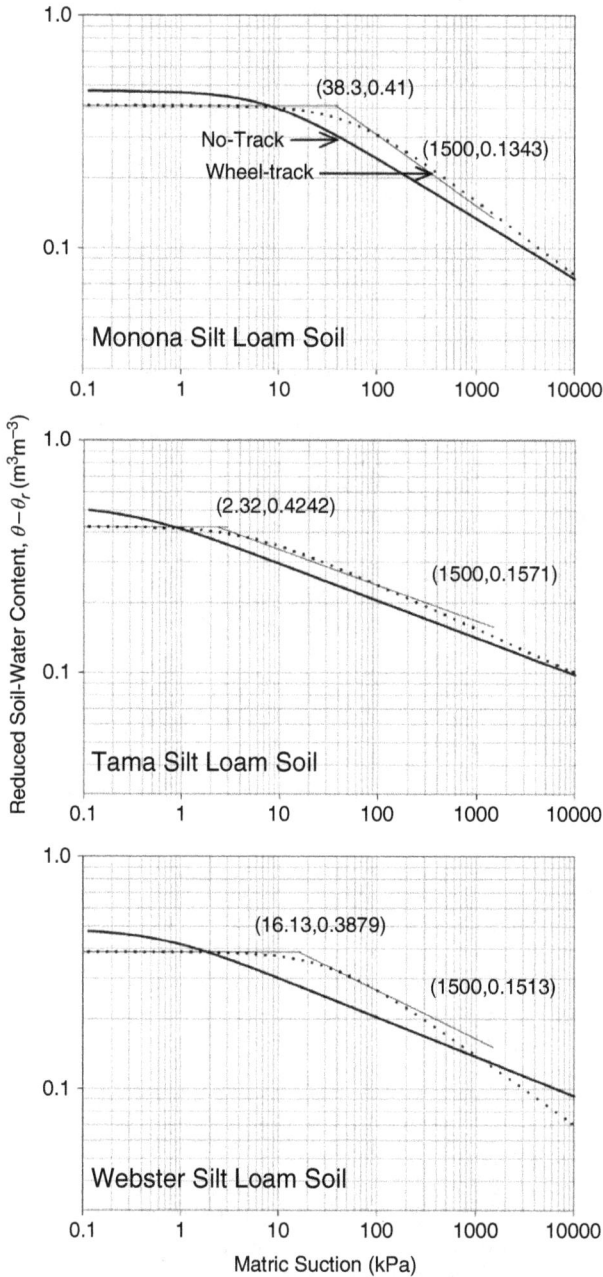

Figure 11.5 Soil water retention curves for three Iowa soils for no-track interrows (solid curves) and for wheel-track interrows (dotted curves). Solid lines are estimates of the track curves based on a known value for 100-kPa water content and a theoretical estimate for 1,500-kPa water content. Figure from Green et al. (2003).

increase in bulk density and precompression stress and reduction of macroporosity in the topsoil of a long-term conventionally tilled soil.

Coquet et al. (2005) measured water and bromide transport in wheel-tracked and plowed parts of a field plot under simulated rainfall. Very little water and bromide penetrated the large compacted zones under wheel tracks, whereas the highly heterogeneous plowed soil between the wheel tracks produced a much higher bromide dispersivity. Filipovic et al. (2019) applied a one-dimensional dual-permeability model to the data of Coquet et al. (2005) in which the compacted soil was considered a soil matrix domain and the uncompacted plowed soil as the fracture domain with an immobile water region. Results from this model using effective parameters for each domain agreed well with a two-dimensional model simulation. Still, the calibration of the effective parameters remains arbitrary and challenging. Silva et al. (2018) used the finite element method for numerical simulation of soil compaction in a sugarcane field in 0–20- and 20–100-cm soil layers, using the Modified Cam Clay elastoplastic constitutive model. The simulation provided the regions where soil compaction occurred, such as the higher soil water content areas.

These results indicate the value of knowing the long-term effects of soil compaction for continued improved management.

Long-Term No-Till and Crop Residue Effects

Macropores

Comparisons of no-till (NT) and minimum tillage with various CT practices during different time periods have not been consistent across soils, climates, and experiments (e.g., Ahuja & Nielsen, 1990; Hill, 1990; Hines, 1986; Logsdon et al., 1999). In general, long-term NT with crop residues has been observed to increase macropores and their connectivity with depth in the profile, with only small changes in bulk density and total porosity, even though NT may initially increase the soil bulk density in some soils (Gozubuyuk et al., 2014). No-till soils often show higher pesticide concentrations in percolate, shallow groundwater, or drainage than tilled soils (Elliott et al., 2000; Isensee et al., 1990; Kanwar et al., 1997; Masse et al., 1998). With NT, decayed root channels serve as continuous macropores, and residue cover associated with NT helps increase the number of continuous earthworm channels (Edwards et al., 1990; Meek et al., 1989; Shipitalo et al., 2000). Tillage is expected to disrupt the continuity or increase the tortuosity of macropores and also increase the K_s of the tilled-zone soil matrix (soil between macropores), which tends to decrease macropore flow (e.g., Petersen et al., 2001; Vervoort et al., 2001). However, some studies have shown no difference in hydraulically active macropores between different management practices (Azevedo et al., 1998). Malone et al. (2003) reanalyzed the macropores estimated from tension infiltrometer data for 20 different structured soils (Kaspar et al., 1995; Logsdon et al., 1993; Logsdon & Kaspar, 1995) to quantify the number of surface (0–1 cm) and subsurface (15–35 cm) macropores as affected by tillage treatments (moldboard plow, disk, chisel, and ridge tillage and NT). They found no clear trend for the number of macropores among tillage treatments and soils. Of course, using the tension infiltrometer method to estimate active macropores has limitations; the infiltrometers wet only a small and ill-defined depth of soil (Logsdon, 1997), approximately 5–7 cm, and thus do not represent continuous macroporosity for the soil profile, and they include the dead-end macropores that may be continuous only in the measurement depth. More recently, van Schaik et al. (2014) found that the abundance and biomass of earthworms were well correlated to different sizes of macropores in different soil depths. However, this was the case for mainly vertical macropores >6 mm in diameter, generally open to the soil surface, and often with a semi-pervious wall coating, which increased infiltration.

In a global review of NT effects on soil properties, Blanco-Canqui and Ruis (2018) concluded that NT had mixed or inconsistent effects on soil bulk density and K_s compared with CT. They also highlighted the variable adoption of NT by country, where most of the global area of NT cultivation is in North and South America. Blanco-Canqui and Wortmann (2020) addressed the practice of occasional tillage in long-term NT systems. Their synthesis of limited existing studies of this topic identified the potential benefits of tillage every 5–10 year, with little detected effect on the long-term effects of NT on soil hydraulic properties. Such mixed tillage practices complicate analyses of limited experimental data and point to the need for improved mechanistic modeling of dynamic soil hydraulic properties.

Further research and new methods are needed to: (a) characterize hydraulically active macropores, with and without wall coating, that are continuous in the whole soil profile (Timlin et al., 1994), as well as noncontinuous dead-end macropores in different soil horizons; and (b) quantify the number and size of macropores as functions of tillage, crop root systems, crop residue mass, climate, and soil type. With this information, we will be able to model the effects of macropores on preferential water and chemical transport in the soil profile, using the explicit Green–Ampt approach, described in detail in Chapter 1, or other approaches such as the one-dimensional dual-permeability model, based on the Richards equation, used by Filipovic et al. (2019) noted above and by Coppola et al. (2012) noted below in this chapter.

Residue Cover Impacts on Infiltration

Crop residue cover has long been recognized to increase infiltration by preventing surface crusting or sealing. Duley (1940) was the leader in showing dramatic effects of surface covers on infiltration by preventing the development of a thin surface seal or crust at the soil surface due to rainfall impact. Van Doren and Allmaras (1978) related the saturated hydraulic conductivity of a seal or crust (K_{sc}) to the rainfall kinetic energy applied (E_{sc}, J cm^{-2}) on the surface:

$$K_{sc} = K_f + (K_0 - K_f)\exp(-CE_{sc}) \tag{11.11}$$

where K_f is the final steady value of K_{sc}, K_0 is the K_s of the uncrusted soil, and C is an empirical coefficient. Lang and Mallett (1984) and Baumhardt and Lascano (1996) presented experimental data for the effects of different levels of residue cover on infiltration. In a theoretical study, Ruan et al. (2001) elucidated the effects of the different levels of residue cover and incomplete surface sealing in a field using a two-dimensional infiltration model based on numerical solution of the Richards equation. Crop residue was assumed to be distributed in regular patches. Beneath the patches, the surface soil was assumed to retain its original K_s value, whereas the bare soil areas between the patches were assumed to form a seal with a K_s value equal to known fractions of the original unsealed value. Interestingly, the results were sensitive to the degree of sealing and percentage of area covered by residue but not to the patch geometry. Ruan et al. (2001) also found that the use of a weighted average K_s of the surface soil for the whole area (covered with residue and uncovered) in a one-dimensional infiltration version of the numerical model reproduced the two-dimensional model results very well, except

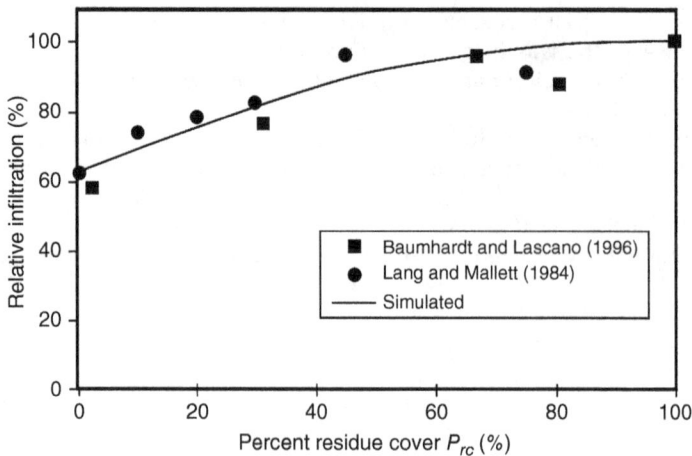

Figure 11.6 Comparison of experimental data and one-dimensional simulation results (Ruan et al., 2001) showing relative infiltration versus residue cover percentage. The square symbols are the data for cotton residue, whereas round dots are for corn residue. The data for the two crop residues merge. Figure from Green et al. (2003).

where the seal was assumed to have zero K_s. Numerical modeling reproduced the results from field studies for corn and cotton residues (Baumhardt & Lascano, 1996; Lang & Mallett, 1984), where the corn and cotton data coalesced when expressed as a function of residue cover percentage (Figure 11.6). Model results also compared well with the graph of precipitation storage efficiency versus residue level during fallow periods of a wheat–fallow rotation from three different locations in the Great Plains of the United States (Greb et al., 1967; Nielsen, 2002; Figure 11.7). Interactions between soil type, residue level (see below), and tillage intensity were apparent. Further work is certainly needed to test these initial findings and quantify the predictive ability of Equation 11.11 and the Richards equation or its simpler versions. In Chapter 1, the one-dimensional Green–Ampt model of infiltration is described in detail. This model can include the presence of a soil surface seal or crust related to surface residue cover. The Richards equation presented in Chapter 1 can also be used for one-dimensional infiltration with a surface seal.

Crop Residues, Manures, and Soil Properties

Frequent additions and decomposition of crop residues at the surface increase the organic matter content of the surface soil and modify its structure and physical properties in the long term (Shaver et al., 2002; Shaver et al., 2003; Sherrod et al., 2003). Heuscher et al. (2005) showed that soil organic C was the strongest contributor to bulk density prediction in regression relations among the soil properties investigated, including water content, silt content, and depth. Shaver et al. (2002, 2003) showed that 12 years of NT and residues in several dryland cropping systems decreased bulk density, increased effective porosity, and increased organic C and macroaggregates in the top 2.5 cm of the surface soil. The magnitude of changes in these properties was linearly related to the amount of residue biomass produced and added to the surface in a given rotation.

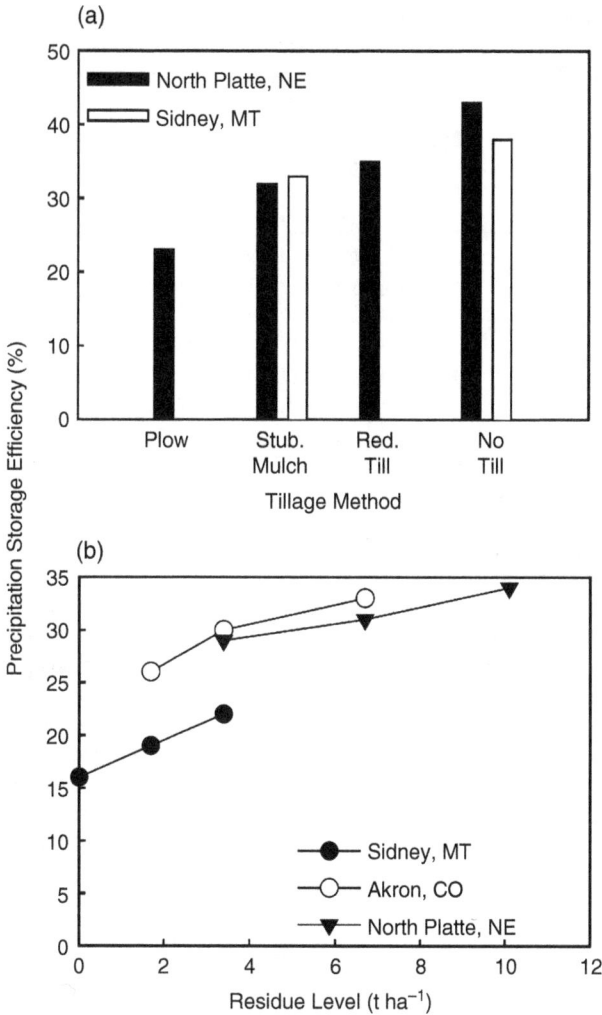

Figure 11.7 Precipitation storage efficiency versus (a) tillage method based on a wheat–fallow system in Nebraska and Montana and (b) residue level in Montana, Colorado, and Nebraska (Nielsen, 2002). Figure from Green et al. (2003).

Ma et al. (1999) compared several models to simulate the decomposition of surface residue in the field at three locations in Colorado as a function of the C/N ratio, air temperature, and rainfall amount. They concluded that crop residue decomposition was adequately simulated by the Douglas–Rickman model (Douglas & Rickman, 1992), which was better than other models because of its consistency in model parameters among the three experimental locations. The residue decay is assumed first order with respect to degree days:

$$M_i = M_{i-1}(1 - K_d DGD_i) \tag{11.12}$$

where M_i and M_{i-1} are residue mass at the current and previous days (kg ha^{-1}), respectively; DGD_i is the degree days for the current day, which is taken as the average temperature above $0\,°C$ of that day; and K_d is a first-order decomposition constant (DGD^{-1}), which is calculated as

$$K_d = kf_N f_W \tag{11.13}$$

where f_N and f_W are factors accounting for initial crop residue C/N ratio and soil moisture effects, respectively, and k is a rate coefficient (DGD^{-1}). Different empirical parameters are used for flat and standing residues. The fitted decomposition rate constants in the study by Ma et al. (1999) matched those originally published by Douglas and Rickman (1992). Quemada (2004) used corrected degree days to improve the effect of moisture and weather conditions on decomposition.

Animal manure, with or without bedding, is mixed with the existing flat residue. Tillage mixes part of the surface residue–manure mass present at the time of the tillage with the upper soil layers. The decomposing residue–manure material within the soil may be divided into more than one residue pool depending on the C/N ratio, which makes it easier to compute decomposition further. Decomposing roots in the soil may be added to these pools. The dynamics of surface residue (standing vs. flat) pools and their mixing with manures and soil use the simple algorithms as described by Rojas and Ahuja (2000). Decomposition and transformations in these pools are linked, commonly, to those of the three soil organic matter pools: fast, slow, and very slow (humus pool). The detailed equations of decomposition and transformations are described in detail in Chapter 5.

The humus component of soil organic matter that forms complexes with clay and silt to create water-stable aggregates thus improves soil structure, increases soil porosity, decreases soil bulk density, and improves its hydraulic properties of soil water retention and hydraulic conductivity (Rawls et al., 2003, 2004). A commonly used method to obtain the change in soil bulk density with the addition of stable organic matter (Maharjan et al., 2018) is

$$BD = \frac{100}{SOM\%/BD_{SOM} + (100 - SOM\%)/BD_{mineral}} \tag{11.14}$$

where BD is the new soil bulk density, $BD_{mineral}$ is the bulk density of the soil mineral component, and BD_{SOM} is the bulk density of the soil organic matter. Both $BD_{mineral}$ (2.65; 2.6–2.75 g cm^{-3}) and BD_{SOM} (1.2; 1.1–1.3 g cm^{-3}) are assumed to stay constant.

The effect of changes in the stable soil organic matter (humus) on porosity can be obtained by using the new BD and the volume-averaged particle density for mineral and humus components. Of course, there is an upper limit to the increase in humus content of a soil, controlled by climatic conditions. Rawls et al. (2003) reported that the relationship between soil water retention and soil C was affected by both the proportions of textural components and the amount of organic C, using regression trees and a group method of data handling on a large soil characterization database. Based on this approach, they developed the following equations to estimate the effects of soil organic C on water content at field capacity (−33 kPa pressure head) and the wilting point (−1,500 kPa), denoted respectively as θ_{33} and $\theta_{1,500}$:

$$\theta_{33} = 29.7528 + 10.3544 \times$$
$$(0.0461615 + 0.290955x - 0.0496845x^2 + 0.00704802x^3 +$$
$$0.269101y - 0.176528xy + 0.0543138x^2y + 0.1982y^2 -$$
$$0.060699y^3 - 0.3220249z - 0.0111693x^2z + 0.14104yz +$$
$$0.0616676xyz - 0.102026y^2z - 0.04012z^2 + 0.160838xz^2 -$$
$$0.121392yz^2 - 0.0616676z^3)$$
(11.15)

$$\theta_{1,500} = 14.2568 + 7.36318 \times$$
$$(0.06865 + 0.108713x - 0.0157225x^2 + 0.00102805x^3 +$$
$$0.886569y - 0.226581xy + 0.0126379x^2y - 0.017059y^2 +$$
$$0.135266xy^2 - 0.0334434y^3 - 0.0535182z - 0.0354271xz -$$
$$0.00261313x^2z - 0.154563yz - 0.0160219xyz - 0.0400606y^2z -$$
$$0.104875z^2 + 0.0159857xz^2 - 0.0671656yz^2 - 0.0260699z^3)$$
(11.16)

where $x = -0.837531 + 0.430183C_{org}$ (soil C content), $y = -1.40744 + 0.0661969C_{clay}$ (soil clay content), and $z = -1.51866 + 0.0393284C_{sand}$ (soil sand content) for $0.02 < C_{org} < 28.44\%$, $0.0 < C_{clay} < 90\%$, and $0.7 < C_{sand} < 95\%$.

Note that the form of Equations 11.15 and 11.16 matches the form as originally published (Rawls et al., 2003). In practice, one would multiply out the coefficients prior to applying repeated calculations. At low soil C contents (C_{org}), these equations give higher water retention in sandy soils due to the addition of organic matter and a decrease in fine-textured soils. At high levels of C_{org}, water retention increases in all soils but more in sandy and silty soils than in clayey soils. The change in field capacity (θ_{33}), along with the change in bulk density and porosity due to soil C (Equation 11.14), allows estimation of K_s due to the change in C_{org}. The change in the 15-bar value ($\theta_{1,500}$) permits estimation of the unsaturated conductivity.

The amount of root mass added at different soil depths in different crop rotations should also have an effect on soil organic matter over the long term (Sherrod et al., 2003). Generally, the root mass added to soil is proportional to the aboveground biomass produced. This is an important issue regarding C sequestration with respect to global climate change, but this will also change the soil properties at different depths.

Dynamics of Soil Structure and Water and Mass Transport

The effects of soil morphology, structure, and aggregation on soil hydraulic properties were addressed by Lilly and Lin (2004) and Guber et al. (2004). The dynamics of soil structure in response to soil mechanical disturbances have been expressed in terms of stress–strain relations by numerous investigators (Horn, 2003; Lipiec & Hatano, 2003; Sillon et al., 2003) and has been noted above for wheel-track compaction. Here, we emphasize the dynamics of soil cracks that cause substantial preferential transport of water and chemicals from the soil surface to deeper layers or the groundwater and the effects of roots on soil structural changes.

Dynamics of Soil Cracks

Swelling and shrinking clayey soils develop cracks on drying that cause rapid movement of water and chemicals to deeper layers and the groundwater. The cracks close on soil rewetting, and crack surface area and depth vary with soil water content. Preferential flow through dynamic cracks can be important but has not been studied much.

Modeling of the flow requires quantification of the opening and closing of cracks (dynamics). While many investigators have measured swelling and shrinking of small soil cores, several of them have also measured shrinkage and subsidence of soils in the field (Bronswijk, 1988, 1989, 1990, 1991a, 1991b; Jamison & Thompson, 1967; Woodruff, 1937; Yule & Ritchie, 1980). In an application of the RZWQM, Hua (1995) assigned crack volume to be either a linear or a quadratic function of soil moisture, which showed success in simulating water table, NO_3, and pesticide concentrations. In his extensive work, Bronswijk (1988, 1989, 1990, 1991a, 1991b) developed and used the following equation to calculate subsidence and shrinkage in soils:

$$1 - \frac{\Delta V}{V} = \left(1 - \frac{\Delta z}{z}\right)^{r_g} \tag{11.17}$$

where V (cm^3) is the initial volume of a soil cube, ΔV is the volume decrease on shrinking, z (cm) is the initial depth, Δz is the vertical subsidence, and r_g is a dimensionless shrinkage geometry factor. In the case of subsidence without cracking, $r_g = 1$, whereas in case of isotropic shrinkage $r_g = 3$. In the case of cracking without subsidence, r_g becomes infinity. With a known value of r_g, Bronswijk (1989) used this equation to estimate ΔV from measured subsidence. The subsidence volume (ΔV_{sub}) and total crack volume (ΔV_{cr}) are then given as

$$\Delta V_{sub} = z^2 \Delta z \tag{11.18a}$$

$$\Delta V_{cr} = \Delta V - \Delta V_{sub} \tag{11.18b}$$

Bronswijk (1988, 1989) used these estimates in modeling the role of continuously changing cracks on water transport in the soil matrix and cracks. Subsequently, Chertkov et al. (2004) improved on Bronswijk's concepts and suggested equations for estimating crack volume. Braudeau et al. (2004a) and Braudeau et al. (2004b) have also addressed these issues. Peng and Horn (2005) presented a simple model of the soil shrinkage curve.

Coppola et al. (2012) used a one-dimensional dual-permeability approach to model flow in shrinking–cracking soil with some simplifying assumptions. The soil was assumed to consist of two dynamic interacting pore domains: fractured or cracked volume and aggregate soil matrix. The shrinkage dynamics were represented by the inversely proportional volume changes of the two domains, while the overall porosity and thickness of the total soil remained constant. The one-dimensional Richards equation was used to compute water flow in both domains. The coupled domain-specific hydraulic properties were changed according to the swell–shrink characteristics of the matrix and partly by changing the fractional contribution of two domains with pressure head. It was also assumed that the aggregate and its hydraulic properties depend only on the aggregate water content. The model predicted the spatial and temporal evolution of soil water contents measured in the field. Coppola et al. (2015) later showed that, during infiltration, the simulated water and solute transfer from cracks to the soil matrix was faster, larger, and deeper when the hydraulic properties and relative fractions of the two domains were changed. By contrast, very similar profiles evolved during soil drying from the wetted condition. Further work is needed to verify these estimates and their applications in modeling transport through dynamic cracks.

Remarks on the Influence of Roots on Soil Structure and Biochannels

Roots and root hairs enter soil pores as they grow into the soil. As such, they change the soil density, porosity, and hydraulic properties adjacent to the roots. Even though the volume of soil occupied by roots is generally <1% of the soil volume (Jungk, 1996), increased resistance to water flow from soil to roots at the root–soil interface may be important for water and nutrient uptake. On the other hand, roots exude ions, organic acids, enzymes, and other substances in their vicinity (the rhizosphere), which increases nutrient availability from adsorbed phases and may also affect soil structure and water uptake. Decayed roots create biochannels or soil macropores that cause a substantial amount of preferential flow and transport, as has already been discussed. Research is needed to characterize and quantify these soil structural changes caused by roots for different rooting systems, crop rotations, soils, climates, and management systems.

Conclusions and Outlook

This chapter describes, in our judgment, the most important knowledge gaps that have been encountered by developers of agricultural systems models. The development of new knowledge in these areas and its quantification will require both innovative experimental research and the development of new concepts, theories, and models. Exciting and potentially high-impact areas of further research lie on the interfacial boundaries of soil physics and other disciplines. Integrated in agricultural systems research, this soil physics research will create breakthroughs in knowledge that will help solve major practical problems that agriculture is facing in the 21st century. The soil physics–agricultural systems models will also be excellent tools for teaching systems modeling to graduate students. These accomplishments will be a source of great personal satisfaction for a new generation of transdisciplinary soil scientists.

References

Ahuja, L. R., Benjamin, J. G., Dunn, G. H., Ma, L., Green, T. R., & Peterson, G. A. (2000a). Quantifying wheel track effects on soil hydraulic properties for agricultural system modeling. In T. W. Riley & J. M. A. Desbiolles (Eds.), *Proceedings of the 4th International Conference on Soil Dynamics, Adelaide, South Australia* (pp. 407–414). School of Advanced Manufacturing & Mechanical Engineering, University of South Australia.

Ahuja, L. R., Rojas, K. W., Hanson, J. D., Shaffer, M. D. J., & Ma, L. (Eds.) (2000b). *Root Zone Water Quality Model: Modeling management effects on water quality and crop production.* Water Resources Publications.

Ahuja, L. R., Cassel, D. K., Bruce, R. R., & Barnes, B. B. (1989). Evaluation of spatial distribution of hydraulic conductivity using effective porosity data. *Soil Science, 148*(6), 404–411.

Ahuja, L. R., Fiedler, F., Dunn, G. H., Benjamin, J. G., & Garrison, A. (1998). Changes in soil water retention curves due to tillage and natural reconsolidation. *Soil Science Society of America Journal, 62,* 1228–1233. https://doi.org/10.2136/sssaj1998.03615995006200050011x

Ahuja, L. R., Ma, L., & Timlin, D. J. (2006). Trans-disciplinary soil physics research critical to synthesis and modeling of agricultural systems. *Soil Science Society of America Journal, 70,* 311–326. https://doi.org/10.2136/sssaj2005.0207

Ahuja, L. R., Naney, J. W., Green, R. E., & Nielsen, D. R. (1984). Macroporosity to characterize spatial variability of hydraulic conductivity and effects of land management. *Soil Science Society of America Journal, 48,* 699–702. https://doi.org/10.2136/sssaj1984.03615995004800040001x

Ahuja, L. R., Naney, J. W., & Williams, R. D. (1985). Estimating soil water characteristics from simpler properties or limited data. *Soil Science Society of America Journal, 49,* 1100–1105. https://doi.org/10.2136/sssaj1985.03615995004900050005x

Ahuja, L. R., & Nielsen, D. R. (1990). Field water relations. In B. A. Stewart & D. R. Nielsen (Eds.), *Irrigation of agricultural crops.* ((Agronomy Monograph 30) (pp. 143–190). ASA, CSSA, & SSSA.

Ahuja, L. R., & Williams, R. D. (1991). Scaling water characteristics and hydraulic conductivity based on Gregson–Hector–McGowan approach. *Soil Science Society of America Journal, 55,* 308–319. https://doi.org/10.2136/sssaj1991.03615995005500020002x

Alakukku, L. (1996). Persistence of soil compaction due to high axle load traffic: 2. Long-term effects on the properties of fine-textured and organic soils. *Soil and Tillage Research, 37*(4), 223–238.

Alberts, E. E., Laflen, J. M., Rawls, W. J., Simanton, J. R., & Nearing, M. A. (1995). Soil components. In D. C. Flanagan & M. A. Nearing (Eds.), *USDA-Water Erosion Prediction Project, hillslope profile and watershed model documentation* ((NSERL Report 10) (pp. 7.1–7.47). USDA–ARS National Soil Erosion Research Laboratory.

Allmaras, R. R., Burwell, W. E., & Holt, R. F. (1967). Plow-layer porosity and surface roughness from tillage as affected by initial porosity and soil moisture at tillage time. *Soil Science Society of America Proceedings, 31,* 550–556. https://doi.org/10.2136/sssaj1967.03615995003100040033x

Allmaras, R. R., Burwell, R. E., Larson, W. E., & Holt, R. F. (1966). *Total porosity and random roughness of the interrow zone as influenced by tillage* (USDA–ARS Conservation Research Report 7). U.S. Government Printing Office.

Allmaras, R. R., Hallaur, E. A., Nelson, W. W., & Evans, S. E. (1977). *Surface energy balance and soil thermal property modification by tillage-induced soil structure* (Technical Bulletin 306). Minnesota Agricultural Experiment Station. The University of Minnesota Digital Conservancy. https://hdl.handle.net/11299/109055

Allmaras, R. R., Nelson, W. W., & Hallaur, E. A. (1972). *Fall versus spring plowing and related soil heat balance in the western Corn Belt* (Technical Bulletin 283). Minnesota Agricultural Experiment Station. The University of Minnesota Digital Conservancy. https://hdl.handle.net/11299/182441

Assouline, S. (2006a). Modeling the relationship between soil bulk density and the water retention curve. *Vadose Zone Journal, 5,* 599–609. https://doi.org/10.2136/vzj2005.0083

Assouline, S. (2006b). Modeling the relationship between soil bulk density and the hydraulic conductivity function. *Vadose Zone Journal, 5,* 697–705. https://doi.org/10.2136/vzj2005.0084

Azevedo, A. S., Kanwar, R. S., & Horton, R. (1998). Effect of cultivation on hydraulic properties of an Iowa soil using tension infiltrometers. *Soil Science, 163,* 22–28.

Basso, B., Ritchie, J. T., Grace, P. R., & Sartori, L. (2006). Simulation of tillage systems impact on soil biophysical properties using the SALUS model. *Italian Journal Agronomy, 1,* 677–688.

Bauer, T., Strauss, P., Grims, M., Kamptner, E., Mansberger, R., & Spiegel, H. (2015). Long-term agricultural management effects on surface roughness and consolidation of soils. *Soil and Tillage Research, 151,* 28–38.

Baumgartl, T., & Kock, B. (2004). Modeling volume change and mechanical properties with hydraulic models. *Soil Science Society of America Journal, 68,* 57–65. https://doi.org/10.2136/sssaj2004.5700

Baumhardt, R. L., & Lascano, R. J. (1996). Rain infiltration as affected by wheat residue amount and distribution in ridged tillage. *Soil Science Society of America Journal, 60,* 1908–1913. https://doi.org/10.2136/sssaj1996.03615995006000060041x

Benjamin, J. G. (1993). Tillage effects on near-surface soil hydraulic properties. *Soil and Tillage Research, 26,* 277–288.

Blanco-Canqui, H., & Ruis, S. J. (2018). No-tillage and soil physical environment. *Geoderma, 326,* 164–200.

Blanco-Canqui, H., & Wortmann, C. S. (2020). Does occasional tillage undo the ecosystem services gained with no-till? A review. *Soil and Tillage Research, 198,* 104534. https://doi.org/10.1016/j.still.2019.104534

Braudeau, E., Frangi, J. P., & Mohtar, R. H. (2004a). Characterizing non-rigid aggregated soil-water medium using its shrinkage curve. *Soil Science Society of America Journal, 68,* 359–370.

Braudeau, E., Mohtar, R. H., & Chahinian, N. (2004b). Estimating soil shrinkage parameters. *Developments in Soil Science, 30,* 225–240.

Bronswijk, J. J. B. (1988). Modeling of water balance, cracking and subsidence of clay soils. *Journal of Hydrology, 97,* 199–212.

Bronswijk, J. J. B. (1989). Prediction of actual cracking and subsidence in clay soils. *Soil Science, 148,* 87–93.

Bronswijk, J. J. B. (1990). Shrinkage geometry of a heavy clay soil at various stresses. *Soil Science Society of America Journal, 54,* 1500–1502. https://doi.org/10.2136/sssaj1990.03615995005400050048x

Bronswijk, J. J. B. (1991a). Relation between vertical soil movements and water-content changes in cracking clays. *Soil Science Society of America Journal, 55,* 1220–1226. https://doi.org/10.2136/sssaj1991.03615995005500050004x

Bronswijk, J. J. B. (1991b). Drying, cracking, and subsidence of a clay soil in a lysimeter. *Soil Science, 152*(2), 92–99.

Brooks, R. H., & Corey, A. T. (1964). *Hydraulic properties of porous media* (Hydrology Paper 3). Colorado State University.

Campbell, G. S. (1974). A simple method for determining unsaturated conductivity from moisture retention data. *Soil Science*, 117, 311–314.

Cassel, D. K. (1983). Spatial and temporal variability of soil physical properties following tillage of Norfolk loamy sand. *Soil Science Society of America Journal*, 47, 196–201. https://doi.org/10.2136/sssaj1983.03615995004700020004x

Chen, Y., Tessier, S., & Rouffignat, J. (1998). Soil bulk density estimation for tillage systems and soil textures. *Transactions of the ASAE*, 41, 1601–1610.

Chertkov, V. Y., Ravina, I., & Zadoenko, V. (2004). An approach for estimating the shrinkage geometry factor at a moisture content. *Soil Science Society of America Journal*, 68, 1807–1817. https://doi.org/10.2136/sssaj2004.1807

Cogo, N. P. (1981). *Effect of residue cover, tillage-induced roughness, and slope length on erosion and related parameters.* Doctoral dissertation, Purdue University. Purdue e-Pubs. https://docs.lib.purdue.edu/dissertations/AAI8210172/

Coppola, A., Comegna, A., Dragonetti, G., Gerke, H. H., & Basile, A. (2015). Simulated preferential water flow and solute transport in shrinking soils. *Vadose Zone Journal*, 14. https://doi.org/10.2136/vzj2015.02.0021

Coppola, A., Gertke, H. H., Comegna, A., Basile, A., & Comegna, V. (2012). Dual-permeability model for flow in shrinking soil with dominant horizontal deformation. *Water Resources Research*, 48, W08527.

Coquet, Y., Coutadeur, C., Labat, C., Vachier, P., van Genuchten, M. T., Roger-Estrade, J., & Šimůnek, J. (2005). Water and solute transport in a cultivated silt loam soil: 1. Field observations. *Vadose Zone Journal*, 4, 573–586. https://doi.org/10.2136/vzj2004.0152

Croney, D., & Coleman, J. D. (1954). Soil structure in relationship to soil suction. *Journal of Soil Science*, 5, 75–84.

Cruse, R. M., Linden, D. R., Radke, J. K., Larson, W. E., & Larntz, K. (1980). A model to predict tillage effects on soil temperature. *Soil Science Society of America Journal*, 44, 378–383. https://doi.org/10.2136/sssaj1980.03615995004400020034x

Culley, J. L. B., Larson, W. E., & Randall, G. W. (1987). Physical properties of a Typic Haplaquoll under conventional and no-tillage. *Soil Science Society of America Journal*, 51, 1587–1593. https://doi.org/10.2136/sssaj1987.03615995005100060033x

Darboux, F., & Huang, C. (2005). Does soil surface roughness increase or decrease water and particle transfer? *Soil Science Society of America Journal*, 69, 748–756. https://doi.org/10.2136/sssaj2003.0311

Défossez, P., Richard, G., Boizard, H., & O'Sullivan, M. F. (2003). Modeling change in soil compaction due to agricultural traffic as function of soil water content. *Geoderma*, 116, 89–105.

Défossez, P., & Richard, G. (2002). Models of soil compaction due to traffic and their evaluation. *Soil and Tillage Research*, 67, 41–64.

Destain, M.-F., Roisin, C., Dalck, A. S., & Mercatoris, B. C. N. (2016). Effect of wheel traffic on the physical properties of a Luvisol. *Geoderma*, 262, 276–284.

Dorner, J., Dec, D., Zunga, F., Sandoval, P., & Horn, R. (2011). Effect of land use change on Andosol's pore functions and their functional resilience after mechanical and hydraulic stresses. *Soil and Tillage Research*, 115–116, 71–79.

Douglas, C. L., Jr., & Rickman, R. W. (1992). Estimating crop residue decomposition from air temperature, initial nitrogen content, and residue placement. *Soil Science Society of America Journal*, 56, 272–278. https://doi.org/10.2136/sssaj1992.03615995005600010042x

Duley, F. L. (1940). Surface factors affecting the rate of intake of water by soils. *Soil Science Society of America Proceedings*, 9, 60–64. https://doi.org/10.2136/sssaj1940.036159950004000C0011x

Edwards, W. M., Shipitalo, M. J., Owens, L. B., & Norton, L. D. (1990). Effect of *Lumbricus terrestris* L. burrows on hydrology of continuous no-till corn fields. *Geoderma*, 46, 73–84. https://doi.org/10.1016/0016-7061(90)90008-W

Elliott, J. A., Cessna, A. J., Nicholaichuk, W., & Tollefson, L. C. (2000). Leaching rates and preferential flow of selected herbicides through tilled and untilled soil. *Journal of Environmental Quality*, 29, 1650–1656. https://doi.org/10.2134/jeq2000.00472425002900050036x

Filipovic, V., Coquet, Y., & Gerke, H. H. (2019). Representation of plot scale soil heterogeneity in dual-domain effective flow and transport models with mass exchange. *Vadose Zone Journal*, 18, 180174. https://doi.org/10.2136/vzj2018.09.0174

Flanagan, D. C., & Nearing, M. A. (Eds.) (1995). *USDA-Water Erosion Prediction Project: Hill slope profile and watershed model documentation* (NSERL Report 10). USDA–ARS National Soil Erosion Research Laboratory.

Franzmeier, D. P. (1991). Estimation of hydraulic conductivity from effective porosity data for some Indiana soils. *Soil Science Society of America Journal*, 55, 1801–1803. https://doi.org/10.2136/sssaj1991.03615995005500060050x

Ghezzehei, T. A. (2019). *What do we miss by treating dynamic soils as rigid porous media?* Paper presented at ASA-CSSA-SSSA International Annual Meetings, San Antonio, TX. https://scisoc.confex.com/scisoc/2019am/meetingapp.cgi/Paper/120682

Ghezzehei, T. A., & Or, D. (2001). Rheological properties of wet soils and clays under steady and oscillatory stresses. *Soil Science Society of America Journal*, 65, 624–637. https://doi.org/10.2136/sssaj2001.653624x

Gozubuyuk, Z., Sahin, U., Ozturk, I., Celik, A., & Adiguzel, M. C. (2014). Tillage effects on certain physical and hydraulic properties of a loamy soil under a crop rotation in a semi-arid region with a cool climate. *Catena*, 118, 195–205.

Greb, B. W., Smika, D. E., & Black, A. L. (1967). Effect of straw mulch rates on soil water storage during summer fallow in the Great Plains. *Soil Science Society of America Journal*, 31, 556–559. https://doi.org/10.2136/sssaj1967.03615995003100040034x

Green, T. R., Ahuja, L. R., & Benjamin, J. G. (2003). Advances and challenges in predicting agricultural management effects on soil hydraulic properties. *Geoderma*, 116, 3–28. https://doi.org/10.1016/S0016-7061(03)00091-0

Green, T. R., Erskine, R. H., Kipka, H., Lighthart, N., Nezat, C., Edmunds, D., David O, Barnard D, Douglas-Mankin, K. (2019). *Would it help? Distributed measurements in a small agricultural catchment may help support spatial complexity of soil hydrologic modeling.* Paper presented at ASA-CSSA-SSSA International Annual Meeting, San Antonio, TX. https://scisoc.confex.com/scisoc/2019am/meetingapp.cgi/Paper/120818

Guber, A., Pachepsky, Y. A., Shein, E., & Rawls, W. J. (2004). Soil aggregates and water retention. *Developments in Soil Science*, 30, 143–150.

Gupta, S. C., & Larson, W. E. (1979). Estimating soil water retention characteristics from particle size distribution, organic matter percent, and bulk density. *Water Resources Research*, 15, 1633–1635.

Gupta, S. C., & Raper, R. L. (1994). Prediction of soil compaction under vehicle. *Developments in Agricultural Engineering*, 11, 53–71.

Gupta, S. C., Sharma, P. P., & DeFranchi, S. A. (1989). Compaction effects on soil structure. *Advances in Agronomy*, 42, 311–338.

Hamblin, A. P., & Tennant, D. (1981). The influence of tillage on soil and water behavior. *Soil Science*, 132, 233–234.

Hansen, B. (2000). Estimation of surface runoff and water-covered area during filling of surface micro-relief depressions. *Hydrological Processes*, 14, 1235–1243.

Heuscher, S. A., Brandt, C. C., & Jardine, P. M. (2005). Using soil physical and chemical properties to estimate bulk density. *Soil Science Society of America Journal*, 69, 51–56. https://doi.org/10.2136/sssaj2005.0051a

Hill, R. L., & Meza-Montalvo, M. (1990). Long-term wheel traffic effects on soil physical properties under different tillage systems. *Soil Science Society of America Journal*, 54, 865–870. https://doi.org/10.2136/sssaj1990.03615995005400030042x

Hill, R. L. (1990). Long-term conventional and no-tillage effects on selected soil physical properties. *Soil Science Society of America Journal*, 54, 161–166. https://doi.org/10.2136/sssaj1990.03615995005400010025x

Hill, J. N. S., & Sumner, M. E. (1967). Effect of bulk density on moisture characteristics of soils. *Soil Science*, 103, 234–238.

Hines, J. W. (1986). *Measurement and modeling of soil hydraulic conductivity under different tillage systems.* Master's thesis, University of Minnesota–St. Paul.

Horn, R. (2003). Stress–strain effects in structured unsaturated soils on coupled mechanical and hydraulic processes. *Geoderma*, 116, 77–88.

Horn, R., van den Akker, J. H., & Arvidsson, J. (2000). *Subsoil compaction: Distribution, processes, and consequences* (Advances in GeoEcology 32). Catena.

Hua, Y. (1995). *The role of variable cracking on agrichemical transport at the Missouri MSEA site using the Root Zone Water Quality Model.* Master's thesis, University of Missouri–Columbia. MOspace. https://doi.org/10.32469/10355/16355

Huang, C., & Bradford, J. M. (1990). Depressional storage for Markov–Gaussian surfaces. *Water Resources Research*, 26, 2235–2242.

Isensee, A. R., Nash, R. G., & Helling, C. S. (1990). Effect of conventional vs. no-tillage on pesticide leaching to shallow groundwater. *Journal of Environmental Quality*, 19, 434–440. https://doi.org/10.2134/jeq1990.00472425001900030014x

Jamison, V. C., & Thompson, G. H. (1967). Layer thickness changes in a clay-rich soil in relation to water content changes. *Soil Science Society of America Proceedings*, 31, 441–444. https://doi.org/10.2136/sssaj1967.03615995003100040010x

Jungk, A. O. (1996). Dynamics of nutrient movement at the soil–root interface. In Y. Waisel, A. Eshel, & U. Kafkafi (Eds.), *Plant roots:The hidden half* (2nd ed., pp. 529–556). Marcel Dekker.

Kanwar, R. S., Colvin, T. S., & Karlen, D. L. (1997). Ridge, moldboard, chisel, and no-till effects on tile water quality beneath two cropping systems. *Journal of Production Agriculture, 10,* 227–234.

Kaspar, T. C., Logsdon, S. D., & Prieksat, M. A. (1995). Traffic patterns and tillage system effects on corn root and shoot growth. *Agronomy Journal, 87,* 1046–1051.

Koolen, A. J., Lerink, P., Kurstjens, D. A. G., van den Akker, J. J. H., & Arts, W. B. M. (1992). Prediction of aspects of soil–wheels systems. *Soil and Tillage Research, 24,* 381–396.

Kosugi, K. (1996). Lognormal distribution model for unsaturated soil hydraulic properties. *Water Resources Research, 32,* 2697–2703.

Kreiselmeier, J., Chandrasekhar, P., Weninger, T., Schwen, A., Julich, S., Feger, K.-H., & Schwärzel, K. (2019). Quantification of soil pore dynamics during a winter wheat cropping cycle under different tillage regimes. *Soil and Tillage Research, 192,* 222–232. https://doi.org/10.1016/j.still.2019.05.014

Lang, P. M., & Mallett, J. B. (1984). Effect of the amount of surface maize residue on infiltration and soil loss from a clay loam soil. *South African Journal of Plant and Soil, 1,* 97–98.

Leij, F. J., Ghezzehei, T. A., & Or, D. (2002a). Analytical models for soil pore-size distribution after tillage. *Soil Science Society of America Journal, 66,* 1104–1114. https://doi.org/10.2136/sssaj2002.1104

Leij, F. J., Ghezzehei, T. A., & Or, D. (2002b). Modeling the dynamics of the soil pore-size distribution. *Soil and Tillage Research, 64,* 61–78.

Liebig, M. A., Jones, A. J., Mielke, L. N., & Doran, J. W. (1993). Controlled wheel traffic effects on soil properties in ridge tillage. *Soil Science Society of America Journal, 57,* 1061–1066. https://doi.org/10.2136/sssaj1993.03615995005700040030x

Lilly, A., & Lin, H. (2004). Using soil morphological attributes and soil structure in pedotransfer functions. *Developments in Soil Science, 30,* 115–141.

Linden, D. R., van Doren, D. M., & Jr. (1987). Simulation of interception, surface roughness, depression storage, and soil settling. In M. J. Schaffer & W. E. Larson (Eds.), *NTRM: A soil crop simulation model for nitrogen, tillage, and crop-residue management* ((Conservation Research Report 31-1) (pp. 90–93). USDA–ARS.

Lindstrom, M. J., & Voorhees, W. B. (1995). Soil properties across a landscape continuum as affected by planting wheel traffic. In P. C. Robert, R. H. Rust, & W. E. Larson (Eds.), *Site-specific management for agricultural systems* (pp. 351–363). ASA, CSSA, & SSSA.

Lindstrom, M. J., & Onstad, C. A. (1984). Influence of tillage systems on soil physical parameters and infiltration after planting. *Journal of Soil and Water Conservation, 39,* 149–152.

Lipiec, J., & Hatano, R. (2003). Quantification of compaction effects on soil physical properties and crop growth. *Geoderma, 116,* 107–136.

Logsdon, S. D. (1997). Transient variation in the infiltration rate during measurement with tension infiltrometers. *Soil Science, 162,* 233–241.

Logsdon, S. D., Allmaras, R. R., Nelson, W. W., & Voorhees, W. B. (1992). Persistence of subsoil compaction from heavy axle loads. *Soil and Tillage Research, 23,* 95–110.

Logsdon, S. D., Jordahl, J. L., & Karlen, D. L. (1993). Tillage and crop effects on pond and tension infiltration rates. *Soil and Tillage Research, 28,* 179–189.

Logsdon, S. D., Kaspar, T. C., & Cambardella, C. A. (1999). Depth-incremental soil properties under no-till or chisel management. *Soil Science Society of America Journal, 63,* 197–200. https://doi.org/10.2136/sssaj1999.03615995006300010028x

Logsdon, S. D., & Kaspar, T. C. (1995). Tillage influences as measured by pond and tension infiltration. *Journal of Soil and Water Conservation, 50,* 571–575.

Ma, L., Peterson, G. A., Ahuja, L. R., Sherrod, L., Shaffer, M. J., & Rojas, K. W. (1999). Decomposition of surface crop residues in long-term studies of dryland agroecosystems. *Agronomy Journal, 91,* 401–409. https://doi.org/10.2134/agronj1999.00021962009100030008x

Maharjan, G. R., Prescher, A. K., Nendel, C., Ewert, F., Mboh, C. M., Gaiser, T., & Seidel, S. J. (2018). Approaches to model the impact of tillage implements on soil physical and nutrient properties in different agroecosystem models. *Soil and Tillage Research, 180,* 210–221.

Malone, R. W., Logsdon, S., Shipitalo, M. J., Weatherington, J., Ahuja, L. R., & Ma, L. (2003). Tillage effects on macroporosity and herbicide transport in percolate. *Geoderma, 116,* 191–216.

Mankin, K. R., Ward, A. D., & Boone, K. M. (1996). Quantifying changes in soil physical properties from soil and crop management. *Transactions of the ASAE, 39,* 2065–2074.

Mapa, R. B., Green, R. E., & Santo, L. (1986). Temporal variability of soil hydraulic properties with wetting and drying subsequent to tillage. *Soil Science Society of America Journal, 50,* 1133–1138. https://doi.org/10.2136/sssaj1986.03615995005000050008x

Masse, L., Patni, N. K., Jui, P. Y., & Clegg, B. S. (1998). Groundwater quality under conventional and no tillage: II. Atrazine, deethylatrazine, and metolachlor. *Journal of Environmental Quality*, 27, 877–883. https://doi.org/10.2134/jeq1998.00472425002700040023x

Meek, B. D., Rechel, E. R., Carter, L. M., & DeTar, W. R. (1989). Changes in infiltration under alfalfa as influenced by time and wheel traffic. *Soil Science Society of America Journal*, 53, 238–241. https://doi.org/10.2136/sssaj1989.03615995005300010042x

Messing, I. (1989). Estimation of saturated hydraulic conductivity in clay soils from soil moisture retention data. *Soil Science Society of America Journal*, 53, 665–668. https://doi.org/10.2136/sssaj1989.03615995005300030003x

Mitchell, J. K., & Jones, B. A. (1976). Micro-relief surface depression storage: Analysis of models to describe the depth of storage function. *AWRA Water Resources Bulletin*, 12, 1205–1222.

Mohanty, B. P., Bowman, R. S., Hendricks, J. M. H., & van Genuchten, M. T. (1997). New piecewise-continuous hydraulic functions for modeling preferential flow in an intermittent-flood-irrigated field. *Water Resources Research*, 33, 2049–2063.

Moore, I. D., & Larson, C. L. (1979). Estimating micro-relief surface storage from point data. *Transactions of the ASAE*, 22, 1073–1077.

Nielsen, D. C. (2002). Enhancing water use efficiency. In R. Lal (Ed.), *Encyclopedia of soil science* (pp. 635–642). Marcel Dekker.

O'Sullivan, M. F., Henshall, J. K., & Dickson, J. W. (1999). A simplified method for estimating soil compaction. *Soil and Tillage Research*, 49, 325–335.

O'Sullivan, M. F., & Simota, C. (1995). Modeling the environmental impacts of soil compaction: A review. *Soil and Tillage Research*, 35, 69–84.

Onstad, C. A. (1984). Depressional storage on tilled soil surfaces. *Transactions of the ASAE*, 27, 729–732.

Onstad, C. A., Wolfe, J. L., Larson, C. L., & Slack, D. C. (1984). Tilled soil subsidence during repeated wetting. *Transactions of the ASAE*, 27, 733–736.

Or, D., & Ghezzehei, T. A. (2002). Modeling post-tillage soil structural dynamics: A review. *Soil and Tillage Research*, 64, 41–59.

Or, D., Leij, F. J., Snyder, V., & Ghezzehei, T. A. (2000). Stochastic model for post tillage soil pore space evolution. *Water Resources Research*, 36, 1641–1652.

Pachepsky, Y., & Rawls, W. J. (Eds.) (2004). *Development of pedotransfer functions in soil hydrology* (Developments in Soil Science 30). Elsevier.

Pelak, N., & Porporato, A. (2019). Dynamic evolution of soil pore size distribution and its connection to soil management and biogeochemical processes. *Advances in Water Resources*, 131, 103384. https://doi.org/10.1016/j.advwatres.2019.103384

Peng, X., & Horn, R. (2005). Modeling soil shrinkage curve across a wide range of soil types. *Soil Science Society of America Journal*, 69, 584–592. https://doi.org/10.2136/sssaj2004.0146

Petersen, C. T., Jensen, H. E., Hansen, S., & Bender Koch, C. (2001). Susceptibility of a sandy loam soil to preferential flow as affected by tillage. *Soil and Tillage Research*, 58, 81–89.

Powers, W. L., Baer, J. U., & Skopp, J. (1992). Alternative soil-water release parameters for distinguishing tillage effects. *Soil Science Society of America Journal*, 56, 873–878. https://doi.org/10.2136/sssaj1992.03615995005600030032x

Quemada, M. (2004). Predicting crop residue decomposition using moisture adjusted time scales. *Nutrient Cycling in Agroecosystems*, 70, 283–291.

Rawls, W. J., Brakensiek, D. L., & Saxton, K. E. (1982). Estimation of soil water properties. *Transactions of the ASAE*, 25, 1316–1320.

Rawls, W. J., Gimenez, D., & Grossman, R. (1998). Use of soil texture, bulk density, and slope of the water retention curve to predict saturated hydraulic conductivity. *Transactions of the ASAE*, 41, 983–988.

Rawls, W. J., Nemes, A., Pachepsky, Y., & Saxton, K. E. (2007). Using NRCS National Soil Survey Soil Information System (NASIS) to provide soil hydraulic properties for engineering applications. *Transactions of the ASAE*, 50, 1715–1718.

Rawls, W. J., Nemes, A., & Pachepsky, Y. A. (2004). Effect of soil organic matter on soil hydraulic properties. *Developments in Soil Science*, 30, 95–111.

Rawls, W. J., Pachepsky, Y. A., Ritchie, J. C., Sobecki, T. M., & Bloodworth, H. (2003). Effect of soil organic carbon on soil water retention. *Geoderma*, 116, 61–76.

Rojas, K. W., & Ahuja, L. R. (2000). Management practices. In L. R. Ahuja, K. W. Rojas, J. D. Hanson, M. D. Shaffer, & L. Ma (Eds.), *Root Zone Water Quality Model: Modeling management effects on water quality and crop production*. Water Resources Publications.

Rousseva, S. S., Ahuja, L. R., & Heathman, G. C. (1988). Use of a surface gamma-neutron gauge for in-situ measurement of changes in bulk density of the tilled zone. *Soil and Tillage Research*, 12, 235–251.

Ruan, H., Ahuja, L. R., Green, T. R., & Benjamin, J. G. (2001). Residue cover and surface-sealing effects on infiltration: Numerical simulations for field applications. *Soil Science Society of America Journal*, 65, 853–861. https://doi.org/10.2136/sssaj2001.653853x

Saxton, K. E., Rawls, W. J., Romberger, J. S., & Papendick, R. I. (1986). Estimating generalized soil water characteristics from texture. *Soil Science Society of America Journal*, 50, 1031–1036. https://doi.org/10.2136/sssaj1986.03615995005000040039x

Saxton, K. E., & Rawls, W. J. (2006). Soil water characteristic estimates by textures and organic matter for hydrologic solutions. *Soil Science Society of America Journal*, 70, 1569–1578. https://doi.org/10.2136/sssaj2005.0117

Schaap, M. G., & Leij, F. J. (2000). Improved prediction of unsaturated hydraulic conductivity with the Mualem–van Genuchten model. *Soil Science Society of America Journal*, 64, 843–851. https://doi.org/10.2136/sssaj2000.643843x

Shaver, T. M., Peterson, G. A., Ahuja, L. R., Westfall, D. G., Sherrod, L. A., & Dunn, G. (2002). Surface soil physical properties after twelve years of dryland no-till management. *Soil Science Society of America Journal*, 66, 1296–1303. https://doi.org/10.2136/sssaj2002.1296

Shaver, T. M., Peterson, G. A., & Sherrod, L. A. (2003). Cropping intensification in dry land systems improves soil physical properties: Regression relations. *Geoderma*, 116, 149–164.

Sherrod, L. A., Peterson, G. A., Westfall, D. G., & Ahuja, L. R. (2003). Cropping intensity enhances soil organic carbon and nitrogen in a no-till agro-ecosystem. *Soil Science Society of America Journal*, 67, 1533–1543.

Shipitalo, M. J., Dick, W. A., & Edwards, W. M. (2000). Conservation tillage and macropore factors that affect water movement and the fate of chemicals. *Soil and Tillage Research*, 53, 167–183.

Sillon, J. F., Richard, G., & Cousin, I. (2003). Tillage and traffic effects on soil hydraulic properties and evaporation. *Geoderma*, 116, 29–46.

Silva, B. O., Moitinho, M. R., Santos, G. A. A., Teixeira, D. B., Fernandes, C., & La Scala, N., Jr. (2019). Soil CO_2 emission short-term soil pore class distribution after tillage operations. *Soil and Tillage Research*, 186, 224–232.

Silva, R. P., Rolim, M. M., Gomes, I. F., Pedrosa, E., Tavares, U. E., & Santos, A. N. (2018). Numerical modeling of soil compaction in a sugarcane crop using the finite element method. *Soil and Tillage Research*, 181, 1–10.

Smith, D. L. O. (1985). Compaction by wheels: A numerical model for agricultural soils. *Journal of Soil Science*, 36, 621–632.

Strudley, M. W., Green, T. R., & Ascough, J. C., II. (2008). Tillage effects on soil hydraulic properties in space and time: State of the science. *Soil and Tillage Research*, 99, 4–48.

Timlin, D. J., Ahuja, L. R., & Ankeny, M. D. (1994). Comparison of three field methods to characterize apparent macropore conductivity. *Soil Science Society of America Journal*, 58, 278–284.

Timlin, D. J., Ahuja, L. R., Pachepsky, Y., Williams, R. D., Gimenez, D., & Rawls, W. J. (1999). Use of Brooks–Corey parameters to improve estimates of saturated conductivity from effective porosity. *Soil Science Society of America Journal*, 63, 1086–1092. https://doi.org/10.2136/sssaj1999.6351086x

van den Akker, J. J. H., & van Wijk, A. L. M. (1987). A model to predict subsoil compaction due to field traffic. In G. Monnier & M. J. Gross (Eds.), *Soil compaction and regeneration* (pp. 69–84). A. A. Balkema.

van Doren, D. M., & Allmaras, R. R. (1978). Effect of residue management practices on the soil physical environment, micro-climate, and plant growth. In *Crop residue management systems* (pp. 49–84). (ASA Special Publication 31)). ASA, CSSA, and SSSA.

van Es, H. M., Ogden, C. B., Hill, R. L., Schindelbeck, R. R., & Tsegaye, T. (1999). Integrated assessment of space, time, and management-related variability of soil hydraulic properties. *Soil Science Society of America Journal*, 63, 1599–1608. https://doi.org/10.2136/sssaj1999.6361599x

van Schaik, L., Palm, J., Klaus, J., Zehe, E., & Schroder, B. (2014). Linking spatial earthworm distribution to macropore numbers and hydrologic effectiveness. *Ecohydrology*, 7, 401–408. https://doi.org/10.1002/eco.1358

Vanderlinden, K., Pachepsky, Y., Pedrera-Parrilla, A., Martinez, G., Espejo-Perez, A., Petrea, F., & Giraldez, J. (2021). Water retention and field water states in a Vertisol under long-term direct and conventional tillage. *European Journal of Soil Science*, 72, 667–678. https://doi.org/10.1111/ejss.12967

Vervoort, R. W., Dabney, S. M., & Römkens, M. J. M. (2001). Tillage and row position effects on water and solute infiltration characteristics. *Soil Science Society of America Journal*, 65, 1227–1234. https://doi.org/10.2136/sssaj2001.6541227x

Warkentin, B. P. (1971). Effects of compaction on content and transmission of water in soils. In K. Barnes (Ed.), *Compaction of agricultural soils* ((ASAE Monograph) (pp. 126–153). American Society of Agricultural Engineers.

Williams, J. R., Jones, C. A., & Dyke, P. T. (1984). A modeling approach to determining the relationship between erosion and soil productivity. *Transactions of the ASAE, 27*, 129–142.

Wilson, G. V., Jardine, P. M., & Gwo, J. P. (1992). Modeling the hydraulic properties of a multiregion soil. *Soil Science Society of America Journal, 56*, 1731–1737. https://doi.org/10.2136/sssaj1992.03615995005600060012x

Wilson, G. V., Zhang, T., Wells, R. R., & Liu, B. (2020). Consolidation effects on relationships among soil erosion properties and soil physical quality indicators. *Soil and Tillage Research, 198*, 104550. https://doi.org/10.1016/j.still.2019.104550

Woodruff, C. M. (1937). Linear changes in the Selby loam profile as a function of soil moisture. *Soil Science Society of America Proceedings, 1*, 65–69. https://doi.org/10.2136/sssaj1937.03615995000100000010x

Wosten, J. H. M., & van Genuchten, M. T. (1988). Using texture and other soil properties to predict unsaturated soil hydraulic functions. *Soil Science Society of America Journal, 52*, 1762–1770. https://doi.org/10.2136/sssaj1988.03615995005200060045x

Wu, L., Swan, J. B., Paulson, W. H., & Randall, G. W. (1992). Tillage effects on measured soil hydraulic properties. *Soil and Tillage Research, 25*, 17–33.

Young, R. N., & Fattah, E. A. (1976). Prediction of wheel–soil interaction and performance using the finite element method. *Journal of Terramechanics, 13*, 227–240.

Yule, D. F., & Ritchie, J. T. (1980). Soil shrinkage relationships of Texas Vertisols: II. Large cores. *Soil Science Society of America Journal, 44*, 1291–1295. https://doi.org/10.2136/sssaj1980.03615995004400060032x

Zhang, M., Lu, Y., Heitman, J., Horton, R., & Ren, T. (2017). Temporal changes of soil water retention behavior as affected by wetting and drying following tillage. *Soil Science Society of America Journal, 81*, 1288–1295. https://doi.org/10.2136/sssaj2017.01.0038

Zhang, S., Harald, G., & Lovdahl, L. (2006). Effect of soil compaction on hydraulic properties of two loess soils in China. *Soil and Tillage Research, 90*, 117–125.

Zobeck, T. M., & Onstad, C. A. (1987). Tillage and rainfall effects on random roughness: A review. *Soil and Tillage Research, 9*, 1–20.

12

Gary Feng and
Saseendran S. Anapalli

Integrating Models with Field Experiments to Enhance Research: Cover Crop, Manure, Tillage, and Climate Change Impacts on Crops in a Humid Region

Abstract

Agricultural system models help integrate, synthesize, and extrapolate data collected in location-specific field experiments with limited spatial and temporal representation to other soils, climates, and locations. Collection of required data that is representative of the soil, water, crop, and climate where the experiments are conducted is key to modeling. Integrating field data into state-of-the-science agricultural system models with long-term climate data can help us simulate the experiments on a computer and answer "what if" questions given soil–water–crop–climate changes over time and location. The RZWQM2 is an agricultural system simulation model that integrates the physical, chemical, and biological processes in agriculture systems and simulates soil–water–crop management impacts on crop productivity and environmental quality. This chapter presents some examples of the performance and application of RZWQM2 in simulating staple crop production systems of the Southeast United States, a region that is characterized by a humid climate with warm summers and mild winters. Model applications in managing cover crops, soil amendments (poultry litter), tillage, and crop-rotations are reported. Examples of system

---

Modeling Processes and Their Interactions in Cropping Systems: Challenges for the 21st Century, First Edition.
Edited by Lajpat R. Ahuja, Kurt C. Kersebaum, and Ole Wendroth.
© 2022 American Society of Agronomy, Inc. / Crop Science Society of America, Inc. / Soil Science Society of America, Inc.
Published 2022 by John Wiley & Sons, Inc.

models-assisted investigations of climate variability and change impacts on planting windows, crop production, water demand and water availability in the region are also presented.

Introduction

Biophysical system models integrated with field research and long-term climate information can help advance agricultural research and technology in multiple ways (Ahuja et al., 2017):

1. Help quantify the field research results in terms of the fundamental theory and concepts that are broadly applicable beyond the site-specific empirical relationships;

2. Foresee experimental outcomes from knowledge of the fundamental factors that determine the environment and plant growth under different climates;

3. Extend experimental results to long-term climate conditions and soils beyond field experimental plots or regions;

4. Assist field experimentation to determine fate and mass balance components of water, nitrogen, carbon, and other chemicals that field experimentation alone is not capable of measuring;

5. Use extended results to develop broad-based management decision support tools or simple management guidelines for producers and other users, which may include linkages to economic and social considerations;

6. Promote a systems approach to agricultural research that looks at all major components and interactions, including the measurement of key parameters to explain results, thus reducing repetition in research and focusing on knowledge gaps;

7. Help optimize site-specific management of spatially variable soil, appropriately selected crops and varieties, and available water resources in the landscape to achieve both environmental and production objectives;

8. Make sense of and assimilate real-time information about soil conditions and crop growth status obtained via remote sensing, which, combined with measured or predicted near-term weather, can be utilized to develop a whole new level of site-specific management;

9. Analyze problems in management, environmental quality, global climate change effects, and other emerging issues;

10. Explore new ideas and alternative strategies under different weather and climatic conditions before assessing them in expensive field experiments, as guides for planning and policies; and

11. Raise more questions for field experimentation to answer, which assist design of field experiments and determine what should be measured to make field research more productive and efficient with limited resources.

There are numerous examples of model applications for enhancing field research all over the world. Ahuja et al. (2017) presented some simple examples of such applications based on their work on critical resource areas in the US Central Great Plains, the US Midwest, and China with the goals of: (a) optimizing the use of limited water in dryland cropping systems through crop selection; (b) optimizing the scheduling of limited irrigations; (c) guiding deficit irrigations based on crop water production functions; (d) minimizing N and pesticide leaching to subsurface drainage (tile flow) and groundwater; and (e) projecting climate change effects and identifying adaptation strategies. Most of these applications utilized the Root Zone Water Quality Model (RZWQM2), developed by USDA-ARS, which contains its own generic crop growth module and incorporated the DSSAT4.6 crop growth model. Importantly, these applications were possible in collaboration with field research scientists at several locations in the United States (Akron, CO; Sidney, NE; Sidney, MT; Greeley, CO; Bushland, TX; Ames, IA), China, and other countries), by making them part of the team. Pertinent applications of the DSSAT model by itself and other agricultural system models were cited. For the dryland crop rotation of wheat–summer crop–fallow, a decision support tool to select the best summer crop was developed based on the soil water content at planting (Nielsen et al., 2012). Charts were developed for the different N balance components at different rate and timing of water and N applications as well as their effects on the crop production function (Anapalli et al., 2008, 2014).

Recently, numerous applications of RZWQM and other models have been regionally extended to the humid southeast United States (Anapalli et al., 2019, 2021; Feng et al., 2005, 2015, 2016, 2018; Gao et al., 2017, 2019a, 2019b; Ouyang et al., 2016, 2017; Tang et al., 2017; Yang et al., 2019a, 2019b, 2020, 2021; Zhang et al., 2018; 2016a, 2016b). Some useful applications of RZWQM2 involving a cover crop, poultry litter, crop water requirement, climate change, and tillage are summarized in this chapter.

Modeling Case Studies
Model Calibration and Validation

The US Southeast is characterized by high spatial variation in soil types, great temporal and spatial variations in rainfall, crop water requirements, soil moisture contents, runoff, drainage, and irrigation demand, all of which are affected by the changing climate patterns. Conducting field experiments to study the climate variability and change impacts on cropping systems and crop water and nitrogen demand–supply scenarios under a wide range of soils and environmental conditions would not be practical. However, a well calibrated and tested model can be used for simulating such impacts with reasonable degree of accuracy for practical applications.

In this regard, the RZWQM2 model was extensively calibrated then tested using three comprehensive field measurement data sets in the region where it was applied (Anapalli et al., 2018; Feng et al., 2015; Yang et al., 2019a, 2019b, 2021). The common statistical criteria for model test or validation were used to test the agreement of measured and simulated results: Nash–Sutcliffe efficiency (NSE) (Nash & Sutcliffe, 1970), relative root mean squared error (RRMSE), coefficient of determination (R^2), and percent error (PE). The NSE value ranges from $-\infty$ to1, where NSE = 1 indicates a perfect match between observed and simulated values, and NSE < 0 indicates that the average of measured values is a better predictor than the simulation. Moriasi et al. (2007) indicated that model performance is acceptable as $R^2 > 0.5$, NSE > 0.5, and PE ±15%. Liu et al. (2011) suggested that model simulations can be considered as "satisfactory" if RRMSE is <30%. The measured and simulated soil water contents (SWC) were compared in two fields at different locations with poultry litter application and cover crops. Observed values matched the predicted trend, the PE was within ±15%, and R^2 was greater than 0.5. Simulated SWC from 0–30 cm was in good agreement with measured values, with a RRMSE < 15% and EF > 0.5. There was good agreement between measured and observed long-term soil organic carbon (SOC) (R^2 = 0.70–0.83, PE = 11%). The observed versus simulated SOC concentration in the 0–30 cm soil depth averaged 7.91 versus 7.98 g kg^{-1} for the L18F treatment, 7.11 versus 6.91 g kg^{-1} for the N202S treatment, and 7.10 versus 7.07 g kg^{-1} for the N202F treatment. Simulated SOC concentration at depths 0–15, 15–30, 30–60, and 60–90 cm at the end of the experiment also compared satisfactorily with corresponding measured SOC concentrations. Differences in SOC concentration at 0–30 cm soil depth between treatments that received poultry litter versus synthetic N were effectively simulated with the RZWQM2 model. Considering the agreement between observed and simulated SOC and the similarities of our results with those of Franzluebbers et al. (2001), the calibrated and validated RZWMQ2 could be considered as satisfactory to model soil C dynamics in the southeastern United States.

The difference between measured and simulated soil nitrate concentrations averaged for all treatments of inorganic and poultry litter was 7%. The maximum difference was 0.95 kg ha^{-1}, which is acceptable considering that the average measured soil nitrate (±standard error) across all fertilizer treatments was 6.7 ± 1.7 kg ha^{-1}. Averaged across all treatments, the RMSE of the simulated soil nitrate and plant N uptake was 47 and 41 kg ha^{-1}. The R^2 and NSE values of all simulated results were >0.5. Although RMSE values across all treatments in three years appeared high, they were comparable to the experimental errors of the measured data. Simulated soil nitrate concentration at a depth of 0–15 and 15–30 cm for winter cover crop (CC) and winter fallow (WF) plots over a four-year period were generally close to measured values. Validation showed "satisfactory" performance with NSE > 0.65, R^2 > 0.75, and RRMSE < 25%. RZWQM2-estimated crop N uptake generally agreed with measured N uptake. The simulated corn N uptake (194 kg N ha^{-1}) in the dry year of 2016 lowered to within the range of measured mean and standard deviation (161 ± 19 kg N ha^{-1}). Qi et al. (2011) reported that, under dry climate conditions, there was 13% higher simulated N uptake than the measured values for the CC plots, and 8% for WF plots on a silty loam soil in a corn–soybean rotation. On average, the simulated N uptake by wheat was overestimated by 11% for CC plots, which was similar to simulation results with 14% overestimation as experienced by Li et al. (2008). However, the NSE and R^2 values were still <16%, and >0.87. Therefore, the model was considered to be acceptable to simulate soil nitrogen dynamics in both poultry litter application and the corn–soybean rotations with and without wheat CC.

The model simulated well in predicting leaf area index (LAI) for corn and soybean in 2016 and 2017. Across two years, the RRMSE, NSE and PE values for measured and simulated LAI were <15%, >0.8, and <15%, respectively. The maximum averaged relative errors of simulated biomass and grain yield across all fertilizer treatments over three year were within ±15%. The PE between observed and simulated yields for the L18F, N202S, and N202F treatments was within a range of ±3 to 14%. The PEs of simulated and measured biomass were 17, 13, and 10% for the Control, L18F, N202F, respectively. In the CC system, the PE of simulated aboveground biomass for corn and soybean were 11.8 and 22.1%, respectively. The RRMSE values were nearly 15% and less than 10%. Corn yield was underestimated by 7.2% and the soybean yield was overestimated by 8.5% when across two years. The simulated yield was within range of measured values and one standard deviation, with a RRMSE of less than 15%, and a NSE of more than 0.8 in the CC system. The largest difference in measured and simulated biomass of wheat cover crop was 16%, which was less than the standard deviation among observations at the same time in different replicates of the same treatment. In summary, the RZWQM2 model provided satisfactory results in terms of the values of statistical criteria. Therefore, the model can be applied to study soil water, N balance and their impacts on environmental quality, water and N crop productivity, crop growth, yield, as affected by poultry litter application and winter wheat cover crop in corn and soybean cropping system.

Long-Term Effects of Cover Crops on Water Availability in Corn–Soybean Rotation

Knowledge of the probability of soil water deficits in rainfed corn and soybean production systems is essential for developing soil-crop-water management decision-support information to improve crop water productivity (CWP) in those cropping systems. In this direction, Tang et al. (2017), Yang et al. (2019a), (2020) integrated RZWQM2 with field research experiments and 100-year climate data to determine crop water requirements, irrigation demands, and rainwater deficits in meeting those demands in soybean and corn production systems in the region (Figure 12.1). These crops are typically planted from late March to May and harvested between mid-September and late August. Rainfall during the simulated growing seasons for soybean and corn was 400 and 510 mm, and accounted for about 31 and 40% of annual rainfall, respectively, over the 100 years. The simulated average crop evapotranspiration (crop water demand) of rainfed soybean and corn was 546 and 588 mm. Weekly soil water (effective rainfall, the portion of rainfall that can be used by the plant naturally) deficits greater than 20 mm were found in Weeks 7 through 16 after planting (22–29 mm) for soybean and in Weeks 11 through 20 (20–26 mm) after planting corn.

The cover cropping is considered one strategy to reduce rainwater loss and mitigate rainfed crop water stress. The impact of a cover crop (CC) on soil water balance and agricultural production is closely dependent on rainfall amount and timing. RZWQM2 was applied to investigate the impact of wheat CC on rainwater balance and CWP in rainfed corn and soybean rotation under different rainfall patterns in eastern Mississippi (Yang et al., 2020). Winter wheat cultivar Terral 8861 was planted as cover crop at the seeding rate of 2.67×10^6 seeds ha^{-1} in mid-October each year. CC were desiccated with glyphosate application four weeks before planting cash crops. Frequency analysis was used to classify three types of rainfall patterns, "wet," "normal," and "dry" years during 1938–2017, based on accumulated rainfall during the growing season of each crop (Figure 12.2). Compared with a no CC scenario (i.e., bare soil), planting CC reduced drainage by 69, 53, and 51 mm (Figure 12.2) and increased evapotranspiration by 79, 81, and 73 mm in the wet, normal, and dry years, respectively (Figure 12.3). Compared with the no CC scenario, the CC-induced decrease in surface evaporation was 64 mm for the corn growing period and 38 mm for soybean growth period, when averaged across 80 years (Figure 12.3). Regardless of rainfall pattern, CC generally improved soil water storage over the cash crop growing season, especially during the early soybean crop growth period under normal rainfall years (Figure 12.4). Yield of corn and soybean in all rainfall patterns was similar between CC and no CC scenarios. CWP was improved after planting CC, especially for corn growing season with above-normal rainfall (Figure 12.3). The increase in CWP can be attributed to the decrease in evaporation while maintaining acceptable yields. Further study found that the best annual agricultural production and minimal loss due to drainage was achieved when wheat CC was planted under conditions of a

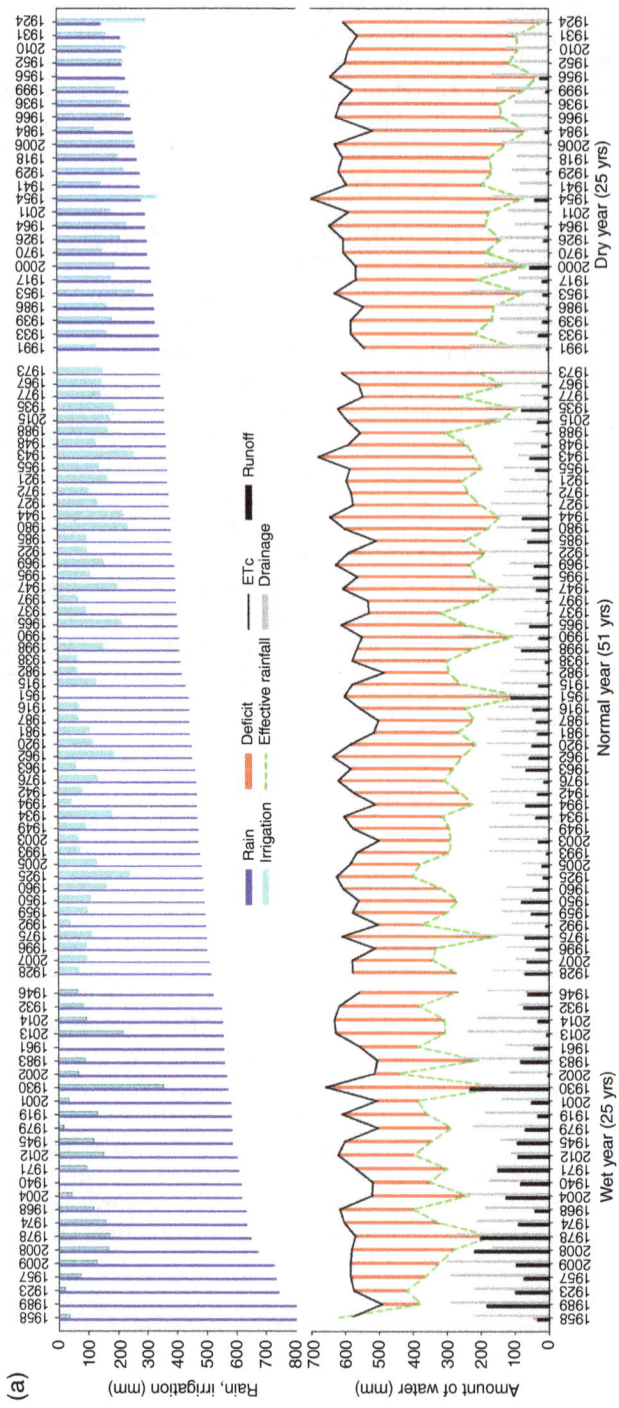

Figure 12.1 Water balance components sorted by rainfall amount during (a) soybean and (b) corn growing seasons in Mississippi (Tang et al., 2017 / with permission of American Society of Agricultural and Biological Engineers (ASABE)).

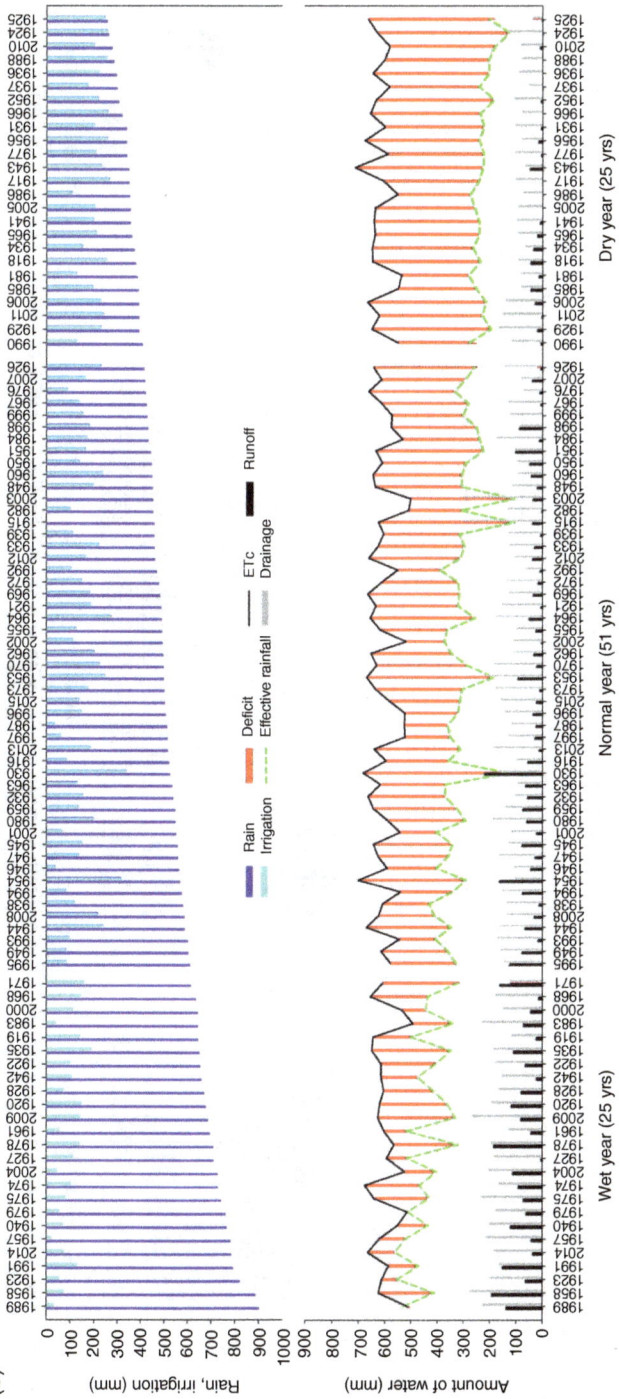

Figure 12.1 (Continued)

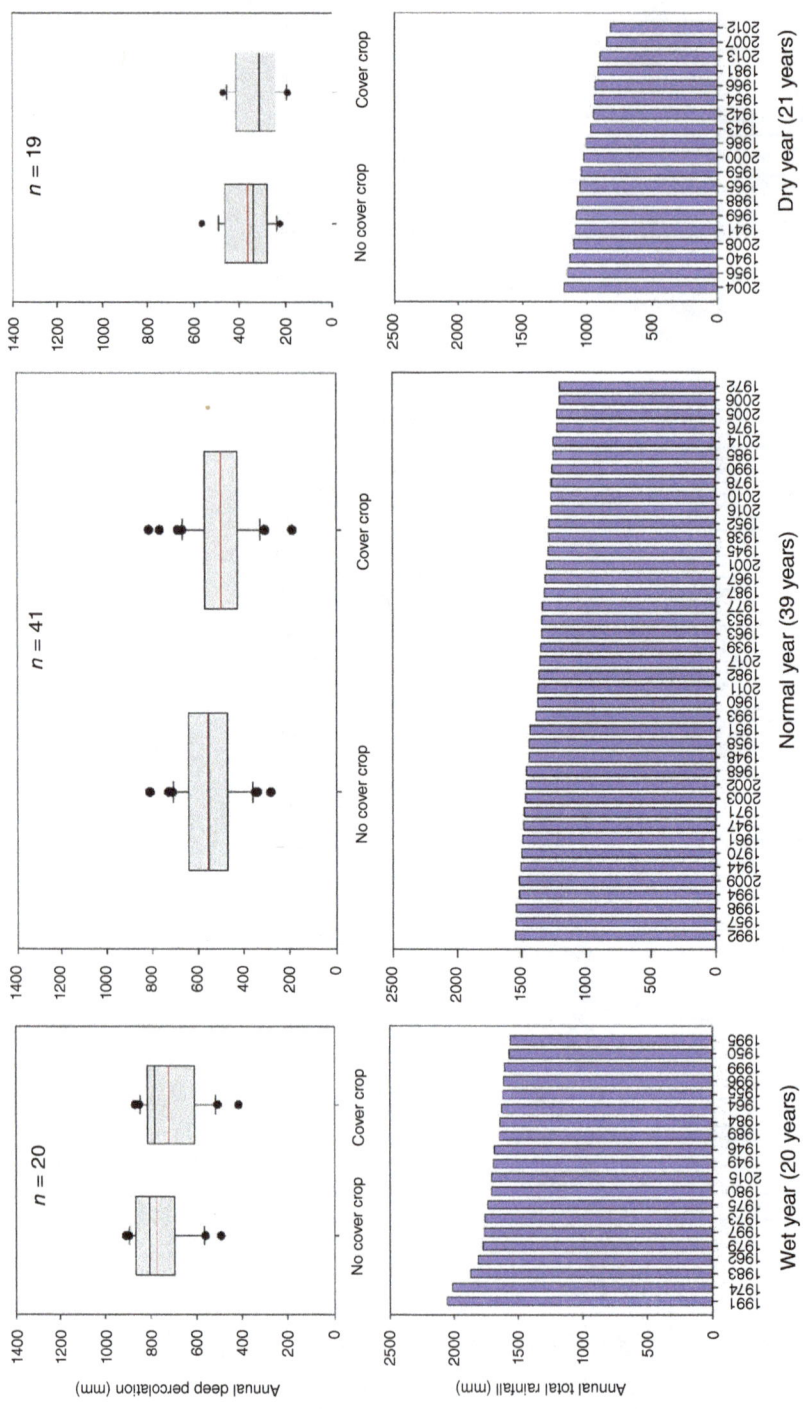

Figure 12.2 Simulated annual deep percolation in a rained corn–soybean rotation with or without wheat cover crop in 80 seasonal years during 1938–2017 in the Mississippi Blackland Prairie. A seasonal year was defined as a period from planting a cover crop to ending 1 day before planting cover crop next year. The red solid line represents mean values from each scenario for individual rainfall patterns (Yang et al., 2019a).

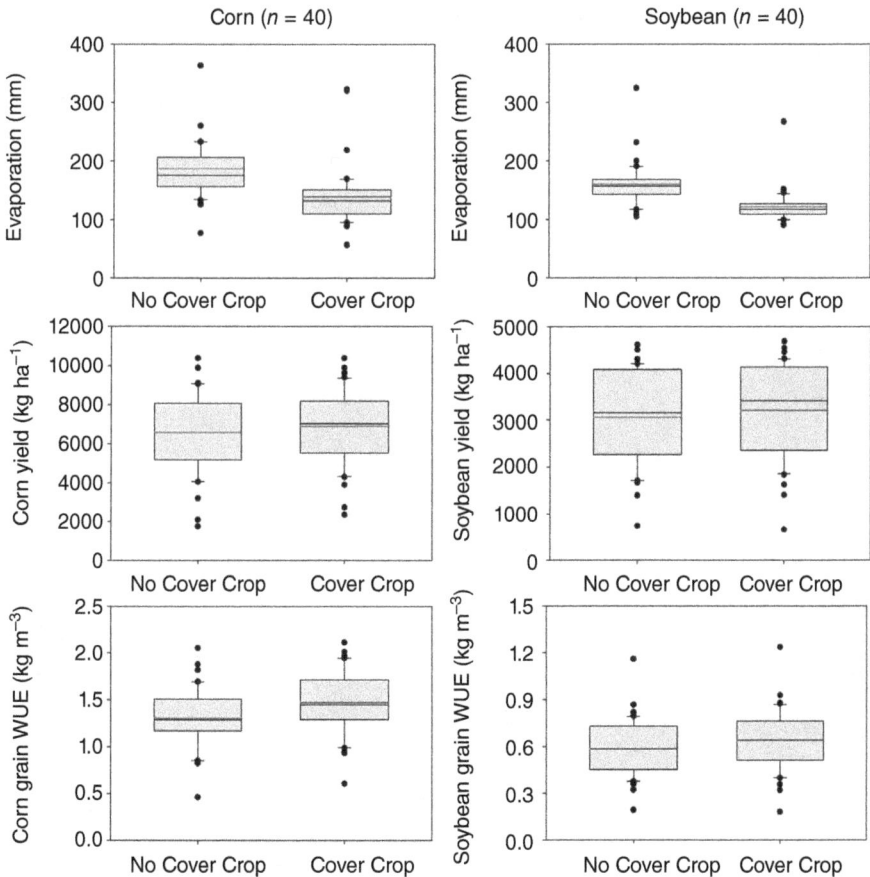

Figure 12.3 Simulated evaporation (E), actual evapotranspiration (ET), grain yield, and water use efficiency (WUE) for rainfed corn and soybean growth periods with a cover crop or no cover crop during 1938–2017 in the Blackland Prairie, Mississippi. The red and blue solid lines represent the mean and medium values of each scenario (Yang et al., 2019a).

"dry" winter fallow season and "normal" summer rainfall (Yang et al., 2020). Blanco-Canqui et al. (2015) also reported that the favorable roles of CC in reducing subsurface drainage beyond root zone and in increasing soil water storage rely on rainfall conditions. Tribouillois et al. (2018) applied the STICS model and found that incorporating CC reduced mean annual subsurface drainage by 20 mm yr^{-1} as simulated for 45 years under temperate climate with dry summers. Dietzel et al. (2016) used 28-year historical precipitation data and APSIM-model to simulate and categorize the optimum growing season rainfall, runoff, and drainage for maintaining optimum system CWP for corn and soybean in the northwestern United States.

Carbon and Nitrogen Contribution of Soil Amended with Poultry Litter

Knowledge of the fate of manure-N is important to effectively manage it and minimize its environmental impact. Conventional methods to track mineralized

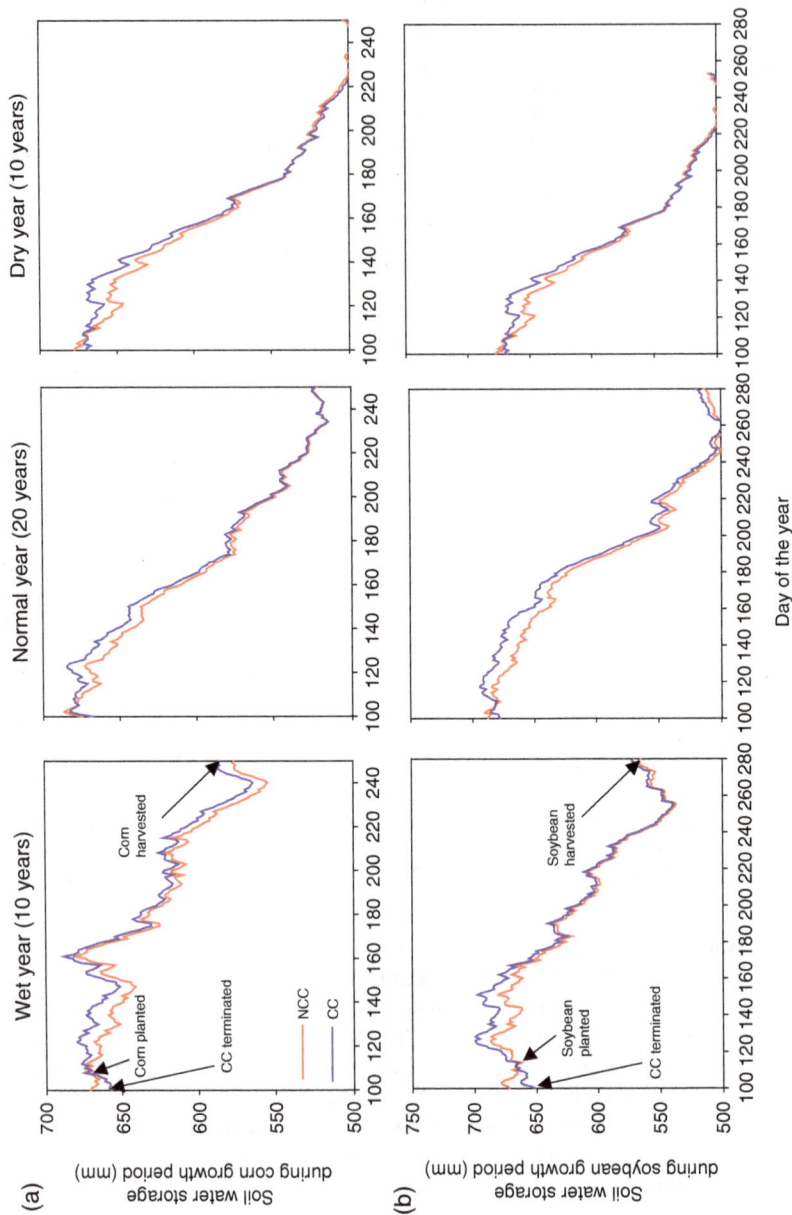

Figure 12.4 Simulated daily average soil water storage in the 1.8-m profile during 40 years of corn (a) and 40 years of soybean (b) growth periods under the no wheat cover crop (NCC) and cover crop (CC) scenarios and different rainfall patterns, based on seasonal rainfall from chemical termination of wheat in spring to grain harvest of summer crop in fall (Yang et al., 2020 / with permission of John Wiley & Sons.).

N from soil-applied manures and quantify the amount lost due to various causes over the years are difficult to implement, inaccurate, and expensive. Yang et al. (2019b) calibrated and evaluated RZWQM2 using three years of data and simulated the mineralization and fate of fall- and spring-applied poultry litter-N (Feng et al., 2015). Litter (18 Mg ha^{-1}, equivalent to 502 kg N ha^{-1}), for comparison, NH_4NO_3-N (202 kg ha^{-1}) were applied in fall and spring in a cornfield near Starkville, MS. The model estimated 57% (279 kg ha^{-1}) of the total litter N applied in the fall and 51% (249 kg ha^{-1}) applied in the spring mineralized by the end of the first year in November. The loss of N to mineralization of litter N by the end of the first year was 24 versus 9% of the total applied for the fall versus spring applications, respectively. At the end of the experiment in November of the third year, 88% of the total 1,507 kg N ha^{-1} litter N applied in the previous three falls, and 72% of that applied in the previous three springs was mineralized. The simulated average daily N mineralization rate during the growing season for the spring-applied litter was 1.3, 1.8, and 2.7 kg ha^{-1} d^{-1} in the three years, respectively. The peak values during the three years ranged from 3 to 4 kg N ha^{-1} d^{-1} under favorable soil moisture and temperature. The loss of mineralized litter N averaged across the three years was 162 kg ha^{-1} yr^{-1} (37% of the total mineralized) if applied in the fall and only 55 kg ha^{-1} yr^{-1} (15% of the total mineralized) the spring. The primary avenue of simulated litter N loss was leaching if applied in the fall and denitrification if applied in the spring. These results demonstrate spring is the best time to apply litter in the southeastern United States.

The soil initially had 102,277 kg C ha^{-1} total C in a 1.8 m soil profile. After three years of growing corn, without any fertilizer applied, the control had the least total C of 102,508 kg C ha^{-1}, which is only 200 kg C ha^{-1} more than the initial C. As only synthetic N was applied each year, the soil had a total C of 103,397 kg C ha^{-1} (Figure 12.5). Applying 18,000 kg ha^{-1} yr^{-1} (corresponding to 4,632 kg C ha^{-1}) poultry litter in either spring or fall for three years resulted in 109,401 kg C ha^{-1}, which is 7,124 kg C ha^{-1} (7.0%) more than the initial and 6,147 kg C ha^{-1} (6.0%) more than the soil that received

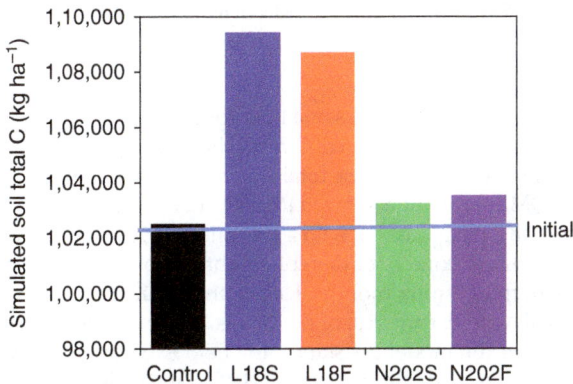

Figure 12.5 Simulated soil total C to a 1.8-m depth from cornfield with the application of 18,000 kg ha^{-1} yr^{-1} (13,896 kg C ha^{-1} yr^{-1} and 490 kg N ha^{-1} yr^{-1}) poultry litter (L18) or 202 kg N ha^{-1} yr^{-1} NO_4NH_3-N (N202) in the fall (F) or spring (S) after three years in Oktibbeha County, Mississippi, USA. L18S, 18,000 kg ha^{-1} poultry litter applied in the spring; L18F, 18,000 kg ha^{-1} poultry litter applied in the fall; N202S, 202 kg ha^{-1} NH_4NO_3-N applied in the spring; N202F, 202 kg ha^{-1} NH_4NO_3-N applied in the fall (Yang et al., 2019a / with permission of John Wiley & Sons.).

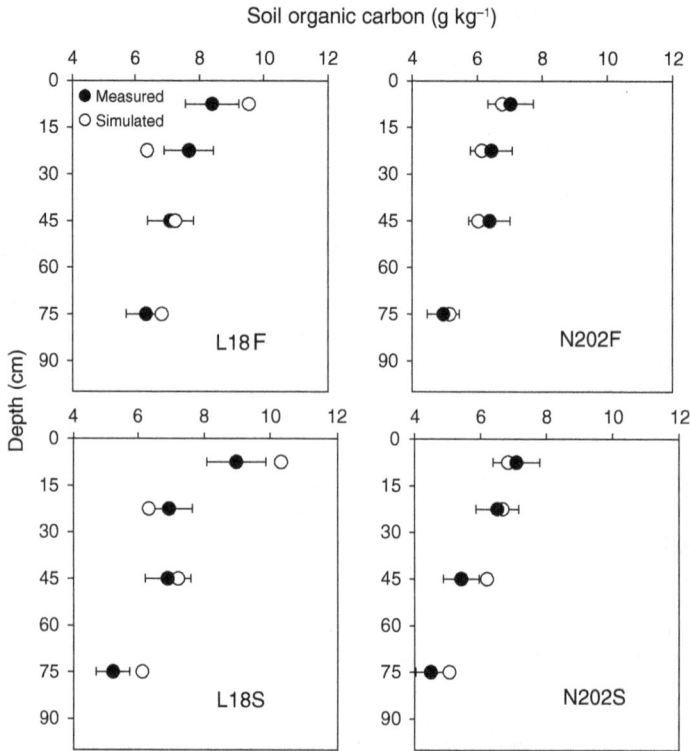

Figure 12.6 Measured and simulated soil organic carbon at soil depths of 0–15, 15–30, 30–60, and 60–90 cm at the end of a three-year (30 Nov. 2008) study in which corn was grown with 18,000 kg ha^{-1} poultry litter or 202 kg ha^{-1} NO$_4$NH$_3$-N applied in the fall or spring from 2006 to 2008 in Oktibbeha County, Mississippi, USA. L18S, 18,000 kg ha^{-1} poultry litter applied in the spring; L18F, 18,000 kg ha^{-1} poultry litter applied in the fall; N202S, 202 kg N ha^{-1} NH$_4$NO$_3$-N applied in the spring; N202F, 202 kg ha^{-1} NH$_4$NO$_3$-N applied in the fall (Yang et al., 2019a / with permission of John Wiley & Sons.).

202 kg C ha^{-1} yr^{-1} NH$_4$NO$_3$-N in the same time for three years. In other words, 6,147 kg ha^{-1} C is the amount of sequestered C from the litter applied over the three-year period, which accounted for 41% of the total C applied in the litter over the three years. Overall, the RZWQM2 simulation showed that 59% of the litter C applied was lost over a three-year period (8,249 kg C ha^{-1}), with a loss rate of 7.53 kg C ha^{-1} d^{-1}. It suggested raising the organic matter content of this soil might be possible in the humid and subtropical region with mild winter months. Half of the sequestered C was located in the top 15 cm of the soil, whereas the rest of the C was leached to depths as great as 90 cm (Figure 12.6). Without the model assistance, the field experiment alone would not be able to determine the detailed fate and transport of N and C derived from poultry litter applied to the soil over time at any given time steps.

Applied litter C was mainly lost as CO$_2$ emissions. Application in the fall resulted in 6,890 kg ha^{-1} yr^{-1} (18.8 kg ha^{-1} d^{-1}) C loss as CO$_2$ emission, 13% of the amount was reduced if the same amount of litter was applied in spring. CO$_2$ emission from the manured field over three years was 79% larger than for the cornfield fertilized with NH_4NO_3-N. Soil CO_2 emission averaged over the three years simulations was 3,592 kg C ha^{-1} yr^{-1} (9.8 kg C ha^{-1} d^{-1}) if synthetic fertilizers were applied instead of litter. Surprisingly, almost half of the total emission occurred during less than 32% of the year from corn maturity to the next litter application (Yang et al., 2019b). Much of the CO_2 emission during this phase is derived from the decomposition of soil organic matter. Therefore, the effort should be taken for sequestration of CO_2 emission during this period by litter application at right time in right form using better method, for example, poultry litter is pelletized and subsurface banded (Feng et al., 2019). Only 19–30% of the average yearly emission occurred in the period between litter application and corn planting. The remaining 25–30% CO_2 was emitted during the 100 days of the growing season. In typical cropland that receives manure, CO_2 emissions show high temporal and spatial variability because of uneven distribution of environmental factors such as soil type, soil moisture, soil temperature, irrigation, and nutrient availability (Henault et al., 2005). Accurate assessment of CO_2 is thus relatively difficult in typical soil systems (Feng et al., 2015; Johnson et al., 2007). In addition, the quantification of long-term CO_2 emissions often requires advanced gas collection and analysis systems under specific experimental conditions (Feng et al., 2015; Wang et al., 2016). Such frequent field trips and efforts are time-consuming, laborious, and costly (Jiang et al., 2017; Li et al., 2005a, 2005b). The calibrated and validated RZWQM2 with continuous three-year comprehensive field data made it possible to simulate and quantify soil CO_2 emission and gain and lose litter-derived C in a corn production system in northeastern Mississippi, USA.

Soil Mineral Nitrogen Distribution in Soil Under Cover Crop

It is also vital to understand soil nitrogen dynamics affected by the interaction of cover crops and rainfall amount as well as distribution in rainfed agricultural production systems. RZWQM2 was employed to investigate the long-term effect of a winter wheat cover crop on soil nitrogen balance in an 80-year no-tillage and rainfed corn–soybean rotations (Yang et al., 2021). In humid northeast Mississippi, planting a winter wheat cover crop into a corn–soybean system increased annual nitrogen mineralization, improved annual denitrification, and reduced annual nitrate loss to deep percolation in any type of the three seasonal rainfall patterns during wet, normal, and dry years (Figure 12.7). The winter wheat cover crop grown from early October to early April led to a 24% (15 kg N ha^{-1}) reduction in nitrate leaching as averaged across three types of rainfall patterns. The difference in annual nitrate leaching between the cover crop-based and fallow cropping systems is equivalent to 17% of the total fertilizer nitrogen-applied annually for the corn–soybean system. Compared with the fallow system, the winter cover crop system from early October to early April reduced annual nitrate leaching losses by 20, 19, and 8 kg N ha^{-1} in wet, normal, and dry years, respectively. Simulation indicated that mean annual net nitrogen mineralization and annual denitrifications were 15 and 9% larger, respectively, under a wheat cover crop than under winter fallow across the past 80 years (162 versus 141 kg N ha^{-1}, 9 versus 8 kg N ha^{-1}). Long-term integration of a wheat cover crop into a no-till corn–soybean rotation is a promising agronomic practice for reducing nitrate leaching and fertilizer-nitrogen input in humid northeastern Mississippi, especially in wetter fallow seasons from early October to early April.

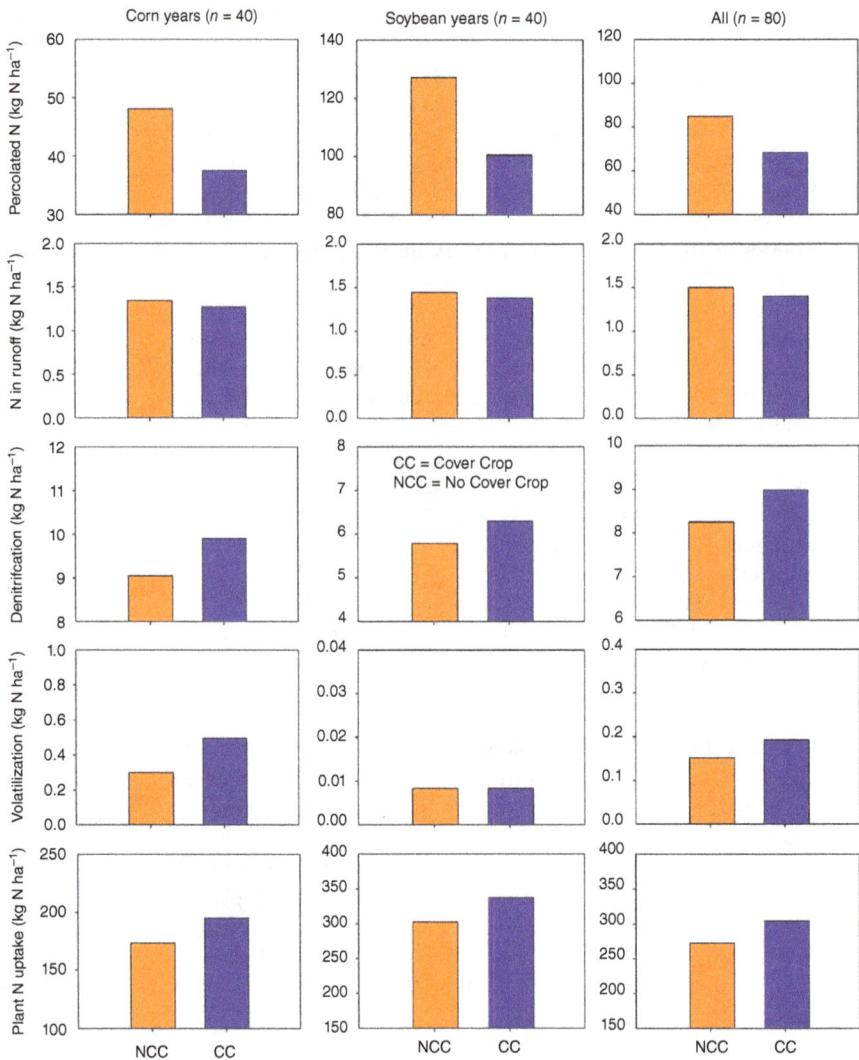

Figure 12.7 Simulated mean annual percolated nitrogen, nitrogen in runoff, denitrification, and volatilization, and plant nitrogen uptake for a corn and soybean rotation with or without wheat cover crop during an 80-year period (1938–2017), Pontotoc County, Mississippi, USA. WCC, wheat cover crop; WF, winter fallow (Yang et al., 2021).

Correct assessment of nitrate load to deep percolation in the soil is relatively difficult under long-term field observations. However, an agricultural system simulation model can help predict the transfer of N and nitrate leaching in the cropping system. The model combined with four years of consecutive field data helped quantify the long-term impact of a cover crop on mitigating N leaching and improving cash crops N uptake in an upland soil for a corn and soybean rotation system under weather variations over seven decades.

Long-Term Effect of Integrated Cover Crop, Poultry Litter, and Tillage on Soil Organic Carbon Under Rainfed Conditions

The unique advantage of modeling research is that models are capable of investigating the long-term effects of integrated management practices under diverse weather conditions. Further RZWQM2 simulations were conducted to identify what management practices combined could greatly improve SOC (Yang et al., 2019b). In the simulations, under the tillage system with 18,000 kg ha^{-1} poultry litter applied in the spring (18S) each year as shown in Tables 1 and 2 published by Feng et al. (2016), followed by corn–wheat CC–soybean fertilized based on the P test, SOC was greatly increased in the rainfed corn–wheat CC (Figure 12.8). The continuous tillage system improved SOC more than the no-till system which removed all disturbing soil tillage activities in Table 2 reported by Feng et al. (2016) as 18,000 kg ha^{-1} poultry litter was applied in each spring in this humid region characterized by hot summer and mild winter (18S) (Figure 12.8). It is in agreement with some of the previous similar studies (Anapalli et al., 2018).

Tillage Study in the Mississippi Delta

A long-term tillage experiment was started in 2008 on 1.25-ha farm-scale fields to assess the production effects of a no-till with full residue retention (NT) system compared with a conventional tillage (CT) system, both under irrigated corn production on a Dundee silt loam soil in a humid climate (Anapalli et al., 2018). Data collected from 2009 to 2015 were confined to grain yield at harvest, which showed that in six out of seven years, the NT tillage treatment system resulted in less corn yield which is in line with findings of a meta-analysis by Pittelkow et al. (2015). To examine the possible reasons for the yield declines in NT in 2016 and 2017, the experiment was integrated with the RZWQM2 to synthesize information on the various components in the system. To run the model, additional data were collected that included soil bulk

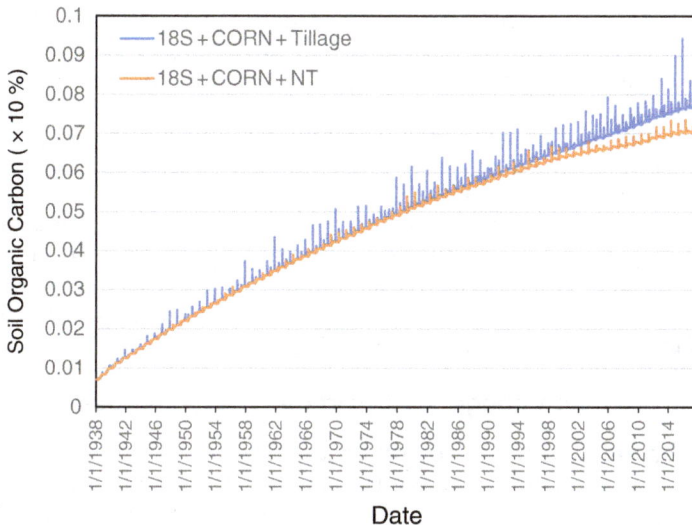

Figure 12.8 Impact of tillage and poultry litter on soil organic carbon at 0–30 cm depth in tillage and no-till (NT) system. Poultry litter 18,000 kg ha^{-1} (18S) was applied to a corn field in the spring.

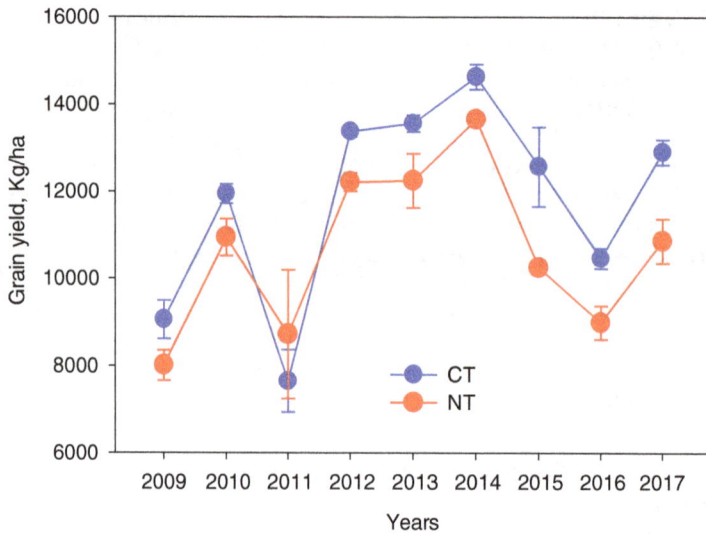

Figure 12.9 Measured corn grain yield in NT and CT systems from 2009 to 2017. Error bars indicate one standard deviation of the measured value from its mean value.

density, carbon, and nitrogen contents, soil surface crop-residue mass (Rm) and crop-cover (Rc); corn LAI and biomass (BM); soil water content (SW), temperature (Ts); and the corn phenology. The RZWQM estimated the soil water retention and hydraulic conductivity from bulk density and texture and changes in these properties due to tillage and reconsolidation changes. In addition, it accounted for the effect of surface crop residues on reducing soil evaporation and surface sealing/crusting. It also estimated plant parameters based on a database.

In seven out of nine years, the measured corn yield in NT was significantly less than the yield harvested in CT (Figure 12.9). The measured 0–60 cm soil bulk density was greater in NT, whereas soil C and N contents were less in NT than in CT (Figure 12.10a–c). In the simulations, compared with CT, low N mineralization rate due to lower soil temperatures under the NT (Figures 12.11a and 12.12), larger N loss to runoff due to larger BD, greater denitrification and deep percolation loss of N due to larger soil water content (Figures 12.11a–d and 12.12), potentially contributed to the observed grain yield decline under this treatment. Simulations indicated that an added N fertilizer at $40\,kg\,ha^{-1}$ at planting or a split application of $50\,kg\,N\,ha^{-1}$ at planting and the remaining $174\,kg\,N\,ha^{-1}$ in the second week of May could enhance the yield return under NT compared with CT (Figure 12.13). It also indicated that the higher soil water contents in the absence of enough N for plant uptake do not help maintain grain yield.

Climate Change Studies in the Mississippi Delta

Climate change impacts on cotton production

After an agricultural system model has been carefully tested for one study, for example, the tillage study described above, its use can be extended to field research involving other management practices, such as the effect of management on crop water use efficiency (Anapalli et al., 2018): simulations of LAI across CT and NT in 2016 and 2018

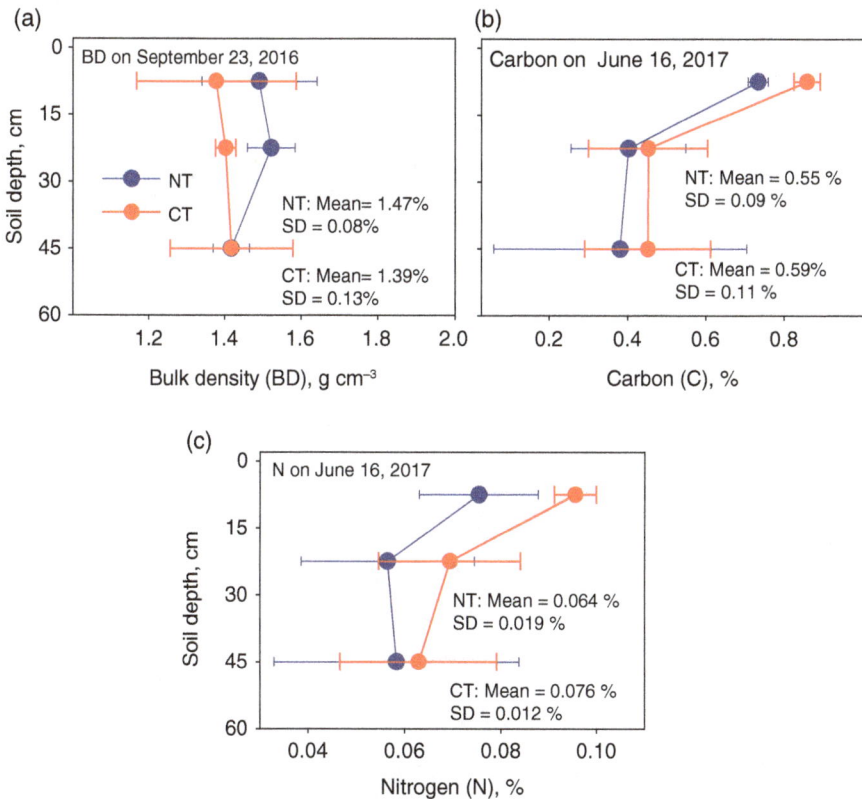

Figure 12.10 Measured soil (a) bulk density (BD) on September 23, 2016, (b) total carbon on 16 June 2017, and (c) total nitrogen (N) on 16 June 2017 in 0–15, 15–30, and 30–60 cm depths under no-till (NT) and conventional till (CT) systems. Mean and SD are averaged (weighted by the respective soil layer thickness) across soil depths (Anapalli et al., 2018 / with permission of John Wiley & Sons.).

yielded RMSD between 0.49 and 0.70 (Figure 12.4a–d in Anapalli et al., 2018); RMSD of simulations of biomass ranged between 6 and 12% across seasons of available measurement; simulated silking and physiological maturity stages deviated from the measured data between −3 and +4 days; grain yields were simulated with an RMSD of 9% under conventional tillage and 12% under no-tillage treatments.

Climate change impacts on agricultural production at different spatial and temporal scales are of great current interest. In this context, foreseeing the possible impacts of climate change on food and fiber production systems in the future is essential for formulating adaptations to sustain production and environmental quality. Anapalli et al. (2016a) used the CSM-CROPGRO-cotton module within the RZWQM2 model for predicting the possible effects of climate change on cotton (*Gossypium hirsutum*) production in the lower Mississippi Delta (MS Delta) region of the United States. The climate change scenarios were based on an ensemble of climate projections of multiple GCMs [Global Climate Models (GCMs)] for furture climate under the CMIP5 (Climate Model Inter-comparison and Improvement Program 5) that were bias-corrected and spatially downscaled (BCSD) at Stoneville in the MS Delta for the years 2050 and 2080. Four RCP

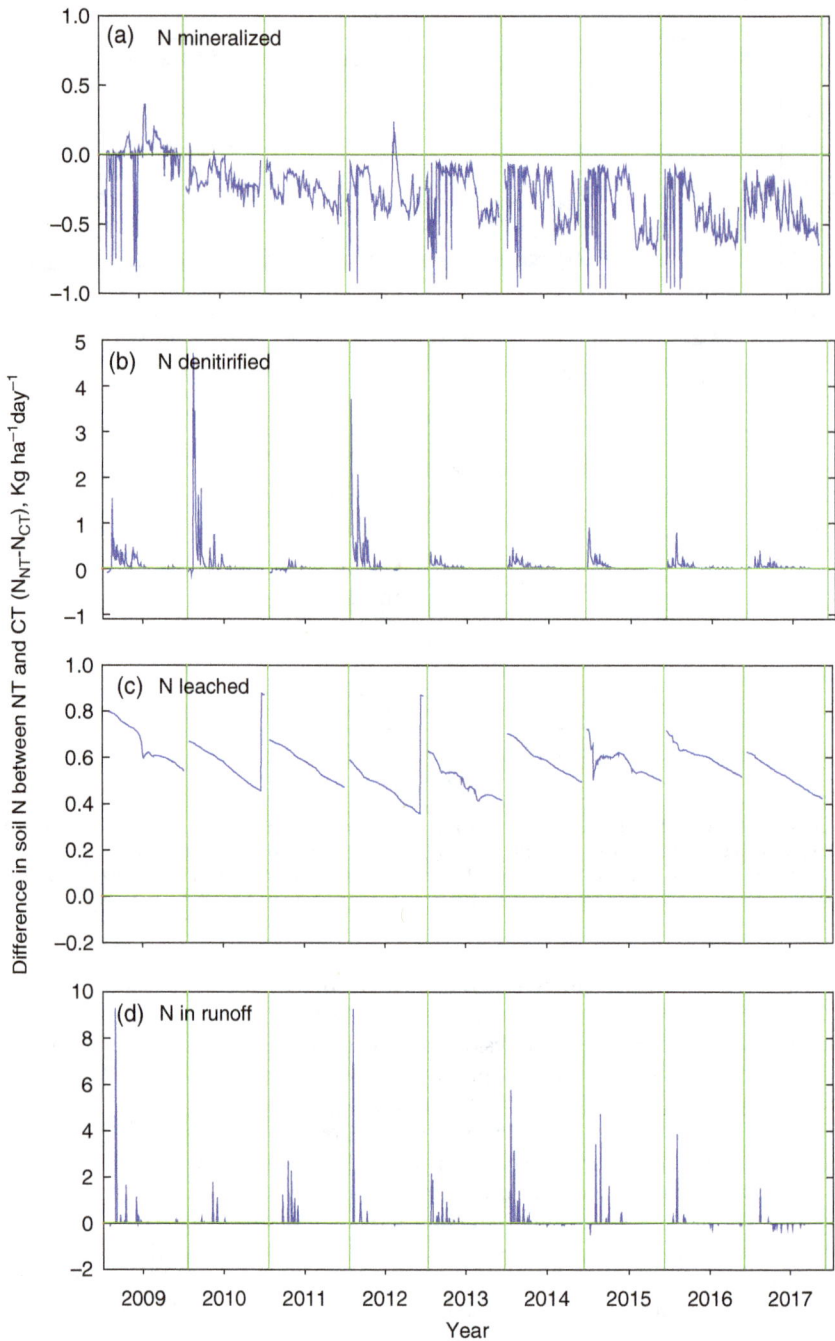

Figure 12.11 Simulated differences between NT and CT (N_{NT}-N_{CT}, N_{NT} is N under NT and N_{CT} is N under CT) in nitrogen (N) in the soil under corn during the corn growth periods from 2009 to 2017. (a) N mineralized, (b) N denitrified, (c) N leached, and (d) N in the runoff (Anapalli et al., 2018 / with permission of John Wiley & Sons.).

integrating models with field experiments

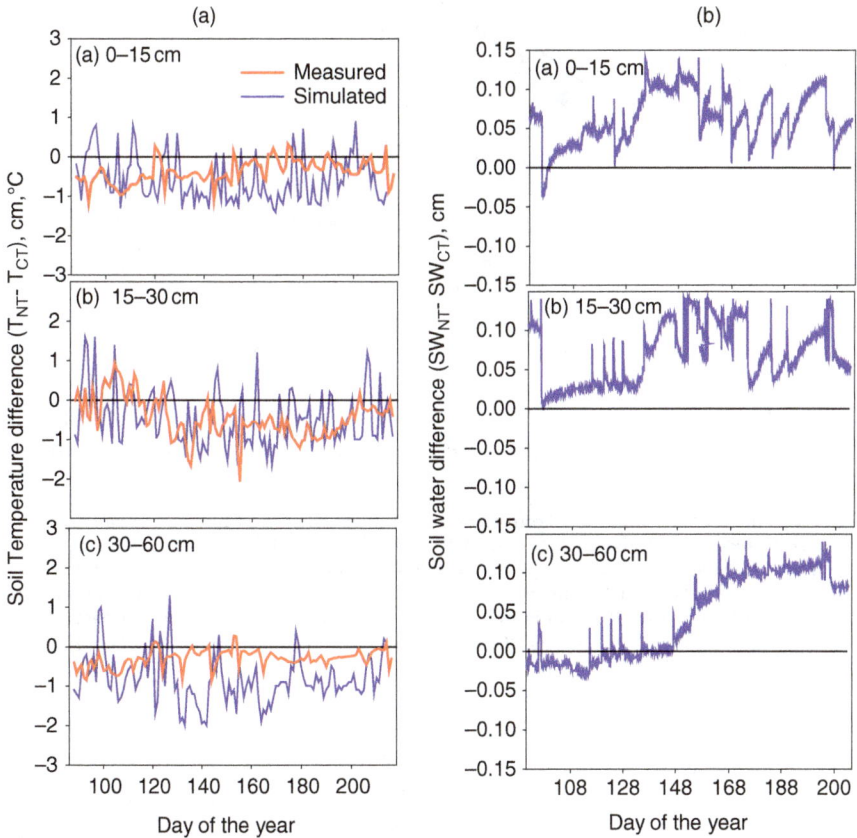

Figure 12.12 Measured and simulated soil temperature (a) and water difference (b) between NT and CT treatments. NT is no-tillage, CT is conventional tillage, T_{NT} is soil temperature under NT, T_{CT} is soil temperature under CT, SW_{NT} is measured soil water under NT, and SW_{CT} is measured soil water under CT treatments at depths 0–15, 15–30, and 30–60 cm in 2017 (Anapalli et al., 2018 / with permission of John Wiley & Sons.).

(Representative Concentration Pathways) drove these climate change projections: 2.6, 4.5, 6.0, and 8.5 (these numbers refer to radiative forcing levels in the atmosphere of 2.6, 4.5, 6.0, and 8.5 W m^{-2}), in lieu of the future levels of the greenhouse gas (GHG) emission scenarios, as used in the Intergovernmental Panel on Climate Change-Fifth Assessment Report (IPCC, 2014). The cotton model within RZWQM2, calibrated and validated for simulating cotton production at Stoneville, was used for simulating production under these climate change scenarios (Anapalli et al., 2016a). Under irrigated conditions, yields increased significantly under the climate change scenarios driven by the low to moderate emission levels of RCP 2.6, 4.5, and 6.0 in the years 2050 and 2080 using more water to meet the higher evapotranspiration demand. Even under the highest emission scenario of RCP 8.5, the cotton yield increased in 2050 but declined appreciably in 2080 (Figure 12.14). Under rainfed conditions, the yield declined in both 2050 and 2080 under all four RCP scenarios; however, the yield still increased when enough rainfall was received to meet the crop's water requirements (in

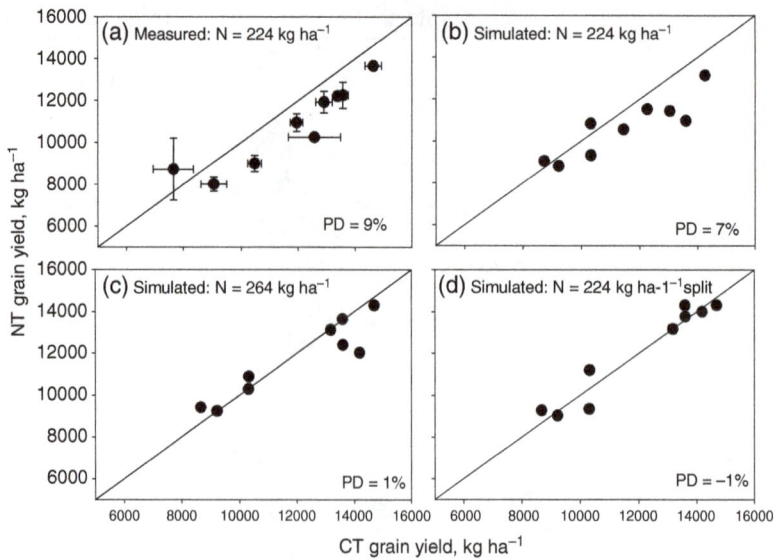

Figure 12.13 Measured and simulated corn grain yields between NT and CT systems (1:1 plot) from 2009 and 2017. (a) Measured yields (N inputs were at 224 kg ha^{-1}), (b) simulated yield with N inputs same as measured, (c) simulated grain yields with N application at 264 kg ha^{-1}, and (d) simulated grain yield in response to split N (50 kg ha^{-1} at planting and 174 kg ha^{-1} on May 14). Horizontal and vertical error bars indicate one standard deviation in the measured corn yield under CT and NT, respectively. PD, percent decrease in yield in NT (Anapalli et al., 2018 / with permission of John Wiley & Sons.).

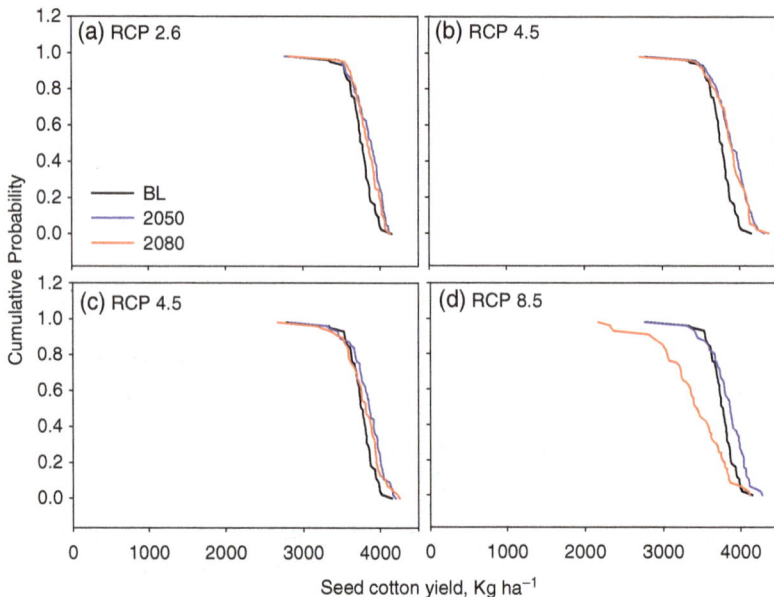

Figure 12.14 Simulated impacts of climate change (changes in temperature, rainfall, and CO_2) on irrigated seed cotton yield in 2050 and 2080 under the four emission scenarios of RCP 2.6 (a), 4.5 (b), 6.0 (c), and 8.5 (d) (Anapalli et al., 2016a / Public Domain CC BY 4.0).

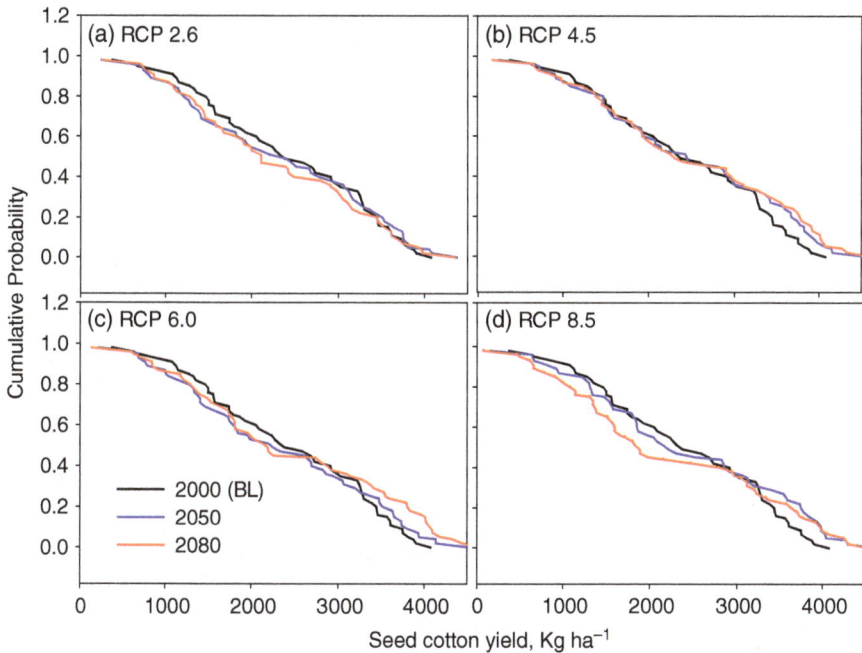

Figure 12.15 Simulated impacts of climate change (changes in temperature, rainfall, and CO_2) on rainfed seed cotton yield in 2050 and 2080 for RCP 2.6 (a), 4.5 (b), 6.0 (c), and 8.5 (d) emission scenarios compared with baseline (BL; measured weather data from 1960 to 2015) conditions (Anapalli et al. 2016a / Public Domain CC BY 4.0).

about 25% of the cases) (Figure 12.15). Mostly, planting cotton six weeks earlier than the normal planting date compensated for the lost yields in all the climate change scenarios (Figures 12.16 and 12.17). This planting strategy only partially compensated for the rainfed yield reductions under all the climate change scenarios; yet, supplemental irrigations up to about 10 cm compensated for the yield losses (Figure 12.18).

Climate change impacts on corn production

Anapalli et al. (2021) investigated the likely impacts of climate change projected by multiple GCMs on rainfed and irrigated corn, a C4 plant, in the Lower Mississippi Delta region (LMD), USA. They used the CSM-CROPGRO-Maize v4.6 module in the RZWQM2 model that was calibrated and tested for modeling corn at Stoneville, Mississippi, a representative location in the region. The climate change scenarios considered in this study were 97 ensemble members of climate projections of multiple GCMs participated in the Climate Model Inter-comparison and Improvement Program 5. These scenarios were BCSD for 2050 and 2080. Four RCP (2.6, 4.5, 6.0, and 8.5) pathways drove these climate change scenarios. Under both irrigated and rainfed conditions, corn yield responses to enhanced CO_2 were weak; thus, yield decreased significantly responding to the enhanced air temperatures under all the RCP scenarios in 2050 and 2080. The irrigated corn yield reductions across RCPs were between 10 and 62%, but between 9 and 60% under rainfed conditions, mainly due to increased frequency of extreme temperatures and reduced crop durations but not from changes in water availability.

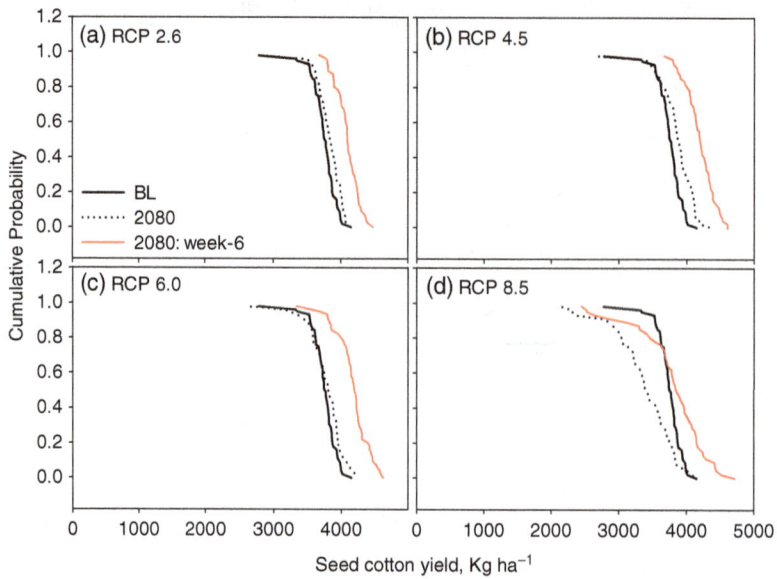

Figure 12.16 Comparison between the simulated irrigated seed cotton yield under baseline (BL) climate and climate change (CC; changes in CO_2, temperature, and rainfall) in 2080 for RCP 2.6 (a), 4.5 (b), 6.0 (c), and 8.5 (d) for the normal day of planting (2 May) and earlier than normal-day plantings of six weeks (week-6) (Anapalli et al., 2016a / Public Domain CC BY 4.0).

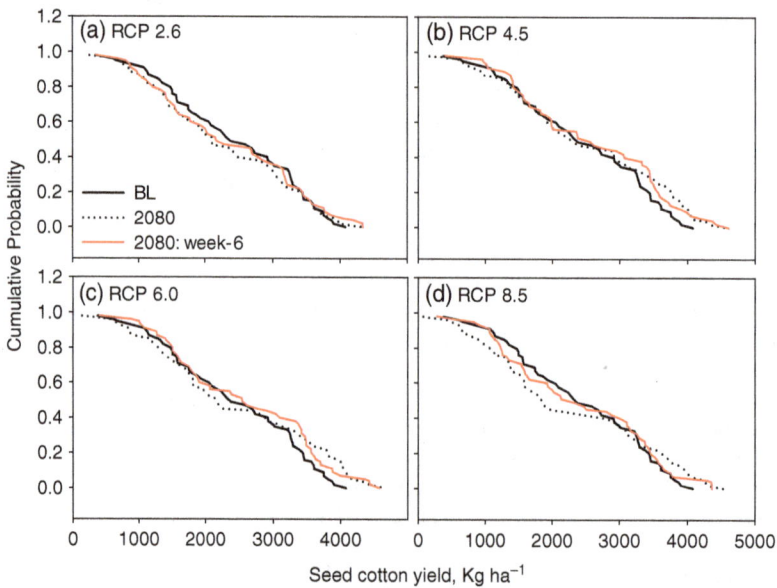

Figure 12.17 Comparison between the simulated rainfed seed cotton yield under baseline (BL) climate and climate change (CC; changes in CO_2, temperature, and rainfall) in 2080 under RCP 2.6 (a), 4.5 (b), 6.0 (c), and 8.5 (d) for normal day of planting (2 May) and earlier than normal-day plantings of six weeks (week-6) (Anapalli et al., 2016a / Public Domain CC BY 4.0)

integrating models with field experiments

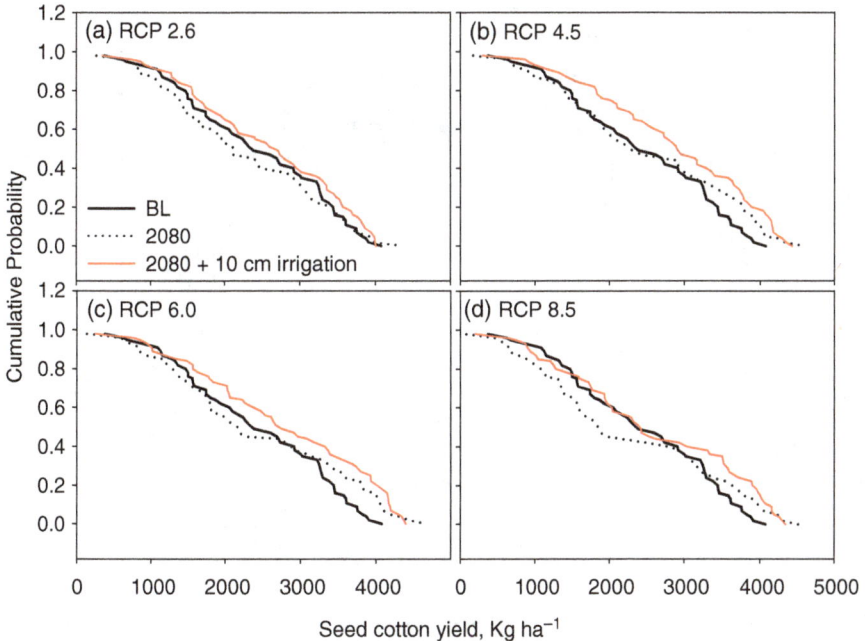

Figure 12.18 Comparison between the simulated rainfed seed cotton yield under baseline (BL) climate and climate change (CC; changes in CO_2, temperature, and rainfall) in 2080 under RCP 2.6 (a), 4.5 (b), 6.0 (c), and 8.5 (d) for normal day of planting (2 May) with normal planting plus 10 cm irrigation (Anapalli et al., 2016a / Public Domain CC BY 4.0)

For simulating corn, the minimum air temperature below and upper air temperature above which the corn plant growth stops were assumed as 8 and 38 °C, respectively, while the temperature for optimum growth was 34 °C (Hatfield & Preuger, 2015; Jones & Kiniry, 1986; Kiniry & Bonhomme, 1991). Water use efficiency declined between 22 and 150% and 8 and 54%, respectively, under irrigated and rainfed conditions. In general, increasing crop duration and reducing chances of encountering upper temperature extremes by planting up to nine weeks earlier in the season, failed to compensate for the yield loss as incidences of lower extreme temperatures also increased simultaneously with this early planting. Cultivars that are more heat tolerant and produce higher grain yields under extreme temperatures would be required to compensate corn yield loss in the region caused by climate change.

Mitigation of climate change impacts by cover crops in a corn–soybean rotation system

Climate change is projected to increase temperature and uncertainty in amount and distribution of rainfall (Anapalli et al., 2016a; Li et al., 2021). Therefore, Li et al. (2021) applied the RZWQM2 model and investigated whether and to what extent cover crops can mitigate the adverse impacts of climate change in the state of Mississippi. The future climate projections for 2020–2079 under RCP 4.5 and 8.5 scenarios were obtained from 10 statistically downscaled and bias corrected GCMs outputs (https://tntcat.

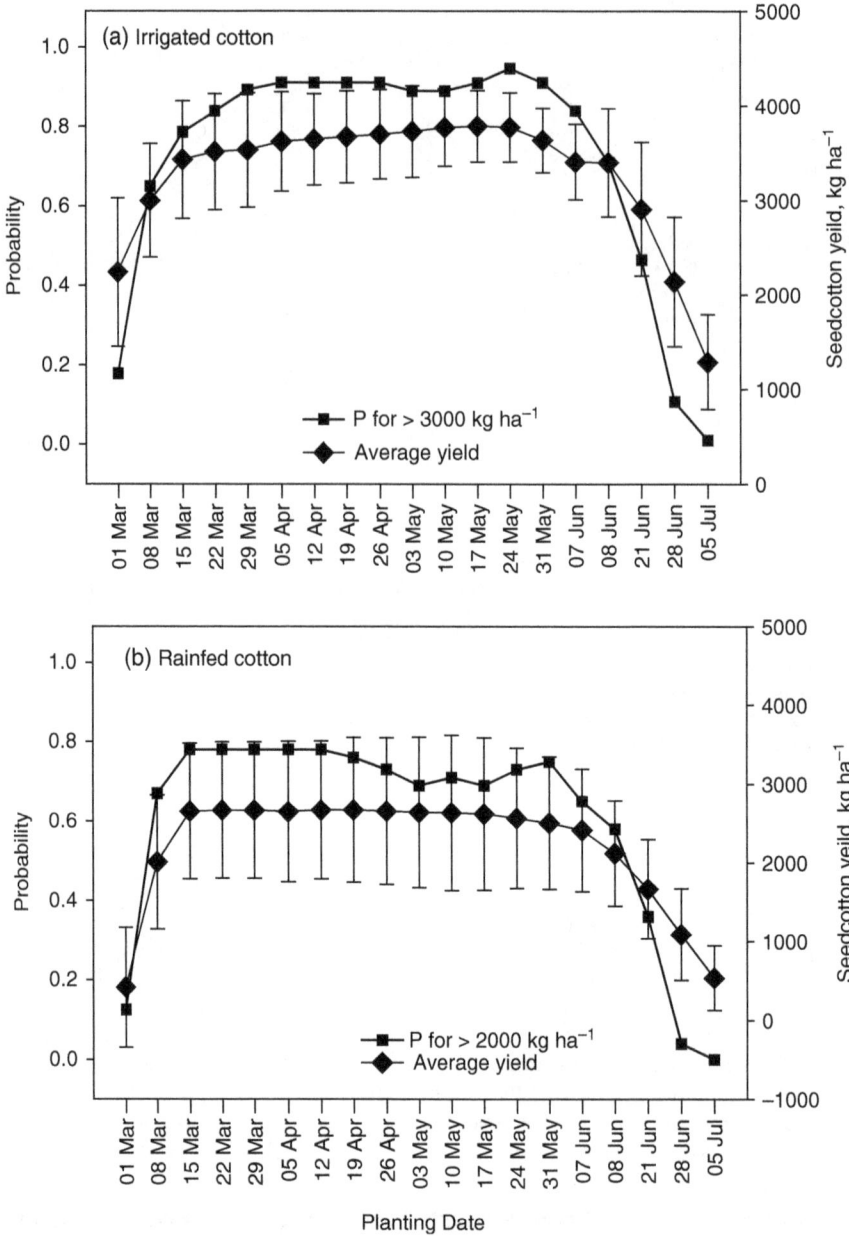

Figure 12.19 Mean simulated irrigated (a) and rainfed (b) seed cotton yields and the probabilities of achieving breakeven seed cotton yields (2,000 kg ha^{-1} for rainfed and 3,000 kg ha^{-1} for irrigated conditions) at Stoneville, MS. The crops were planted at weekly intervals from 1 March through 5 July (Anapalli et al., 2016b).

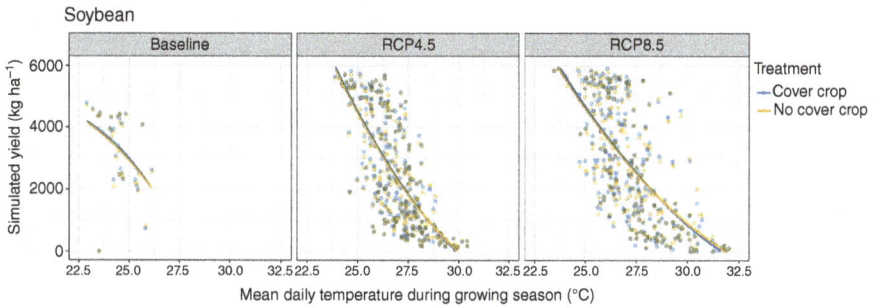

Figure 12.20 Relationships between mean growing season temperature and simulated yield of soybean under the historical baseline, and the future RCP4.5 and RCP8.5 scenarios for the cover crop and no cover crop treatments (Li et al., 2021).

iiasa.ac.at/RcpDb). The CO_2 concentration for baseline was obtained from the National Ocean and Atmospheric Administration (NOAA) Earth System Research Laboratory (ESRL) (https://www.esrl.noaa.gov/gmd/ccgg/trends/data.html) during 1956–2015. The average CO_2 concentrations over 60 years were 350, 480, and 560 µmol mol^{-1} for the historical baseline (1956–2015), and the future RCP4.5 and RCP8.5 scenarios (2020–2079), respectively.

Compared with the historical baseline, future temperature and precipitation were projected to be significantly increased under both RCP4.5 and RCP8.5 climate change scenarios in MS (Li et al., 2021). GCM projections showed the possibility of warmer and wetter conditions in the region, especially under the RCP8.5 scenario. Yang et al. (2020) applied RZWQM2 and determined the roles of cover crops in reducing the historical adverse weather impacts. Li et al. (2021) investigated cover crop benefits under climate change scenarios, they revealed that cover crops increased soybean yields by 1.4 and 1.3% (Figure 12.20), and improved CWP by 2.5% for corn and 1.0% for soybean (Figure 12.21) under the future RCP4.5 and RCP8.5 scenarios. As Figure 12.22 shows, soil evaporation, as the largest source of water loss from the cropping system, soil evaporation was reduced by 28.4% (52.2 mm), 24.4% (53.3 mm), and 26.8% (58.8 mm) during the corn growing season, and by 26.6% (42.7 mm), 25.8% (43.1 mm) and 27.2% (46.5 mm) during the soybean growing season under the baseline, RCP4.5 and RCP8.5, respectively. The annual drainage below 1.8-m soil depth was decreased by 38 mm (8.0%), 53 mm (11.8%), and 67 mm (15.4%) for the cover crop treatment compared with the no cover crop treatment under the baseline, RCP4.5 and RCP8.5 (Figure 12.22). As a result, cover crops increased soil water storage (0–1.8 m depth of soil profile) during early stages of main crop growing period under historical and future climate scenarios (Figure 12.23). The Figure 12.23 shows that wheat cover crop enhanced the soil water storage by as much as 14.8 mm (on 133 day of the year, DOY), 7.6 mm (on 133 DOY) and 4.6 mm (134 DOY) under the baseline, RCP4.5 and RCP8.5 during corn growing season, respectively. For soybean, soil water storage could be improved by as much as 16.1 mm (149 DOY), 13.2 mm (150 DOY) and 11.7 mm (150 DOY) under the baseline, RCP4.5 and RCP8.5 during the soybean growing season (Figure 12.23).

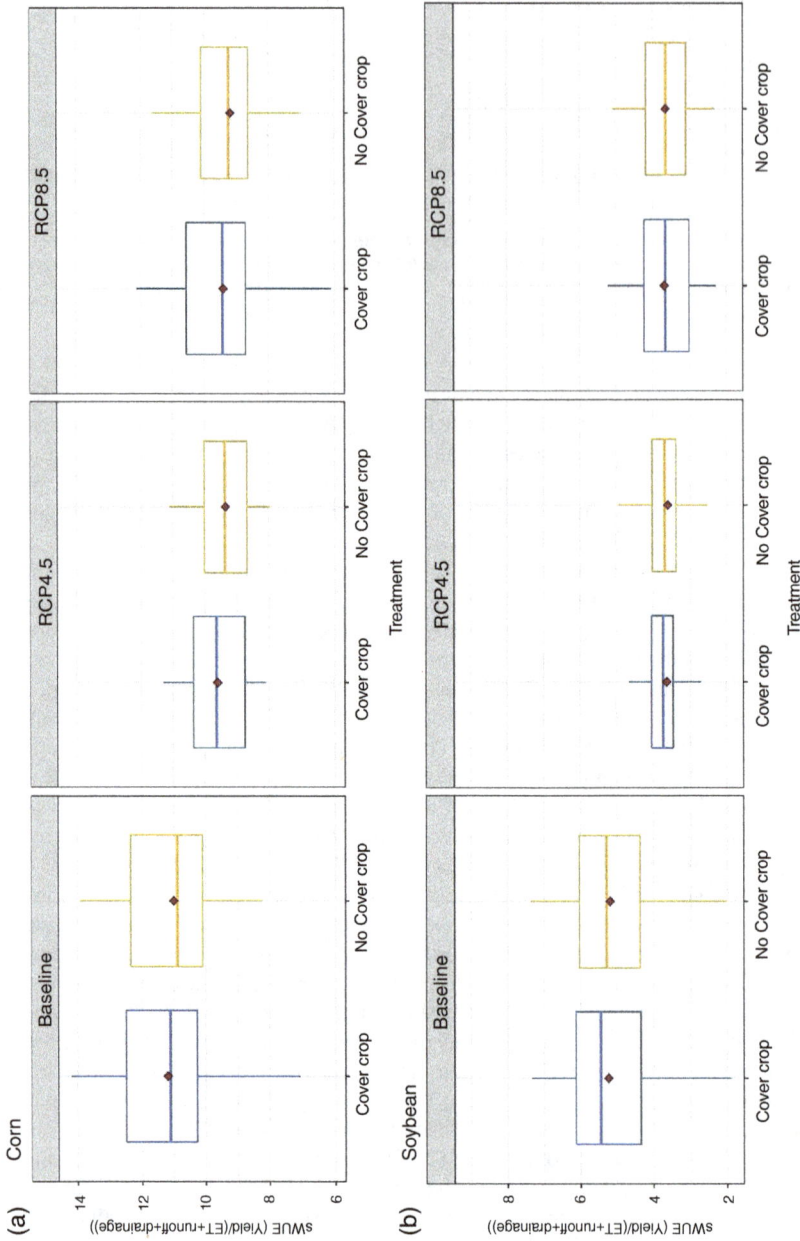

Figure 12.21 Crop water productivity (CWP) of (a) corn and (b) soybean under the baseline, RCP4.5 and RCP8.5 scenarios for the cover crop and no cover crop treatments in a corn–soybean rotation system. The solid red dots in the figure represent the average values starting from the termination date of cover crop to the harvest of main crops over 30 even-numbered years (corn growing period) or 30 odd-numbered years (soybean growing period) under different scenarios (Li et al., 2021).

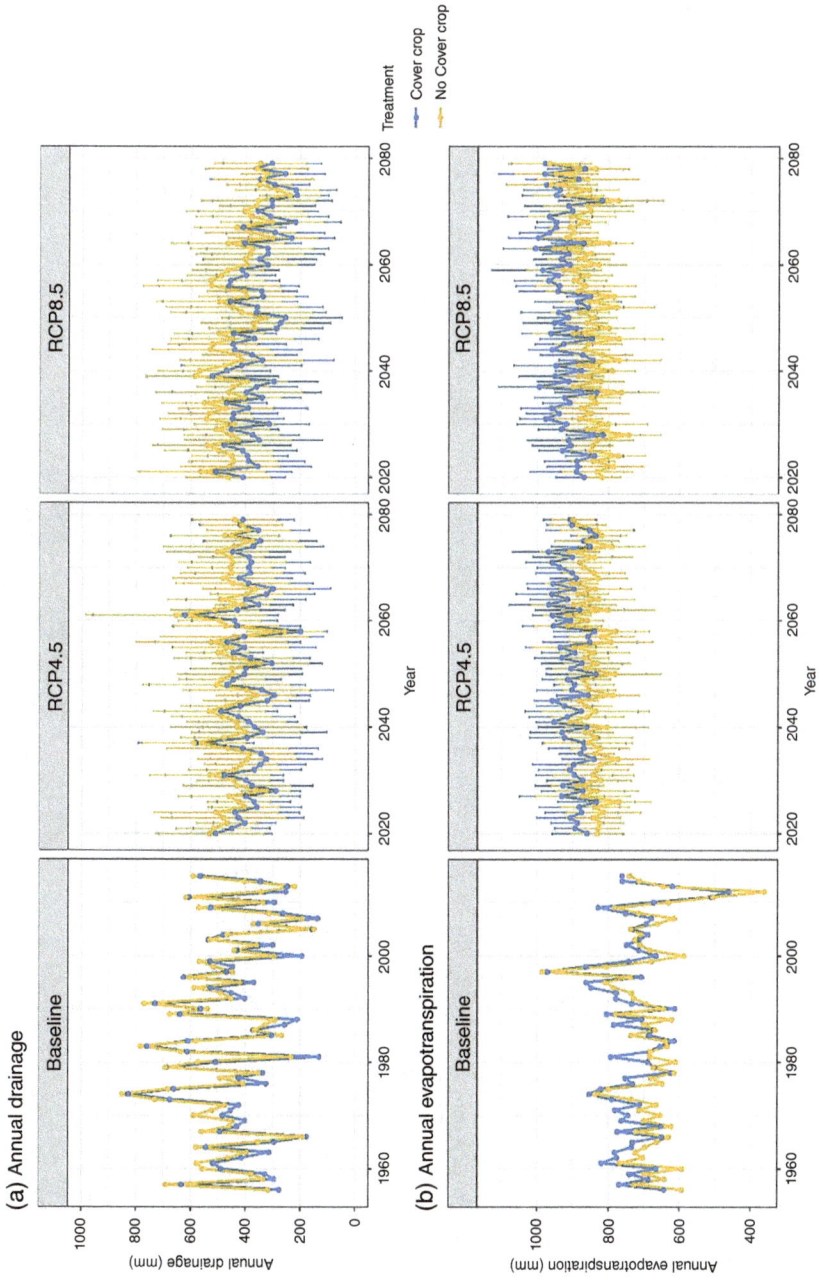

(a) Annual drainage

(b) Annual evapotranspiration

Figure 12.22 Average daily simulated soil water storage (at 0–1.8 m depth of soil profile) over (a) corn and (b) soybean growing seasons for cover crop (CC) and no cover crop (NCC) treatments under the historical baseline, and the future RCP4.5 and RCP8.5 scenarios in a corn–soybean rotation system (Li et al., 2021).

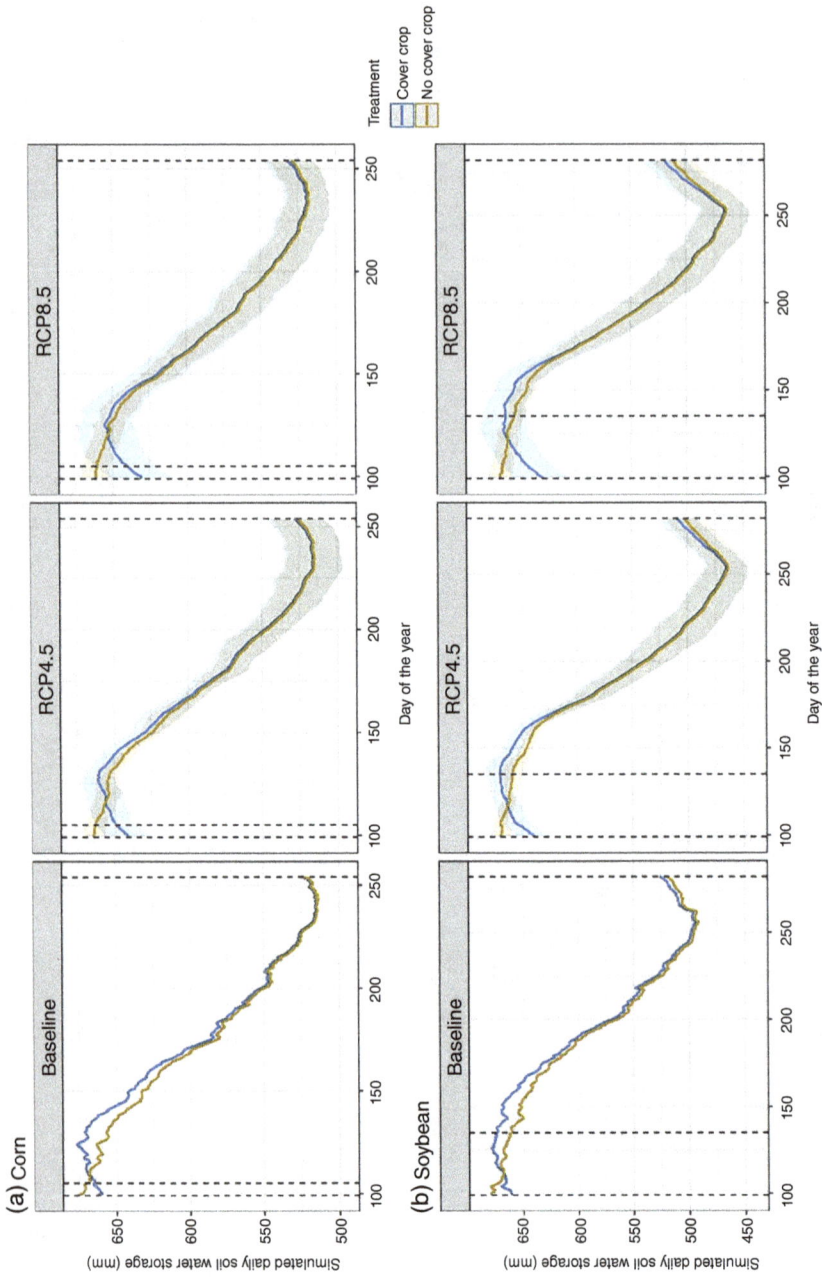

Figure 12.23 Average daily simulated soil water storage (at 0–1.8 m depth of soil profile) over (a) corn and (b) soybean growing seasons for cover crop (CC) and no cover crop (NCC) treatments under the historical baseline, and the future RCP4.5 and RCP8.5 scenarios in a corn–soybean rotation system (Li et al., 2021).

Growing cover crop, buffered impacts of climate extremes on crop yields and water loss, is an effective means to mitigate the adverse impacts of climate changes and extremes in a humid environment. With models long-term simulation studies, can be conducted that assist or extend field experimental research and evaluate existing and alternative crop management systems on different soil types under historical, current, and projected weather conditions. Combining a model with experimental data (to calibrate the model) made it possible to obtain the above greatly useful information, in particular, the information on long-term impacts of cover crop on water balance under century-long historical weather conditions and projected future climate conditions.

Planting Windows for Cotton in Response to Long-Term Climate Variability

Exceptionally variable summer rainfalls of the lower Mississippi (MS) Delta region are a serious challenge to cotton producers deciding when to plant the crop to maximize yield returns. Old-style two- to four-year agronomic field trials conducted in this area failed to capture the effects of long-term climate variabilities in the location for developing reliable planting windows for producers. Anapalli et al. (2016b) integrated a four-year study at Stoneville, MS during 2005–2008, which investigated impacts of planting dates on cotton production, with long-term climate data in an agricultural system model and developed optimum planting windows for cotton under both irrigated and rainfed conditions (Fig. 12.19). Weather data collected at this location from 1960 to 2015 and the CSM-CROPGRO-Cotton v4.6 model within the RZWQM2 were used. The cotton model was able to simulate both the variable planting date and variable water regimes reasonably well: relative errors of seed cotton yield, aboveground biomass, and LAI were 14, 12, and 21% under rainfed conditions and 8, 16, and 15% under irrigated conditions, respectively. Planting windows under both rainfed and irrigated conditions with targeted yields of 2,000 and 3,000 kg ha^{-1}, respectively, extended from mid-March to mid-June: windows from mid-March to the last week of May under rainfed conditions, and from the last week of April to the end of May under irrigated conditions were better suited for optimum yield returns (Figure 12.24). Rainfed cotton tends to lose yield from later plantings within these windows, but irrigated cotton benefits; however, irrigation requirements increase as the planting windows advance in time. Irrigated cotton produced about 1,000 kg ha^{-1} seed cotton more than rainfed cotton, with irrigation water requirements averaging 15 cm per season (Figure 12.24). Under rainfed conditions, there is a 5, 14, and 27% chance that the seed cotton production is below 1,000, 1,500, and 2,000 kg ha^{-1}, respectively. Information developed in this study can help MS farmers in decision support for cotton planting.

Summary and Conclusions

Water withdrawals exceeding recharge rates are causing the MS River Valley Alluvial Aquifer (MSRVAA) underlying the Lower Mississippi Delta region to decline at an alarming rate. To ease the pressure on the MSRVAA for irrigations and sustainable water management, it is critical to base water withdrawals on exact crop consumptive water use (ET, evapotranspiration) in response to enhanced climate variability and change under the current observed warming trends in global climate. In this chapter we presented case studies for integrating short term experiments and long-term past

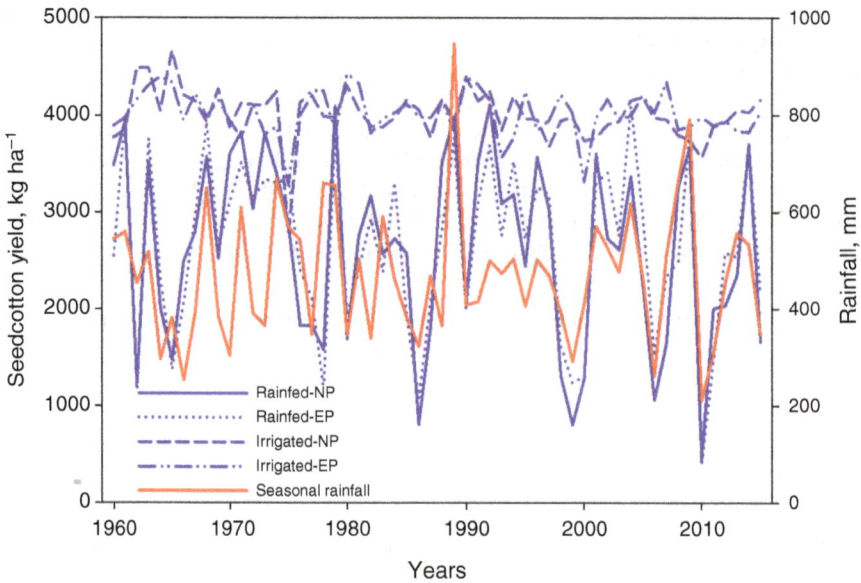

Figure 12.24 Simulated seed cotton yields of normal (NP) and early plantings (EP) under rainfed and irrigated conditions from 1960 to 2015 at Stoneville, MS. The EP was on 30 March, and NP was on 2 May (Based on Anapalli et al., 2016a).

and future climate data in agricultural system models for investigating and generating alternative crop–soil–water management strategies for sustainable irrigated and rainfed agriculture: (a) cover crops in corn–soybean systems for soil water management, (b) poultry litter for soil N, carbon and water management in corn–soybean systems, (c) N distribution management under cover crop in corn–soybean systems, (d) integrated effects of cover crop, poultry litter, and tillage on soil carbon, N and water management, (e) tillage and no-tillage in corn for soil water and N management, (f) investigating impacts of climate variability and change impacts and managing adaptation strategies in cotton, corn, and soybean systems under rainfed and irrigated conditions, and (g) developing cotton planting windows in response to long-term climate variability and change. Further studies are being conducted for accurately quantifying crop growth and water use (evapotranspiration, ET) responses to irrigated and rainfed crop management in relation to various soil amendments and tillage and cover crop management systems. The state-of-the-art micrometeorological theory-based eddy covariance (EC) method is being used for accurately quantifying evapotranspiration from cropping systems. Efforts are underway to develop and improve ET modules in the RZWQM USDA ARS for simulating cropping systems in the region in response to above management strategies for developing farmer decision support systems for sustainable agriculture in the region. The improved ET modules, crop coefficients for linking actual crop ET with a reference crop ET for irrigation scheduling, and cropping system models will be available online to the public, farmers, and crop consultants for crop, soil, and irrigation water management coping with climate variability and change in the region.

References

Ahuja, L. R., Ma, L., & Anapalli, S. S. (2017). Biophysical system models advance agricultural research and technology: Some examples and further research needs. In O. Wendroth, R. J. Lascano, & L. Ma (Eds.), *Bridging among disciplines by synthesizing soil and plant processes*. Advances in Agricultural Systems Modeling 8 (pp. 1–32). ASA, CSSA, and SSSA.

Anapalli, S. S., Ahuja, L. R., Ma, L., & Trout, T. J. (2014). Modeling the effects of irrigation frequencies, initial water, and nitrogen on corn yield responses for best management. *Advances in Agricultural Systems Modeling*, 5, 25–52.

Anapalli, S. S., Ahuja, L. R., Nielsen, D. C., Trout, T., & Ma, L. (2008). Use of crop simulation models to evaluate limited irrigation management options for corn in a semiarid environment. *Water Resources Research*, 44, W00E02. http://dx.doi.org/10.1029/2007WR006181

Anapalli, S. S., Fisher, D. K., Reddy, K. N., Pettigrew, W. T., Sui, R., & Ahuja, L. R. (2016a). Vulnerability and adaptation of cotton to climate change in the Mississippi Delta. *Climate*, 4(4), 55. 1–20; http://dx.doi.org/ http://dx.doi.org/10.3390/cli4040055

Anapalli, S. S., Pettigrew, W. T., Reddy, K. N., Ma, L., Fisher, D. K., & Sui, R. (2016b). Climate optimized planting windows for cotton in the lower Mississippi Delta region. *Agronomy*, 6(4), 46. 1–15

Anapalli, S. S., Fisher, D. K., Reddy, K. N., Rajan, N., & Pinnamaneni, S. R. (2019). Modeling evapotranspiration for irrigation water Management in a Humid Climate. *Agricultural Water Management*, 225, 1–11.

Anapalli, S. S., Pinnamaneni, S. R., Fisher, D. K., & Reddy, K. N. (2021). Vulnerabilities of irrigated and rainfed corn to climate change in a humid climate. *Climatic Change*, 164, 5.

Anapalli, S. S., Reddy, K. N., & Sindhu, J. (2018). Conservation tillage impacts and adaptations in irrigated corn (*Zea mays* L.) production in a humid climate. *Agronomy Journal*, 110, 1–14. 2018.

Blanco-Canqui, H., Shaver, T. M., Lindquist, J. L., Shapiro, C. A., Elmore, R. W., Francis, C. A., & Hergert, G. W. (2015). Cover crops and ecosystem services: Insights from studies in temperate soils. *Agronomy Journal*, 107, 2449–2474. http://dx.doi.org/10.2134/agronj15.0086

Dietzel, R., Liebman, M., Ewing, R., Helmers, M., Horton, R., Jarchow, M., & Archontoulis, S. (2016). How efficiently do corn- and soybean-based cropping systems use water? A systems modeling analysis. *Global Change Biology*, 22, 666–681. http://dx.doi.org/10.1111/gcb.13101

Feng, G., Adeli, A., Read, J. J., McCarty, J. C., & Jenkins, J. N. (2019). Consequences of pelletized poultry litter applications on soil physical and hydraulic properties in reduced tillage, continuous cotton system. *Soil & Tillage Research*, 194(104309), 1–7. http://dx.doi.org/10.1016/j.still.2019.104309

Feng, G., Cobb, S., Abdo, Z., Fisher, D. K., Ouyang, Y., Adeli, A., & Jenkins, J. (2016). Trend analysis and forecast of precipitation, reference evapotranspiration and rainfall deficit in the Blackland prairie of eastern Mississippi. *Journal of Applied Meteorology and Climatology*, 55, 1425–1439. http://dx.doi.org/10.1175/JAMC-D-15-0265.1

Feng, G., Letey, J., Chang, A., & Mathews, M. (2005). Simulating dairy liquid waste management options as a nitrogen source for crops. *Agriculture, Ecosystems, and Environment*, 110, 219–229.

Feng, G., Ouyang, Y., Adeli, A., Read, J., & Jenkins, J. (2018). Rainwater deficit and irrigation demand for major row crops in Mississippi Blackland prairie. *Soil Science Society of America Journal*, 82, 423–435. http://dx.doi.org/10.2136/sssaj2017.06.0190

Feng, G., Tewolde, H., Ma, L., Adeli, A., Sistani, K., & Jenkins, J. (2015). Simulating the fate of fall- and spring-applied poultry litter nitrogen in corn production. *Soil Science Society of America Journal*, 79, 1804–1814. http://dx.doi.org/10.2136/sssaj2015.06.0211

Franzluebbers, A. J., Stuedemann, J. A., & Wilkinson, S. R. (2001). Bermuda grass management in the Southern Piedmont USA: I. Soil and surface residue carbon and sulfur. *Soil Science Society of America Journal*, 65, 834–841.

Gao, F., Feng, G., Han, M., Dash, P., Jenkins, J., & Liu, C. (2019a). Assessment of surface water resources in the big sunflower river watershed using coupled SWAT–MODFLOW model. *Water*, 11(528), 1–21. http://dx.doi.org/10.3390/w11030528

Gao, F., Feng, G., Ouyang, Y., Jenkins, J., & Liu, C. (2019b). Simulating potential weekly stream and pond water available for irrigation in the big sunflower river watershed of Mississippi delta. *Water*, 11(1271), 1–15. http://dx.doi.org/10.3390/w11061271

Gao, F., Feng, G., Ouyang, Y., Wang, H., Fisher, D., Adeli, A., & Jenkins, J. (2017). Evaluation of reference evapotranspiration methods in arid, semiarid and humid regions. *Journal of the American Water Resources Association (JAWRA)*, 53(4), 791–808. http://dx.doi.org/10.1111/1752-1688.12530

Hatfield, J. L., & Preuger, J. H. (2015). Temperature extremes: The effect on plant growth and development. *Weather and Climate Extremes*, 10, 4–10.

Henault, C., Bizouard, F., Laville, P., Gabrielle, B., Nicoullaud, B., Germon, J. C., & Cellier, P. (2005). Predicting in situ soil N₂O emission using NOE algorithm and soil database. *Global Change Biology*, 11, 115–127.

IPCC. (2014). Summary for policymakers. In C. B. Field, V. R. Barros, D. J. Dokken, K. J. Mach, M. D. Mastrandrea, T. E. Bilir, M. Chatterjee, K. L. Ebi, Y. O. Estrada, R. C. Genova, B. Girma, E. S. Kissel, A. N. Levy, S. MacCracken, P. R. Mastrandrea, & L. L. White (Eds.), *Climate change: Impacts, adaptation, and vulnerability. Part a: Global and sectoral aspect. Contribution of Working Group II to Fifth Assessment Report of the Intergovernmental Panel on Climate Change* (pp. 1–32). Cambridge University Press.

Jiang, Q., Qi, Z. M., & Madramootoo, C. A. (2017). *Simulating agronomic management effects on greenhouse gas emissions in a subsurface drainage field in Southern Quebec using RZWQM2*. In: 2017 ASABE Annual International Meeting 1701223.

Johnson, J. M. F., Franzluebbers, A. J., Weyers, S. L., & Reicosky, D. C. (2007). Agricultural opportunities to mitigate greenhouse gas emissions. *Environmental Pollution*, 150, 107–124.

Jones, C. A., & Kiniry, J. R. (1986). *CERES-maize: A simulation model of maize growth and development*. Texas A&M University Press.

Kiniry, J. R., & Bonhomme, R. (1991). Predicting maize phenology. In T. Hodges (Ed.), *Predicting crop phenology* (pp. 115–131). CRC Press.

Li, C., Frolking, S., Xiao, X., Moore, B., III, Boles, S., Qiu, J., Huang, Y., Salas, W., & Sass, R. (2005a). Modeling impacts of farming management alternatives on CO₂, CH₄, and N₂O emissions: A case study for water management of rice agriculture of China. *Global Biogeochemical Cycles*, 19, GB301.

Li, Y., Chen, D., Zhang, Y., Edis, R., & Ding, H. (2005b). Comparison of three modeling approaches for simulating denitrification and nitrous oxide emissions from loam-textured arable soils. *Global Biogeochemical Cycles*, 19, 1–15.

Li, L., Malone, R. W., Ma, L., Kaspar, T. C., Jaynes, D. B., Saseendran, S. A., Thorp, K. R., Yu, Q., & Ahuja, L. R. (2008). Winter cover crop effects on nitrate leaching in subsurface drainage as simulated by RZWQM-DSSAT. *Transactions of the ASABE*, 51, 1575–1583.

Li, Y. D., Tian, G., Feng, W. Y., & Feng, L. (2021). Climate change and cover crop effects on water use efficiency of a corn-soybean rotation system. *Agricultural Water Management*, 255(2021), 107042. http://dx.doi.org/10.1016/j.agwat.2021.107042

Liu, H. L., Yang, J. Y., Tan, C. S., Drury, C. F., Reynolds, W. D., Zhang, T. Q., Bai, Y. L., Jin, J., He, P., & Hoogenboom, G. (2011). Simulating water content, crop yield and nitrate N loss under free and controlled tile drainage with subsurface irrigation using the DSSAT model. *Agricultural Water Management*, 98, 1105–1111.

Moriasi, D. N., Arnold, J. G., Van Liew, M. W., Gingner, R. L., Harmel, R. D., & Veith, T. L. (2007). Model evaluation guidelines for systematic quantification of accuracy in watershed simulation. *Transactions of the ASABE*, 50, 885–900. http://dx.doi.org/10.13031/2013.23153

Nash, J. E., & Sutcliffe, J. V. (1970). River flow forecasting through conceptual models Part I: A discussion of principles. *Journal of Hydrology*, 10, 282–290.

Nielsen, D. C., Anapalli, S. S., Ma, L., & Ahuja, L. R. (2012). Simulating the production potential of dryland spring canola in the central Great Plains. *Agronomy Journal*, 104, 1182–1188.

Ouyang, Y., Feng, G., Read, J., Leininger, T., & Jenkins, J. (2016). Estimating the ratio of pond size to irrigated soybeans land in Mississippi: A case study. *Water Science and Technology: Water Supply*, 16(6), 1639–1647. http://dx.doi.org/10.2166/ws.2016.087

Ouyang, Y., Paz, J. O., Feng, G., Read, J. J., Adeli, A., & Jenkins, J. N. (2017). A model to estimate hydrological processes and water budget in an irrigation farm pond. *Water Resources Management*, 31(7), 2225–2241. http://dx.doi.org/10.1007/s11269-017-1639-0

Pittelkow, C., Liang, X., Linquist, B., Jan, K., van Groenigen, J., Lee, M. E., Lundy, N., van Gestel, J., Six, R. T., Venterea, C., & van Kessel (2015). Productivity limits and potentials of the principles of conservation agriculture. *Nature*, 517, 365–368.

Qi, Z., Helmers, M. J., Malone, R. W., & Thorp, K. R. (2011). Simulating long-term impacts of winter rye cover crop on hydrologic cycling and nitrogen dynamics for a corn-soybean crop system. *Transactions of the ASABE*, 54, 1575–1588.

Tang, Q., Feng, G., Fisher, D., Zhang, H., Ouyang, Y., Adeli, A., & Jenkins, J. (2017). Rainwater deficit and irrigation demand of major row crops in the Mississippi Delta. *Transactions of the ASABE*, 61(3), 927–935. http://dx.doi.org/10.13031/trans.12397

Tribouillois, H., Constantin, J., & Justes, E. (2018). Cover crops mitigate direct greenhouse gases balance but reduce drainage under climate change scenarios in temperate climate with dry summers. *Global Change Biology*, 24, 2513–2519. http://dx.doi.org/10.1111/gcb.14091

Wang, Z., Qi, Z., Xue, L., & Bukovsky, M. (2016). RZWQM2 simulated management practices to mitigate climate change impacts on nitrogen loss and corn production. *Environmental Modelling and Software*, 84, 99–111.

Yang, W., Feng, G., Adeli, A., Kersebaum, K. C., Jenkins, J., & Li, P. (2019a). Long-term effect of cover crop on rainwater balance components and use efficiency in a no-till and rainfed corn and soybean rotation system. *Agricultural Water Management*, 219, 27–39. http://dx.doi.org/10.1016/j.agwat.2019.03.022

Yang, W., Feng, G., Tewolde, H., & Li, P. (2019b). CO_2 emission and soil carbon sequestration from spring- and fall-applied poultry litter in corn production as simulated by RZWQM2. *Journal of Cleaner Production*, 209, 1285–1293. http://dx.doi.org/10.1016/j.agwat.2019.03.022

Yang, W., Feng, G., Adeli, A., Tewolde, H., & Qu, Z. (2021). Simulated long-term effect of wheat cover crop on soil nitrogen losses from no-till corn-soybean rotation under different rainfall patterns. *Journal of Cleaner Production*, 280. http://dx.doi.org/10.1016/j.jclepro.2020.124255

Yang, W., Feng, G., Read, J., Ouyang, Y., & Li, P. (2020). Impact of cover crop on corn-soybean productivity and soil water dynamics under different seasonal rainfall patterns. *Agronomy Journal*, 112, 1–15. http://dx.doi.org/10.1002/agj2.20110

Zhang, B., Feng, G., Ahuja, L., Kong, X., Ouyang, Y., Adeli, A., & Jenkins, J. (2018). Soybean crop-water production functions in a humid region across years and soils determined with APEX model. *Agricultural Water Management*, 204, 180–191. http://dx.doi.org/10.1016/j.agwat.2018.03.024

Zhang, B., Feng, G., Kong, X., Rattan, L., Ouyang, Y., Adeli, A., & Jenkins, J. (2016a). Simulating yield potential by irrigation and yield gap of rainfed soybean using APEX model in a humid region. *Agricultural Water Management*, 177, 440–453.

Zhang, B., Feng, G., Read, J., Kong, X., Ouyang, Y., Adeli, A., & Jenkins, J. (2016b). Simulating soybean productivity under rainfed conditions for major soil types using APEX model in East Central Mississippi. *Agricultural Water Management*, 177, 379–391.

Index

abscisic acid (ABA) signaling, 130–131
accelerated degradation, 226
acid soils, 193
adsorption models, 196–197
aerial parenchyma, 145
aerobic biomass growth
  autotrophs, 166
  heterotrophs, 165–166
aerobic decay of organic matter, 161–162
aerodynamic resistance, 312
aging effect, 220
AgMip (Agricultural Model Improvement), 140
agricultural applications, of ET models, 70–77
Agricultural Model Inter-Comparison and
    Improvement Project (AgMIP), 71
Agricultural Production System sIMulator (APSIM),
    132, 297, 299, 307, 314
  comparison of performance, 267, 268
  hybrid modeling, 269
    APSIM + MLR, 261–264
    APSIM + RF, 255, 261–268, 270
  wheat module, 257–258
    performance assessment, 262, 263
    simulations, 259–260
Agricultural Reference Index for Drought
    (ARID), 260
agro-ecosystem models, 292
agrometeorological models, 124
Al–organic matter complex, 195–196
animal manure, 348
anisohydric plants, 137–138
APSIM see Agricultural Production System
    sIMulator (APSIM)
APSoil database, 256, 259
AQUACROP model, 299, 301, 302, 306
aquatic pesticide exposure, 232
ASCE-EWRI-PM models, 91, 98
atmospheric CO_2 concentrations
  crop models, 296–298
  effect on crop transpiration, 295–298
  uncertainty from model ensembles, 298–299
autotrophs, 166
axial root resistance, 133

Ball–Berry equation, 94
Ball–Berry–Leuning model, 131
bare soil evaporation, 123–124

basal crop coefficient (K_{cb}), 63, 84
Bayesian method, 91
bicarbonate buffer system, 186–187
bioavailability, 222
biomass production, 300–301
bound residues, 222
Bowen ratio energy balance system (BREBS)
    ET measurements, 72–74, 87
Bowen's ratio, 119
bromide tracer, 23–27
Brooks–Corey parameters, 4
Brunetti–Trapp model, 209
buffer zones, 230–231
bulk density of soil, 335–336, 348
bulk surface resistances, 70

calcite crystal growth models, 189
calibration, 197
canopy resistance, 297
canopy temperature, 310–314
capacitance, storage, 128
carbon and nitrogen pools, 159–161
cardinal temperature, 302
CCM see Constant capacitance model (CCM)
CD-MUSIC model, 197
century model, 158
CERES-Maize model, 278
  crop N demand, 285
  crop yield, 284–285
  developmental phases, 279
  DTT, 279
  growth partitioning, 282–283
  leaf area, 280–281
  light capture and growth rate, 281–282
  parameters, 286
  performance, 287, 288
  senescence, 281
chemical equilibrium approach, 181–184
chemical transport in soil matrix see Water and
    chemical transport in soil matrix
chemo-denitrification, 159
climate change, 293
  and ECEs, 252–253
  impact assessments, 292
    atmospheric CO_2 concentrations, 295–298
    biomass production, 300–301
    canopy temperature, 310–313

Modeling Processes and Their Interactions in Cropping Systems: Challenges for the 21st Century, First Edition.
Edited by Lajpat R. Ahuja, Kurt C. Kersebaum, and Ole Wendroth.
© 2022 American Society of Agronomy, Inc. / Crop Science Society of America, Inc. / Soil Science Society of America, Inc.
Published 2022 by John Wiley & Sons, Inc.

climate change (*cont'd*)
    crop growth and yield, 299–300
    crop models, 296–298
    ET, 294
    frost, 307–308
    heat stress, 304
    heavy rainfall, 318–319
    potential ET, 294–295
    temperature and drought relationships, 309–314
    temperature effects, 301–308
    uncertainty from model ensembles, 298–299
    underrepresented processes in, 318–319
    projected climate trends, 293–294
    studies in Mississippi Delta
        CC benefits, 381, 383–387
        corn production, 379, 381
        cotton production, 374, 375, 377, 379–382
        planting windows for cotton, 387, 388
climatic demand, 120–123, 126
climatic evapotranspiration, 122–123
C–N dynamics, 156–159
C–N models, 173–174
CO_2 *see also* Elevated CO_2
    emission, 370–371
    sensitivity of stomata, 295–296
coefficient of determination (R_2), 361–362
combination-based ET models, 56, 57
COMPISOIL model, 342
conductance
    hydraulic, 129, 338–340
    stomatal, 129–132
constant capacitance model (CCM), 196–197
convection–dispersion equation, 2
conventional tillage (CT), 339, 345
    NT *vs.*, 373–378
    study in Mississippi Delta, 373–378
corn production, 379, 381
corn–soybean rotation system, CC in
    long-term effects on water availability, 363–368
    mitigation of climate change impacts, 381, 383–387
cotton production
    Mississippi Delta, climate change studies in, 374, 375, 377, 379–382
    planting windows for, 387, 388
Coupled Model Intercomparison Project Phase 5 (CMIP5), 90, 252
cover crop (CC), in corn–soybean rotation system
    long-term effects on water availability, 363–368
    mitigation of climate change impacts, 381, 383–387
cracks, soil, 349–350
crop coefficient (K_c)
    basal, 63, 84

dual, 69–70, 77, 83
    time-dependent, 74
    values, 85
        database of, 98
        transferability of, 82–83
crop growth
    climate change impact assessment, 299–300
    elevated CO_2 impact on, 313–318
crop models, 117–118, 125 *see also specific crop models*
    process-based, 254–255
    RUE concept in, 140, 300
crop nitrogen demand, 285–286
crop phenology, 304
Cropping System Model (CSM), 278
crop potential evapotranspiration, 55
crop residue
    decomposition, 347
    long-term NT with
        changes in soil properties, 346–349
        infiltration, 345–346
        macropores, 344–345
        manures, 348
        precipitation storage efficiency *vs.* residue level, 346, 347
CropSyst model, 132–133, 297–298, 300, 306, 307, 309, 311, 314
crop transpiration, 125
    atmospheric CO_2 concentration effect on, 295–298
    plant-based estimation of, 94–96
crop water productivity (CWP), 363, 367, 383
crop water requirements, 68–69
crop yield, 284–285 *see also* Yield
        loss/reduction
    climate change impact assessment, 299–300, 318
    frost, 307
    heat stress, 304
    high temperature, 304, 306
    water deficit, 308–309
CSM-IXIM model, 278
    crop N demand, 285–286
    crop yield, 285
    developmental phases, 279, 280
    DTT, 279–280
    growth partitioning, 283–284
    leaf area, 281
    light capture and growth rate, 282
    parameters, 285, 286
    performance, 287, 288
    senescence, 281
C use efficiency (CUE), 158

daily thermal time (DTT)
    CERES-Maize model, 279
    CSM-IXIM model, 279–280

Darcy's equation, 41
Darcy's law, 119
Davies equation, 183
DayCent model, 158–159
Debye–Huckel equation, 182–183
decayed roots, 351
Decision Support System for Agrotechnology
    Transfer (DSSAT), 132, 255, 278, 361
  CERES-Maize model, 278
    crop N demand, 285
    crop yield, 284–285
    developmental phases, 279
    DTT, 279
    growth partitioning, 282–283
    leaf area, 280–281
    light capture and growth rate, 281–282
    parameters, 286
    performance, 287, 288
    senescence, 281
  crop models, 278
  crop-related coefficients, 278
  CSM-IXIM model, 278
    crop N demand, 285–286
    crop yield, 285
    developmental phases, 279, 280
    DTT, 279–280
    growth partitioning, 283–284
    leaf area, 281
    light capture and growth rate, 282
    parameters, 285, 286
    performance, 287, 288
    senescence, 281
  future research needs, 287
denitrification, 157, 159, 163, 171
depression storage (DS), 341
desertification, 146
Dexter ratio, 234
direct drill (DD), 339
disk-till (DT), 57–60
dissolved organic C (DOC), 158
DNDC (DeNitrification–DeComposition) model, 159
dolomite, 192–193
Douglas–Rickman model, 347
drained upper limit (DUL), 132
drought, 145, 253, 265–270
  temperature and, 309–314
  yield reduction, 310
dry mass partitioning, 282
DS see depression storage (DS)
dual crop coefficient method, 69–70, 77, 83
dynamic carbon processes, 118
dynamic plant uptake (DPU) models, 209

EBN models, 310
ECEs see Extreme climate events (ECEs)
ECOSYS model, 159, 312

elevated CO_2, 61, 97, 294, 296
  effect on canopy temperature, 313
  failure to account for, 87–90
  growth response to, 314–318
  RUE vs., 314
embolism, xylem, 138–139
emerging environmental contaminants (EECs), 233
energy balance, 116
  approaches, 310–312
  equation, 119, 120
  and water balance, coupling of, 120–126
energy budget models, 125
EPIC model, 297, 307–308, 315, 335
equilibrium constant, 181–182
EU BROWSE project, 206
European Enlightenment, 119
evaporation
  bare soil, 123–124
  potential, 122
evaporative losses, 54
evapotranspiration (ET), 45, 54–55
  BREBS measurements, 72–74, 87
  climate change effects on, 294
  climatic, 122–123
  crop potential, 55
  ET, E and T measurements, 86–87, 98
  metadata for, 96
  modeling biological control on, 90–92
    Ball–Berry equation, 94
    Jarvis–Stewart equation, 92
    KP model, 92
    Li et al. model, 93
    Medlyn equation, 94
    Shuttleworth one-step model, 93
    Todorovic model, 92–93
  models
    agricultural applications, 70–77
    applications in natural landscapes, 78–80
    combination-based, 56, 57
    daily and hourly approaches, 80–81
    elevated CO_2, 61, 87–90
    extended M–S, 69–70
    extended SW, 66–68
    horticultural applications, 77
    Matt–Shuttleworth model, 68–69
    multiple layer, 79
    NT vs. DT, 56–60
    Penman–Monteith model, 62–64
    Shuttleworth–Wallace model, 64–66
    suitability of, 80–86, 96–97
  under non-water-stressed conditions, 89
  potential, 55
  from residue-covered system, 66–68
  sensitivity analysis, 90–91
  in SPAC model, 120–123
  from sparse canopy, 64–66

evapotranspiration (ET) (*cont'd*)
  terrestrial, 54
  two-step approach, 63, 68, 74, 81
  under water-stressed environments, 75–77
extended Debye–Huckel equation, 183
extended Matt–Shuttleworth (M–S) model, 69–70
extended Shuttleworth–Wallace (SW) model
  agricultural applications of, 72–74
  of evapotranspiration, 66–68
  via RZWQM2, 74–76
extreme climate events (ECEs)
  climate change and, 252–253
  defined, 252
  hybrid model, 255
  impact on crops, 253–254
  impact on wheat yields (*see* New South Wales
    wheat belt)
  process-based crop models, 254–255
  statistical models, 255
  yield losses, 252
    factors, 254
    frost-induced, 254
    heat stress and, 253

FAO-56 approach, 98
  crop coefficient (K_c)
    basal, 63
    dual, 69–70, 77, 83
    time-dependent, 74
    transferability of K_c values, 82–83
  K_c curve, 64, 83, 84
  strengths of, 80
  two-step, 83, 97
Farquhar photosynthesis model, 141–142
FASSET model, 311
Feddes model, 133
fenpropidin, 212
field capacity (FC), 132
FOCUS, 224, 232
FOOTPRINT EU projects, 232
Free-Air-Carbon-Enrichment (FACE) experiments,
  295, 296
frost, 254, 265–269
  crop damages, 307–308

Gapon convention, 194
Gibbs free energy, 181–182
global climate models (GCMs), 253, 258–259, 268,
  271, 379
global warming, 294
grain number per plant (GPP), 284, 285
GRDC-NVT, 259
Green–Ampt approach, 3, 8
greenhouse gas (GHG) emissions, 292
ground heat flux, 37
growing degree days (GDD), 283

growth partitioning, 282–284
growth rate, 282
gypsum, 190

heat fluxes
  ground, 37
  latent, 36–37, 121, 124
  modeling, 48–51
  sensible, 36–37, 121
heat stress, 253, 304
heavy rainfall, 318–319
HERMES model, 298, 299, 302, 304, 312, 315–317
heterotrophs, 165–166
Hoffmann approach, 316, 317
Hooghoudt's steady-state equation, 17
horticultural applications of ET models, 77
HUMEWHeat model, 311
hybrid models, 255, 261–262
hydraulic conductivity, 338–340
  effective, 18
  saturated, 6–8
  vs. matric suction, 5
hydraulic failure, 138–139
hydraulic system, in plants, 116–117, 126
  hydraulic conductance, 129
  plant structure, 127
  storage and capacitance, 127–129
  xylem and phloem, 117
hydrolysis of urea, 165
HydroShoot, 127
HYDRUS-1D soil hydrology model, 209, 212
HYDRUS model, 198

IDEFICS drift model, 206
infiltration, 345–346, 350
  chemical transport within soil matrix, 10–12
  flow into tile drains during, 9–10
  rates, calculation of, 8
  redistribution of water and chemicals, 15–16
  time step for, 9
  vertical, 8–9
integrated assessment and modeling (IAM)
    approaches, 292
integrating models with field experiments, 360–361
  climate change studies in Mississippi Delta, 374
    CC benefits, 381, 383–387
    corn production, 379, 381
    cotton production, 374, 375, 377, 379–382
    planting windows for cotton, 387, 388
  cover crop
    effects on water availability, 363–368
    soil nitrogen dynamics, 371–372
  investigating long-term effects, 373
  litter-derived N and C, 367, 369–371
  model calibration and validation, 361–363
  RZWQM2 model, 361

tillage studies, 373–378
ion activity product (IAP), 188–190
ion exchange, 193–195
ionic mobility, 182
ionizable pesticides, 218–219
ion pairs and complexes, 185
island effect, 123
isohydric plants, 137–138
isotope exchange experiments, 222

Jarvis–Stewart equation, 92

Katerji and Perrier (KP) model, 92
K_c curve, 64, 83, 84

lacunar parenchyma, 145
LAI *see* Leaf area index (LAI)
land surface models (LSMs), 118
Langmuir isotherm, 196
latent heat flux, 36–37, 121, 124
latent heat of fusion
 snow, 37
 soil, 39–40
latent heat of sublimation, 38
latent heat of vaporization, 40
leaf area, 280–281
leaf area index (LAI), 58, 74, 79, 82, 95, 362
leaf number (LN), 280–281, 283
leaf senescence, 281
leaf water potential regulation, 136–137
Leistra pesticide volatilization model, 233
Li et al. model, 93
light capture, 281–282
light response curve, 300
litter-derived N and C, 367, 369–371
lodging, 318

machine learning algorithms, 255
macropore flow, 3, 12–15
macropores, 344–345
magnesium carbonate precipitation, 191
maize crop processes in DSSAT
 CERES-Maize model, 278
  crop N demand, 285
  crop yield, 284–285
  developmental phases, 279
  DTT, 279
  growth partitioning, 282–283
  leaf area, 280–281
  light capture and growth rate, 281–282
  parameters, 286
  performance, 287, 288
  senescence, 281
 CSM-IXIM model, 278
  crop N demand, 285–286

crop yield, 285
 developmental phases, 279, 280
 DTT, 279–280
 growth partitioning, 283–284
 leaf area, 281
 light capture and growth rate, 282
 parameters, 285, 286
 performance, 287, 288
 senescence, 281
 future research needs, 287
MAIZSIM model, 134–135
manure, 348
Matt–Shuttleworth (M–S) approach
 evapotranspiration, 68–69
 suitability of, 96–97
 vs. FAO-56 two-step approach, 83–84
Medlyn equation, 94
mesotrione, 227
methane gas (CH_4) production, 164–165
methyl bromide volatilization, 210
Michaelis–Menten constants, 141
micrometeorology, 119
Mississippi Delta
 climate change studies in
  CC benefits, 381, 383–387
  corn production, 379, 381
  cotton production, 374, 375, 377, 379–382
 planting windows for cotton, 387, 388
 tillage studies in, 373–378
Mitchell approach, 315–317
MLRs *see* Multiple linear regressions (MLRs)
MONICA model, 298, 306
Monin-Obukhov Similarity Theory (MOST), 313
Monte Carlo techniques, 79
multiple linear regressions (MLRs), 255, 261
 APSIM + MLR hybrid model, 261–264

Na–organic matter complex, 196
Nash–Sutcliffe efficiency (NSE), 361–362
National Water Quality Assessment (NAWQA)
 Pesticide National Synthesis Project, 233
natural landscapes, application of ET models, 78–80
NER *see* non-extractable residues (NER)
net photosynthesis, 142
net radiation, 35–36
New South Wales wheat belt, 255–256
 adaptation strategies, 270
 adjusting sowing dates, 270
 APSIM wheat module, 257–258
  performance assessment, 262, 263
  simulations, 259–260
  vs. hybrid model, 267, 268
 ARID, 260
 climate data, 258–259
 comparison of results, 267–268
 ECEs effects, 270

New South Wales wheat belt (*cont'd*)
    heat, drought and frost events, 265–270
    indicators of, 260–261
    projected changes and, 265–267
    yield loss, 261, 265, 267–269
    GCMs, 253, 258–259, 268, 271
    hybrid-modeling approach, 269
        APSIM + MLR, 261–264
        APSIM model *vs.*, 267, 268
        APSIM + RF, 261–268, 270
    in-situ trial data, 259
    limitations, 270–271
    MLR model, 261
    RF model, 261
    study sites in, 256–257
    yield monitoring, 270–271
N factor (NF), 286
nitrification, 157, 162–163
nitrogen and carbon pools, 159–161
nitrogen demand, 285–286
nitrogen dynamics, 371–372
nitrous oxide (N_2O) emission, 157, 164
NOE model, 164
nonequilibrium kinetic soil sorption, 219–220
non-extractable residues (NER), 221, 223
nonhebel approach, 316, 317
normalized difference vegetation index (NDVI), 76
no-till (NT), 56–60, 97
    CT *vs.*, 373–378
    long-term NT with crop residues
        changes in soil properties, 346–349
        infiltration, 345–346
        macropores, 344–345
        manures, 348
        precipitation storage efficiency *vs.* residue level, 346, 347
    study in Mississippi Delta, 373–378

O_2 saturation solubility, 168
oasis effect, 123
Ohm's law, 117
OMNI
    aerobic biomass growth
        autotrophs, 166
        heterotrophs, 165–166
    aerobic decay of organic matter, 161–162
    anaerobic biomass growth, 166–168
    biomass death, 167–168
    carbon and nitrogen pools, 159–161
    correction factors, 168–169
    denitrification, 163
    described, 155–159
    field tests of, 172–173
    flow diagrams, 156–157
    hydrolysis of urea, 165
    methane gas production, 164–165

N_2O emission, 164
    nitrification, 162–163
    saturation O_2 solubility, 168
    state variables, 159, 160
    testing, verification, and calibration, 169–172
    transition-state rate equations, 156
one-parameter model, 7
osmotic potential, 39–40
overland flow, 12
oxygen stress, 319

parenchyma, aerial, 145
PEARL model, 204, 227
PEC model, 204, 227
Penman–Monteith (PM) model *see also* PM equation
    applications of, 70, 78–79
    of evapotranspiration, 62–64
    FAO-56, 71
    one-step, 86
    PET estimation, 87, 89, 294–295
    two-step, 74
percent error (PE), 361–362
perennial plants, 146
permafrost, 34
pesticides, 203
    application efficiency, 206–207
    aquatic exposure, 232
    communication with toxicologists, 233
    Dexter ratio, 234
    dissipation in top/less millimeter of soil, 234
    environmental parameters, 212–213
    foliar retention, 207
    IDEFICS drift model, 206
    in leachate, 228–229
    losses, 228–232
    PEC model, 204, 227
    penetration, 208
    photodegradation of, 208, 232
    remediation of
        buffer zones and water catchments, 230–231
        tillage practices, 230
    root uptake and plant metabolism, 208–209
    runoff, 229–230, 234
    soil degradation, 224–225
        accelerated, 226
        soil sorption and, 227–228
        spatial variability in, 226–227
        temperature, moisture and tropical environments, 225–226
        two-stage degradation model, 228
    soil sorption, 215–216
        aging, 220
        bioavailability, 222
        bound residues, 222
        and degradation, 227–228
        of ionizable pesticides, 218–219

NER, 221, 223
nonequilibrium kinetic, 219–220
reviews, 216
SOM and, 217–218
spatial variability, 223–224
time-dependent, 220, 228
spray drift, 205–206
studies and trends, 232–233
TSCF, 208
volatilization
effects of formulation, deposit site and foliar
penetration, 213–214
losses, 209, 210
methyl bromide, 210
modeling, 210–212, 233
prediction of, 212–213
recent progress in, 214
vapor pressure vs., 209, 210, 212–213
wind tunnel experiment for, 211
wash-off of residues, 214–215
phloem, 117
photoperiod, 278, 280
photosynthesis, 117, 119
modeling, 139–142
net, 142
photosynthesis-transpiration relationship, 139–142
photosynthetic active radiation (PAR), 281, 282, 300
photosynthetic photon flux density (PPFD), 92–94
Pitzer approach, 184
plants
hydraulic system, 126–129
regulation of water use
hydraulic failure, 138–139
isohydric and anisohydric plants, 137–138
leaf water potential regulation, 136–137
response to water stress, 142–144
structure, 127
transpiration, 124–126
water storage compartments in, 127–129
PM equation, 91, 97, 295 see also Penman–Monteith
(PM) model
daily vs. hourly time step calculation, 80
ET estimation, 125
surface resistance in, 125
uses of, 123
PM model see Penman–Monteith (PM) model
pond food web structure, 232
pore size distribution (PSD), 340
porosity of soil, 336
potential evaporation, 122
potential evapotranspiration (PET)
climate impact assessment, 294–295
described, 55
failure to account for elevated CO_2, 87–90
models, 78, 79
PM-based, 87, 89, 294–295

potential growth (PG) rate, 282
potential transpiration, 55
poultry litter, N and C from, 367, 369–371
PPFD see Photosynthetic photon flux density (PPFD)
precipitation change, 252–253
precipitation models, 189–191
predicted environmental concentrations (PECs), 204
preferential flow, 228–229
primary tillage, 334
process-based crop models, 254–255, 292
protodolomite, 193

QSAR, 232–233

radial root resistance, 133–134
radiation absorption, of snow, 38
radiation use efficiency (RUE), 95, 140, 300, 314
rainfall
heavy, 253
interception modeling, 233
Random Forest (RF), 261
APSIM + RF hybrid model, 255, 261–268, 270
random roughness (RR), 341
redistribution
fluctuating water tables and tile drainage, 17–18
of water and chemicals, 15–16
reduced-till system, 56–60, 97
regression equations, 337
regression methods, 255
relative root mean squared error
(RRMSE), 361–362
representative concentration pathway (RCP) 4.5 and
8.5, 381, 383–386
residual soil water content, 337
Richards equation, 2, 3, 15, 119, 350
Riparian Ecosystem Management Model (REMM),
220, 230, 231
root mean square error (RMSE), 262, 263
roots
decayed, 351
influence on soil structure, 351
water uptake, 16–20, 120, 132–136, 308, 309, 319
Root Zone Water Quality Model (RZWQM), 3–4,
20–24, 203–204, 215, 229, 374
Root Zone Water Quality Model 2 (RZWQM2), 74–76,
89–90, 361–363 see also Integrating models with
field experiments

saturated hydraulic conductivity, 6–8
saturation O_2 solubility, 168
secondary tillage, 334
senescence, 281
sensible heat flux, 36–37, 121
sensitivity analysis, 90–91
sepiolite, 191–192
Shuttleworth one-step model, 93

Shuttleworth–Wallace–Hu (SWH) model, 79
Shuttleworth–Wallace (SW) model, 91
  applications
    agricultural, 70–71
    horticultural, 77
    in natural landscapes, 78–79
  of evapotranspiration, 64–66
  extended, 66–68
  PET, 79
  thermal stress index, 76
  under water-stressed environments, 76–77
similar-media scaling, 6–7
Simultaneous Heat and Water (SHAW) model, 33,
    34, 40
snow dynamics
  density changes, 38–39
  energy balance, 37
  illustration of
    site characteristics, 42
    site with contrasting elevation, 42–46
    site with contrasting slope or aspect, 46–50
  latent heat of fusion, 37
  latent heat of sublimation, 38
  outflow of water, 39
  radiation absorption, 38
  specific heat, 37
  thermal conduction, 38
sodium–organic matter complex, 196
soil
  bulk density, 335–336, 348
  compactness, 342
  cracks, 349–350
  drought, 137
  effects of wheel track compaction, 341–344
  evaporation, 123–124
  hydraulic properties, 3, 6–7
  influence of roots, 351
  latent heat of fusion, 39–40
  latent heat of vaporization, 40
  nitrogen dynamics, 371–372
  porosity, 336
  profile, 3–6
  resistance, 133
  shrinkage dynamics, 350
  solid phases in, 187–193
  specific heat, 39
  thermal conductivity, 40
  vapor transfer in, 40–41
  water flow, 41
  water retention characteristics, 336–340
soil chemistry equations
  adsorption–desorption, 196–197
  Al and Na complexes with SOM, 195–196
  bicarbonate buffer system, 186–187
  ion exchange, 193–195
  ion pairs and complexes, 185

solid phases, 187–193
soil matrix
  porosity, 10
  suction
    hydraulic conductivity vs., 5
    soil water content vs., 4
  vertical infiltration of water, 8–9
soil organic carbon (SOC), 362, 373
soil organic matter (SOM), 157 see also OMNI
  Al and Na complexes, 195–196
  dynamics models, 158–159
  heterogeneity of, 158
  nature of, 217–218
soil–plant–atmosphere continuum (SPAC), 116, 120
  energy balance–water balance coupling
    bare soil evaporation, 123–124
    climatic demand, potential evaporation and ET,
      120–123
    plant transpiration, 124–126
  leaf water potential, 136–138
  plant hydraulic system, 116–117, 126
    hydraulic conductance, 129
    plant structure, 127
    storage and capacitance, 127–129
    xylem and phloem, 117
  root water uptake, 132–136
  stomatal conductance, 129–132
soil properties, temporal variability in, 333–334
soil solution composition, 179–180
soil–vegetation–atmosphere transfer (SVAT)
    model, 34
soil-water modeling, 180
solid phases, 187–193
sorption by soil, pesticides, 215–216
  aging, 220
  bioavailability, 222
  bound residues, 222
  of ionizable pesticides, 218–219
  NER, 221, 223
  nonequilibrium kinetic, 219–220
  reviews, 216
  soil degradation and, 227–228
  SOM and, 217–218
  spatial variability, 223–224
  time-dependent, 220, 228
space discretization, 19–20
spatial variability
  soil degradation, 226–227
  soil sorption, 223–224
specific adsorption, 196
specific heat
  of snow, 37
  of soil, 39
specific leaf area (SLA), 283–284
spray drift task force, 205
statistical models, 255

Index

Stefan–Boltzmann constant, 121
STICS model, 297, 308, 311, 367
stomata response to CO_2 concentration, 295–296
stomatal conductance, 87–88, 125, 129–132, 295, 298
stomatal resistance, 81, 87, 89–90, 93–94, 130, 296
storage capacitance, 128
stress-driven senescence, 281
sublimation, latent heat of, 38
SUCROS model, 302
supersaturation, 189, 190
surface-applied bromide tracer, 23–27
surface energy balance, 35
surface resistance, 64, 68–70, 83, 86, 90–94, 123
surface roughness, 340–341
SW model *see* Shuttleworth–Wallace (SW) model

temperature
  canopy, 310–314
  cardinal, 302
  and drought, 309–314
  effects on crops, 301–308
  heat stress, 304
  response curves, 302–305
thermal conduction
  snow, 38
  soil, 40
thermodynamic equilibrium approach, 187
tile drainage, 17–18
tillage
  effects and reconsolidation, 335
    under DD *vs.* CT, 339
    depression storage, 341
    hydraulic conductivity, 338–340
    soil bulk density and porosity, 335–336
    soil water retention, 336–340
    surface roughness, 340–341
  intensity, 335
  long-term NT with crop residues
    changes in soil properties, 346–349
    infiltration, 345–346
    macropores, 344–345
    manures, 348
    precipitation storage efficiency *vs.* residue level, 346, 347
  NT *vs.* DT, 56–60
  primary, 334
  remediation of pesticides, 230
  secondary, 334
  wheel-track compaction effects, 341–344
time-dependent sorption of pesticides, 220, 228
time-driven senescence, 281
Todorovic model, 92–93
transpiration *see also* Evapotranspiration (ET)
  crop
    atmospheric CO_2 concentration effect on, 295–298

plant-based estimation of, 94–96
  metadata for, 96
  models, 119
  plant, 124–126
  potential, 55
transpiration-photosynthesis relationship, 139–142
transpiration stream concentration factor (TSCF), 208
transpiration use efficiency (TUE), 140, 300–301, 315
transport of water and chemicals, 1–3
  bromide tracer, 23–27
  estimation of soil hydraulic properties
    one-parameter model, 7
    similar-media scaling, 6–7
  evaluation of, 24, 28
  infiltration
    flow into tile drains during, 9–10
    rates, calculation of, 8
    sequential partial-displacement and mixing approach, 10–12
    time step for, 9
    vertical, 8–9
  macropore flow, 12–15
  overland flow, 12
  redistribution process, 15–16
  root water uptake, 16–20
  RZWQM, 3–4
    preferential macropore transport studied with, 20–24
  soil profile, 3–6
transport system in plants *see* Hydraulic system, in plants
tropical environments, 225–226
TURFP model, 212
two-site models, 219–220
two-step approach, 63, 68, 74, 81

uncertainties, 298–299, 317
underground transfer of water, 119
UNSATCHEM model, 181, 184, 187, 190, 192, 195
urea, hydrolysis of, 165
USDA–ARS OM3 Object Modeling System, 233

validation, 197
van den Honert's equation, 127
van Genuchten model, 5–6
Vanselow convention, 194
vapor flux, 40–41
vapor pressure deficit (VPD), 69, 95, 297
  in PM equation, 80
  sensitivity analyses, 91
vapor pressure (VP) *vs.* volatilization, 209, 210, 212–213
vegetation, importance of, 116, 145 *see also* Water fluxes in vegetation
vertical infiltration of water, 8–9

volatilization, pesticides
  effects of formulation, deposit site and foliar
      penetration, 213–214
  losses, 209, 210
  methyl bromide, 210
  modeling, 210–212, 233
  prediction of, 212–213
  recent progress in, 214
  vapor pressure *vs.*, 209, 210, 212–213
  wind tunnel experiment for, 211

WATEQ, 180
water and chemical transport model, 1–3
  bromide tracer, 23–27
  estimation of soil hydraulic properties
    one-parameter model, 7
    similar-media scaling, 6–7
  evaluation of, 24, 28
  infiltration
    flow into tile drains during, 9–10
    rates, calculation of, 8
    sequential partial-displacement and mixing
        approach, 10–12
    time step for, 9
    vertical, 8–9
  macropore flow, 12–15
  overland flow, 12
  redistribution process, 15–16
  root water uptake, 16–20
  RZWQM, 3–4
    preferential macropore transport studied
        with, 20–24
  soil profile, 3–6
water balance–energy balance, coupling of, 120–126
water balance models, 308
water cycle, 119
water deficit, 308–309
water excess, 319
water-filled pore space (WFPS), 163, 168–169
water fluxes, 127
  modeling, 48–51
  surface energy and, 35–37
water fluxes in vegetation, 115
  historical overview of, 118–120
  photosynthesis-transpiration relationship, 139–142

plant regulation of water use
  hydraulic failure, 138–139
  isohydric and anisohydric plants, 137–138
  leaf water potential regulation, 136–137
plant response to water stress, 142–144
SPAC model, 116, 120
  bare soil evaporation, 123–124
  climatic demand, potential evaporation and ET,
      120–123
  hydraulic conductance, 129
  plant hydraulic system, 126–129
  plant structure, 127
  plant transpiration, 124–126
  root water uptake, 132–136
  storage and capacitance, 127–129
water potential gradients, 128
water retention characteristics, of soil, 336–340
water storage compartments, in plants, 127–129
water stress
  factor, 76
  plant response to, 142–144
water table, fluctuating, 17–18
water use efficiency (WUE), 209
WEPS model, 234
wheat plants, 253–254
wheel-track compaction effects, 341–344
Willis equation, 214–215
wilting point (WP), 132
WOFOST model, 298, 304, 315

xylem, 117
  embolism, 138–139
  hydraulic resistance, 137

yield loss/reduction
  drought, 310
  factors, 254
  frost-induced, 254, 307
  heat stress and, 253, 304
  high temperatures, 304, 306
  lodging, 318
  New South Wales wheat belt, 261, 265, 267–269
  water deficit, 308–309
  for wheat, 318

CPSIA information can be obtained
at www.ICGtesting.com
Printed in the USA
JSHW010548110822
28786JS00001B/2

9 780891 183853